The Lymphatic Continuum Revisited

ANNALS OF THE NEW YORK ACADEMY OF SCIENCES
Volume 1131

The Lymphatic Continuum Revisited

Edited by
STANLEY G. ROCKSON

Published by Blackwell Publishing on behalf of the New York Academy of Sciences
Boston, Massachusetts
2008

Library of Congress Cataloging-in-Publication Data

The lymphatic continuum revisited/editor, Stanley G. Rockson.
    p.; cm. – (Annals of the New York Academy of Sciences; 1131)
    Includes index.
    ISBN 978-1-57331-699-6
    1. Lymphatics–Physiology–Congresses. 2. Lymphatics–
Pathophysiology–Congresses. I. Rockson, Stanley G. II. Series.
    [DNLM: 1. Lymphatic System–Congresses. 2. Lymphatic Diseases–Congresses.
W1 AN626YL v. 1131 2008/WH700 L98464 2008]

    QP115.L96 2008
    612.4'2–dc22

2008014699

The *Annals of the New York Academy of Sciences* (ISSN: 0077-8923 [print]; ISSN: 1749-6632 [online]) is published 28 times a year on behalf of the New York Academy of Sciences by Blackwell Publishing with offices at (US) 350 Main St., Malden, MA 02148-5020, (UK) 9600 Garsington Road, Oxford, OX4 2ZG, and (Asia) 165 Cremorne St., Richmond VIC 3121, Australia. Blackwell Publishing was acquired by John Wiley & Sons in February 2007. Blackwell's program has been merged with Wiley's global Scientific, Technical, and Medical business to form Wiley-Blackwell.

MAILING: *Annals* is mailed Standard Rate. Mailing to rest of world by IMEX (International Mail Express). Canadian mail is sent by Canadian publications mail agreement number 40573520. POSTMASTER: Send all address changes to *Annals of the New York Academy of Sciences*, Blackwell Publishing Inc., Journals Subscription Department, 350 Main St., Malden, MA 02148-5020.

Disclaimer: The Publisher, the New York Academy of Sciences and Editors cannot be held responsible for errors or any consequences arising from the use of information contained in this publication; the views and opinions expressed do not necessarily reflect those of the Publisher, the New York Academy of Sciences and Editors.

Blackwell Publishing is now part of Wiley-Blackwell.

Information for subscribers: For ordering information, claims, and any inquiry concerning your subscription, please contact your nearest office:

UK: Tel: +44 (0)1865 778315; Fax: +44 (0) 1865 471775
USA: Tel: +1 781 388 8599 or 1 800 835 6770 (toll free in the USA & Canada); Fax: +1 781 388 8232 or Fax: +44 (0) 1865 471775
Asia: Tel: +65 6511 8000; Fax: +44 (0)1865 471775,
Email: customerservices@blackwellpublishing.com

Subscription prices for 2008: Premium Institutional: US$4265 (The Americas), £2370 (Rest of World). Customers in the UK should add VAT at 7%; customers in the EU should also add VAT at 7%, or provide a VAT registration number or evidence of entitlement to exemption. Customers in Canada should add 5% GST or provide evidence of entitlement to exemption. The Premium institutional price also includes online access to the current and all online back files to January 1, 1997, where available. For other pricing options, including access information and terms and conditions, please visit www.blackwellpublishing.com/nyas.

Delivery Terms and Legal Title: Prices include delivery of print publications to the recipient's address. Delivery terms are Delivered Duty Unpaid (DDU); the recipient is responsible for paying any import duty or taxes. Legal title passes to the customer on despatch by our distributors.

Membership information: Members may order copies of *Annals* volumes directly from the Academy by visiting www.nyas.org/annals, emailing membership@nyas.org, faxing +1 212 298 3650, or calling 1 800 843 6927 (toll free in the USA), or +1 212 298 8640. For more information on becoming a member of the New York Academy of Sciences, please visit www.nyas.org/membership. Claims and inquiries on member orders should be directed to the Academy at email: membership@nyas.org or Tel: 1 800 843 6927 (toll free in the USA) or +1 212 298 8640.

Printed in the USA. Printed on acid-free paper.

Annals is available to subscribers online at Blackwell Synergy and the New York Academy of Sciences Web site. Visit www.blackwell-synergy.com or www.annalsnyas.org to search the articles and register for table of contents e-mail alerts.

The paper used in this publication meets the minimum requirements of the National Standard for Information Sciences Permanence of Paper for Printed Library Materials, ANSI Z39.48 1984.

ISSN: 0077-8923 (print); 1749-6632 (online)
ISBN 978-1-57331-699-6; ISBN 1-57331-699-7

A catalogue record for this title is available from the British Library.

ANNALS OF THE NEW YORK ACADEMY OF SCIENCES
Volume 1131

The Lymphatic Continuum Revisited

Editor
STANLEY G. ROCKSON

CONTENTS

Part III. Biological Principles in Lymphedema Diagnosis and Management

Part IV. Lymphatic Biology of Cancer and Metastasis

Preface

The Lymphatic Continuum Revisited

STANLEY G. ROCKSON

Stanford Center for Lymphatic and Venous Disorders, Division of Cardiovascular Medicine, Stanford University School of Medicine, Stanford, California, USA

In 2002, a seminal conference was held at the Natcher Center of the National Institutes of Health in Bethesda, Maryland. This conference, devoted to the subject of lymphatic research and biology, was entitled The Lymphatic Continuum, and provided an interdisciplinary forum for leading international biomedical investigators to discuss current and anticipated developments in lymphatic research. The proceedings of that conference were published in a dedicated volume of this *Annals* series.[1]

By design, the conference was intended to address both the compelling motives for intensified research efforts in lymphatic structure, function, and disease, as well as the importance of lymphatic investigation to the expansion of fundamental comprehension of the mechanisms of human disease, including cancer, infection, metabolism, wound healing and fibrosis, immune disorders, and vascular and developmental biology. At the time of publication of the conference proceedings, it was anticipated that clear directions for future research had emerged and that this field of investigation would advance substantially within the next 5 years.

Those predictions have indeed been proven to be accurate. There has, in fact, been exponential growth in the technology that is applied to questions in lymphatic research, and a similar exponential growth in our comprehension of the molecular regulatory processes that govern the normal development and function of the lymphatic system. Awareness of the importance of lymphatic mechanisms to the continuum of human biology and disease is growing.[2] This "lymphatic continuum" now easily encompasses cardiovascular disease,[3] respiratory inflammation,[4] obesity,[5] autoimmune disease,[6]

and chronic transplant rejection,[7–9] among many other expressions of human pathology.

This unprecedented emphasis upon lymphatic mechanisms in health and disease has produced a parallel, remarkable growth in biomedical publications devoted to these topics. Once a virtually neglected field, lymphatic research has experienced a near explosion. Interrogation of the National Library of Medicine database discloses the fact that, from 1987 to 2006 (20 years), the annual number of publications that cite the term "lymphatic" has grown from 1699 to 3598, reflecting a growth of 112%; as a point of reference, the search term "cardiac" yields a much more modest growth of 73% over the corresponding interval (FIG. 1). Calling up "lymphangiogenesis" as an investigative topic reveals a corresponding 15,800% growth over just the last 6 years (FIG. 2).

Thus, 5 years of biomedical investigation have produced extraordinary progress. It is therefore timely to revisit the lymphatic continuum. The purpose of this volume is to review and document the wonderful growth of insight into the tools for lymphatic investigation, biological principles of the lymphatic system, and biological principles in lymphedema diagnosis and management, as well as the lymphatic biology of cancer and metastasis.

It has been a pleasure to participate as the editor of this current compendium of recent advances in this exciting field. It is with great sense of anticipation that I look ahead to the next five years of growth of studies within the lymphatic continuum.

Address for correspondence: Stanley G. Rockson, M.D., Division of Cardiovascular Medicine, Falk Cardiovascular Research Center, Stanford University School of Medicine, Stanford, CA 94306. Voice: +1-650-725-7571; fax: +1-650-725-1599.
srockson@cvmed.stanford.edu

References

1. ROCKSON, S.G., ED. 2002. The Lymphatic Continuum: Lymphatic Biology and Disease. Ann. N.Y. Acad. Sci. **979:** 1–235.

Ann. N.Y. Acad. Sci. 1131: ix–x (2008). © 2008 New York Academy of Sciences.
doi: 10.1196/annals.1413.000

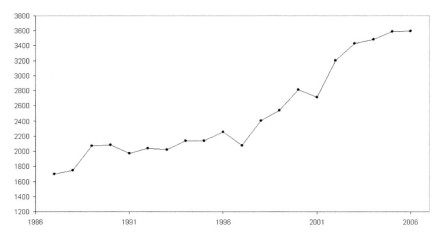

FIGURE 1. Publications citing the "lymphatic" search term within the National Library of Medicine database. The number of publications/year on lymphatics is shown for the period 1987–2006. During this period, the 112% increase eclipsed the corresponding growth of "cardiac" citations (73%).

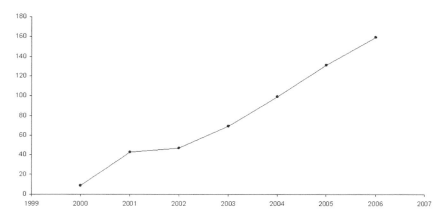

FIGURE 2. Publications citing the "lymphangiogenesis" search term within the National Library of Medicine database. The number of publications/year is shown for the period 2000–2006. During this period, there was a 15,800% increase.

2. ROCKSON, S.G. 2006. The lymphatic continuum continues. Lymphat. Res. Biol. **4:** 1–2.

3. NAKAMURA, K. & S.G. ROCKSON. 2008. The role of the lymphatic circulation in the natural history and expression of cardiovascular disease. Int. J. Cardiol. In press.

4. BALUK, P., T. TAMMELA, E. ATOR, *et al.* 2005. Pathogenesis of persistent lymphatic vessel hyperplasia in chronic airway inflammation. J. Clin. Invest. **115:** 247–257.

5. HARVEY, N.L., R.S. SRINIVASAN, M.E. DILLARD, *et al.* 2005. Lymphatic vascular defects promoted by Prox1 haploinsufficiency cause adult-onset obesity. Nat. Genet. **37:** 1072–1081.

6. ZHANG, Q., Y. LU, S. PROULX, *et al.* 2007. Increased lymphan-

giogenesis in joints of mice with inflammatory arthritis. Arthritis. Res. Ther. **9:** R118.

7. BOCK, F., J. ONDERKA, T. DIETRICH, *et al.* 2007. Bevacizumab as a potent inhibitor of inflammatory corneal angiogenesis and lymphangiogenesis. Invest. Ophthalmol. Vis. Sci. **48:** 2545–2552.

8. KERJASCHKI, D., N. HUTTARY, I. RAAB, *et al.* 2006. Lymphatic endothelial progenitor cells contribute to de novo lymphangiogenesis in human renal transplants. Nat. Med. **12:** 230–234.

9. KERJASCHKI, D., H.M. REGELE, I. MOOSBERGER, *et al.* 2004. Lymphatic neoangiogenesis in human kidney transplants is associated with immunologically active lymphocytic infiltrates. J. Am. Soc. Nephrol. **15:** 603–612.

Markers for Microscopic Imaging of Lymphangiogenesis and Angiogenesis

PETER BALUK AND DONALD M. MCDONALD

Cardiovascular Research Institute, Comprehensive Cancer Center, and Department of Anatomy, University of California, San Francisco, San Francisco, California, USA

Imaging of lymphangiogenesis and angiogenesis requires robust and unambiguous markers of lymphatic and blood vessels. Although much progress has been made in recent years in identifying molecules specifically expressed on lymphatic and blood vessels, no perfect marker has been found that works reliably in all species, tissues, vascular beds, and in all physiological and pathologic conditions. The heterogeneity of expression of markers in both blood and lymphatic vessels reflects underlying differences in the phenotype of endothelial cells. Use of only one marker can lead to misleading interpretations, but these pitfalls can usually be avoided by use of multiple markers and three-dimensional whole-mount preparations. LYVE-1, VEGFR-3, Prox1, and podoplanin are among the most useful markers for microscopic imaging of lymphatic vessels, but, depending on histologic location, each marker can be expressed by other cell types, including vascular endothelial cells. Other markers, including CD31, junctional proteins, and receptors, such as VEGF-2, are shared by lymphatic and blood vessels.

Key words: imaging of angiogenesis; imaging of lymphangiogenesis; LYVE-1; podoplanin; Prox1; VEGFR-3

Introduction

Angiogenesis and lymphangiogenesis are essential components of normal development and physiological and pathologic processes in adults, and consequently have been recognized as potential targets for therapy. Accordingly, the need has arisen for reliable imaging of the vascular and lymphatic systems to determine their normal extent and to understand their changes in pathologic situations and response to therapy. Progress in functional imaging of lymphatics and lymphangiogenesis has lagged behind imaging of the vascular system and angiogenesis, but has recently received a boost with the explosion in knowledge of molecular markers and the application of contemporary technological advances. The goal of this article is to assess the advantages, limitations, and pitfalls of markers for microscopic imaging of lymphatics and lymphangiogenesis in comparison to markers of blood vessels and angiogenesis.

Microscopic Imaging of Lymphangiogenesis

An ideal cell marker for microscopic imaging studies has several desirable features: It should be expressed only on the cell type of interest, should be absent from other cells, and should have a good signal-to-noise ratio with surrounding tissues. A good lymphatic marker should preferably be expressed at a uniform level on all vessel segments during all stages of development and disease, in all species studied, and be resistant to a wide range of fixation conditions. Although few markers satisfy all of these conditions, multiple robust markers are available for microscopic imaging studies of lymphatics (TABLE 1) and blood vessels (TABLE 2).

It has only been relatively recently realized that lymphatic endothelial cells are not homogenous, but instead show phenotypic variation according to their anatomic location and functional state (e.g., during growth, regression, or inflammation). Moreover, lymphatic endothelial cells share many markers in common with vascular endothelial cells,[1] and even some with other hematopoietic progenitor cells. The intensity of staining with particular markers can increase or decrease when cells are grown in tissue-culture conditions or when tissues are exposed to certain physiological or pathologic conditions. Caution is

Address for correspondence: Peter Baluk, Ph.D., Cardiovascular Research Institute, Box 0130, University of California, 513 Parnassus Avenue, San Francisco, CA 94143. Voice: +1-415-476-2118 or +1-415-476-5285; fax: +1-415-476-4845.

peter.baluk@ucsf.edu

Ann. N.Y. Acad. Sci. 1131: 1–12 (2008). © 2008 New York Academy of Sciences.
doi: 10.1196/annals.1413.001

TABLE 1. Selected markers for microscopic imaging of lymphatic vessels

Marker	Alternative name	Species[a]	Notes	References
LYVE-1	Hyaluronan receptor	H, M, R	Also expressed on some blood vessels and macrophages	2, 5, 7, 22, 75, 76
VEGFR-3	Flt-4 (fms-like tyrosine kinase 4)	H, M, R	Also expressed on some blood vessels and stem cells	10, 77
Prox1	Prospero-related homeobox transcription factor	H, M, R	Expressed in nuclei only	78, 79
Podoplanin	Gp38, T1α, AGGRUS, D2–40,	H, M, R	Also expressed on some epithelial cells, podocytes	17
Neuropilin-2	VEGFR-3 co-receptor	M	Also expressed on veins	22, 23, 80
FOXC2	Forkhead transcription factor C2	H, M, R	Expressed in nuclei only	24, 25
CCL21	6Ckine, Exodus-2, SLC	H, M, R	Also on high endothelial veins	31, 81
D6	Decoy chemokine receptor	H, M, R	Binds chemokines, but does not signal	35
5′-nucleotidase	5′-nucleotidase, 5′-nase, CD73	H, M, R	Observed best by enzyme cytochemistry	31, 82, 83
Aquaporin-1	AQ-1, CHIP28	R	Also on some blood vessels	39, 41
LA102	Mouse lymphatic EC marker	M	Does not correspond to any of above markers	44
B27	Rat lymphatic EC marker	R	Also on some blood vessels	43

[a]H, human; M, mouse; R, rat.

therefore needed to avoid coming to false conclusions on the basis of single markers. Good morphologic studies should strive to use multiple markers and experimental conditions. For example, use of whole-mount specimens or thick sections can help to prevent misinterpretations common in examination of histologic thin sections alone.

LYVE-1 (lymphatic vessel endothelial hyaluronan receptor 1) is one of the first and best-characterized markers of lymphatic endothelial cells.[2] It is strongly expressed on the entire luminal and abluminal surface of lymphatic endothelial cells, even on fine filopodia of growing vessels during lymphangiogenesis (FIG. 1A and B). A mutant mouse with a beta-galactosidase reporter knocked into the LYVE-1 locus is useful for whole-mount histochemical studies of lymphatics.[3] Expression of LYVE-1 is downregulated by cytokines, such as TNFα,[4] in some inflammatory conditions (FIG. 1C and D). LYVE-1 also is less strongly expressed on collecting lymphatics than on initial lymphatics. However, LYVE-1 is expressed by some blood vessels, notably liver sinusoids,[5] pulmonary capillaries,[6] and pulmonary arteries and their branches in the lung (FIG. 1E and F). Furthermore, some macrophages also express LYVE-1 (FIG. 2A and B).[7–9]

VEGFR-3, the receptor for the lymphatic growth factors VEGF-C and VEGF-D, is expressed by lymphatic endothelial cells from an early stage of development (E12.5 in mouse).[10] VEGFR-3 is absent from the LYVE-1-immunoreactive macrophages (FIG. 2C and D) and from pulmonary vessels. On the other hand, VEGFR-3 is expressed by some blood vessels, including those during early development,[10] fenestrated cap-

illaries in normal organs,[11] angiogenic blood vessels in tumors and in inflammation,[12,13] and by some stem cells.[14]

The transcription factor Prox1 is expressed in lymphatic endothelial cells from early stages of embryonic development (E9.5 in mouse), that is, even earlier than VEGFR-3 becomes largely restricted to lymphatics.[15,16] Prox1 is believed to be a master control gene inducing the expression of other lymphatic markers[15,16] and is a convenient marker of the nuclei of lymphatic endothelial cells (FIG. 2E).

Podoplanin, a protein first detected in podocytes,[17] labels most lymphatic vessels (from E11.0 onward in mice), where it may play a role in cell adhesion and tube formation.[18] Podoplanin-deficient mice die at birth on account of respiratory failure and lymphatic defects.[18] However, podoplanin is also expressed in type I epithelial cells of the lung,[19] mesothelial cells, myoepithelial cells, ependymal cells, osteocytes, ovarian granulosa cells, and stromal reticular cells and follicular dendritic cells in lymph nodes. Podoplanin is found in skin carcinomas[20] and some other in pathologic situations, thus reducing its usefulness as an unambiguous marker of lymphatics. The monoclonal antibody D2–40, widely used for staining human lymphatics, recognizes podoplanin.[20,21]

Neuropilin 2, which is a co-receptor for VEGF-C and VEGF-D, is found on lymphatic vessels, but is also present on some veins[22,23] (FIG. 3A and B).

FOXC2 is a transcription factor essential for the normal development of lymphatic vessels.[24,25] The lymphatics of mice deficient in FOXC2 lack properly formed valves and have an investment of

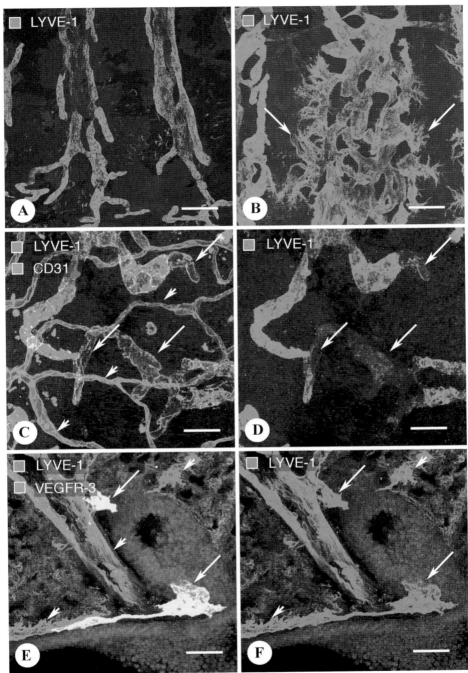

FIGURE 1. (A, B) LYVE-1-immunoreactivity on mouse tracheal lymphatics at baseline **(A)** and during lymphangiogenesis induced by airway infection by *Mycoplasma pulmonis* **(B)**. Lymphatic vessels at base line have smooth contours, but during lymphangiogenesis have numerous lymphatic sprouts (*arrows*). Bar = 200 μm. (From Baluk et al.[61] Reproduced by permission.) **(C, D)** LYVE-1 downregulation in mouse airways during slow regression of lymphatics after elimination of *M. pulmonis* infection by antibiotic treatment. Some individual lymphatic endothelial cells (*arrows*) are still clearly part of lymphatic vessels outlined by CD31 staining, but have little or no LYVE-1 immunoreactivity. CD31 also marks a network of blood vessels (*arrowheads*). Bar = 50 μm. **(E, F)** LYVE-1 on pulmonary arteries and lymphatics of mouse lung. **(E)** Lymphatic vessels (*arrows*) stain for VEGFR-3 (green) and LYVE-1 (red), whereas **(F)** branches of the pulmonary artery (*arrowheads*) stain for LYVE-1 only. Bar = 50 μm. (In color in *Annals* online.)

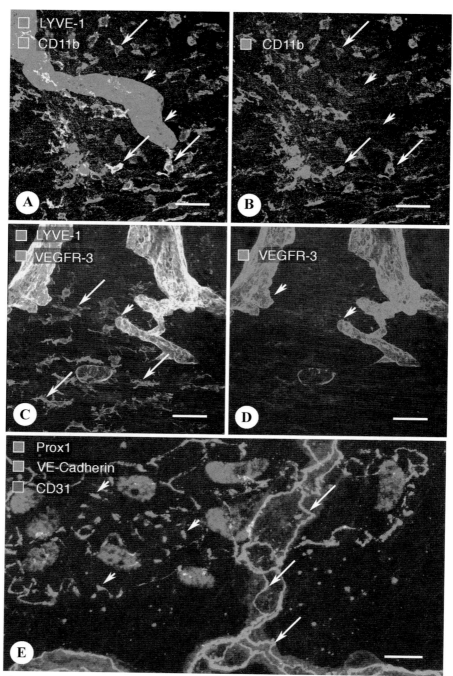

FIGURE 2. (A, B) LYVE-1 and CD11b on macrophages in mouse trachea. LYVE-1 (red) and CD11b (green) are both found on a population of macrophages (*arrows*), but only LYVE-1 is expressed on lymphatics (*arrowheads*). Bar = 50 μm. **(A)** LYVE-1 (red) is found on both lymphatics (arrowheads) and on macrophage-like cells (arrows), but **(B)** VEGFR-3 marks only lymphatics. Bar = 50 μm. **(C, D)** VEGFR-3 expressed on lymphatics, but not on macrophages. **(E)** Prox1 staining of nuclei of lymphatic endothelial cells. An initial lymphatic of mouse trachea has endothelial cell nuclei marked by Prox1 (red) and with discontinuous button-like junctions (*arrowheads*) labeled by VE–cadherin (green) and CD31 (blue). In contrast, a vascular capillary that crosses over the lymphatic vessels does not have nuclei marked by Prox1 and has continuous endothelial junctions (*arrows*) strongly labeled by VE–cadherin and CD31. Bar = 10 μm. (In color in *Annals* online.)

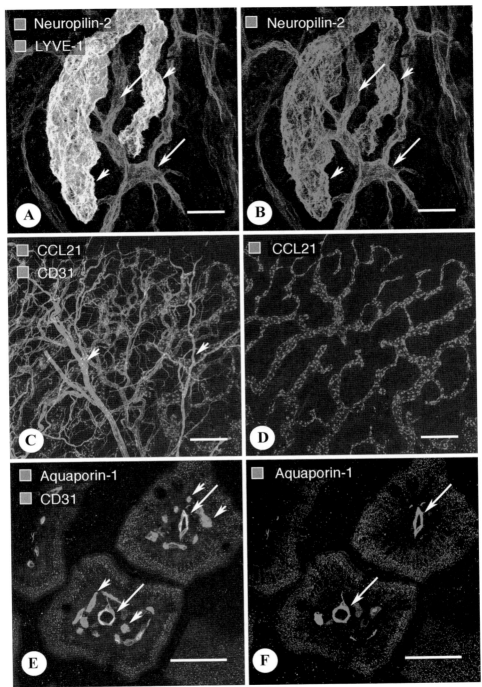

FIGURE 3. (A, B) Neuropilin-2 expression on lymphatics and venules of mouse trachea. **(A)** LYVE-1 (green) labels lymphatic vessels only (*arrowheads*), whereas **(B)** neuropilin-2 also labels some venules (*arrows*). Bar = 50 μm. **(C, D)** Chemokine ligand 21 (CCL21) expression on lymphatics in whole-mount specimen of mouse ear. **(C)** CCL21 (red) is expressed in lymphatic vessels, but not in plexus of blood vessels (*arrowheads*) marked by CD31 (green). **(D)** Staining for CCL21 appears strongest in the perinuclear regions of lymphatic endothelial cells. Bar = 200 μm. **(E, F)** Aquaporin-1 expression on lacteals of mouse intestinal villi. Aquaporin-1 (red) labels central lymphatic vessels (*arrows*) that are surrounded by a meshwork of blood capillaries (*arrowheads*) labeled by CD31 (green). Bar = 50 μm. (In color in *Annals* online.)

TABLE 2. Selected markers for microscopic imaging of blood vessels

Marker	Alternative name	Species[a]	Notes	References
CD31	PECAM-1, endocam	H, M, R	Also expressed on platelets, some leukocytes and weakly on lymphatics. Human CD31 recognized by antibody EN4	51, 53, 64
CD34	Sialomucin, mucosialin,	H	Good antibodies available for human, but not rodent tissues	64
CD105	Endoglin, TGF-β receptor	H, M, R	May be upregulated in tumors	55, 56
MECA-32	Mouse endothelial antigen-32	M	Identical to PV-1 Absent from brain vessels	59, 84, 85
PAL-E	*Pathologische anatomie Leiden-endothelium*	H	Identical to PV-1	58 59
VEGFR-2	Flk-1, Kdr	H, M, R	Also expressed by lymphatics	61, 62
von Willebrand factor	Factor VIII–related antigen	H, M, R	Variably expressed in blood vessels; also found in lymphatics	63, 64, 67, 86
Acetylated LDL uptake	Scavenger receptor	H, M, R	Also taken up by lymphatic endothelial cells	67, 87, 88
Endosialin	TEM1	H, M	Upregulated in tumor vessels	89–91
CD146	MUC18, Mcam, P1H12, S-endo-1	H, M, R	Good antibody available for human but not rodent tissues	92, 93
VE–cadherin	Cadherin-5, CD144	H, M, R	Expressed also by lymphatics	52
Integrin αvβ3	LM609	H	Good antibody available for human but not rodent tissues	94, 95
Integrin αvβ5	CD51, vitronectin receptor	H	Upregulated in angiogenesis	96
Integrin α5β1	CD49e, fibronectin receptor	H, M	Upregulated in tumor vessels	97–99
Delta-like ligand 4	Dll4	H, M, R	Acts as negative regulator of tumor angiogenesis	100, 101
ROBO4	Roundabout homologue 4	H, M	Restricted to blood vessels	48, 102
RECA-1	Rat endothelial cell antigen	R	Useful marker for rat endothelial cells	86
OX43	MRC OX43	R	Useful marker for rat endothelial cells	103, 104
Ulex lectin	*Ulex europeus* agglutinin lectin	H	Stains human and bovine endothelium, but not rodent endothelium; present on some lymphatics	67, 71
Griffonia lectin	*Bandeiraea simplicifolia* lectin	M, R	Stains rodent endothelial cells, but not human endothelial cells	68, 69
Tomato lectin	*Lycopersicon esculentum* lectin	M, R	Stains rodent endothelial cells, but not human endothelial cells	70
Collagen-IV	Basement membrane marker	H, M, R	Found in basement membrane of blood vessels and larger lymphatics	60, 73

[a]H, human; M, mouse; R, rat.

pericytes. FOXC2, like Prox1, is largely expressed in nuclei of lymphatic endothelial cells.[25] FOXC2 is also expressed in developing heart, kidney, bone, eyelid (all tissues associated with the lymphatic disorder known as lymphedema–distichiasis), and in adipose tissue.[24]

Chemokine ligand 21 (CCL21), also known as secondary lymphoid chemokines (SLC), 6Ckine, or Exodus-2, is a chemotactic ligand for chemokine receptor 7 (CCR7) present on dendritic cells, monocytes, and lymphocytes that enter lymphatics during inflammation.[26,27] CCL21 is found on afferent lymphatics (FIG. 3C and D) and it appears to be upregulated in inflammation.[28–32] It is also expressed on stromal cells within lymph nodes.[33]

D6 is a silent non-signaling chemokine receptor that can competitively bind chemokine ligands. It dampens inflammation by mopping up chemokine ligands and preventing them from binding to their receptors.[34] D6 is expressed by some lymphatic endothelial cells.[35,36] 5′-Nucleotidase is an enzyme that is preferentially on the surface of lymphatic endothelial cells, but not vascular endothelial cells. It has been particularly useful for demonstrating the overall anatomy of lymphatic networks by enzyme histochemical methods,[37,38] but good antibodies to it are lacking. 5′-Nucleotidase corresponds to CD73.

Aquaporin-1 is a water-transporter molecule that is strongly expressed in some subtypes of lymphatics, for example, in lacteals of intestinal villi (FIG. 3E and F) and in lymph nodes.[39,40] However, aquaporin-1 does not appear to be uniformly present in all lymphatics. Moreover, it is also expressed in some blood vessels,[41] including nonfenestrated blood vessels, high endothelial venules,[40] some tumor blood vessels,[41] and in other nonendothelial cell types.[42]

Two monoclonal antibodies, LA102 and B27, that recognize mouse and rat lymphatic vessels, respectively, and are useful for immunohistochemical studies have been produced recently.[43,44] The antigens recognized by these antibodies do not appear to correspond to known lymphatic markers, but have not yet been identified.

Microscopic Imaging of Angiogenesis

Many markers of vascular endothelial cells have been identified (TABLE 2) and reviewed in detail.[45–50] However, the overlap and lack of overlap of markers of vascular endothelial cells and lymphatic endothelial cells have received less attention.

CD31, also known as PECAM-1 or endocam, has long been considered a marker of blood vessels.[51] CD31 is a pan-endothelial marker, present in blood vessels and lymphatics, albeit at a lower level of expression on lymphatic endothelial cells.[52] CD31 is also expressed by platelets, granulocytes, and monocytes.[51] The antigen detected by monoclonal antibody EN4 to human endothelial cells is CD31.[53]

CD34 is expressed on mature vascular endothelial cells and circulating hematopoietic progenitor cells. The monoclonal antibody QBEnd/10 stains human endothelial cells robustly,[54] but the homologous markers for rodent CD34 are more problematic.

CD105 or endoglin, the receptor for transforming growth factor-β (TGF-β), is expressed by vascular endothelial cells (FIG. 4A and B) and has been reported to be upregulated in angiogenic vessels and accumulate preferentially in tumors.[55] This property allows radiolabeled antibodies to CD105 to be used as probes for imaging tumors in the clinical setting.[55,56]

The antibody MECA-32, originally described as a marker of endothelial cells in rodents, and the *pathologische anatomie Leiden-endothelium* (PAL-E) antibody to human endothelium, bind to PV-1 in the thin diaphragm that covers endothelial fenestrations and caveolae.[57–59] PV-1 has not been reported in lymphatics.

VEGFR-2 is strongly expressed on most blood vessels, with some differences in staining intensity among different vessel types. It is upregulated on angiogenic vessels (FIG. 4C–E) and is downregulated after anti-VEGF anti-angiogenic therapy.[60] However, some lymphatic vessels express VEGFR-2 at levels comparable to those of blood vessel (FIG. 4F and G).

See also the work of Baluk *et al.*[61] and Wirzenius *et al.*[62]

Von Willebrand factor, also known as factor VIII–related antigen, is a marker present in endothelial cell organelles known as Weibel–Palade bodies. However, vascular endothelial cells do not all express the von Willebrand factor uniformly. Expression is greater in arterial and venular endothelial cells than in capillary endothelial cells, and is also found in lymphatic endothelial cells.[63,64]

VE–cadherin, which is selectively expressed at the borders of endothelial cells, is a member of the large cadherin family of junctional proteins. Although VE–cadherin is strongly expressed in vascular endothelial cells, it is also present in lymphatic endothelial cells. In initial lymphatics, VE–cadherin is located at button-like junctions that are sites of fluid uptake.[52]

Integrins, including αvβ3, αvβ5, and α5β1, are upregulated on angiogenic blood vessels, particularly in tumors, and are attractive targets for anti-angiogenic therapy.[65,66] Their distribution on blood vessels is beginning to be mapped, but their distribution on lymphatic vessels is unknown.

Lectins can be used in several interesting ways for imaging studies of blood vessels. First, *Ulex europaeus* agglutinin-1 lectin has been very useful in labeling human and bovine endothelium,[67] and *Lycopersicon esculentum*, *Griffonia simplicifolia* and *Ricin communis* aggutinin-1 mark rodent vascular endothelial cells.[68–70] *Ulex europaeus* lectin also labels human lymphatic endothelial cells.[71] When injected intravenously, lectins can be used to mark blood vessels with a patent lumen, because they only have access to vessels that are perfused with blood. When used in combination with CD31 as a marker of endothelial cells, the occluded segments of blood vessels can be identified.[72] Last, *Griffonia simplicifolia* isolectin B4 binds to the delicate endothelial filopodia of angiogenic blood vessels much more effectively than most antibodies.[68]

Type IV collagen is a prominent component of the basement membrane of blood vessels and of larger collecting lymphatics. Tumor blood vessels also express type IV collagen.[73] When endothelial cells undergo rapid regression, as after inhibition of VEGF signaling, empty basement membrane sleeves persist, thus providing a historical record, marking "ghost" vessels that have regressed.[60]

Summary

Recent studies have revealed a considerable heterogeneity in the expression of endothelial markers

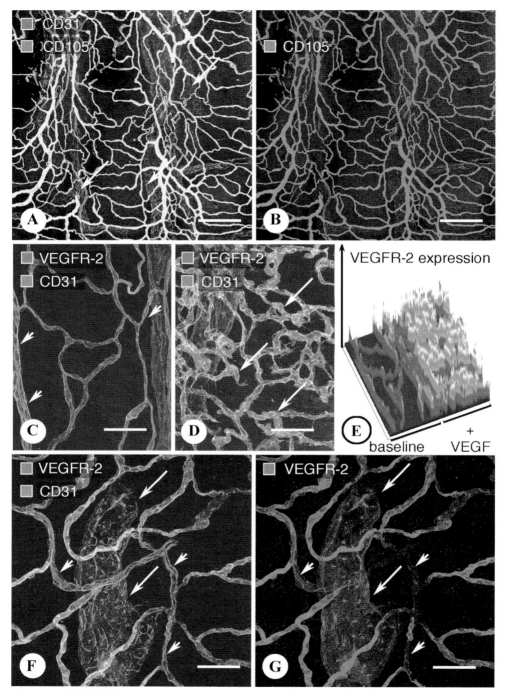

FIGURE 4. (A, B) CD105 (endoglin) expression on blood vessels but not on lymphatics of mouse trachea. **(A)** CD31 (red) labels blood vessels and lymphatics (*arrows*), but **(B)** CD105 (green) labels blood vessels only. Bar = 200 μm. **(C–E)** VEGFR-2 upregulation on angiogenic blood vessels. **(C)** In baseline conditions most endothelial cells of blood vessels of mouse trachea have modest expression of VEGFR-2, and some (*arrowheads*) have only weak staining. **(D)** In contrast, after transgenic overexpression of the ligand VEGF, most newly formed blood vessels (*arrows*) have strong VEGFR-2 staining. **(E)** A contour plot shows the relative vessel density and VEGFR-3 staining intensity of regions **(A)** and **(B)**. (With permission from Ref. 74. Bar = 50μm.) **(F, G)** VEGFR-2 expression on lymphatics. VEGFR-2 (red) labels most endothelial cells of blood vessels of mouse trachea, but some only weekly (*arrowheads*). In addition, VEGFR-2 is expressed on lymphatics (*arrows*) at levels similar to those on blood vessels. Bar = 50 μm. (In color in *Annals* online).

of lymphatic and blood vessels in different species, tissues, developmental stages, and physiological and pathologic conditions. Numerous markers have been discovered that are selective for either lymphatic or blood vessels. Other markers are shared between lymphatic and blood endothelial cells, but are expressed to a different degree, while yet others are shared with other cells types, notably epithelial cells and inflammatory cells. Studies using two or more markers and three-dimensional whole-mount preparations invariably reveal a greater complexity for either vessel type than studies using single markers and thin sections alone. The existence of such hitherto unrecognized endothelial heterogeneity implies that the endothelial phenotype is specified on a cell-by-cell basis. The challenge for the future is to understand the factors that regulate the phenotype of individual endothelial cells of lymphatic and blood vessels.

Acknowledgments

This work was supported in part by grants from the National Heart, Lung, and Blood Institute (HL-24136 and HL-59157) and from the National Cancer Institute (CA-82923).

Conflicts of Interest

The authors declare no conflicts of interest.

References

1. WICK, N. *et al.* 2007. Transcriptomal comparison of human dermal lymphatic endothelial cells ex vivo and in vitro. Physiol. Genom. **28:** 179–192.

2. JACKSON, D.G. 2004. Biology of the lymphatic marker LYVE-1 and applications in research into lymphatic trafficking and lymphangiogenesis. APMIS **112:** 526–538.

3. GALE, N.W. *et al.* 2007. Normal lymphatic development and function in mice deficient for the lymphatic hyaluronan receptor LYVE-1. Mol. Cell. Biol. **27:** 595–604.

4. JOHNSON, L.A., R. PREVO & D.G. JACKSON. 2007. Inflammation-induced uptake and degradation of the lymphatic endothelial hyaluronan receptor LYVE-1. J. Biol. Chem. **282:** 33671–33680.

5. MOUTA CARREIRA, C. *et al.* 2001. LYVE-1 is not restricted to the lymph vessels: expression in normal liver blood sinusoids and down-regulation in human liver cancer and cirrhosis. Cancer Res. **61:** 8079–8084.

6. FAVRE, C.J. 2003. Expression of genes involved in vascular development and angiogenesis in endothelial cells of adult lung. Am. J. Physiol. Heart Circ. Physiol. **285:** H1917–H1938.

7. CHO, C.H. *et al.* 2007. Angiogenic role of LYVE-1-positive macrophages in adipose tissue. Circ. Res. **100:** e47–e57.

8. MARUYAMA, K. *et al.* 2005. Inflammation-induced lymphangiogenesis in the cornea arises from CD11b-positive macrophages. J. Clin. Invest. **115:** 2363–2372.

9. XU, H. *et al.* 2007. LYVE-1-positive macrophages are present in normal murine eyes. Invest. Ophthalmol. Vis. Sci. **48:** 2162–2171.

10. KAIPAINEN, A. *et al.* 1995. Expression of the fms-like tyrosine kinase 4 gene becomes restricted to lymphatic endothelium during development. Proc. Natl. Acad. Sci. USA **92:** 3566–3570.

11. KAMBA, T. *et al.* 2006. VEGF-dependent plasticity of fenestrated capillaries in the normal adult microvasculature. Am. J. Physiol. Heart Circ. Physiol. **290:** H560–H576.

12. CLARIJS, R. *et al.* 2002. Induction of vascular endothelial growth factor receptor-3 expression on tumor microvasculature as a new progression marker in human cutaneous melanoma. Cancer Res. **62:** 7059–7065.

13. WITMER, A.N. *et al.* 2004. In vivo angiogenic phenotype of endothelial cells and pericytes induced by vascular endothelial growth factor-A. J. Histochem. Cytochem. **52:** 39–52.

14. SALVEN, P. *et al.* 2003. VEGFR-3 and CD133 identify a population of CD34 +lymphatic/vascular endothelial precursor cells. Blood **101:** 168–172.

15. WIGLE, J.T. *et al.* 2002. An essential role for Prox1 in the induction of the lymphatic endothelial cell phenotype. EMBO J. **21:** 1505–1513.

16. WILTING, J. *et al.* 2002. The transcription factor Prox1 is a marker for lymphatic endothelial cells in normal and diseased human tissues. FASEB J. **16:** 1271–1273.

17. BREITENEDER-GELEFF, S. *et al.* 1999. Angiosarcomas express mixed endothelial phenotypes of blood and lymphatic capillaries: podoplanin as a specific marker for lymphatic endothelium. Am. J. Pathol. **154:** 385–394.

18. SCHACHT, V. *et al.* 2003. T1alpha/podoplanin deficiency disrupts normal lymphatic vasculature formation and causes lymphedema. EMBO J. **22:** 3546–3556.

19. DOBBS, L.G., M.C. WILLIAMS & R. GONZALEZ. 1988. Monoclonal antibodies specific to apical surfaces of rat alveolar type I cells bind to surfaces of cultured, but not freshly isolated, type II cells. Biochim. Biophys. Acta **970:** 146–156.

20. SCHACHT, V. *et al.* 2005. Up-regulation of the lymphatic marker podoplanin, a mucin-type transmembrane glycoprotein, in human squamous cell carcinomas and germ cell tumors. Am. J. Pathol. **166:** 913–921.

21. SONNE, S.B. *et al.* 2006. Identity of M2A (D2–40) antigen and gp36 (Aggrus, T1A-2, podoplanin) in human developing testis, testicular carcinoma in situ and germ-cell tumours. Virchows Arch. **449:** 200–206.

22. YUAN, L. *et al.* 2002. Abnormal lymphatic vessel development in neuropilin 2 mutant mice. Development **129:** 4797–4806.

23. KARPANEN, T. *et al.* 2006. Functional interaction of VEGF-C and VEGF-D with neuropilin receptors. FASEB J. **20:** 1462–1472.

24. DAGENAIS, S.L. *et al.* 2004. Foxc2 is expressed in developing lymphatic vessels and other tissues associated with

lymphedema-distichiasis syndrome. Gene Expr. Patterns **4:** 611–619.

25. PETROVA, T.V. *et al.* 2004. Defective valves and abnormal mural cell recruitment underlie lymphatic vascular failure in lymphedema distichiasis. Nat. Med. **10:** 974–981.

26. LIRA, S.A. 2005. A passport into the lymph node. Nat. Immunol. **6:** 866–868.

27. RANDOLPH, G.J., V. ANGELI & M.A. SWARTZ. 2005. Dendritic-cell trafficking to lymph nodes through lymphatic vessels. Nat. Rev. Immunol. **5:** 617–628.

28. JI, R.C., K. KURIHARA & S. KATO. 2006. Lymphatic vascular endothelial hyaluronan receptor (LYVE)-1- and CCL21-positive lymphatic compartments in the diabetic thymus. Anat. Sci. Int. **81:** 201–209.

29. JIN, Y. *et al.* 2007. The chemokine receptor CCR7 mediates corneal antigen-presenting cell trafficking. Mol. Vis. **13:** 626–634.

30. ODAKA, C. *et al.* 2006. Distribution of lymphatic vessels in mouse thymus: immunofluorescence analysis. Cell Tissue Res. **325:** 13–22.

31. QU, P., R.C. JI & S. KATO. 2005. Expression of CCL21 and 5′-Nase on pancreatic lymphatics in nonobese diabetic mice. Pancreas **31:** 148–155.

32. EBERHARD, Y. *et al.* 2004. Up-regulation of the chemokine CCL21 in the skin of subjects exposed to irritants. BMC Immunol. **5:** 7.

33. LUTHER, S.A. *et al.* 2002. Differing activities of homeostatic chemokines CCL19, CCL21, and CXCL12 in lymphocyte and dendritic cell recruitment and lymphoid neogenesis. J. Immunol. **169:** 424–433.

34. MARTINEZ DE LA TORRE, Y. *et al.* 2007. Protection against inflammation- and autoantibody-caused fetal loss by the chemokine decoy receptor D6. Proc. Natl. Acad. Sci. USA **104:** 2319–2324.

35. NIBBS, R.J. *et al.* 2001. The beta-chemokine receptor D6 is expressed by lymphatic endothelium and a subset of vascular tumors. Am. J. Pathol. **158:** 867–877.

36. SIRONI, M. *et al.* 2006. Generation and characterization of a mouse lymphatic endothelial cell line. Cell Tissue Res. **325:** 91–100.

37. KATO, S., M. MIURA & R. MIYAUCHI. 1993. Structural organization of the initial lymphatics in the monkey mesentery and intestinal wall as revealed by an enzyme-histochemical method. Arch. Histol. Cytol. **56:** 149–160.

38. JI, R.C. *et al.* 2004. Expression of VEGFR-3 and 5′-nase in regenerating lymphatic vessels of the cutaneous wound healing. Microsc. Res. Tech. **64:** 279–286.

39. GANNON, B.J. & C.J. CARATI. 2003. Endothelial distribution of the membrane water channel molecule aquaporin-1: implications for tissue and lymph fluid physiology? Lymphat. Res. Biol. **1:** 55–66.

40. OHTANI, O. *et al.* 2003. Fluid and cellular pathways of rat lymph nodes in relation to lymphatic labyrinths and Aquaporin-1 expression. Arch. Histol. Cytol. **66:** 261–272.

41. VERKMAN, A.S. 2006. Aquaporins in endothelia. Kidney Int. **69:** 1120–1123.

42. GANNON, B.J. *et al.* 2000. Aquaporin-1 expression in visceral smooth muscle cells of female rat reproductive tract. J. Smooth Muscle Res. **36:** 155–167.

43. EZAKI, T. *et al.* 1990. A new approach for identification of rat lymphatic capillaries using a monoclonal antibody. Arch. Histol. Cytol. **53**(Suppl): 77–86.

44. EZAKI, T. *et al.* 2006. Production of two novel monoclonal antibodies that distinguish mouse lymphatic and blood vascular endothelial cells. Anat. Embryol. (Berl.) **211:** 379–393.

45. MCDONALD, D.M. & P.L. CHOYKE. 2003. Imaging of angiogenesis: from microscope to clinic. Nat. Med. **9:** 713–725.

46. AUGUSTIN, H., ED. 2004. Methods in Endothelial Cell Biology. Springer-Verlag. Berlin–Heidelberg.

47. MILLER, J.C. *et al.* 2005. Imaging angiogenesis: applications and potential for drug development. J. Natl. Can. Inst. **97:** 172–187.

48. NERI, D. & R. BICKNELL. 2005. Tumour vascular targeting. Nat. Rev. Can. **5:** 436–446.

49. OCAK, I. *et al.* 2007. The biologic basis of in vivo angiogenesis imaging. Front Biosci. **12:** 3601–3616.

50. ADAMS, R.H. & K. ALITALO. 2007. Molecular regulation of angiogenesis and lymphangiogenesis. Nat. Rev. Mol. Cell Biol. **8:** 464–478.

51. ALBELDA, S.M. *et al.* 1991. Molecular and cellular properties of PECAM-1 (endoCAM/CD31): a novel vascular cell-cell adhesion molecule. J. Cell Biol. **114:** 1059–1068.

52. BALUK, P. *et al.* 2007. Functionally specialized junctions between endothelial cells of lymphatic vessels. J. Exp. Med. **204:** 2349–2362.

53. BURGIO, V.L. *et al.* 1994. Characterization of EN4 monoclonal antibody: a reagent with CD31 specificity. Clin. Exp. Immunol. **96:** 170–176.

54. RAMANI, P., N.J. BRADLEY & C.D. FLETCHER. 1990. QBEND/10, a new monoclonal antibody to endothelium: assessment of its diagnostic utility in paraffin sections. Histopathology **17:** 237–242.

55. DUFF, S.E. *et al.* 2003. CD105 is important for angiogenesis: evidence and potential applications. FASEB J. **17:** 984–992.

56. FONSATTI, E. *et al.* 2003. Endoglin (CD105): a powerful therapeutic target on tumor-associated angiogenetic blood vessels. Oncogene **22:** 6557–6563.

57. STAN, R.V., M. KUBITZA & G.E. PALADE. 1999. PV-1 is a component of the fenestral and stomatal diaphragms in fenestrated endothelia. Proc. Natl. Acad. Sci. USA **96:** 13203–13207.

58. NIEMELA, H. *et al.* 2005. Molecular identification of PAL-E, a widely used endothelial-cell marker. Blood **106:** 3405–3409.

59. STAN, R.V. 2007. Endothelial stomatal and fenestral diaphragms in normal vessels and angiogenesis. J. Cell Mol. Med. **11:** 621–643.

60. INAI, T. *et al.* 2004. Inhibition of vascular endothelial growth factor (VEGF) signaling in cancer causes loss of endothelial fenestrations, regression of tumor vessels, and appearance of basement membrane ghosts. Am. J. Pathol. **165:** 35–52.

61. BALUK, P. *et al.* 2005. Pathogenesis of persistent lymphatic vessel hyperplasia in chronic airway inflammation. J. Clin. Invest. **115:** 247–257.

62. WIRZENIUS, M. *et al.* 2007. Distinct vascular endothelial growth factor signals for lymphatic vessel enlargement and sprouting. J. Exp. Med. **204:** 1431–1440.

63. DUMONT, A.E. 1993. Factor VIII-related antigen. J. Natl. Can. Inst. **85:** 674–676.

64. PUSZTASZERI, M.P., W. SEELENTAG & F.T. BOSMAN. 2006. Immunohistochemical expression of endothelial markers CD31, CD34, von Willebrand factor, and Fli-1 in normal human tissues. J. Histochem. Cytochem. **54:** 385–395.

65. STUPACK, D.G. & D.A. CHERESH. 2004. Integrins and angiogenesis. Curr. Top. Dev. Biol. **64:** 207–238.

66. HYNES, R.O. 2007. Cell-matrix adhesion in vascular development. J. Thromb. Haemostatis. **5**(Suppl. 1): 32–40.

67. LAURENS, N. & V. VAN HINSBERG. 2004. Isolation, purifaction and culture of human micro- and macrovascular endothelail cells. *In* Methods in Endothelial Cell Biology. H. Augustin, Ed. Springer-Verlag. Berlin–Heidelberg.

68. GERHARDT, H. *et al.* 2003. VEGF guides angiogenic sprouting utilizing endothelial tip cell filopodia. J. Cell Biol. **161:** 1163–1177.

69. LAITINEN, L. 1987. *Griffonia simplicifolia* lectins bind specifically to endothelial cells and some epithelial cells in mouse tissues. Histochem. J. **19:** 225–234.

70. THURSTON, G. *et al.* 1996. Permeability-related changes revealed at endothelial cell borders in inflamed venules by lectin binding. Am. J. Physiol. **271:** H2547–H2562.

71. GARRAFA, E. *et al.* 2006. Isolation and characterization of lymphatic microvascular endothelial cells from human tonsils. J. Cell Physiol. **207:** 107–113.

72. BAFFERT, F. *et al.* 2006. Cellular changes in normal blood capillaries undergoing regression after inhibition of VEGF signaling. Am. J. Physiol. Heart Circ. Physiol. **290:** H547–H559.

73. BALUK, P. *et al.* 2003. Abnormalities of basement membrane on blood vessels and endothelial sprouts in tumors. Am. J. Pathol. **163:** 1801–1815.

74. BALUK, P. *et al.* 2004. Regulated angiogenesis and vascular regression in mice overexpressing vascular endothelial growth factor in airways. Am. J. Pathol. **165:** 1071–1085.

75. BANERJI, S. *et al.* 1999. LYVE-1, a new homologue of the CD44 glycoprotein, is a lymph-specific receptor for hyaluronan. J. Cell Biol. **144:** 789–801.

76. JACKSON, D.G. 2003. The lymphatics revisited: new perspectives from the hyaluronan receptor LYVE-1. Trends Cardiovasc. Med. **13:** 1–7.

77. ALITALO, K., T. TAMMELA & T.V. PETROVA. 2005. Lymphangiogenesis in development and human disease. Nature **438:** 946–953.

78. WIGLE, J.T. & G. OLIVER. 1999. Prox1 function is required for the development of the murine lymphatic system. Cell **98:** 769–778.

79. OLIVER, G. & K. ALITALO. 2005. The lymphatic vasculature: recent progress and paradigms. Annu. Rev. Cell Dev. Biol. **21:** 457–483.

80. STATON, C. *et al.* 2007. Neuropilins in physiological and pathological angiogenesis. J. Pathol. **212:** 237–248.

81. GUNN, M.D. *et al.* 1998. A chemokine expressed in lymphoid high endothelial venules promotes the adhesion and chemotaxis of naive T lymphocytes. Proc. Natl. Acad. Sci. USA **95:** 258–263.

82. KATO, S. *et al.* 1996. Enzyme triple staining for differentiation of lymphatics from venous and arterial capillaries. Lymphology **29:** 15–19.

83. JI, R.C., P. QU & S. KATO. 2003. Application of a new 5′-nase monoclonal antibody specific for lymphatic endothelial cells. Lab. Invest. **83:** 1681–1683.

84. ENGELHARDT, B., F.K. CONLEY & E.C. BUTCHER. 1994. Cell adhesion molecules on vessels during inflammation in the mouse central nervous system. J. Neuroimmunol. **51:** 199–208.

85. MADDEN, S.L. *et al.* 2004. Vascular gene expression in nonneoplastic and malignant brain. Am. J. Pathol. **165:** 601–608.

86. ULGER, H., A.K. KARABULUT & M.K. PRATTEN. 2002. Labelling of rat endothelial cells with antibodies to vWF, RECA-1, PECAM-1, ICAM-1, OX-43 and ZO-1. Anat. Histol. Embryol. **31:** 31–35.

87. VOYTA, J.C. *et al.* 1984. Identification and isolation of endothelial cells based on their increased uptake of acetylated-low density lipoprotein. J. Cell Biol. **99:** 2034–2040.

88. OHTANI, Y. & O. OHTANI. 2001. Postnatal development of lymphatic vessels and their smooth muscle cells in the rat diaphragm: a confocal microscopic study. Arch. Histol. Cytol. **64:** 513–522.

89. CHRISTIAN, S. *et al.* 2001. Molecular cloning and characterization of endosialin, a C-type lectin-like cell surface receptor of tumor endothelium. J. Biol. Chem. **276:** 7408–7414.

90. NANDA, A. & B. ST CROIX. 2004. Tumor endothelial markers: new targets for cancer therapy. Curr. Opin. Oncol. **16:** 44–49.

91. RETTIG, W.J. *et al.* 1992. Identification of endosialin, a cell surface glycoprotein of vascular endothelial cells in human cancer. Proc. Natl. Acad. Sci. USA **89:** 10832–10836.

92. ST CROIX, B. *et al.* 2000. Genes expressed in human tumor endothelium. Science **289:** 1197–1202.

93. ANFOSSO, F. *et al.* 2001. Outside-in signaling pathway linked to CD146 engagement in human endothelial cells. J. Biol. Chem. **276:** 1564–1569.

94. BROOKS, P.C., R.A. CLARK & D.A. CHERESH. 1994. Requirement of vascular integrin alpha v beta 3 for angiogenesis. Science **264:** 569–571.

95. VARNER, J.A., P.C. BROOKS & D.A. CHERESH. 1995. Review: the integrin alpha V beta 3: angiogenesis and apoptosis. Cell Adhes. Commun. **3:** 367–374.

96. FRIEDLANDER, M. *et al.* 1996. Involvement of integrins alpha v beta 3 and alpha v beta 5 in ocular neovascular diseases. Proc. Natl. Acad. Sci. USA **93:** 9764–9769.

97. MAGNUSSEN, A. *et al.* 2005. Rapid access of antibodies to alpha5beta1 integrin overexpressed on the luminal surface of tumor blood vessels. Can. Res. **65:** 2712–2721.

98. PARSONS-WINGERTER, P. *et al.* 2005. Uniform overexpression and rapid accessibility of alpha5beta1 integrin on blood vessels in tumors. Am. J. Pathol. **167:** 193–211.

99. YAO, V.J. *et al.* 2006. Antiangiogenic therapy decreases integrin expression in normalized tumor blood vessels. Can. Res. **66:** 2639–2649.

100. NOGUERA-TROISE, I. *et al.* 2006. Blockade of Dll4 inhibits tumour growth by promoting non-productive angiogenesis. Nature **444:** 1032–1037.

101. RIDGWAY, J. *et al.* 2006. Inhibition of Dll4 signalling inhibits tumour growth by deregulating angiogenesis. Nature **444:** 1083–1087.

102. OKADA, Y. *et al.* 2007. A three-kilobase fragment of the human Robo4 promoter directs cell type-specific expression in endothelium. Circ. Res. **100:** 1712–1722.

103. ROBINSON, A.P., T.M. WHITE & D.W. MASON. 1986. MRC OX-43: a monoclonal antibody which reacts with all vascular endothelium in the rat except that of brain capillaries. Immunology **57:** 231–237.

104. ANDERSON, C.R., A.M. PONCE & R.J. PRICE. 2004. Absence of OX-43 antigen expression in invasive capillary sprouts: identification of a capillary sprout-specific endothelial phenotype. Am. J. Physiol. Heart Circ. Physiol. **286:** H346–H353.

New Horizons for Imaging Lymphatic Function

RUCHI SHARMA,[a,b] JULIET A. WENDT,[a,b] JOHN C. RASMUSSEN,[b] KRISTEN E. ADAMS,[b] MILTON V. MARSHALL,[b] AND EVA M. SEVICK-MURACA[a,b]

[a]Translational Biology and Molecular Medicine, and the

[b]Division of Molecular Imaging, Department of Radiology, Baylor College of Medicine, Houston, Texas, USA

In this review, we provide a comprehensive summary of noninvasive imaging modalities used clinically for the diagnosis of lymphatic diseases, new imaging agents for assessing lymphatic architecture and cancer status of lymph nodes, and emerging near-infrared (NIR) fluorescent optical imaging technologies and agents for functional lymphatic imaging. Given the promise of NIR optical imaging, we provide example results of functional lymphatic imaging in mice, swine, and humans, showing the ability of this technology to quantify lymph velocity and frequencies of propulsion resulting from the contractility of lymphatic structures.

Key words: lymph flow; NIR optical imaging; fluorescence; indocyanine green

Introduction

Many of the recent advances in the field of lymphovascular research can be ascribed to the discovery of lymphatic endothelium–specific biomarkers that enable identification of lymphatics. Yet there are few noninvasive, *in vivo* imaging modalities to visualize and quantitatively assess functional lymphatic vasculature within subjects, whether animal or human. The ability to noninvasively image functional lymphatics and markers of lymphangiogenesis could have an impact on our understanding of a number of diseases in which the lymphatic system is implicated, such as cancer metastases,[1] asthma,[2] diabetes,[3] and obesity.[4] Imaging could further play an important role in elucidating the role of lymphatics, not only in disease, but in health as well. In comparison to the hemovascular system, comparatively little is known about the lymphatic system, probably because of the inability to readily image the lymphatics. For example, while arteriovenous malformations (AVMs) are readily assessed with angiographic techniques, there are few available imaging techniques to assess lymphovascular malformations or even to discern whether lymphovascular malformations present with AVMs. In short, there is a paucity of imaging techniques to assess lymphatic function *in vivo*.

Here we review the imaging modalities used clinically and in animal studies as well as those under development in order to provide a "state of the art" review of new horizons in noninvasive, lymphatic imaging. In the first section, we begin with conventional imaging modalities of X-ray, nuclear, magnetic resonance imaging (MRI), and ultrasound (US) and focus upon their clinical use as well as their associated new experimental imaging agents currently under development. Next, the emerging developments in optical imaging modality are summarized and the enhanced sensitivity of near-infrared (NIR) optical imaging for imaging lymphatic function, lymph flow, and propulsion from contractile lymphangions is described. Examples of NIR fluorescence optical imaging within mice, swine, and humans are provided, demonstrating its unique ability to show lymphatic function. Finally, we describe work using labeled hyaluronan as a molecular "stain" that targets the LYVE-1 receptor and that may provide useful images of lymphatic architecture for therapy, surgical planning, and use in an intraoperative setting.

It is hoped that with an emerging armamentarium that allows the noninvasive imaging of lymphatic function and architecture, a new understanding will be gained of lymphatic biology, new means of therapeutic modulation of lymphatic function will be developed, and more effective diagnoses of lymphatic diseases will be enabled.

Modalities for Lymph Imaging

Imaging of the hemovascular system typically requires the administration of milliliters of contrast

Address for correspondence: Eva M. Sevick-Muraca, Division of Molecular Imaging, Baylor College of Medicine, One Baylor Plaza, BCM 360, Houston, TX 77030. Voice: +1-713-798-3684; fax: +1-713-798-2749.

evas@bcm.edu

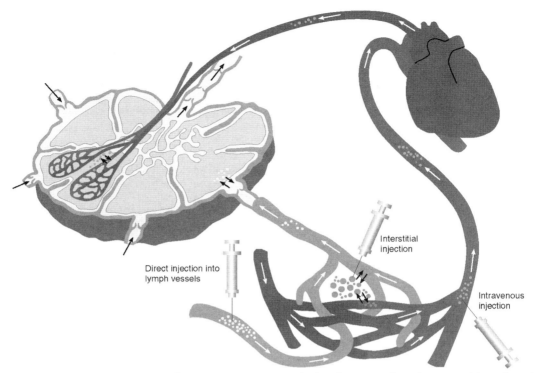

FIGURE 1. Injection routes to introduce contrast agents into the lymphatics are through interstitial (intradermal or subcutaneous) administration, direct administration into a cannulated lymphatic vessel, or intravenous injection.

agents that are injected intravenously for direct imaging of the arteriovenous systems. However, there is no simple route for administration of milliliter volumes of contrast agents into the lymphatic system. The dermis of the skin is made up of blind-ended initial lymphatics that serve as an entry portal for particles, cells, and interstitial fluid. Under a positive pressure gradient, these initial lymphatics distend, creating openings between loosely anchored endothelial cells that further allow influx of fluid and particulate matter from extracellular space.[5] Lymph drainage from these initial lymphatics is received by deeper collecting vessels that contain valves to maintain unidirectional flow of lymph. These collecting vessels are surrounded by smooth muscle cells that rhythmically contract to propel lymph to lymph nodes. As described below and in FIGURE 1, there are four distinct routes to introduce imaging agents into the lymphatic space: (i) intradermally for delivery into the lymphatic vessels via the lymph plexus in the dermis; (ii) subcutaneously into the interstitium for permeation into the lymphatic capillaries and vessels; (iii) directly introducing agents into the lymphatic space through lymphatic vessel cannulation; and (iv) intravenously for transit from the vascular space to the interstitium for permeation into the lymphatics or for direct depo-

sition via the microcirculation found within the lymph nodes. Particles less than a few nanometers (~11 nm) in size can pool into the lymph nodes via blood circulation, but can also diffuse into lymphatic vessels via the gap junctions between the endothelial cells under a hydrostatic pressure gradient.[6] Particles as large as 100 nm in diameter extravasate into the interstitial space, where they are phagocytosed by macrophages and are then transported to lymph nodes. Particles larger than 100 nm typically remain trapped in the interstitium.[7]

The following sections in this chapter are limited to lymphatic imaging modalities employed for whole-animal or patient imaging that employs contrast agents. While microscopy studies of lymphatic function in animal preparations has enabled significant understanding of the lymphovascular system, their translation to clinical use may be doubtful because of the restricted depth of tissue studied. In the following, we focus upon X-ray, nuclear, MRI, US, and NIR optical imaging modalities.

X-Ray-Based Techniques

X-ray imaging provides the most common form of hemovascular imaging. High-energy X-rays, which span the electromagnetic spectrum from 0.001 to 10 nm,[8] have a low soft-tissue attenuation coefficient. By introducing a contrast agent into the vasculature which absorbs X-rays, two-dimensional projections containing transmission shadows of vascular architecture are acquired. A series of projection images can then be tomographically rendered, using computed tomography (CT) to produce a three-dimensional image of the vasculature. Vascular imaging of blood vessels, or angiography, is established in the clinic, while lymphangiography is more complex.

Lymphangiography involves the direct administration of an iodinated contrast agent into a cannulated lymph vessel for radiography or CT of lymphatic architecture. Hultén *et al.*[9] and Kinmonth[10,11] first reported the use of blue dye administered interdigitally in hands or in feet to find, dissect, and cannulate lymphatic vessels for subsequent infusion of a contrast agent to detect lymphatic vasculature by serial radiography. Hultén *et al.*[9] also utilized subcutaneous injection of a blue dye to identify and cannulate lymphatic vessels in the arms of patients with breast carcinoma. The cannulated lymphatic vessels were then infused with 4–7 mL of Lipiodol (Guerbet, Aulnay-Sous-Bois, France), an oily contrast agent used to detect lymph nodes. Recently, lymphangiography was employed to detect postoperative lymphatic leakage in patients with known or suspected lymphocele, chylothorax, or lymphatic fistulas.[12] FIGURE 2 shows an X-ray scan taken after injection of up to 14 mL of Lipidiol. The pedal lymphangiography depicts a lymphatic fistula (arrows) in a patient who underwent resection of melanoma in the right thigh and inguinal lymphadenectomy.[12] Although lymphangiography enables imaging of lymph vessels and lymph nodes,[13,14] it has been largely abandoned because of such complications as infection, lung embolization, pulmonary edema, respiratory distress, and damage to the lymphatics, as well as the technical skill required to cannulate lymph vessels.[15–17]

As an alternative to direct injection into the lymphatic space, water-soluble iodinated contrast agents can be delivered intradermally for subsequent lymphatic uptake in an indirect lymphangiography technique. Partsch and co-workers[18,19] administered up to 20 mL of Iotasul (Schering AG, Berlin, Germany), a nonionic water-soluble contrast agent, into the interdigital or intradermal space of patients with lymphedema in order to visualize the dermal lymphatics that were too small to be cannulated. The technique avoids the complexity and invasiveness of direct injection into the lymphatic space,[20] but is restricted by the small volume (~0.2 mL) of fluid that can occupy the dermal tissues at each injection site.

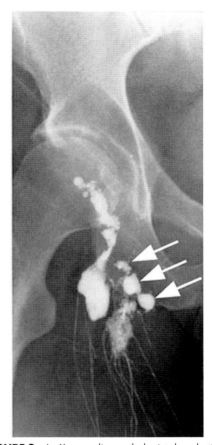

FIGURE 2. An X-ray radiograph depicts lymphatics and lymphatic fistula in a patient who underwent resection of melanoma of right thigh and inguinal lymphadenectomy. The *arrow* indicates Lipidiol leakage. Pedal lymphangiography was performed after injection of 14 mL of Lipidiol iodinated glycerol ester. (From Kos *et al.*[12] Reproduced by permission.)

Iopamidol, another water-soluble nonionic contrast agent, has also been employed to visualize breast and esophagogastric tract lymph drainage pathways. CT lymphangiography using subcutaneous injections of up to 1 mL of undiluted iopamidol in 10 female dogs enabled visualization of breast lymph pathways, 20 sentinel lymph nodes, and 110 distant nodes.[21] In addition, CT lymphangiography following endoscope-guided administration of 2 mL of Iopamidol (Nihon Schering, Osaka, Japan) into the esophageal or gastric submucosal space of nine dogs enabled visualization of lymphatic channels draining into regional lymph nodes (possibly sentinel lymph nodes), hence demonstrating the technique's potential to aid surgeons in the surgical planning of the extent of lymph node

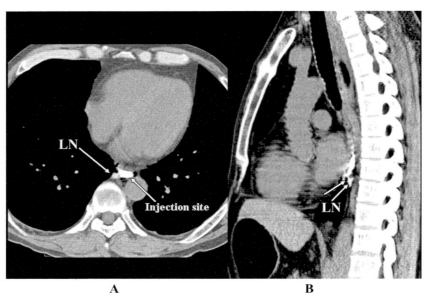

FIGURE 3. Transverse (**A**) and sagittal (**B**) endoscopic contrast-enhanced CT lymphangiograms of a patient with esophageal cancer injected peritumorally with 2 mL of iopamidol. The images identify short lymphatic vessel (*unlabeled arrow* in B) and a paraesophageal lymph node. (From Suga *et al.*[22] Reproduced by permission.)

dissection. This imaging agent has also been demonstrated for esophageal cancer staging by Suga and co-workers,[22] as depicted in FIGURE 3, which illustrates transverse and sagittal CT lymphangiograms of a patient with middle-thoracic esophageal cancer. An endoscope was used to administer 2 mL of iopamidol in the peritumoral submucosal space. The CT lymphangiogram identified a short lymphatic vessel (unlabeled arrow in FIG. 3B) at a paraesophageal region draining into a middle-thoracic paraesophageal lymph node. The patient underwent esophagectomy and regional lymphadenectomy, including the unenhanced lymph nodes, which were later found to be positive for micrometastases. These images demonstrate the ability of CT lymphangiography to detect sentinel lymph nodes from the lymphatic drainage routes around the tumor. The potential of CT lymphangiography to possibly detect malignant lymph nodes was demonstrated by Wisner *et al.*,[23] who administered 2 mL per subcutaneous injection (6–12 mL total) of a 15% wt/vol of an iodinated suspension in swine with melanoma. Reviews of CT scans revealed differences in iodine concentration and lymph node morphology in diseased as compared to the normal lymph nodes, suggesting that the use of the iodinated contrast enhanced CT to identify malignant lymph nodes. However, once within the lymphatic compartment, the iodinated contrast agents approved for human use drain rapidly to the blood vascular compartment and clear from the body, limiting the time for imaging and their clinical usefulness. In addition, the millimolar tissue agent concentrations needed to produce X-ray image contrast make their use difficult when administered in limited volumes within the dermal space.

A number of experimental CT agents are currently under development as a result of these limitations. Novel bismuth sulfide nanoparticles with a longer hemovascular circulation half-life may address the clearance issues of current iodinated CT contrast agents In an experimental study by Rabin *et al.*,[24] 50 μL of polymer-coated bismuth sulfide nanoparticles with a hydrodynamic diameter of 30 ± 10 nm and corresponding to 11.4 μmol Bi^{3+} were subcutaneously injected and imaged for up to 140 h in mice. The images depicted contrast-enhanced lymph nodes. Although CT imaging of long-circulating bismuth sulfide nanoparticles can be achieved at low volumes and with efficacies comparable to that of iodinated agents, concerns regarding fate and long-term toxicity may hinder future clinical translation.

Because of the difficulties of delivering sufficient amounts of agents necessary to produce X-ray contrast, lymphangiography is not routinely used for diagnosis of lymphatic disorders and does not have the clinical diagnostic impact of hemovascular angiography.

Nuclear Imaging

Nuclear imaging techniques employ radionuclides as contrast agents which, upon decay, generate either gamma photons of ~0.0001-nm wavelength[8] or positrons that annihilate to produce two 511-kEV photons at 22-nm wavelength. Gamma photons and annihilation photons can propagate efficiently and have low attenuation in tissues. Gamma scintigraphy is the most common nuclear imaging in which gamma photons are detected for the generation of a two-dimensional planar image detailing the position of the radionuclide. While scintigraphy provides two-dimensional images, single-photon emission computed tomography (SPECT) reconstructs three-dimensional positions of radionuclide sources within the body from a series of images acquired by a gamma camera that is rotated around the subject. However, since the technique requires pinhole rejection of scattered gamma photons, a large number of photons remain undetected in SPECT, and scanning times can be long. Conversely, positron emission tomography (PET) depends upon coincident detection of the annihilation photons for three-dimensional reconstruction of the position of positron annihilation. Unlike X-ray-based contrast agents, nuclear techniques, such as gamma scintigraphy, SPECT, and PET, are associated with exquisite sensitivity because of the ability to image tissues with pico- to femtomolar tissue concentrations of radionuclides. The enhanced sensitivity enables intradermal administration of small volumes and concentrations of radiopharmaceuticals for lymphatic delivery.

Lymphoscintigraphy

The most common nuclear method of imaging the lymphatics is lymphoscintigraphy. In the United States, lymphoscintigraphy is performed with filtered 99m-technetium (Tc-99m) sulfur colloid (40–100 nm in diameter), whereas in Europe it is performed with Tc-99m albumin nanocolloid (5–80 nm), Tc-99m rhenium colloid (2–15 nm), Tc-99m antimony colloid (3–30 nm), and other soluble noncolloids, such as Tc-99m human serum albumin, 99mTc human immunoglobin, and Tc-99m dextran.[25–27] Typically. intradermal or subcutaneous injections of 11.1–111 MBq (0.3–3 mCi) of filtered Tc-99m sulfur colloid in up to 4 mL of saline solution are administered between 30 min to 2 h prior to imaging; however, administration of radiocolloid can be performed 1 day prior to imaging. The scintigram requires anywhere from 20 min up to 2 h to be acquired. For nodal staging in breast cancer and melanoma, intradermal or subcutaneous injections of radiocolloid are generally done peritumorally or subareolarly (in breast cancer), and gamma scintigraphy is used to identify the draining nodal basins for subsequent resection for pathologic examination for cancer.

Because of the slow depot clearance and incomplete lymph node extraction in sentinel lymph node mapping (SLNM) using Tc-99m-labeled sulfur colloid, several radiopharmaceuticals are under investigation to stage cancer. In Japan, Tc-labeled rhenium colloid (Tc-99m Re, 2–15 nm in diameter) and soluble Tc-99m human serum albumin diethylene-triamine-pentaacetic acid (Tc-99m HSA-D; molecular weight greater than 50,000) have been more widely used than Tc-99m sulfur colloid. Sato and co-workers[28] compared the usefulness of both radiopharmaceuticals for nodal staging. In their study, 65 patients with oral cancer received injections of 0.25 mL of 37 MBq (1 mCi) Tc-99m Re, whereas while 20 received 0.1 mL of 74 MBq (2 mCi) Tc-99m HSA-D for nodal staging. When compared to the soluble Tc-99m HSA-D, Tc-99m Re was found to provide better scintigrams of lymph nodes, presumably because of the preferential uptake of radiocolloids into the lymph nodes while the soluble Tc-99m-HSA-D passed through the lymphatics. However, the radiocolloid required longer times for transit to the lymph nodes than the soluble radiopharmaceutical. Recently, in the United States, Wallace and colleagues[29–31] introduced "Lymphoseek," which consists of a dextran backbone covalently bonded to mannose units and linked to Tc-99m. The mannose units target the mannose-binding protein present on reticuloendothelial cells (phagocytic cells, such as macrophages) found in lymph nodes. The success rate of SLNM following intradermal and subcutaneous administration of Lymphoseek has been mixed, and further clinical trials are required to demonstrate improved injection site clearance and lymph node extraction as compared to Tc-99m-sulfur colloid.

In clinical practice, if a sentinel lymph node is found to be positive for tumor cells, then further lymph node dissection is typically performed to acquire adequate information for prognosis. In women with advanced cancer, complete axillary lymph node dissection (ALND) may also be performed. As a consequence of these lymph node dissections, 5–80% of the breast cancer patients undergoing SLNM and ALND develop breast cancer–related lymphedema (BCRL) in the arm.[32] To reduce this occurrence, Thompson *et al.*[33] recently proposed a novel technique of axillary reverse-mapping (ARM) that seeks to identify and preserve the lymphatics that drain the arm as opposed to those that drain the breast. In their study, 40 breast cancer patients who were scheduled for ALND and/or SLNM were injected with 37 MBq (1 mCi) of unfiltered Tc-99m sulfur colloid in 4 mL within the subareolar

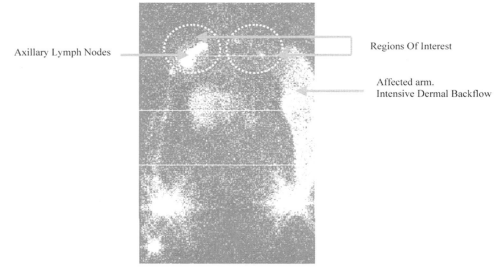

FIGURE 4. Lymphoscintigram of a postmastectomy lymphedema patient. The affected arm depicts dermal backflow as compared to the unaffected arm that shows a channel draining the injection site to the axillary lymph nodes. (From Szuba *et al.*[32] Reproduced by permission.)

plexus for sentinel lymph node detection and 2.5 mL of blue dye subcutaneously or intradermally within the inner upper arm to map arm lymphatic drainage. In all patients, 100% of the resected sentinel lymph nodes were radioactive but not blue. In addition, pathologic examination revealed the negative status of the blue lymph nodes that drained the arm in the seven patients who had positive axillary lymph nodes. ARM may be an effective technique to prevent resection of arm lymphatics and subsequent occurrence of BCRL. However, a long-term prospective study is needed to ensure that along with a reduction in the risk of BCRL, the addition of ARM does not lower the accuracy of nodal staging.

Lymphoscintigraphy is also used to diagnose lymphedema and to assess lymphedema treatment efficacy on the basis of radiocolloid clearance rates, dermal backflow, and ratios of radioactivity in affected and unaffected regions. To investigate the potential of lymphoscintigraphy to evaluate lymphedema and to monitor response to therapy, Szuba *et al.*[32] administered 0.2 mL of 0.25 mCi Tc-99m filtered sulfur colloid subcutaneously in the first and second interdigital space of postmastectomy lymphedema patients. FIGURE 4 depicts a lymphoscintigram of a postmastectomy lymphedema patient with dermal backflow in the affected arm. Scintigraphy enabled quantitative estimation of dermal backflow and radioactivity ratio in the axilla of the affected arm to that in the healthy arm. In addition, the authors developed an empiri-

cal scoring system for postmastectomy lymphedema based upon visualization of lymph nodes and dermal backflow. Their results showed correlations of both the radioactivity ratio in the axilla of the affected arm to that in the healthy arm and the percent reduction in edema post treatment to both the lymphoscintigraphic score and excess initial arm volume. These results demonstrate the utility of lymphoscintigraphy to assess BCRL. Because the incidence of BCRL and the time at which the condition strikes varies from patient to patient, O'Mahony[34] and co-workers designed a lymphoscintigraphic study to identify lymphatic functional and morphologic changes in 19 patients who had undergone ALND. The authors performed intradermal injections of 0.1 mL of soluble Tc-99m human immunoglobin (40 MBq) in the second interdigital space of the hand to trace protein kinetics. Their rationale was that Tc-99m human immunoglobin is more stable than Tc-99m human serum albumin, has a slower blood clearance rate, and can provide information of lymphatic transport from its accumulation in blood. Although lymphatic architecture may be expected to change after ALND, there was a general tendency of more rapid appearance of radiolabeled proteins within the vascular space in patients who had undergone surgery than in those who had not. However, soluble radioisotopes such as Tc-99m human immunoglobin and Tc-99m human serum albumin may clear from injection sites by resorption into blood capillaries as well as by transport through the lymphatics. In order to link incidence of BCRL to more rapid appearance of protein in the bloodstream, a better understanding of the mechanism of transport from interstitium to the circulatory system is needed.

In addition to studies that involve interstitial injections to assess lymphedema, Peer *et al.*[35] recently reported on the use of intravenous injections of soluble Tc-99m hexakis-2-methoxy isobutyl isonitrile (Tc-99m MIBI) to detect AIDS-related cutaneous/subcutaneous Kaposi's sarcoma (KS) lesions and the associated condition of lymphadenopathy. KS is thought to arise from the vascular and lymphatic endothelium, and there are cases which report lymphadenopathy, and associated lymphedema can arise without the appearance of cutaneous/subcutaneous lesions. In these investigators' studies, the lymphoscintigrams of patients injected with 740 MBq (20 mCi) of Tc-99m MIBI showed accumulation in KS lesions and enabled detection of abnormal lymph nodes, suggesting its use for KS staging and for assessing response to therapy.

Lymphoscintigraphy has also been employed to assess the efficacy of emerging therapies for treatment of lymphedema. In order to investigate the success of transplanted autologous lymphatic vessels in patients with secondary lymphedema, Weiss *et al.*[36] conducted preoperative baseline and 8-year postoperative follow-up lymphoscintigraphy studies. To map the lymphatics of the affected limb, patients were given approximately 74 MBq (2 mCi) of Tc-99m nanocolloid injected subcutaneously into the first interdigital space. Lymphoscintigraphy scans provided lymph drainage maps that assisted in the planning of microsurgical interventions. Postoperative scintigraphy studies successfully depicted significant improvement of lymph drainage and visualization of lymph nodes. Thus, lymphoscintigraphy proved the potential of autologous lymph vessel transplant as a therapeutic option to treat lymphatic obstruction in lymphedema.[36,37] The recent discovery of lymphangiogenesis-inducing growth factor, VEGF-C (vascular endothelial growth factor C), has provided another possible avenue to treat lymphedema. Administration of VEGF-C in the mouse skin[38] and rabbit ear[39] model of postsurgical lymphatic insufficiency has been shown to induce lymph vessel growth. However, the function of regenerating lymphatics is still unknown. In order to understand the physiology of regenerating lymphatics in an experimental animal model, Blacker and co-workers[6] employed lymphoscintigraphy after subdermal injections of 1.5 MBq in 10 μL of Tc-99m antimony trisulfide colloid (Tc-99m ATC) (~10 nm diameter), Tc-99m tin fluoride colloid (Tc-99mTFC) (~2000 nm diameter), and

water-soluble, Tc-99m diethylenetriaminepentaacetic acid (Tc-99m DTPA) in order to examine depot clearance rates and lymph velocity in Australian geckos with original tails (OTs), fully regenerated tails (FRTs), and regenerating tails (RTs) at 6, 9, and 12 weeks after autotomy. Tc-99m ATC had a greater clearance and faster lymph velocity in OT geckos than in RT geckos. However, Tc-99m TFC had better clearance and faster velocity in RT than in FRT or OT animals, implying that larger particles were impeded as tail regeneration progressed. Last, water-soluble molecules immediately cleared from the injection site via venous capillaries in all tails, indicating that the slower clearance of colloids occurred via the lymphatics. The variable transit patterns for different types of radiopharmaceutical agents imply possible changes in vessel density and tissue porosity in the regenerating lymphatic model.

In summary, lymphoscintigraphy is a radionuclide-based imaging technique that enables two-dimensional visualization of the lymphatic network in cancer patients for lymph node staging and in lymphedema patients for evaluating response to therapy. Colloidal particle size and route of administration have an impact on the distribution of radiopharmaceuticals and the capacity for lymphatic imaging. For example, the imaging of lymphatic vessels was better enabled in BCRL patients by intradermal administration of Tc-99m human immunoglobin than subcutaneous administration because of the direct access to initial dermal lymphatics.[40] Nonetheless, the clinical evaluation of lymphedema remains qualitative or semiquantitative on the basis of clearance rates and dermal backflow parameters obtained from lymphoscintigraphy. Gamma camera integration times are typically long, preventing the dynamic imaging needed to quantitatively assess lymph flow, which, as we have found, is an inherent characteristic of healthy lymph vessels.

Positron Emission Tomography

Because of long gamma camera integration times, three-dimensional SPECT has not made a significant impact upon lymphatic imaging, whereas three-dimensional PET imaging promises to improve nodal staging in some kinds of cancer. PET involves intravenous injection of positron-emitting molecularly targeted radiopharmaceuticals to monitor biological processes and has been used to detect diseased lymph nodes.[30,41,42] PET images are often combined with X-ray CT to gain the additional advantage of structural information.[42] The contrast agents are typically analogues of naturally existing compounds or targeting agents that are either (1) isotopically substituted with a beta-emitting isotope (e.g.,

C-11,[43] N-13,[44] O-15,[45] F-18, I-124, I-125, or I-131)[30,42,43,46] or (2) conjugated to a chelating unit for sequestration of a beta-emitting substance, such as Cu-64 or Cu-61. 18-F-fluoro-2-deoxy-D-glucose (FDG) is the most common PET agent currently employed in the clinic. FDG is a glucose analogue with an –OH group replaced with F-18. The molecule is taken up by cells via Glut-1 transporter and, as in the case of glucose, undergoes phosphorylation to FDG-6-phophate, reflecting the hexokinase enzyme activity that is a rate-limiting step in glucose metabolism. However, FDG-6-phosphate is not further metabolically processed and becomes trapped within the cell. Since tumor cells exhibit increased glucose metabolism partly because of increased levels of glucose transporters and hexokinase enzyme activity, FDG-PET imaging may provide a means to localize sites of increased glucose metabolism, as in the case of reactive or cancer-positive lymph nodes.

PET has high sensitivity with a potential of signal detection at picomolar tissue concentrations of tracer, but falls short of conventional imaging techniques in terms of spatial resolution (∼2 mm). A combination of the complementary roles of functional imaging by PET and anatomic imaging by CT has improved staging of malignancies.[14,47] FIGURE 5 depicts a sagittal CT scan (A), PET scan (B), and a coregistered PET-CT scan (C) of a patient with non–small cell lung cancer after intravenous injection of 350–400 MBq of FDG.[48] No enlarged lymph nodes are identified on the unenhanced CT scan. The PET scan identifies an area of increased FDG uptake, but it is difficult to identify the precise anatomic location of the abnormality. Integrated PET-CT, however, enables identification of a normally sized supraclavicular lymph node that was later excised and histologically confirmed for metastasis.[48] While 18-FDG PET-CT has been proposed to improve the diagnostic accuracy in nodal staging, not all kinds of cancer have higher glucose metabolism. Hence specificity and sensitivity for cancer nodal staging may prevent 18-FDG PET-CT from ubiquitous clinical use. However, more tumor-specific nuclear imaging agents are under development to more accurately perform nodal staging with PET.

Magnetic Resonance Imaging

Unlike nuclear imaging techniques, MRI is a technique that does not involve exposure to ionizing radiation. MRI utilizes magnetic properties of nuclei containing an odd number of protons and/or neutrons (such as the hydrogen atom) and their interaction with an external magnetic field and radiofrequency pulse to generate high-resolution images of the hu-

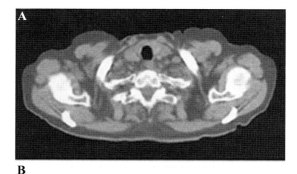

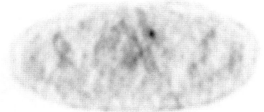

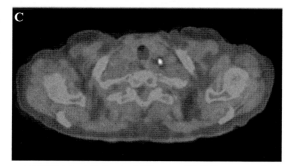

FIGURE 5. (**A**) Sagittal CT scan, (**B**) PET scan, and (**C**) coregistered PET-CT scan of a patient with non–small cell lung cancer after intravenous injection of 350–400 MBq of 18-FDG. (From Lardinois *et al.*[48] Reproduced by permission.)

man body. T1-weighted images result from the relaxation of excited nuclei due to energy exchange with its surroundings, while T2 relaxation occurs from the exchange of energy between nuclei. MR lymphangiography involves interstitial or intravenous injection of contrast agents including gadolinium-labeled diethylene-triaminepentaacetic acid (Gd-DTPA), Gd dendrimers or liposomes, and iron oxide particles[14,49,50]

Pan and co-workers[51] demonstrated the feasibility of MR lymphangiography to assess lymphatic obstruction in the murine model of surgically induced tail lymphedema. T1-weighted MRI scans were conducted in normal and lymphedematous mice after subcutaneous tail injections of 0.1 mmol/kg Gd-DTPA 5 cm distal to the base of the tail. Unlike the normal mice, in

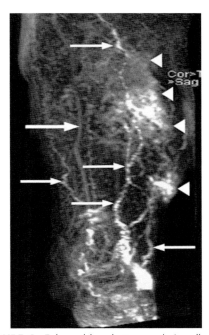

FIGURE 6. Enlarged lymphatic vessels (*small arrows*), dermal backflow (*arrowheads*), and concomitantly enhanced veins (*large arrow*) in the lower left part of the leg of a 69-year-old man with primary lymphedema. The image was acquired 35 min after gadodiamide injection. (From Lohrmann *et al.*[53] Reproduced by permission.)

which the two major bilateral collecting lymph vessels were visible from Gd-DTPA contrast (before rapid clearance), the lymphedematous mice showed accumulation of Gd tracer in the surgical site in the tail (1.6 cm distal from the base of the tail). In addition, lymph flow obstruction was assessed by comparing the mean signal intensity as a function of time in normal and edematous tails. Reduction in signal intensity with time at a region of interest (ROI) provided a measure of agent clearance in normal mice as opposed to the increase in signal intensity that indicated agent accumulation in edematous mice. These results demonstrate the feasibility of MR lymphangiography to assess lymphedema in experimental models. In human studies, Ruehm and co-workers[52] evaluated the potential of gadoterate meglumine, an extracellular paramagnetic agent, as an interstitial agent for visualization of lymphatics and lymph nodes in five healthy volunteers and three patients suffering from lymphatic disorders. After a total of 4.5 mL of contrast agent was injected into the dorsum of the foot, lymph vessels and lymph nodes could be clearly visualized from MR (indirect) lymphangiography. Lohrmann and colleagues[53–55] demonstrated the feasibility of detecting

lymphatic vessels in patients with lymphedema using gadodiamide or gadoteridol, which are nonionic water-soluble paramagnetic contrast agents. A total amount of 4.5 mL of gadodiamide or 9 mL of gadoteridol at a dose of 0.1 mmol/kg of body weight was administered intracutaneously into the dorsal aspect of foot in the four interdigital spaces and medial to both first proximal phalanges. MR lymphangiography depicted lymphatic vessels, collateral vessels with dermal backflow, and inguinal lymph nodes. FIGURE 6 depicts enlarged lymphatic vessels (small arrows), dermal backflow (arrowheads), and concomitantly enhanced veins (large arrow) in the left lower part of the leg of a 69-year-old man with primary lymphedema. The image was acquired 35 min after gadodiamide administration. Inguinal lymph nodes with external iliac lymphatics were also visualized in these studies. Recently, Matsushima *et al.*[56] reported unenhanced noninvasive visualization of the lymphatics of the trunk that included the thoracic duct, cisterna chyli, and lumbar lymphatics. In their studies, they employed heavily T2-weighted imaging with the addition of respiratory triggering to enable imaging of lymph separately from venous flows. FIGURES 7A to C depict coronal maximum intensity projection images of a 28-year-old male, with arrows identifying the cisterna chyli, and arrowheads indicating lumbar lymphatics and para-aortic lymphatic trunks. While other slowly flowing fluids may contaminate the images, the opportunity to assess lymphatic architecture without the administration of a contrast agent could have a significant impact on how we diagnostically assess the lymphatics.

The MRI techniques described above do not enable cancer staging of lymph nodes. MRI with superparamagnetic iron oxide contrast may depict salient features of normal and cancer-positive lymph nodes.[53] Superparamagnetic contrast agents are composed of a water-insoluble crystalline magnetic core (4–6 nm in diameter), usually magnetite (Fe_3O_4) or magnemite ($\gamma\text{-}Fe_2O_3$), surrounded by a layer of dextran or starch derivatives. The particles shorten proton relaxation times, giving a signal distinct from tissue without particles. To detect diseased lymph nodes, dextran-coated ultra-small superparamagnetic iron oxide (USPIO) particles of less than 50 nm in diameter, known as ferumoxtran, have been intravenously injected into patients in investigational studies.[7] USPIO particles disseminate to lymph nodes via two mechanisms: (*a*) direct transcapillary migration across blood capillaries into the medullary sinuses within the lymph node; and (*b*) extravasation from the venules into the interstitium for uptake by lymph vessels draining to lymph nodes. The iron oxide particles may be subsequently

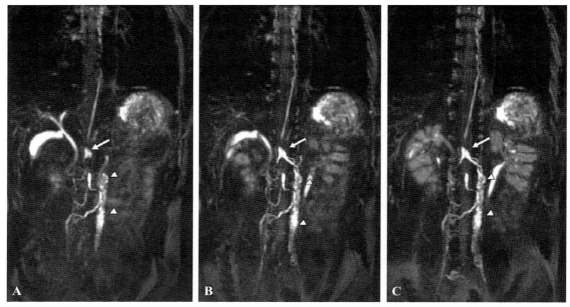

FIGURE 7. Coronal maximum intensity projection images of a 28-year-old male. Three dimensional heavily T2-weighted images were obtained in the expiratory phase when the diaphragm was at the highest level. (**A** to **C**) A high intensity linear structure clearly identifies thoracic duct. *Arrows* identify a saccular structure to be cysterna chili, and *arrowheads* indicate lumbar lymphatics and para-aortic lymphatic trunks. The lymphatics were solely identified on the basis of the knowledge of systemic anatomy. (From Matsushima *et al.*[56] Reproduced by permission.)

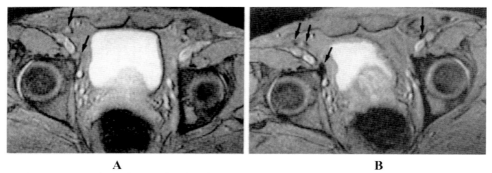

FIGURE 8. MR images depict benign iliac lymph nodes in a patient with prostatic carcinoma. (**A**) Precontrast image of external iliac lymph nodes. *Arrows* indicate iliac lymph nodes that appear bright on account of high signal intensity. (**B**) Postcontrast image of iliac lymph nodes (*arrows*) that appear to be dark because of the accumulation of USPIO. Contrast-enhanced MR images are acquired after injection of 1.7 Fe/kg USPIO. (From Bellin *et al.*[57] Reproduced by permission.)

phagocytosed by macrophages within the lymph nodes, introducing field inhomogeneities that reduces signal. Bellin *et al.*[57] studied the uptake of USPIO particles in normal and diseased lymph nodes. The observed T2-weighted images in FIGURE 8 show inhomogeneities in benign lymph nodes, which are produced by accumulated iron oxide particles. FIGURE 8A depicts a precontrast image of external iliac lymph nodes of a patient with prostate cancer. The arrows

indicate iliac lymph nodes that have high signal intensity. FIGURE 8B depicts the image of normal iliac lymph nodes (arrows) acquired after injection of 1.7 Fe/kg USPIO that appear to be dark on account of agent accumulation. Malignant lymph nodes that lack reticuloendothelial cells cannot accumulate USPIO and thus retain the bright MR signal.[14,57,58]

The high spatial resolution of MRI has the greatest potential for assessing the morphology of the lymphatic system, including lymph vessels and nodes. USPIOs provide indirect evidence of nodal cancer involvement through negative contrast and until sensitivity and specificity for nodal detection can be determined it is unlikely that they will supplant current biopsy procedures. While enhanced MR lymphangiography shows great promise for limb imaging, the opportunity for unenhanced lymphangiography for imaging the lymphatic tract holds unprecedented promise for diagnosis and treatment planning of lymphatic disorders.

Ultrasound

US employs sound waves at frequencies ranging from 1 MHz to 20 MHz (or 1.57 mm to 0.08 mm in wavelengths within tissue) that exceed the frequency range audible to humans. The US device consists of a transducer that emits US waves that pass through tissue and, upon encountering tissues of varying acoustic properties, generate echoes that are registered by piezodetectors placed at the tissue surface.[59] US is the imaging modality most frequently used for evaluation of a parasitic form of lymphedema called *filariasis*. Suresh *et al.*[60] reported the use of US to detect dilated lymph vessels in patients with filariasis as well as to measure worm motion within lymphatics using Doppler US. Used typically to quantify the velocity of blood flow toward and away from a US probe, Doppler US is unable to assess acellular lymph flow. However, Mellor *et al.*[61] reported the use of Doppler US to examine venous flow of patients with lymphedema–distichiasis (LD) who had known mutations in the *FOXC2* gene. While FOXC2 mutations are implicated in defective lymphatic valves, Mellor and co-workers found venous reflux and venous valvular incompetence in the proximal veins in LD patients. The result may not be surprising considering the close genetic relationship between lymphatic and hemovascular disorders and that many of the venous vasculopathologies clinically evaluated may also have an unidentified lymphatic component.

In addition to lymphedema, US may be utilized to characterize metastatic lymph nodes on the basis of their size, shape, and echogenicity using US. A US image of a normal lymph node appears ellipsoid in shape, with hypoechoic cortex and hyperechoic hilum. A malignant lymph node can appear round with loss of hyperechoic hilum and noticeable hypoechogenicity of cortex.[14] Since it does not employ a contrast agent, US imaging of lymph nodes may prove to be useful for detecting tumor burden in lymph nodes that obstruct the drainage of radiotracer to sentinel lymph nodes, thereby providing false negative SLNM results. While nodal staging by ultrasonography is reported to have

sensitivity ranging from 50% to 87%,[62,63] and specificity of up to 92%[62]; and the addition of fine-needle aspiration is reported to have a controversial sensitivity ranging from 18% to 78.5% and specificity of 93.5%–100%,[62,63] it is unlikely that US can detect occult micrometastases, which are increasingly thought to be clinically relevant. US is also restricted from studying deep lymph nodes.

Contrast-enhanced ultrasonography includes the administration of submicron-sized microbubbles to image the microcirculation as well as the lymphatics and lymph nodes in animals.[13,49,64,65] The echogenic microbubbles are phagocytosed by macrophages or reticuloendothelial cells and appear bright on an ultrasonogram. Malignant lymph nodes that lack reticuloendothelial cells cannot take up the microbubbles and remain hypoechoic. Goldberg *et al.*[65] injected a milliliter volume of 2.4–3.5 μm diameter, lipid-stabilized perflurobutane microbubbles subcutaneously or intradermally in the peri-tumoral region of swine. FIGURES 9A and B illustrate US scans of a normal lymph node at pre- and postcontrast in a swine model of melanoma. The postcontrast US scans depict hyperechoic lymph nodes. FIGURE 9C represents a contrast-enhanced lymphatic channel draining into a sentinel lymph node (N) after an intradermal injection of microbubbles. Last, FIGURE 9D indicates a post-contrast US scan of a lymph node (arrows), mostly composed of hypoechoic region depicting buildup of melanoma tumor cells (T), with a small region of normal tissue (N) that demonstrates contrast enhancement. Hauff and co-workers[66] developed and utilized L-selectin ligand-specific polymer-stabilized air-filled microparticles (MPs) for active targeting of peripheral lymph nodes in mice and dogs. The feasibility of detecting lymph nodes by molecularly targeting the L-selectin ligand present on high endothelial venules was demonstrated by intravenous injection of MPs (10–14 million particles per kg) in mice or dogs by means of stimulated acoustic emission (SAE) ultrasonography, in which the applied US caused bursting of microbubbles and generation of harmonic US frequencies for detection. The technique provides yet another potential tool to detect cancer-positive lymph nodes with specificity depending upon the targeting, which in this case was a nonspecific L-selectin ligand in a cancerous node. Unfortunately, the microbubble agent is destroyed in the process of SAE imaging.

Optical Imaging

Optical imaging is based upon the administration of contrast agents that, when excited by tissue

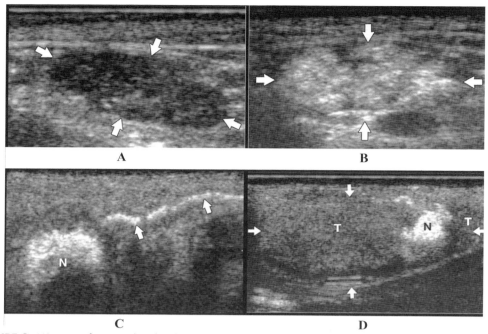

FIGURE 9. US scans of a normal and malignant lymph nodes pre- and postcontrast subcutaneous administration in a swine model of melanoma. (**A**) Precontrast US image depict hypoechoic ellipsoid lymph node; (**B**) postcontrast US scans depict hyperechoic lymph nodes; (**C**) contrast-enhanced lymph channel draining into a sentinel lymph node (N) subsequent to an intradermal injection of microbubbles; (**D**) postcontrast US scan of a lymph node (*arrows*) composed of hypoechoic region depicting buildup of melanoma tumor cells (T), with a small region of normal tissue (N) that demonstrates contrast enhancement. (From Goldberg *et al.*[65] Reproduced by permission.)

propagating excitation light, exhibit fluorescence. The process of fluorescent emission occurs when an incident photon of resonant energy is absorbed by a molecule residing at equilibrium in a ground state. The molecule, upon absorption of the photon, is subsequently excited to a higher electronic energy level. Fluorescence occurs when the molecule radiatively relaxes, releasing the energy required for it to return back to its ground state. The energy released is in the form of photons that either have wavelengths in the visible range (400–700 nm) or in the NIR region (700–900 nm). Blood, melanin, and water have a low absorption of light in the NIR region[67] that enables NIR light to penetrate deeply in tissues.

Fluorescence microlymphangiography utilizes contrast agents, such as fluorescein, which is excited and emits at visible wavelengths and, as a result, provides information on superficial lymphatics. The technique utilizes interstitial injections of approximately $10\,\mu L$ of fluorescein isothiocyanate dextran and 494-nm excitation illumination of the superficial lymphatics.[68,69] FIGURE 10 compares fluorescence microlymphangiograms after the administration of $10\,\mu L$ of 25%

(w/v) fluorescein isothiocyanate into the dermis on the ventral aspect of the normal (panel A) and affected forearm (panel B) of a BCRL patient. The superficial fluorescent lymphatic channels are visualized by fluorescent video microscopy usually across a quarter-sized area, and demonstrate the denser lymphatic network in the dermis of the edematous as opposed to the unaffected arm. In addition to mapping the superficial lymphatics, microlymphangiography may enable noninvasive estimation of the width of lymph channels.[69] Fluorescence imaging of lymph nodes in the visible wavelength range was also demonstrated by McGreevy *et al.*,[70] who utilized a contrast agent that was constituted of a red fluorescent dye, Cy5, conjugated to vitamin B_{12} for excitation at 633–650 nm and emission at 668 nm. Increased sequestration of cobalamin (vitamin B_{12}) by primary tumor cells is hypothesized to provide contrast for cancer-positive lymph nodes. Its feasibility to detect lymph nodes was demonstrated in normal swine following intradermal injection of 1 mL containing 1.5 mg of the conjugate. However, because of the limited tissue penetration depth of the visible light, the investigators were not

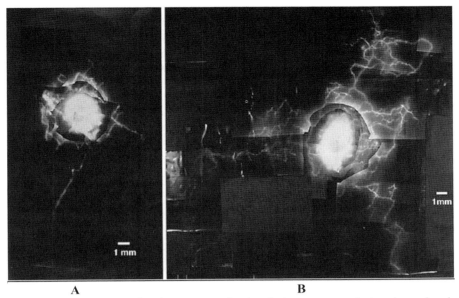

FIGURE 10. Fluorescence lymphangiogram of a lymphedematous arm depicts denser lymphatic network as opposed to that in normal lymphatics in a postmastectomy lymphedema patient. (**A**) Lymphatics in affected limb; (**B**) lymphatics of affected arm of a postmastectomy patient with edema. Ten microliters of 25% (w/v) fluorescein isothiocyanate was injected into the dermis on the ventral aspect of the forearm. (From Mellor *et al.*[69] Reproduced by permission of S. Karger AG.)

able to image deep lymphatics, such as pre- and postcollector vessels.

The NIR emission from quantum dots (QDs) offers somewhat better tissue penetration, but QDs are mostly excited by visible light and are restricted from deep tissue imaging. Type II core/shell QDs are semiconductor nanocrystals that can comprise of a core of cadmium telluride, a shell of cadmium selenide, and an outer coating of soluble phosphines.[71] QDs have a hydrodynamic diameter of 15–20 nm and a peak emission of 840 nm.[71] Injections of type II QDs within the footpad of rodents,[72] the intraparenchymal[73,74] and submucosal esophageal[71,73] spaces, have demonstrated the feasibility of intraoperative mapping of lymph nodes for resection and subsequent biopsy. Even though the use of high quantum yield type II QDs enables visualization of lymph nodes reportedly as deep as 1 cm within an intraoperative field, concerns about the fate and toxicity of QDs due to their heavy metal content may impede their translation into the clinic.

NIR imaging involves intradermal, subcutaneous, or intravenous injections of fluorescently labeled soluble agents to map lymphatic vasculature. Previously, Gurfinkel *et al.*[75] reported the use of intravenous injections of 1.0 mg kg^{-1} of an NIR dye, indocyanine green (IC-Green), for excitation at 778 nm and col-

lection of emission at 830 nm and 0.3 mg kg^{-1} of carotene-conjugated 2-devinyl-2-(1-hexyloxyethyl) pyropheophorbide (HPPH-car) with excitation at 660 nm and collection of fluorescence at 710 nm to image reactive lymph nodes within a canine population with spontaneous mammary disease. HPPH-car is a photodynamic agent modified with a carotene moiety to eliminate the cytotoxic triplet state upon excitation. Unfortunately, the suboptimal red excitation and emission of HPPH-car restricted the tissue penetration of both excitation and emission light in comparison to the NIR dye, IC-Green. Recently, Sharma and co-workers[76] reported quantitative optical imaging of lymph function in anesthetized swine after intradermal injections of about 200 μL of 3–32-μM IC-Green and showed for the first time, dynamic lymph propulsion of "packets" that traveled on the order of 0.1–1 cm/s. The unique opportunities enabling dynamic fluorescent imaging and emerging results in humans are described in more detail below. However, prior to this work, lymph flow was reported in animal preparations to be on the order of μm/s,[77] instead of the cm/s that was quantitatively and noninvasively imaged using NIR optical imaging.

NIR optical imaging technique has also been employed for mapping lymphatics in humans. Kitai *et al.*[78] and Motomura *et al.*[79] demonstrated mapping of lymphatic trunks after subdermal administration of 25 mg of IC-Green. Kitai *et al.*[78] used fluorescence, while Motomura *et al.*[79] visually tracked the green dye. In addition, Unno and co-workers[80] used fluorescence from IC-Green to map lymphatics of secondary lymphedema patients and healthy subjects. Subcutaneous injections of 1 mg IC-Green in the dorsum of the foot of lymphedema patients depicted dilated lymph channels, dermal backflow, and accumulation of dye in the dorsal and plantar region of the foot as opposed to discrete lymphatic channels in healthy subjects. Recently, Sevick-Muraca *et al.*[81] reported the results from quantitative optical imaging of lymph flow in breast cancer patients following microdose (i.e., less than $100 \mu g$) administration of IC-Green. The first report of optical lymph imaging in humans at trace amounts of IC-Green demonstrates the potential to administer an NIR-labeled, targeting contrast agent for the noninvasive nodal staging of cancer-positive lymph nodes as proposed by Sampath *et al.*[82]

Unique Opportunities for Imaging Lymphatic Function

Evaluation of functional status of lymph channels will be crucial for developing an understanding of lymphatic biology, evaluating the role of the lymphatics in health and disease, as well as diagnosing patients with disorders of the lymphatic system. Once a quantifiable imaging approach for assessing lymphatic function is developed and made available for clinical study, the opportunity to assess new therapeutic strategies and pharmacologic agents may spur the development of new treatments for patients suffering from lymphatic disorders. Fortunately for lymphatic imaging, NIR technology may be simple and economic to employ. As detailed below, NIR fluorescence imaging represents the only imaging modality to date sensitive enough for the direct imaging of propulsive lymph flow through lymphatic vessels deep within tissues.

The reason for the sensitivity arises from the penetration of NIR light, optimally in the wavelength range of 750–850 nm, and the ability to stably and repeatedly re-activate NIR-excitable dyes. To illustrate the sensitivity, compare a radiotracer (whether a gamma- or positron-emitter for the nuclear imaging techniques described above) and a typical NIR fluorophore. Whether produced in a cyclotron or generator, a radiotracer relaxes only once, emitting a single photon event before becoming spent. On the other hand, once a stable NIR fluorophore relaxes in tissues generating a fluorescent photon, it can be subsequently reactivated by propagating excitation light, enabling it

to relax multiple times. Since the lifetime, or the mean time that the fluorophore resides in its activated state, is on the order of a nanosecond, theoretically there can be as many as one billion fluorescent photons emitted from a single fluorescent molecule per second. Realistically, low quantum efficiencies for fluorescent photon generation, high tissue attenuation of propagating excitation and emission photons, and other instrumentation factors prevent such enormous fluorescent photon output. Nonetheless, in a study of a dual-labeled NIR optical and gamma-emitter, Houston and co-workers[83] found that the image quality of NIR optical imaging surpasses gamma scintigraphy in small animals, even though NIR imaging required only 800 ms of integration time while gamma scintigraphy required 15 min.

It is the high sensitivity of NIR optical imaging that allows image integration times to range from 50 ms to 800 ms, enabling rapid pharmacokinetic analysis of dye uptake[75,84] from images as well as imaging of lymph trafficking. While pulsatile lymph flow has been quantified using intravital microscopy techniques in the mesentery of animals, there have been no previous reports of quantifiable lymphatic flow from noninvasive imaging. More recently, the ability to optically image propulsive lymph flow in swine,[76] mice,[84] and humans[81] has been demonstrated. The common feature in these studies is the use of an image-intensifier which is sensitive in the NIR wavelength range, and a frame-transfer, 16-bit CCD camera which enables rapid and sensitive data acquisition with a low noise floor. The noise floor of fluorescence optical image acquisition arises because of autofluorescence as well as excitation light leakage through the rejection filters, which enable selective passage of fluorescent light to the image intensifier. For NIR imaging, the autofluorescence is virtually zero, but as one moves into the red excitation wavelengths using red exciting dyes, natural porphyrins and chlorophylls (from food) create a background that can obscure the signal from the fluorophore. In addition, the use of holographic rejection filters, which block 10 orders of magnitude of collimated, monochromatic excitation light, can also reduce the noise floor. Unfortunately, few commercially available instruments for small-animal imaging employ holographic filters. In addition, the success of holographic filters depends upon collimating the incident light and removing all possible stray light components. Not widely recognized is the fact that the use of broadband excitation sources, such as light-emitting diodes and lamps, reduce the effectiveness of holographic and interference filters, since these light sources are seldom monochromatic enough for effective blockage of excitation light. Finally, the availability of NIR-excitable dyes has increased in the past few years. IC-Green is approved in humans for assessing cardiac and hepatic function and for ophthalmologic studies and can be excited serendipitously at 780 nm, with a significant Stoke's shift enabling fluorescent measurement at 830 nm and greater. Small Stoke's shifts less than 30–40 nm can cause significant problems as they impede the ability to efficiently reject excitation light and collect the fluorescent signal. Unfortunately, IC-Green does not have a functional group and cannot be conjugated with a targeting peptide, protein, polysaccharide, or other molecule, but it can be used as a soluble dye which noncovalently associates with albumin to create a lymphotropic agent. A few commercially available NIR-excitable dyes with functional groups for creating molecular imaging agents are now available with sufficient Stoke's shifts: IRDye800 CW (Licor, Inc., Lincoln NE), AlexaFluor 790 (Invitrogen, Carlsbad, CA), and Cy7 and Cy 7.5 (Amersham/GE Health-Science, Piscataway, NJ). Usually, as a general rule of thumb, it is important to excite NIR organic dyes with as narrow or as monochromatic source as possible in order to maximally reflect excitation light, and to use bandpass filters well away from the excitation line in order to collect as much emission as possible for enhanced sensitivity. With these caveats, as shown below, the use of NIR optical imaging of lymphatic function offers a new, economical, and significant research and diagnostic tool to the lymphatic research community. In the following, we briefly review the progress in dynamic imaging of lymphatic function in mice, swine, and humans as well as highlight the opportunities and challenges specific to each.

Lymph Imaging in Mice

Preclinical investigation of potential pro- and anti-lymphangiogenic agents are typically performed in rodent models, thereby necessitating the imaging of lymphatic function in these small-animal models. While their small size may facilitate imaging, the temporal and spatial resolution provides challenges for functional lymphatic imaging in small animals. As comprehensively reviewed by Kwon and Sevick-Muraca,[84] lymphatic vessels in mice and rats have been imaged using (1) radiography to visualize mercury uptake in the two major lymph vessels in the mouse tail,[85] and (2) micro-MR lymphangiography with Gd-laden dendrimers[86–88] and (3) fluorescence video-microscopy to see the circumferential "honeycomb" lymphatic structure in the mouse tail.[89–91] Lymphatic function has also been inferred from lymphoscintigraphy and microangiography by Slavin *et al.*,[92] who reported that myocutaneous flap transfer restores lymphatic function.

Several groups have demonstrated lymphatic transport of QDs to map out drainage patterns.[72,93,94] Wunderbaldinger *et al.*[95] used an enzyme-sensing optical probe with visible wavelength excited fluorophore, Cy5.5, for the detection of lymph nodes. To date there have been few, if any, studies that demonstrated contractile lymphatic motion that characterizes lymphatic function in larger animals. As shown in FIGURE 11, Kwon and Sevick-Muraca[84] recently demonstrated the ability to image IC-Green trafficking from the lymph plexus, through lymph vessels and lymphangions, to ischial nodes in the tail and to axillary nodes in mice. The intensity profile shows that the peak fluorescence occurs at an average of every 6.56 ± 1.14 s and the lymph flow velocity ranged from 0.28 mm/s to -1.35 mm/s. Lymph flow velocities from the propelled IC-Green packet in the major lymph vessels in the mouse tail ranged from 1.33 mm/s to -3.88 mm/s. While pulsatile lymph flow was detected in the deep lymph vessels, lymph propulsion was not visualized in the superficial lymphatic network of the tail. When imaging lymph flow to the axillary nodes, propulsive lymph flow was also detected.

The ability to image contractile motion in mice provides new opportunities to (1) assess lymphatic function in transgenic mice models to better understand the role that specific gene expression has on the lymphatic function and to (2) investigate pharmacologic agents that stimulate the formation of functional lymphatics as well as stimulate the contractile apparatus of existing lymphatics.

Lymph Imaging in Swine

Yorkshire swine is the commonly preferred animal model for preclinical lymph-imaging investigations, as swine skin and lymph drainage pattern are similar to those of human skin.[70,96] Similar to human dermis, swine dermis is characterized by open-ended lymphatic capillaries that take up particles and lymphotropic agents under a pressure gradient. Passively diffusing agents, such as soluble NIR fluorophores, are taken up in the initial lymphatics and then drain into afferent lymphatic channels that propel lymph to the lymph nodes. Recently, we reported results of quantitative imaging of afferent lymphatic function in swine.[76] In addition to mapping lymphatic vasculature after intradermal injections of 100–200 μL of 32-μM IC-Green near mammary teats or on hindlimb, we conducted quantitative analysis of (1) the lymph flow velocities of transiting "packets" of IC-Green, and (2) their frequency of propulsion caused by lymphangion contractions. The "packets" of dye transited along lymph vessels 2–16 cm in length at velocities of

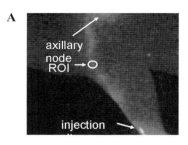

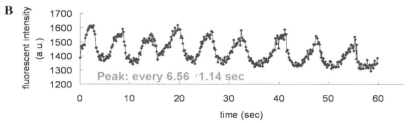

FIGURE 11. NIR optical imaging enables detection of pulsatile lymph flow in mice. (**A**) Fluorescence image; (**B**) intensity profile as a function of time in a specific ROI along a lymph vessel after intradermal injection of 2 μL of 1.29 μM IC-Green in dorsal aspect of paw. The intensity profile shows that the peak fluorescence occurs at an average of every 6.56 ± 1.14 s. (From Kwon and Sevick-Muraca[84] Reproduced by permission.)

2.3–7.5 mm/s and frequencies of 0.5–3.3 pulses per minute. FIGURE 12 illustrates pulsatile lymph flow in a swine lymphatic. The figure represents images of a bolus of IC-Green (circle) transiting along a swine's abdominal lymph channel from the injection site around the mammary teat to the inguinal lymph node at times 0, 2.6, and 5.2 s. The pulsatile lymph flow was also analyzed quantitatively to compute the period between the "packets" of dye transiting along a lymph channel. FIGURE 13 represents a plot of intensity profile at an ROI on an abdominal lymph channel as a function of imaging time after injection of 200 μL of 32-μM IC-Green around the mammary teats of swine. The peaks indicate a "bolus" of fluid transiting from the injection site to the inguinal lymph node on an average at every 72 s. Since manual lymphatic drainage (MLD) is an important component of complete decongestive therapy for lymphedema patients,[97] we employed the noninvasive lymph-imaging technique to measure the efficacy of MLD in a healthy anesthetized swine. For instance, in one case, the NIR optical lymph-imaging technique enabled detection of change in lymph flow velocity in a swine abdominal lymphatic from 6.2 (±1.9) mm/s to 8.0 (±2.5) mm/s and change in the mean period between pulses from 48 (±37) to 69 (±40) s in response to massage. Noninvasive determination of pulsatile lymph flow and transport velocities via imaging can provide a tool to evaluate lymphatic function in response to existing therapies and may help assess func-

FIGURE 12. Fluorescent images depict a bolus of IC-Green (*circle*) transiting along a swine's abdominal lymph channel at (left) $t = 0$, (middle) $t = 2.6$, and (right) $t = 5.6$ s. The lymphatic channel drained to the swine's inguinal region.

tion of newly regenerated lymph channels in response to novel therapeutic interventions, including VEGF-C or gene therapies.

Lymph Imaging in Humans

The preclinical imaging studies in swine demonstrated the feasibility of performing deep tissue lymph

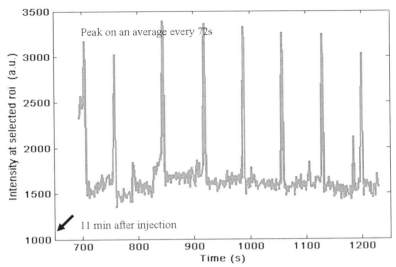

FIGURE 13. Plot of mean NIR fluorescent intensity profile of a ROI selected on a swine's abdominal lymphatic vessel as a function of imaging time. Consistent peaks that appear on average every 72 s are observed 11 min after intradermal injection of 200 μL of 32 μM IC-Green and indicate pulsatile lymph flow in swine abdomen.

imaging with microgram quantities of IC-Green. To translate the technique to the clinic we take advantage of the established safety record of IC-Green given in mg quantities. Sevick-Muraca *et al.*[81] recently reported the feasibility of detecting sentinel lymph nodes with microgram administration of IC-Green in breast cancer patients. FIGURE 14A depicts a white-light and fluorescent image overlay of a 46-year-old African American breast cancer patient intradermally injected with 100 μL of 20 μg IC-Green in each quadrant of the peri-areolar region on her right breast. The injection sites were covered with an opaque plastic to prevent oversaturation of the imaging camera. Arrows indicate lymphatics pooling the dye to axilla. FIGURE 14B represents the pulsatile lymph flow in the breast lymphatics. The peaks in the plot of intensity profile at an ROI (oval in panel A) as a function of duration of image acquisition indicate that the lymph channel conducts "packets" of IC-Green from the injection site to the axilla on an average every 22.9 ± 7.8 s at a velocity of 2.2 (±0.6) mm/s. A fluorescent "hot-spot" in the axilla was resected and found to be a sentinel lymph node. IC-Green is now being used for human lymph imaging in an ongoing feasibility study to compare lymph flow of healthy subjects and lymphedema patients as next described.

Before aberrant lymphatic function can be identified through NIR imaging, baseline lymphatic function was first evaluated in healthy volunteers. FIGURE 15 shows white-light and fluorescent overlay images of a foot of a 46-year-old Caucasian female in a supine position during image acquisition. The subject was injected with 100 μL of 25 μg of IC-Green each in the first and second interdigital spaces (arrows) in the dorsum of the foot. The images are a combination of the white-light photograph of the foot overlain with corresponding fluorescent images at 0, 2, and 15 s. The circle identifies a "packet" of dye transiting at a velocity of 5.0 mm/s between lymphangions along the lymphatic channel, while the arrowheads identify the localized collection of fluorescent dye that demarks lymphangions. The period between two consecutive pulses observed during image acquisition was 164 s. FIGURE 16 is another white-light and fluorescent image overlay that illustrates the lymphatic vasculature of a human hand. The image represents a hand of a 22-year-old Caucasian male in a supine position at the time of imaging. The subject was injected with 100 μL of 25 μg of IC-Green in the four interdigital spaces. The velocity of a single packet transiting in one of the lymph vessels draining an injection site next to the thumb was 16.6 mm/s. The ability to safely and noninvasively map the lymphatic network, locate lymphangions, and quantify lymph function in healthy volunteers can be employed for a comparative assessment of lymph drainage in persons with lymphedema. In addition, the potential to detect pulsatile lymph flow noninvasively using microdose concentrations of IC-Green augurs well for molecular imaging of lymphatics.

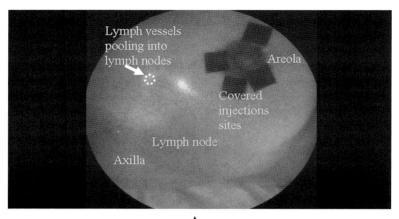

A

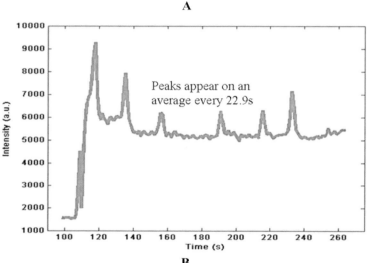

B

FIGURE 14. NIR optical lymph imaging detects pulsatile flow in humans. (**A**) White-light and NIR fluorescent image overlay of a subject with breast cancer. Intradermal injection is performed with 100 μL of 20 μg IC-Green in the peri-areolar region on the right breast. *Arrow* identifies a lymph channel on patient's left breast. (**B**) Plot of intensity profile at a ROI (*oval* in **A**) as a function of duration of image acquisition. The peaks indicate that lymph channel conducts the bolus of IC-Green from the injection site to axilla on an average every 22.9 ± 7.8 s.

Molecular Staining of Lymph Endothelium for Imaging Using Hyaluron

The sudden interest in lymphatic research is in part attributed to the discovery of lymph-specific biomarkers, such as Prox-1, podoplanin, and lymph endothelial-specific hyaluronan receptor 7 (LYVE-1),[1] and regulators of lymphangiogenesis. Of these biomarkers, LYVE-1 is not only expressed on lymphatic endothelium, but is also found on liver sinusoids, tubular epithelial cells of kidney, cells from adrenal glands and pancreas, and tumor-infiltrating macrophages.[98–100] The expression of LYVE-1 is correlated with poor prognosis of breast, endometrial, lung, and head and neck cancer,[101,102] and it is hypothesized that the kinds of cancer that have a tendency to metastasize express lymph markers. LYVE-1 is thought to mediate cell adhesion, uptake of hyaluronan (HA) from tissue for catabolism in lymph endothelial cells, and transport of HA from tissues to lymph across the lymphatic endothelium via transcytosis for its subsequent degradation in the nodes.[100,103,104] Hyaluron, an important component of the extracellular matrix, is an acidic glucosaminoglycan made of repeating disaccharides, and is synthesized by fibroblasts and other

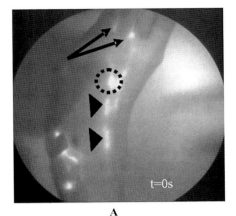

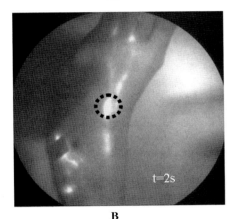

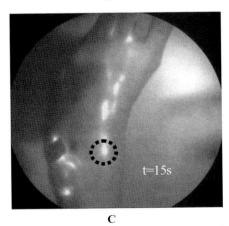

FIGURE 15. White-light overlay with corresponding fluorescent images of lymphatic map of foot of a 46-year-old female who was injected with 100 μL of 25 μg of IC-Green in the first and second interdigital spaces (*arrows*) in the dorsum of the foot. *Circles* indicate a "packet" of dye transiting along the lymphatic channel. *Arrowheads* identify localized collection of fluorescent dye that could be lymphangions of a lymphatic vessel. Panels **A**, **B** and **C** depict images at 0, 2, and 15 s.

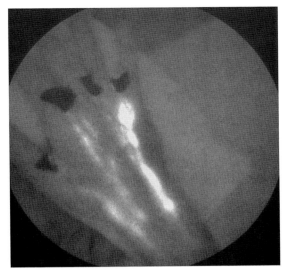

FIGURE 16. A white-light and fluorescent overlay image of lymphatics in a human hand. A 22-year-old male was injected with 100 μL of 25 μg of IC-Green in the four interdigital spaces on the right hand. Five discrete lymphatics appear to drain the dye from the injection site up to the axilla.

mesenchymal cells.[103–105] It is suggested that HA plays a key role in regulating interstitial hydration, plasma protein exclusion, maintaining interstitial fluid volume, supporting cell adhesion and migration in embryogenesis and wound healing, and in inducing inflammation.[103,106,107] While 90% of all HA is degraded in the lymph nodes, the remaining HA is subsequently taken up and broken down in the liver. HA is approved by Food and Drug Administration for use in joint therapy and correction of facial wrinkles.

Recently, we reported swine lymph vessel staining with an NIR dye conjugated to HA that binds to LYVE-1 on the lymphatic endothelium. FIGURE 17 depicts a plot of mean fluorescent intensity of an ROI identified on a swine leg lymph vessel stained with HA-NIR and IC-Green for dynamic imaging as a function of time. FIGURE 17A depicts an ROI selected on a leg lymph vessel stained with HA-NIR dye. FIGURE 17B represents an ROI selected on a leg lymph vessel with trafficking IC-Green. FIGURE 17C compares the intensity profiles in the two lymph vessels. The initial peaks of fluorescent intensity for IC-Green, a nonspecific dye, indicate that the dye is propelled as "packets" of fluid and cleared out of the lymph vessel. On the other hand, the fluorescence intensity due to HA-NIR remains almost unchanged for the duration of image acquisition. The observation supports the hypothesis that HA-NIR binds to LYVE-1 present on the lymph endothelium.

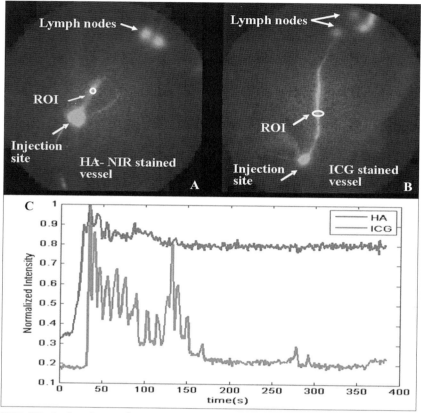

FIGURE 17. A plot of mean fluorescent intensity of leg lymph vessel stained with HA-NIR and IC-Green at ROIs (*ovals*) for duration of image acquisition. (**A**) Oval indicates an ROI selected on leg lymph vessel stained with HA-NIR dye. (**B**) Oval identifies an ROI on leg lymph vessel with trafficking IC-Green, and (**C**) comparison of intensity profile as a function of time for selected ROIs. (From Sharma *et al.*[76] Reproduced by permission.)

Kitayama and colleagues[108] recently reported the use of free HA mixed with patent blue or ferumoxides to map sentinel lymph nodes in patients undergoing gastric cancer surgery. Submucosal injections of one part HA and four parts patent blue or ferumoxides resulted in long tracer retention times in the sentinel nodes (<2 h) as opposed to patent blue or ferumoxides alone, which stained the efferent lymphatics within 20 min. In summary, since HA is a ligand for LYVE-1 present on the lymph endothelium, NIR imaging agents conjugated to HA may enable accurate detection of sentinel lymph nodes in cancer patients and computation of lymphatic vessel architecture and density in patients with lymphatic disorders.

Summary

From the compendium of imaging techniques summarized above, one can see that new opportunities for imaging lymphatic architecture and function are promised primarily from the development of new imaging agents and perhaps by the emergence of fluorescent optical imaging technologies. The translation of these discoveries into clinical practice may first depend upon their successful demonstration in preclinical models of lymphatic impairment and repair. With the ultimate adoption of noninvasive imaging and novel imaging agents in the clinic, it may be possible to obtain more accurate phenotyping for identification of genetic bases of lymphatic disorders, furthering advanced understanding of human disease.

Conflicts of Interest

The authors declare no conflicts of interest.

References

1. ALITALO, K., T. TAMMELA & T.V. PETROVA. 2005. Lymphangiogenesis in development and human disease. Nature **438:** 946–953.

2. BALUK, P., T. TAMMELA, E. ATOR, *et al.* 2005. Pathogenesis of persistent lymphatic vessel hyperplasia in chronic airway inflammation. J. Clin. Invest. **115:** 247–257.

3. JI, R.C. 2005. Characteristics of lymphatic endothelial cells in physiological and pathological conditions. Histol. Histopathol. **20:** 155–175.

4. HARVEY, N.L., R.S. SRINIVASAN, M.E. DILLARD, *et al.* 2005. Lymphatic vascular defects promoted by Prox1 haploinsufficiency cause adult-onset obesity. Nat. Genet. **37:** 1072–1081.

5. CUENI, L.N. & M. DETMAR. 2006. New insights into the molecular control of the lymphatic vascular system and its role in disease. J. Invest. Dermatol. **126:** 2167–2177.

6. BLACKER, H.A., C. TSOPELAS, S. ORGEIG, *et al.* 2007. How regenerating lymphatics function: lessons from lizard tails. Anat. Rec. **290:** 108–114.

7. MOGHIMI, S.M. & A.R. RAJABI-SIAHBOOMI. 1996. Advanced colloid-based systems for efficient delivery of drugs and diagnostic agents to the lymphatic tissues. Progress Biophys. Mol. Biol. **65:** 221–249.

8. ENDERLE, J.D., S.M. BLANCHARD & J. BRONZINO. 2000. Introduction to Biomedical Engineering. Academic Press. San Diego, CA.

9. HULTÉN, L., C. ÅHRÉN & M. ROSENCRANTZ. 1966. Lymphangio-adenography in carcinoma of the breast. Acta Chirurg. Scand. **132:** 261–274.

10. KINMONTH, J.B. 1952. Lymphangiography in man; a method of outlining lymphatic trunks at operation. Clin. Sci. **11:** 13–20.

11. KINMONTH, J.B. 1972. The Lymphatics: Diseases, Lymphography and Surgery. Edward Arnold. London.

12. KOS, S., H. HAUEISEN, U. LACHMUND, *et al.* 2007. Lymphangiography: forgotten tool or rising star in the diagnosis and therapy of postoperative lymphatic vessel leakage. Cardiovasc. Intervent. Radiol. **30:** 968–973.

13. WITTE, C.L., M.H. WITTE, E.C. UNGER, *et al.* 2000. Advances in imaging of lymph flow disorders. Radiographics **20:** 1697–1719.

14. BARRETT, T., P.L. CHOYKE & H. KOBAYASHI. 2006. Imaging of lymphatic system: new horizons. Contrast Media Mol. Imaging **1:** 230–245.

15. SILVESTRI, R.C., J.S. HUSEBY, I. RUGHANI, *et al.* 1980. Respiratory-distress syndrome from lymphangiography contrast-medium. Am. Rev. Resp. Dis. **122:** 543–549.

16. TIWARI, A., K.-S. CHEUNG, M. BUTTON, *et al.* 2003. Differential diagnosis, investigation, and current treatment of lower limb lymphedema. Arch. Surg. **138:** 152–161.

17. VOGL, T.J., M. BARTJES & K. MARZEC. 1997. Contrast-enhanced lymphography. CT or MR imaging? Acta Radiol. Suppl. **412:** 47–50.

18. PARTSCH, H., A. URBANEK & B. WENZEL-HORA. 1984. The dermal lymphatics in lymphoedema visualized by indirect lymphography. Br. J. Dermatol. **110:** 431–438.

19. PARTSCH, H., B.I. WENZEL-HORA & A. URBANEK. 1983. Differential diagnosis of lymphedema after indirect lymphography with Iotasul. Lymphology **16:** 12–18.

20. PECKING, A.P., J.P. DESPREZ-CURLEY & R.V. CLUZAN. 2002. Tests and imaging of the lymphatic system. Rev. Med. Interne. **23:** d391s–397s.

21. SUGA, K., N. OGASAWARA, Y. YUAN, *et al.* 2003. Visualization of breast lymphatic pathways with an indirect computed tomography lymphography using a nonionic monometric contrast medium iopamidol: preliminary results. Invest. Radiol. **38:** 73–84.

22. SUGA, K., K. SHIMIZU, Y. KAWAKAMI, *et al.* 2005. Lymphatic drainage from esophagogastric tract: feasibility of endoscopic CT lymphography for direct visualization of pathways. Radiology **237:** 952–960.

23. WISNER, E.R., K.W. FERRARA, R.E. SHORT, *et al.* 1996. Indirect computed tomography lymphography using iodinated nanoparticles to detect cancerous lymph nodes in a cutaneous melanoma model. Acad. Radiol. **3:** 40–48.

24. RABIN, O., J.M. PEREZ, J. GRIMM, *et al.* 2006. An X-ray computed tomography imaging agent based on long circulating bismuth sulphide nanoparticles. Nat. Mater. **5:** 118–122.

25. YUAN, Z., L. CHEN, Q. LUO, *et al.* 2006. The role of radionuclide lymphoscintigraphy in extremity lymphedema. Ann. Nucl. Med. **20:** 341–344.

26. MCNEILL, G.C., M.H. WITTE, C.L. WITTE, *et al.* 1989. Whole-body lymphangioscintigraphy: preferred method for initial assessment of the peripheral lymphatic system. Radiology **172:** 495–502.

27. SZUBA, A., W.S. SHIN, H.W. STRAUSS, *et al.* 2003. The third circulation: radionuclide lymphoscintigraphy in the evaluation of lymphedema. J. Nucl. Med. **44:** 43–57.

28. SATO, T., K. YAMAGUCHI, Y. MORITA, *et al.* 2000. Lymphoscintigraphy for the interpretation of changes of cervical lymph node function in patients with oral malignant tumors: comparison of Tc-99m-Re and Tc-99m-HSA-D. Oral Surg. Oral Med. Oral Pathol. Oral Radiol. Endod. **90:** 526–536.

29. WALLACE, A.M., C.K. HOH, D.R. VERA, *et al.* 2003. Lymphoseek: a molecular radiopharmaceutical for sentinel node detection. Ann. Surg. Oncol. **10:** 531–538.

30. SCHÖDER, H., E.C. GLASS, A.P. PECKING, *et al.* 2006. Molecular targeting of lymphovascular system for imaging and therapy. Cancer Metastasis Rev. **25:** 185–201.

31. ELLNER, S.J., C.K. HOH, D.R. VERA, *et al.* 2003. Dose-dependent biodistribution of [(99m)Tc]DTPA-mannosyl-dextran for breast cancer sentinel lymph node mapping. Nucl. Med. Biol. **30:** 805–810.

32. SZUBA, A., W. STRAUSS, S.K. SIRSIKAR, *et al.* 2002. Quantitative radionuclide lymphoscintigraphy predicts outcome of manual lymphatic therapy in breast cancer-related lymphedema of the upper extremity. Nucl. Med. Commun. **23:** 1171–1175.

33. THOMPSON, M., S. KOROURIAN, R. HENRY-TILLMAN, *et al.* 2007. Axillary reverse mapping (ARM): a new concept to identify and enhance lymphatic preservation. Ann. Surg. Oncol. **14:** 1890–1895.

34. O'MAHONY, S., T.M. BENNETT BRITTON, C.K. SOLANKI, *et al.* 2007. Lymphatic transfer studies with

immunoglobin scintigraphy after axillary surgery. Eur. J. Surg. Oncol. **33:** 1052–1060.

35. PEER, F.I., M.H. PURI, A. MOSAM, *et al.* 2007. 99mTc-MIBI imaging of cutaneous AIDS-associated Kaposi's sarcoma. Int. J. Dermatol. **46:** 166–171.

36. WEISS, M., R.J.H. BAUMEISTER & K. HAHN. 2002. Post-therapeutic lymphedema: scintigraphy before and after autologous lymph vessel transplantation 8 years of long-term follow-up. Clin. Nucl. Med. **27:** 788–792.

37. WEISS, M., R. BAUMEISTER & K. HAHN. 2003. Dynamic lymph flow imaging in patients with oedema of the lower limb for evaluation of the functional outcome after autologous lymph vessel transplantation: an 8-year follow-up study. Eur. J. Nucl. Med. Mol. Imaging **30:** 202–206.

38. SAARISTO, A., T. TAMMELA, J. TIMONEN, *et al.* 2004. Vascular endothelial growth factor-C gene therapy restores lymphatic flow across incision wounds. FASEB J. **18:** 1707–9.

39. SZUBA, A., M. SKOBE, M.J. KARKKAINEN, *et al.* 2002. Therapeutic lymphangiogenesis with human recombinant VEGF-C. FASEB J. **16:** 1985–1987.

40. O'MAHONY, S., C.K. SOLANKI, R.W. BARBER, *et al.* 2006. Imaging of lymphatic vessel in breast cancer related lymphedema: intradermal versus subcutaneous injection of 99mTc-immunoglobulin. Am. J. Roentgenol. **186:** 1349–1355.

41. DELBEKE, D. 1999. Oncological applications of FDG PET imaging: brain tumors, colorectal cancer lymphoma and melanoma. J. Nucl. Med. **40:** 591–603.

42. HOFFMAN, J.M. & S.S. GAMBHIR. 2007. Molecular imaging: the vision and opportunity for radiology in the future. Radiology **244:** 39–47.

43. LINDQVIST, U., G. WESTERBERG, M. BERGSTROM, *et al.* 2000. [^{11}C]Hyaluronan uptake with positron emission tomography in liver disease. Eur. J. Clin. Invest. **30:** 600–607.

44. HUTCHINS, G.D., M. SCHWAIGAER, K.C. ROSENSPIRE, *et al.* 1990. Noninvasive quantification of regional blood flow in the human heart using N-13 ammonia and dynamic positron emission tomographic imaging. J. Am. Coll. Cardiol. **15:** 1032–1042.

45. BRAUN, A.R., T.J. BALKIN, N.J. WESENTEN, *et al.* 1997. Regional cerebral blood flow throughout the sleep-wake cycle. An $H_2{}^{15}O$ PET study. Brain **120:** 1173–1197.

46. METTLER, F.A. & M.J. GUIBERTEAU. 2006. Essentials of Nuclear Medicine Imaging. Saunders Elsevier. Philadelphia, PA.

47. BAR-SHALOM, R., N. YEFREMOV, L. GURALNIK, *et al.* 2003. Clinical performance of PET/CT in evaluation of cancer: additional value for diagnostic imaging and patient management. J. Nucl. Med. **44:** 1200–1209.

48. LARDINOIS, D., W. WEDER, T.F. HANY, *et al.* 2003. Staging of non–small-cell lung cancer with integrated positron-emission tomography and computed tomography. N. Engl. J. Med. **348:** 2500–2507.

49. CLÉMENT, O. & A. LUCIANI. 2004. Imaging the lymphatic system: possibilities and clinical application. Eur. Radiol. **14:** 1498–1507.

50. MISSELWITZ, B. 2006. MR contrast agents in lymph node imaging. Eur. J. Radiol. **58:** 375–382.

51. PAN, D., Y. SUZUKI, P.C. YANG, *et al.* 2006. Indirect magnetic resonance lymphangiography to assess lymphatic function in experimental murine lymphedema. Lymph. Res. Biol. **4:** 211–215.

52. RUEHM, S.G., T. SCHROEDER & J.F. DEBATIN. 2001. Interstitial MR lymphography with gadoterate meglumine: initial experience in humans. Radiology **220:** 816–821.

53. LOHRMANN, C., E. FOELDI & M. LANGER. 2006. Indirect magnetic resonance lymphangiography in patients with lymphedema: preliminary results in humans. Eur. J. Radiol. **59:** 401–406.

54. LOHRMANN, C., E. FOELDI, J.-P. BARTHOLOMAE, *et al.* 2007. Gadoteridol for MR imaging of lymphatic vessels in lymphoedematous patients: initial experience after intracutaneous injection. Br. J. Radiol. **80:** 569–573.

55. LOHRMANN, C., E. FOELDI, O. SPECK, *et al.* 2006. High-resolution MR lymphangiography in patients with primary and secondary lymphedema. Am. J. Roentgenol. **187:** 556–561.

56. MATSUSHIMA, S., N. ICHIBA, D. HAYASHI, *et al.* 2007. Nonenhanced magnetic resonance lymphoductography: visualization of lymphatic system of the trunk on 3-dimensional heavily T2-weighted image with 2-dimensional prospective acquisition and correction. J. Comput. Assist. Tomography **31:** 299–302.

57. BELLIN, M.F., C. ROY, K. KINKEL, *et al.* 1998. Lymph node metastases: safety and effectiveness of MR imaging with ultrasmall superparamagnetic iron oxide particles–initial clinical experience. Radiology **207:** 799–808.

58. RÉTY, F., O. CLEMENT, N. SIAUVE, *et al.* 2000. MR lymphography using iron oxide nanoparticles in rats: pharmacokinetics in the lymphatic system after intravenous injection. J. Magn. Res. Imaging **12:** 734–739.

59. SUTTON, D. 1998. Radiology and Imaging for Medical Students. Churchill Livingstone. Edinburgh, UK.

60. SURESH, S., V. KUMARASWAMI, I. SURESH, *et al.* 1997. Ultrasonographic diagnosis of subclinical filariasis. J. Ultrasound Med. **16:** 45–49.

61. MELLOR, R.H., G. BRICE, A.W.B. STANTON, *et al.* 2007. Mutations in FOXC2 are strongly associated with primary valve failure in veins of the lower limb. Circulation **115:** 1912–1920.

62. MOTOMURA, K., H. INAJI, Y. KOMOIKE, *et al.* 2001. Gamma probe and ultrasonographically-guided fine-needle aspiration biopsy of sentinel lymph nodes in breast cancer patients. Eur. J. Surg. Oncol. **27:** 141–145.

63. BEDROSIAN, I., D. BEDI, H.M. KUERER, *et al.* 2003. Impact of clinicopathological factors on sensitivity of axillary ultrasonography in the detection of axillary nodal metastases in patients with breast cancer. Ann. Surg. Oncol. **10:** 1025–1030.

64. WISNER, E.R., K.W. FERRARA, R.E. SHORT, *et al.* 2003. Sentinel node detection using contrast-enhanced power Doppler ultrasound lymphography. Invest. Radiol. **38:** 358–365.

65. GOLDBERG, B.B., D.A. MERTON, J-B. LIU, *et al.* 2004. Sentinel lymph nodes in a swine model with melanoma: contrast-enhanced lymphatic US. Radiology **230:** 727–734.

efficiency and accumulation in the tumor tissue in a time-dependent manner, it also harbors cytotoxic activity. When systemically administered to MDA-MB-435 xenograft-bearing mice, the LyP-1 peptide inhibits tumor growth.[37] This effect was not enhanced by conjugating the LyP-1 to the $_D$(KLAKLAK)$_2$ peptide (Laakkonen, Akerman and Ruoslahti; unpublished observations), and could be abolished by introducing a point mutation to the LyP-1 peptide sequence.[37] LyP-1 treatment also significantly reduces the number of lymphatic vessels and increases apoptosis in the tumor. Since the destruction of lymphatic vessels in tumor is not enough to inhibit tumor growth, the anti-tumor effect of LyP-1 is most probably attributable to its ability to recognize and kill other cells in the tumor, rather than the anti-lymphatic effect. This conclusion is also supported by our findings with LyP-1 peptide linked to an anti-angiogenic protein, anastellin.[47,48] This chimeric protein homed to the tumor lymphatic vessels, but did not spread into the tumor tissue like the peptide does. When MDA-MB-435 tumor-bearing mice were treated systemically with this conjugate, the treated tumors had significantly fewer lymphatic vessels compared to the control-treated tumors, while no effect on tumor growth was observed (Laakkonen, Akerman, and Ruoslahti; unpublished observations). We conclude that the lymphatic homing peptides can be used to target therapeutic agents to tumors. Because mere destruction of tumor lymphatics is unlikely to affect tumor growth, the effect of the conjugate has to reach beyond the lymphatics. Destroying tumor lymphatics alone may have an inhibitory effect on the dissemination of tumor to the distant organs.

Concluding Remarks

In this review we have discussed peptides that recognize tumor lymphatic vessels. In our screens to date we have identified five different lymphatic homing peptides. Each of these peptides recognizes lymphatic vessels in a different set of tumors, and some of them also recognize tumor cells. None of the peptides recognizes blood endothelial cells, even though we access the lymphatic vessels via intravenous injections. In addition, none of our peptides home to any normal organ. Whether the lack of homing to normal lymphatic vessels is due to the inaccessibility of these vessels via blood circulation or the lack of the receptor molecules on their surface remains to be studied. These questions can best be addressed after we have identified the receptor molecules these peptides recognize in the lymphatic vessels. Our studies demonstrate that the tumor lymphatic vessels carry molecular markers that are specific for a given tumor type and, in some cases, recognize tumor stage–specific markers. A different set of such markers has been previously shown to exist in the tumor blood vessels. The extraordinary homing efficiency of the LyP-1 peptide, and its ability to recognize metastatic lesions suggest potential diagnostic uses for this peptide, and perhaps some of the other peptides as well. Conjugation of the lymphatic homing peptides to drugs and other therapeutic molecules provides an opportunity to specifically deliver therapeutic agents into tumors using a route not previously exploited. This is another example of drug delivery that depends on specific docking of a drug at its target, rather than specific action of the drug on the target. We refer to this type of drug delivery as *synaphic* (from the Greek root for *affinity*) targeting.

Conflicts of Interest

The authors declare no conflicts of interest.

References

1. HOLASH, J., P.C. MAISONPIERRE, D. COMPTON, *et al*. 1999. Vessel cooption, regression, and growth in tumors mediated by angiopoietins and VEGF. Science **284:** 1994–1998.
2. COUSSENS, L.M., D. HANAHAN & J.M. ARBEIT. 1996. Genetic predisposition and parameters of malignant progression in K14-HPV16 transgenic mice. Am. J. Pathol. **149:** 1899–1917.
3. HATTORI, K., S. DIAS, B. HEISSIG, *et al*. 2001. Vascular endothelial growth factor and angiopoietin-1 stimulate postnatal hematopoiesis by recruitment of vasculogenic and hematopoietic stem cells. J. Exp. Med. **193:** 1005–1014.
4. RAFII, S. 2000. Circulating endothelial precursors: mystery, reality, and promise [comment]. J. Clin. Invest. **105:** 17–19.
5. MCDONALD, D.M. & A.J. FOSS. 2000. Endothelial cells of tumor vessels: abnormal but not absent. Cancer Metast. Rev. **19:** 109–120.
6. CARMELIET, P. & R.K. JAIN. 2000. Angiogenesis in cancer and other diseases. Nature **407:** 249–257.
7. RUOSLAHTI, E. 2002. Specialization of tumour vasculature. Nat. Rev. Cancer. **2:** 83–90.
8. ST. CROIX, B., C. RAGO, V. VELCULESCU, *et al*. 2000. Genes expressed in human tumor endothelium [see comments]. Science **289:** 1197–1202.
9. OH, P., Y. LI, J. YU, *et al*. 2004. Subtractive proteomic mapping of the endothelial surface in lung and solid tumours for tissue-specific therapy [see comment]. Nature **429:** 629–635.

10. RIBATTI, D., B. NICO, A. VACCA, *et al.* 2002. Endothelial cell heterogeneity and organ specificity. J. Hematother. Stem Cell Res. **11:** 81–90.

11. ZHANG, L., J.A. HOFFMAN & E. RUOSLAHTI. 2005. Molecular profiling of heart endothelial cells. Circulation **112:** 1601–1611.

12. ARAP, W., W. HAEDICKE, M. BERNASCONI, *et al.* 2002. Targeting the prostate for destruction through a vascular address. Proc. Natl. Acad. Sci. USA **99:** 1527–1531.

13. RAJOTTE, D., W. ARAP, M. HAGEDORN, *et al.* 1998. Molecular heterogeneity of the vascular endothelium revealed by in vivo phage display. J. Clin. Invest. **102:** 430–437.

14. PASQUALINI, R. & E. RUOSLAHTI. 1996. Organ targeting in vivo using phage display peptide libraries. Nature **380:** 364–366.

15. GERLAG, D.M., E. BORGES, P.P. TAK, *et al.* 2001. Suppression of murine collagen-induced arthritis by targeted apoptosis of synovial neovasculature. Arthritis Res. **3:** 357–361.

16. PORKKA, K., P. LAAKKONEN, J.A. HOFFMAN, *et al.* 2002. A fragment of the HMGN2 protein homes to the nuclei of tumor cells and tumor endothelial cells in vivo. Proc. Natl. Acad. Sci. USA **99:** 7444–7449.

17. LAAKKONEN, P., K. PORKKA, J.A. HOFFMAN, *et al.* 2002. A tumor-homing peptide with a targeting specificity related to lymphatic vessels. Nat. Med. **8:** 751–755.

18. PASQUALINI, R., E. KOIVUNEN & E. RUOSLAHTI. 1997. Alpha v integrins as receptors for tumor targeting by circulating ligands [see comments]. Nat. Biotechnol. **15:** 542–546.

19. BURG, M.A., R. PASQUALINI, W. ARAP, *et al.* 1999. NG2 proteoglycan-binding peptides target tumor neovasculature. Cancer Res. **59:** 2869–2874.

20. JOYCE, J.A., P. LAAKKONEN, M. BERNASCONI, *et al.* 2003. Stage-specific vascular markers revealed by phage display in a mouse model of pancreatic islet tumorigenesis. Cancer Cell. **4:** 393–403.

21. HOFFMAN, J.A., E. GIRAUDO, M. SINGH, *et al.* 2003. Progressive vascular changes in a transgenic mouse model of squamous cell carcinoma. Cancer Cell. **4:** 383–391.

22. SAHARINEN, P., T. TAMMELA, M.J. KARKKAINEN, *et al.* 2004. Lymphatic vasculature: development, molecular regulation and role in tumor metastasis and inflammation. Trends Immunol. **25:** 387–395.

23. MATTILA, M.M., J.K. RUOHOLA, T. KARPANEN, *et al.* 2002. VEGF-C induced lymphangiogenesis is associated with lymph node metastasis in orthotopic MCF-7 tumors. Int. J. Cancer. **98:** 946–951.

24. SKOBE, M., T. HAWIGHORST, D.G. JACKSON, *et al.* 2001. Induction of tumor lymphangiogenesis by VEGF-C promotes breast cancer metastasis [see comments]. Nat. Med. **7:** 192–198.

25. MANDRIOTA, S.J., L. JUSSILA, M. JELTSCH, *et al.* 2001. Vascular endothelial growth factor-C-mediated lymphangiogenesis promotes tumour metastasis. EMBO J. **20:** 672–682.

26. KARPANEN, T., M. EGEBLAD, M.J. KARKKAINEN, *et al.* 2001. Vascular endothelial growth factor C promotes tumor lymphangiogenesis and intralymphatic tumor growth. Cancer Res. **61:** 1786–1790.

27. STACKER, S.A., C. CAESAR, M.E. BALDWIN, *et al.* 2001. VEGF-D promotes the metastatic spread of tumor cells via the lymphatics [see comments]. Nat. Med. **7:** 186–191.

28. LIN, J., A.S. LALANI, T.C. HARDING, *et al.* 2005. Inhibition of lymphogenous metastasis using adeno-associated virus-mediated gene transfer of a soluble VEGFR-3 decoy receptor. Cancer Res. **65:** 6901–6909.

29. HE, Y., K. KOZAKI, T. KARPANEN, *et al.* 2002. Suppression of tumor lymphangiogenesis and lymph node metastasis by blocking vascular endothelial growth factor receptor 3 signaling [see comments]. J. Natl. Cancer Inst. **94:** 819–825.

30. PODGRABINSKA, S., P. BRAUN, P. VELASCO, *et al.* 2002. Molecular characterization of lymphatic endothelial cells [erratum appears in Proc. Natl. Acad. Sci. USA 2003 Apr 15;100(8):4970]. Proc. Natl. Acad. Sci. USA **99:** 16069–16074.

31. PETROVA, T.V., T. MAKINEN, T.P. MAKELA, *et al.* 2002. Lymphatic endothelial reprogramming of vascular endothelial cells by the Prox-1 homeobox transcription factor. EMBO J. **21:** 4593–4599.

32. KRIEHUBER, E., S. BREITENEDER-GELEFF, M. GROEGER, *et al.* 2001. Isolation and characterization of dermal lymphatic and blood endothelial cells reveal stable and functionally specialized cell lineages [see comments]. J. Exp. Med. **194:** 797–808.

33. HOFFMAN, J.A., P. LAAKKONEN, K. PORKKA, *et al.* 2004. In vivo and ex vivo selections using phage-displayed libraries. *In* Phage Display. T. Clackson & H.B. Lowman, Eds.: 171–192. Oxford University Press. Oxford, UK.

34. PREVO, R., S. BANERJI, D.J. FERGUSON, *et al.* 2001. Mouse LYVE-1 is an endocytic receptor for hyaluronan in lymphatic endothelium. J. Biol. Chem. **276:** 19420–19430.

35. JACKSON, D.G., R. PREVO, S. CLASPER, *et al.* 2001. LYVE-1, the lymphatic system and tumor lymphangiogenesis. Trends Immunol. **22:** 317–321.

36. BREITENEDER-GELEFF, S., A. SOLEIMAN, H. KOWALSKI, *et al.* 1999. Angiosarcomas express mixed endothelial phenotypes of blood and lymphatic capillaries: podoplanin as a specific marker for lymphatic endothelium. Am. J. Pathol. **154:** 385–394.

37. LAAKKONEN, P., M.E. AKERMAN, H. BILIRAN, *et al.* 2004. Antitumor activity of a homing peptide that targets tumor lymphatics and tumor cells. Proc. Natl. Acad. Sci. USA **101:** 9381–9386.

38. LANGEL, U. 2006. Handbook of Cell-Penetrating Peptides, 2nd ed. U. Langel, Ed. CRC Press (Taylor & Francis). Boca Raton, FL and New York.

39. RUOSLAHTI, E., T. DUZA & L. ZHANG. 2005. Vascular homing peptides with cell-penetrating properties. Curr. Pharm. Des. **11:** 3655–3660.

40. LAAKKONEN, P. & E. RUOSLAHTI. 2006. Selective delivery to vascular addresses: in vivo applications of cell-type-specific cell-penetrating peptides. *In* Handbook of Cell-Penetrating Peptides. U. Langel, Ed.: 413–422. CRC Press (Taylor & Francis). Boca Raton, FL and New York.

41. ZHANG, L., E. GIRAUDO, J.A. HOFFMAN, *et al.* 2006. Lymphatic zip codes in premalignant lesions and tumors [erratum appears in Cancer Res.] Cancer Res. **66:** 5696–5706.

42. AKERMAN, M.E., W.C. CHAN, P. LAAKKONEN, *et al.* 2002. Nanocrystal targeting in vivo. Proc. Natl. Acad. Sci. USA **99:** 12617–12621.

43. RAE, J.M.C., C.J. MECK, J.M. HADDAD, B.R. JOHNSON, M.D. 2007. MDA-MB-435 cells are derived from M14 melanoma cells: a loss for breast cancer, but a boon for melanoma research. Breast Cancer Res. Treat. **104**: 13–19.

44. RAE, J.M., S.J. RAMUS, M. WALTHAM, *et al.* 2004. Common origins of MDA-MB-435 cells from various sources with those shown to have melanoma properties. Clin. Exp. Met. **21:** 543–552.

45. GREENBERG, N.M., F. DEMAYO, M.J. FINEGOLD, *et al.* 1995. Prostate cancer in a transgenic mouse. Proc. Natl. Acad. Sci. USA **92:** 3439–3443.

46. ELLERBY, H.M., W. ARAP, L.M. ELLERBY, *et al.* 1999. Anticancer activity of targeted pro-apoptotic peptides. Nat. Med. **5:** 1032–1038.

47. YI, M. & E. RUOSLAHTI. 2001. A fibronectin fragment inhibits tumor growth, angiogenesis, and metastasis. Proc. Natl. Acad. Sci. USA **98:** 620–624.

48. PASQUALINI, R., S. BOURDOULOUS, E. KOIVUNEN, *et al.* 1996. A polymeric form of fibronectin has antimetastatic effects against multiple tumor types. Nat. Med. **2:** 1197–1203.

Lymphatic Tissue Engineering

Progress and Prospects

THOMAS HITCHCOCK AND LAURA NIKLASON

Department of Biomedical Engineering, Yale University, New Haven, Connecticut, USA

In the last 5 years major advances have been made in the field of tissue engineering. However, while engineering of tissues from nearly every major system in the body have been studied and improved, little has been done with the engineering of viable lymphatic tissues. Recent advances in understanding of lymphatic biology have allowed the easy isolation of pure lymphatic cell cultures, increasing, in turn, the ability to study lymphatic biology in greater detail. This has allowed the elucidation of lymphatic properties on the structural, cellular, and molecular levels, making possible the successful development of the first lymphatic engineered tissues. Among such advances are the engineering of lymphatic capillaries, the development of a functioning bioreactor designed to culture lymph nodes *in vitro*, and *in vivo* growth of lymphatic organoids. However, there has been no research on the engineering of functional lymphangions. While the advances made in the study of lymphatic biology are encouraging, the complexities of the system make the engineering of certain functional lymphatic tissues somewhat more difficult.

Key words: tissue engineering; lymphatics; lymphangions; vascular; endothelium; smooth muscle

Introduction

Major advances have been made in the field of tissue engineering since the topic of lymphatic engineering was taken up by this journal 5 years ago.[a] The relatively new discipline of tissue engineering has proven successful in using a combination of life sciences and engineering research to produce technologies that allow the assembly of multicellular, complex, and functioning tissues and organs. In the last 5 years, scientists have been able to grow and transplant human bladders, create three-dimensional (3-D) tissues using the techniques of the ink-jet printer, develop transplantable nervous tissue constructs, and advance research in cardiac, cartilage, and bone tissue engineering.[1–6] In comparison, the work done in the area of lymphatic tissue engineering to this point has been more sparse. However, in the last 3 years there has been a small, yet important movement toward research geared to the engineering of lymphatic tissues.

As the critical importance of the lymphatic system in the human body becomes increasingly evident, novel research efforts will drive the need for proper models by which lymphatic biology might be studied. Tissue-engineered lymphatic organs (lymph nodes, spleens, lymphatic vessels, and lymphoid derived cells) would not only serve well as *ex vivo* research models, but also have the potential to be used via implantation for the alleviation of lymphatic deficiencies caused by disease or injury.

Our lab currently has been successful in the engineering of mammalian blood vessels made from cells isolated from the cardiovascular system.[7] Given the similarities between the vessels in the cardiovascular and lymphatic systems, it may be useful to apply arterial tissue engineering techniques to the engineering of lymphatic vessels and organs. In this review we will concentrate on (1) current advances in lymphatic biology research and how such advances may pave the way for the initial phases of lymphatic tissue engineering; (2) differences between lymphatic and vascular vessel biology and their effects on tissue engineering strategies; and (3) recent efforts in lymphatic tissue engineering and their implications and prospects.

Relevant Advances in Lymphatic Biology

The lymphatic and blood vascular systems are interconnected—extracellular fluid flows from vascular capillaries into lymphatic microvessels and is

---

[a]See *The Lymphatic Continuum*, Volume 979 of the *Annals of the New York Academy of Sciences* (2002), edited by Stanley G. Rockson and *Reparative Medicine: Growing Tissues and Organs*, volume 961 of the *Annals of the New York Academy of Sciences*

Address for correspondence: Laura Niklason, 301D Amistad Building, 10 Amistad Street, Yale University, New Haven, CT 06520. Voice: +1-203-737-1422; fax: +1-203737-1484.

laura.niklason@yale.edu

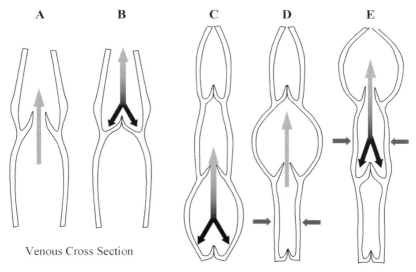

A   B   C   D   E

Venous Cross Section

Lymphangion Cross Section

FIGURE 1. Cross sections of veins and lymphangions.

returned to the vascular system via the thoracic duct. However, while much is known about the biology of the blood vascular system, the biology behind the development and regulation of the lymphatic system remains comparatively poorly understood.[8] Until recently, the number of molecular tools that allowed us to distinguish between blood and lymphatic vessels and cells within tissues was limited. However, in recent years researchers have been able to distinguish quite definitively between the tissues of the two systems at the molecular level.[9] This ability to discriminate between cell types allows for the easy isolation of pure lymphatic cell cultures, and in turn the ability to study lymphatic biology in greater detail.[8–10]

Blood Vessels and Lymphatic Vessels: Comparison and Contrast

A macroscopic look at the vessels of the cardiovascular and lymphatic systems makes it obvious that there are structural and organizational similarities. This is in part due to the connection between the function of the two systems. Both systems contain large networks of vessels, the vessels in both systems are composed of smooth muscle walls lined with a monolayer of endothelium, and they both range in complexity from larger multi-layered vessels to capillaries composed mostly of endothelium. However, the differences between the two also correlate with their function and are evident at the structural, cellular, and molecular

levels. Such similarities and differences make an impact as to how closely the tissue engineering strategies for each system should correlate.

Structure

Current engineering of cardiovascular tissues have yielded vessels that are suitable for arterial implantation and flow.[10,11] The structure of such vessels is fairly straightforward, comprising multiple layers of smooth muscle cells and interlaced extracellular matrix, lined with a monolayer of endothelium. Unlike arteries, the vessels of the venous portion of the cardiovascular system require much less mechanical strength, and many veins incorporate a valve system to prevent the backflow of blood. Lymph vessels (or lymphangions) also require less wall strength than arteries do, and utilize a two-leaflet valve system to prevent the backflow of lymph, making the structural aspects of lymph vessels more similar to venous than to arterial blood vessels. Incorporation of valvular structures into the scaffolding for engineered lymphangions increases the complexity of these structures. However, the lower mechanical requirements for function of venous and lymphangitic structures may make these "easier" to engineer in some respects (FIG. 1).

Cells

The cells that make up the vessels of the lymphatic and cardiovascular system, too, are quite similar. Noting such similarities is very important when considering cell sourcing for tissue engineering strategies. While the origin of the smooth muscle cells that make up the cardiovascular and lymphatic systems is not completely known, it is thought that smooth muscle is recruited by the endothelium that initially constitutes each system.[12] Smooth muscle cells, along with the extracellular matrices that they produce, provide the mechanical strength of the vessels. However, the function of the smooth muscle cells between the two systems differs in their contractile properties. While vascular smooth muscle cells contract in a radial fashion to either increase or decrease the lumen size of the vessel, lymphatic smooth muscle contracts in a rhythmic and unidirectional manner in order to propel lymph toward the thoracic duct.[12,13] It is now thought that native smooth muscle tissues contain specialized pacemaker cells that serve to regulate any spontaneous electrical activity, and that it is these cells that provide the distinct contractile properties of the tissues and not any particular differences in the smooth muscle itself.[14] This implies that current strategies to obtain smooth muscle by the selective culturing of mesenchymal stem cells derived from the peripheral blood or bone marrow may be useful for both systems. However, pacemaker cells would have to be either isolated and cultured, or recruited from surrounding tissues upon engineered graft implantation, adding another consideration to the composition of a working lymphangion.

Developmentally, the lymphatic endothelium (or lymphatic endothelial cells, LECs) originates from a subpopulation of embryonic venous endothelium (or blood endothelial cells, BECs) which expresses the lymphatic-specific homeobox transcriptional factor, prox1.[15] LECs bud off from the venous endothelial populations, and commit to lymphatic lineages. It is debated whether LECs originate directly from the existing venous endothelium or from lymphangioblast precursor cells, which along with angioblasts (vascular endothelial precursors) are derived from mesodermal somites.[15,16] Either way, BECs and LECs are relatively close in origin, function, and structure. When considering cell source for use in tissue engineering, mixtures of BECs and LECs are readily available and easily obtained from harvest of excess dermal tissue.[17] Although the cell types are similar, it is unlikely that they are completely interchangeable on account of the dissimilar sets of genes expressed by each.[18]

Molecular Makeup

Gene array studies have delineated about 300 genes that are differentially expressed between BECs and LECs.[18] Among the genes that are differentially expressed are the known LEC-specific genes (VEGFR-3, LYVE-1, podoplanin, β-chemokine receptor D6, and prox1) as well as an array of proinflammatory cytokines, chemokines and their receptors, cadherins and integrins.[19–23] Interestingly enough, induced overexpression of the PROX1 gene in BECs allows for the upregulation of many of the LEC-specific genes and suppresses expression of ∼40% of BEC-specific genes.[18] This suggests that BECs may be a default lineage, with prox1 inducing lymphangiogenesis of the cells during development. Such insights may allow researchers to use endothelium obtained from less invasive means (i.e., peripheral blood–derived vascular endothelial progenitor cells) along with simple gene therapies in order to supply LEC-like cells for tissue engineering purposes.

Progress of Lymphatic Tissue Engineering

Within the last 5 years, some progress has been made in the area of lymphatic tissue engineering. While the research is still somewhat sparse, it is at the same time relatively diverse and promising. Of the lymphoid tissues, research has been published on the engineering of lymphatic capillaries, lymph nodes, and nonspecific secondary lymphoid tissue-like organoids. Interestingly enough, absent from the list is any research into the engineering of lymphangions.

Lymphatic Capillaries

In the engineering of 3-D tissues, one of the greatest challenges is getting nutrition to all of the cells within a complex structure. For this reason, in correlation with the engineering of larger, more complex tissues, much work is being done in the area of tissue engineering microvasculature. With the ability to reliably differentiate between vascular and lymphatic endothelial cells, some labs are including LECs in their microvasculature research.[24,25] This has led to the development of lymphatic endothelium–lined microstructured extracellular matrices and has allowed investigators to study the function of the lymphatic endothelium in a 3-D *in vitro* system, as opposed to typical 2-D flow systems. In such systems, the endothelium of both the lymphatic and cardiovascular systems can be studied and compared in relation to their response to different stresses and conditions. It is essential to note that research on the microvasculature of lymphatic and blood capillaries should differ in that their functions differ (i.e., exchange of gasses and nutrients with continual flow versus uptake of excessive interstitial fluid).

Researchers have shown that extracellular matrix composition differentially influences the organization of lymphatic and vascular endothelium, with lymphatic endothelium showing the most extensive organization in fibrin-only matrix and vascular endothelium preferring a matrix containing collagen.[24] It is also observed that the types of organizational structures produced are distinct between the two types of endothelium. Lymphatic endothelium tends to produce slender, overlapping networks with fine lumina, while vascular endothelium produces thick, branched networks containing wide lumina.[24] Such differences may be observed *in vivo* in the several days' time lapse that occurs between angiogenesis and lymphangiogenesis during dermal wound healing.[26] It is hypothesized that the change in matrix that occurs during the process of wound-healing produces the optimal environment for the transition from angiogenesis to lymphangiogenesis.

Other research has studied the effects of interstitial and laminar flow on lymphatic endothelium in both 3-D and 2-D environments, respectively.[27] Here we observe that under interstitial flow, lymphatic endothelium tends to form large vacuoles and long extensions. This is in contrast to the multicellular branched lumen-containing networks formed by vascular endothelium under the same conditions. Under interstitial flow, lymphatic endothelium in a 2-D confluent monolayer tends to decrease its cell–cell adhesions in comparison to vascular endothelium. These observations make sense, as they agree with the native function of lymphatic capillaries which provide for looser junctions to allow for the uptake of excess interstitial fluids. This shows that it is not only the matrix that may influence LEC development during tissue culture, but also the mechanical stress that is imparted on the cells. Such observations may further elucidate differences between vascular and lymphatic endothelium *in vivo*, and address key concerns as to what specific cell source, matrix, and environmental stresses are required for lymphatic endothelialization in engineered tissues and creation of lymphatic capillary networks.

Human Lymph Nodes In Vitro

Secondary lymphoid organs are tissues that allow for the development of the immune system within the mammalian body. Some success has been obtained in the development of a bioreactor made for the engineering of a human lymph node *in vitro*.[28] This system was developed to culture human dendritic cells generated from peripheral blood mononuclear cells using an 11-day protocol involving the timed addition and removal of cytokines and growth factors (GM-CSF, IL-4, and TNF-alpha). Such cells were tested for maturation by selecting for a panel of CD-markers and for phagocytic properties.

A small, disposable bioreactor was fashioned that allowed perfusion of cells and medium from a 4-

FIGURE 2. Bioreactor incorporates two external fluidic systems, O-ring sealed caps, and a sample reservoir.[28] Used with permission from ProBioGen AG.

mL outer culture space into a 1-mL central culture space (FIG. 2). The central culture space contained a matrix composed of sheets of agarose and nonwoven polyamide fibers, which were chosen on the basis of their flexibility, cell adhesion properties, porosity, and ease of manipulation. Primed dendritic cells were placed onto two sheets of matrix stabilized by a macroporous membrane and enclosed in the central space, which is supported by a set of microporous hollow fibers that allow medium and gas supply. The porosity of the system would allow any suspended cells within the outer space to freely pass into the central space and have contact with any matrix-bound cells, as well as allow continual medium exchange. After 2 weeks of bioreactor culture, both the nonwoven polyamide sheets and agarose matrix sheets showed evidence of sustainable lymphocyte clusters containing antigen-specific leukocytes. Responses of the cells to IL-2 and LPS showed evidence of early T cell activation and long-term viability.

This bioreactor system allowed for the construction of an environment similar to that of the lymph node, leading to T cell and B cell swarming and T cell clustering. This type of system could be useful for mimicking the effects of drugs and cell therapies within physiologic environments. However, it is not evident that such a system is useful for the construction of implantable tissues.

There has also been success in the area of growing lymphoid tissues *in vivo*. Transplantation of biocompatible "sponge-like" collagenous scaffolds that are seeded with thymus stromal cells (TEL-2-LTα) into mice have yielded "lymphoid tissue-like organoids" that have similar structure to secondary lymphoid organs and are able to produce antibodies in the host.[29] Addition of activated dendritic cells to the stromal cell–containing matrix resulted in the promotion of T cell and B cell cluster formation more than two times as much as without dendritic cells. However, in these experiments it was essential that the TEL-2-LTα stromal cells be present, suggesting that they play a major role in cluster formation. Another intriguing point is that sections of explanted organoids showed evidence of blood vessel and high endothelial venule development, which are typical of secondary lymphoid tissues.

Further research with these "artificial lymph nodes" has yielded promising results in immunodeficient SCID mice. It has been found that implantation of the artificial lymph nodes in the SCID mice led to strong antigen-specific antibody responses, migration of lymphoid cells to the spleen and bone marrow, and maintenance of secondary immune responses over time.[30] Such technology could lead to the successful treatment of immunodeficiency diseases in the future, and allow for the synthesis of engineered lymphoid tissues that can be used for both *in vitro* therapeutic analysis as well as *in vivo* implantation.

Prospects for Lymphatic Tissue Engineering

While the advances made in the study of lymphatic biology are encouraging, the complexities of the system that have been elucidated make the engineering of certain functional lymphatic tissues somewhat more daunting than previously thought. Specifically, such difficulties are apparent in the engineering of functional lymphangions, which may affect how we translate techniques used in engineering blood vessels to engineering lymphatic vessels. Contrary to what has been proposed in the past, it may not be sufficient to simply design a conduit for the passive movement of lymph. A lymphatic vessel must be functional: the vessel wall must be able to respond to lymph flow, leakage

of lymph backwards in the system must be prevented, and there must be active contraction to move lymph forward. While the mechanical requirements that must be met for cardiovascular vessel engineering are much higher, the anatomic design of an arterial blood vessel seems comparatively simpler. Such differences may be one reason why current research in lymphatic tissue engineering has not yet broached the engineering of lymphangions.

When we look at the progress made in the engineering of secondary lymphoid tissues and the prospects of furthering such research, it is easy to become optimistic. Such findings may lead the way to therapies aimed at diseases, such as HIV and leukemia. As with all fields of tissue engineering, there is a long way to go before engineered lymphatic tissues can be implemented as a clinical regimen. However, looking at the progress made in the last several years, we see that it is more a question of timing than plausibility.

Acknowledgments

Work supported by ROI HL076485 and RZI HL081560 (both LEN).

Conflicts of Interest

The authors declare no conflicts of interest.

References

1. ATALA, A. *et al*. 2006. Tissue-engineered autologous bladders for patients needing cystoplasty. Lancet **367:** 1241–1246.
2. MIRONOV, V. *et al*. 2003. Organ printing: computer-aided jet-based 3D tissue engineering. Trends Biotechnol. **21:** 157–161.
3. PFISTER, B.J. *et al*. 2006. Development of transplantable nervous tissue constructs comprised of stretch-grown axons. J. Neurosci. Meth. **153:** 95–103.
4. RISBUD, M.V. & M. SITTINGER. 2002. Tissue engineering: advances in *in vitro* cartilage generation. Trends Biotechnol. **20:** 351–356.
5. SHARMA, B. & J.H. ELISSEEFF. 2004. Engineering structurally organized cartilage and bone tissues. Ann. Biomed. Eng. **32:** 148–159.
6. ZAMMARETTI, P. & M. JACONI. 2004. Cardiac tissue engineering: regeneration of the wounded heart. Curr. Opin. Biotechnol. **15:** 430–434.
7. POH, M. *et al*. 2005. Blood vessels engineered from human cells. Lancet **365:** 2122–2124.
8. CARMELIET, P. 2000. Mechanisms of angiogenesis and arteriogenesis. Nat. Med. **6:** 389–395.
9. EZAKI, T. *et al*. 2006. Production of two novel monoclonal antibodies that distinguish mouse lymphatic and block vascular endothelial cells. Anat. Embryol. **211:** 379–393.

10. NIKLASON, L.E. *et al*. 1999. Functional arteries grown in vitro. Science **284:** 489–493.

11. NIKLASON, L.E. *et al*. 2001. Morphologic and mechanical characteristics of engineered bovine arteries. J. Vasc. Surg. **33:** 628–638.

12. VON DER WEID, P.-Y. & D.C. ZAWIEJA. 2004. Lymphatic smooth muscle: the motor unit of lymph drainage. Int. J. Biochem. Cell Biol. **36:** 1147–1153.

13. GASHEV, A.A. 2002. Physiologic aspects of lymphatic contractile function. Ann. N.Y. Acad. Sci. **979:** 178–187.

14. MCHALE, M. *et al*. 2006. Origin of spontaneous rythmicity in smooth muscle. J. Physiol. **570:** 23–28.

15. KONO, T. *et al*. 2006. Differentiation of lymphatic endothelial cells from embryonic stem cells on OP9 stromal cells. Arterioscler. Thromb. Vasc. Biol. **26:** 2070–2076.

16. WILTING, J. & J. BECKER. 2006. Two endothelial cell lines derived from the somite. Anat. Embryol. **211:** S57–S63.

17. HIRAKAWA, S. *et al*. 2003. Identification of vascular lineage-specific genes by transcriptional profiling of isolated blood vascular and lymphatic endothelial cells. Am. J. Pathol. **162:** 575–586.

18. PETROVA, T.V. *et al*. 2002. Lymphatic endothelial reprogramming of vascular endothelial cells by the PRox-1 homeobox transcription factor. EMBO J. **21:** 4593–4599.

19. GALE, N.W. *et al*. 2007. Normal lymphatic development and function in mice deficient for the lymphatic hyaluronan receptor LYVE-1. Mol. Cell. Biol. **27:** 595–604.

20. GOLDMAN, J. *et al*. 2007. Cooperative and redundant roles of VEGFR-2 and VEGFR-3 signaling in adult lymphangiogenesis. FASEB J. **21:** 1–10.

21. NIBBS, R.J.B. *et al*. 2001. The beta-chemokine receptor D6 is expressed by lymphatic endothelium and a subset of vascular tumors. Am. J. Pathol. **158:** 867–877.

22. SCHACHT, V. *et al*. 2003. T1alpha/podoplanin deficiency disrupts normal lymphatic vasculature formation and causes lymphedema. EMBO J. **22:** 3546–3556.

23. WIGLE, J.T. *et al*. 2002. An essential role for Prox1 in the induction of the lymphatic endothelial cell phenotype. EMBO J. **21:** 1505–1513.

24. HELM, E.-L.E., A. ZISCH & M.A. SWARTZ. 2006. Engineered blood and lymphatic capillaries in 3-D VEGF-fibrin-collagen matrices with interstitial flow. Biotechnol. Bioeng. **96:** 167–176.

25. NELSON, C.M. & J. TIEN. 2006. Microstructured extracellular matrices in tissue engineering and development. Curr. Opin. Biotechnol. **17:** 518–523.

26. RUTKOWSKI, J.M., K.C. BOARDMAN & M.A. SWARTZ. 2006. Characterization of lymphangiogenesis in a model of adult skin regeneration. Am. J. Physiol. Heart Circ. Physiol. **291:** H1402–H1410.

27. NG, C.P., C.E. HELM & M.A. SWARTZ. 2004. Interstitial flow differentially stimulates blood and lymphatic endothelial cell morphogenesis in vitro. Microvasc. Res. **68:** 258–264.

28. GIESE, C. *et al*. 2006. A human lymph node in vitro: challenges and progress. Artif. Organs **30:** 803–808.

29. SUEMATSU, S. & T. WATANABE. 2004. Generation of a synthetic lymphoid tissue-like organoid in mice. Nat. Biotechnol. **22:** 1539–1545.

30. OKAMOTO, M. *et al*. 2007. Artificial lymph nodes induce potent secondary immune responses in naive and immunodeficient mice. J. Clin. Invest. **117:** 997–1007.

Animal Models for the Molecular and Mechanistic Study of Lymphatic Biology and Disease

WILLIAM S. SHIN AND STANLEY G. ROCKSON

Stanford Center for Lymphatic and Venous Disorders, Division of Cardiovascular Medicine, Stanford University School of Medicine, Stanford, California, USA

The development of animal model systems for the study of the lymphatic system has resulted in an explosion of information regarding the mechanisms governing lymphatic development and the diseases associated with lymphatic dysfunction. Animal studies have led to a new molecular model of embryonic lymphatic vascular development, and have provided insight into the pathophysiology of both inherited and acquired lymphatic insufficiency. It has become apparent, however, that the importance of the lymphatic system to human disease extends, beyond its role in lymphedema, to many other diverse pathologic processes, including, very notably, inflammation and tumor lymphangiogenesis. Here, we have undertaken a systematic review of the models as they relate to molecular and functional characterization of the development, maturation, genetics, heritable and acquired diseases, and neoplastic implications of the lymphatic system. The translation of these advances into therapies for human diseases associated with lymphatic dysfunction will require the continued study of the lymphatic system through robust animal disease models that simulate their human counterparts.

Key words: animal model systems; diseases; lymphangiomas; vascular malformations; VEGF-C

Introduction

The lymphatic vascular system mediates tissue fluid homeostasis by providing an important route for fluid and protein transport, and plays a complementary role to the blood vasculature in fluid reabsorption and tissue perfusion.[1] Furthermore, the lymphatics serve as conduits for intestinal lipid absorption and for the transport of lymphocytes and antigen-presenting cells to lymph nodes. A variety of disease processes affect the lymphatic system. Dysfunction or hypoplasia of the lymphatic vasculature can lead to a debilitating form of regional swelling known as lymphedema, whereas abnormal development or hyperplasia of these vessels is associated with lymphangiomas and vascular malformations. Recent evidence suggests that the lymphatic vasculature plays an active role in the nodal and systemic metastasis of cancer; this, among many lines of biomedical investigation, has brought increased scientific attention to the lymphatic system. The character-ization of lymphatic-specific growth factors, receptors, and transcriptional regulators has afforded insight into the mechanisms of lymphangiogenesis; in parallel the generation of increasingly sophisticated animal models has helped to establish the current molecular concepts of embryonic lymphatic development. Studies of these animal models have facilitated the unraveling of the mechanisms of human diseases associated with lymphatic dysfunction, and offer promise for the development of targeted therapies for the treatment of these disorders.

The Lymphatic System: Structure and Function

The lymphatic system is a part of both the circulatory and immune systems.[1,2] In addition to maintaining tissue fluid homeostasis through the evacuation of protein-rich lymph from tissues, with transport to the blood vascular system for recirculation, the lymphatics serve as conduits for the transport of lymphocytes and antigen-presenting dendritic cells to regional lymph nodes.[3] Lymphatic capillaries are blind-ended endothelial tubes lacking pericytes and a continuous basal lamina, but containing large interendothelial openings and anchoring filaments that connect the

Address for correspondence: Stanley G. Rockson, M.D., Stanford Center for Lymphatic and Venous Disorders, Division of Cardiovascular Medicine, Stanford University School of Medicine, 300 Pasteur Drive, Stanford, CA 94305. Voice: +1-650-725-7571; fax: +1-650-725-1599. srockson@cvmed.stanford.edu

Ann. N.Y. Acad. Sci. 1131: 50–74 (2008). © 2008 New York Academy of Sciences.
doi: 10.1196/annals.1413.005

vessels to the extracellular matrix.[4–10] The lymphatic vasculature comprises an open-ended network through which lymph is drained from the interstitial space of tissues. Mediated by pressure gradients emanating from respiration, musculoskeletal movement, smooth muscle contraction, and one-way valves,[11] lymph is transported from initial lymphatic capillaries to larger collecting vessels and, finally, to the inferior vena cava for recirculation.[3]

Lymphatics are present in most vascularized tissues, with the exception of the brain and retina.[2,3] Lymphatic tissue is prevalent in organs that are directly exposed to the external environment, including the skin, the lungs, and the gastrointestinal tract, presumptively reflecting the role that the lymphatics play in the defense against potential pathogens. In the gastrointestinal tract, the lymphatic system mediates the absorption of fat from the intestine and transports the lipids, in the form of chyle, to the liver. As previously described, the lymphatic system also transports fluid excesses, along with cellular debris and metabolic waste products from peripheral sites, back to the systemic circulation.

In addition to its vascular structures, the lymphatic system comprises lymphoid cells and organized lymphoid tissues, including the lymph nodes, tonsils, spleen, thymus, Peyer's patches in the intestine, and lymphoid tissue in the lungs, liver, and parts of the bone marrow.[3,12] These lymphoid organs play a vital role in the immune response. Foreign antigens are concentrated by dendritic cells and presented to lymphocytes in these specialized lymphoid tissues, leading to signaling cascades that result in orchestrated immune responses.[10] It is important to note that lymphocytes circulate between the lymphatic and blood vasculatures. Lymphocytes in circulating lymph enter lymph nodes via the afferent lymphatic vessels, while lymphocytes present in blood may access the lymph node through the wall of special postcapillary venules. Lymphocytes, along with lymph, are subsequently recirculated to the blood circulation through the efferent lymphatic vessels and thoracic duct.

Disease as a Manifestation of Lymphatic Dysfunction

The lymphatic system is essential for homeostasis of the circulatory and immune systems, and lymphatic dysfunction can manifest within a predictable spectrum of sequelae,[11] ranging from isolated immune impairment to a debilitating form of regional swelling termed *lymphedema*, in which inadequate transport of interstitial fluid, impaired immunity, and fibrosis are

observed. Historically, the pathologic expression of lymphedema has been poorly understood.[1] According to the classical model, an imbalance between lymphatic load and transport capacity predicates the accumulation of protein and obligate fluid in the interstitial space,[13–15] which results in further edema formation as a consequence of increases in tissue colloid osmotic pressure. Chronic lymph stasis predisposes to inflammatory and architectural changes that are often profound,[15–18] with an increase in the number of fibroblasts, adipocytes, and keratinocytes in the edematous tissues. The inflammatory response is often demarcated by mononuclear cells.[16,17] Ultimately, skin thickening and subcutaneous fibrosis ensue,[15] although the mechanisms of this transformation are still not well understood.

Lymphedema can be classified as having primary and secondary etiologies. Primary lymphedemas are rare developmental disorders resulting from inborn malformations of the lymphatics.[11,13,18] These conditions most commonly arise at puberty in Meige's disease, or hereditary *lymphedema praecox*. Lymphedema that is present at birth (hereditary congenital lymphedema) has historically been termed *Milroy's disease*. Secondary lymphedema is more prevalent than the primary form and develops as a result of postsurgical, radiotherapeutic, traumatic, or infectious disruption of lymphatic pathways.[13–15,18,19] Edema of the arm after breast cancer interventions is the most common cause of lymphedema in the United States; the global incidence of secondary lymphedema can be ascribed predominantly to filiarisis, which globally afflicts more than 129 million people.[20,21]

Lymphedema is a chronic, unremitting condition that predisposes to substantial morbidity.[13–15] As an example, postsurgical lymphatic dysfunction is estimated to affect approximately one-fourth to one-third of the estimated 2 million American breast cancer survivors.[13] Lymphedema poses a complex therapeutic challenge for the physician.[13,15,22–26] Treatment options are limited to physiotherapeutic interventions that reduce the volume of edema volume, but at best provide only limited relief to afflicted patients.[13,15,27] Furthermore, such conservative measures do not ensure freedom from the advent of irreversible fibrosis.[13] Recent studies in animal models of both primary and secondary lymphatic insufficiency have demonstrated the promise of growth factor–mediated therapies for this debilitating disease.

A newly identified role for lymphatic vessels in the metastasis of cancer cells has brought substantial scientific and clinical interest to the lymphatic system. Whereas the lymphatic system was once believed to play a passive role in the metastatic process, there is now an increasing body of evidence suggesting that lymphatics play an active role in promoting tumor metastasis. As will subsequently be discussed, animal models of growth factor–mediated tumor lymphangiogenesis have provided insight into the mechanisms of tumor mestastasis, and may facilitate the development of targeted molecular therapies for inhibiting cancer spread.

Historical Animal Models of Acquired Lymphatic Insufficiency

The prototypically successful approach to experimental secondary lymphedema was developed in the canine hindlimb.[14,28] It has been demonstrated that chronic resistant lymphedema can be obtained if, in addition to transection of the lymphatics, a circular strip of skin, subcutaneous tissues, fascia, and periosteum is removed from the thigh so as to interpose a scar to close the capillary network that would otherwise retain the ability to restore lymph flow. The specific technique requires a circumferential incision at the midthigh, resection of the main lymphatic trunk with ligation at both ends, excision of popliteal nodes, suturing of the skin edges, and ligation of the afferent lymphatics. Other subsequent canine models represent modifications of this technique, which was initially elaborated by Olszewski.[28] While these attempts to produce clinical lymphedema have met with some measure of success, all have required substantial surgical intervention, with an observed surgical morbidity exceeding that seen in conventional breast cancer operations in humans.[14,28–33]

Surgical approaches have been adapted to smaller laboratory animals, capitalizing on economy of cost and time in obtaining clinically relevant observations.[34] Models of postsurgical lymphedema in the rat hindlimb are examples.[35–37] One such model,[35] which entails a circumferential incision and resection of lymph trunks analogous to that in the canine model, bears superficial resemblance to the human state of postsurgical lymphedema.[14] However, despite early significant edema, there is a late tendency for the volume of ipsilateral and contralateral limb to approximate one another, as the normal limb also becomes edematous. Furthermore, a substantial death rate is observed in treated animals. An alternative rodent hindlimb model[37] has combined radiation with microsurgical ablation of regional groin lymphatics and lymph to successfully produce chronic lymphedema. However, in comparison to human lymphedema, an uncharacteristic spectrum of late morbidity that includes skin breakdown, extensive soft tissue fibrosis, and bone necrosis in the lymphedematous limb is observed.

Inherent obstacles to creating a stable postsurgical model of lymphatic insufficiency are posed by the regenerative capacity of the lymphatics,[38] development of collateral circulations,[39] and formation of lymphaticovenous shunts.[40] Extensive surgical intervention is thus often required to create sustained lymphedema, with a surgical morbidity that is out of proportion to its human clinical counterpart.[14] Some of these problems have been circumvented through the use of the rabbit ear, a structure that provides a comparatively large area of homogenous tissue with its lymphatic drainage through a small conduit that is easily accessible for surgical intervention.[14] In one such rabbit model of experimental lymphedema,[41] a circular strip of skin, subcutaneous tissue, and perichondrium is excised from the base of the ear, with preservation of the central neurovascular bundle and chondrium. Methylene blue 0.5% is injected intradermally into the tip of the ear for visualization of the lymphatics. The several major lymphatic channels that can be visualized to converge around the neurovascular bundle are resected and ligated. With this approach, investigators have been able to create sustained, chronic lymphedema in 47 of 50 rabbits over 6 months of observation.

The mouse tail has proven to be another practical target for the experimental production of postsurgical lymphatic insufficiency.[14] The original report[42] of a successful mouse-tail model of lymphedema describes a circumferential incision through the dermis at the tail base which severs the superficial lymphatic network. The deep lymphatic vessels in the lateral aspects of the tail are then identified with blue dextran and cauterized. A 3-mm gap is created in the skin via cautery for subsequent healing through secondary intent. The study has reported sustained lymphedema over 60 days in all subjects, and over more than 100 days in those that were not sacrificed, with tail diameter increasing from 20% to 50% over 14 weeks as compared to control subjects whose deep lymphatics are not cauterized. The applicability of this model for chronic lymphedema of longer duration has not been published.

While the aforementioned historical animal models represent approaches to the gross simulation of human postsurgical lymphedema, current animal model studies of both primary and secondary lymphatic insufficiency have benefited from powerful molecular tools to shed light on the molecular mechanisms governing embryonic and postnatal lymphatic development. Prior to embarking on a discussion of these new animal models, it is important to review our current understanding of lymphatic development.

Embryonic Lymphatic Development

The lymphatic vascular system has been anatomically recognized for centuries,[11,43] but only recently has the discovery of lymphatic-specific receptors, transcriptional regulators, and growth factors provided insight into the mechanisms of lymphatic development on a molecular level. In the early 20th century, Sabin suggested that initial lymph sacs emerge by budding from embryonic veins, and that these primitive lymphatics then spread out throughout the body to form lymphatic networks.[44,45] Several lymphatic-specific markers including vascular endothelial growth factor receptor-3 (VEGFR-3), lymphatic vessel endothelial hyaluronan receptor (LYVE-1), and the transcription factor prospero-related homeobox 1 (PROX-1) are expressed in endothelial cells that form the budding lymph sacs in mouse embryos, supporting Sabin's theory of lymphatic development.[1,10,46–48]

Detailed analyses of genetic mouse models have to led to a molecular model for early lymphatic vascular development,[1,49,50] in which a subpopulation of embryonic venous endothelial cells becomes programmed for lymphatic differentiation. It is believed that the progression of lymphangiogenesis occurs in four distinct regulatory stages[11]: (1) lymphatic competence; (2) lymphatic commitment; (3) lymphatic specification; and (4) lymphatic maturation.

Lymphatic Competence

Lymphatic endothelial cell (LEC) competence represents the stage in which all embryonic cardinal vein endothelial cells possess independent potential to respond to an as-yet unknown inductive signal and thereby undergo lymphatic differentiation.[11] In mice, LEC competence is observed by embryonic day 8.5–9.5 (E8.5-E9.5)[49,50] and is defined by the surrogate expression of VEGFR-3 and LYVE-1, both of which are later expressed specifically by lymphatic endothelium. While the precise role of LYVE-1 is largely unknown, the vascular endothelial growth factor (VEGF) family of ligands and receptors is integrally involved in the development of blood vascular and lymphatic structures[10] and has benefited from intense investigation.[11]

The VEGF family comprises seven identified growth factors, namely VEGF (VEGF-A), VEGF-B, VEGF-C, VEGF-D, orf virus VEGF (VEGF-E),

VEGF-F, and PIGF, which bind to three receptors, VEGFR-1, VEGFR-2, and VEGFR-3, with varying specificities.[1,11,51] VEGFRs are high-affinity receptor kinases that are believed to mediate VEGF signaling by undergoing dimerization and transphosphorylation upon ligand binding.[10] VEGF-C and VEGF-D have been characterized as pro-lymphangiogenic factors in a number of models of varying developmental and postnatal stages,[9,11,51–55] and their cognate receptor, VEGFR-3, appears to be the most critical of the VEGFRs in the context of lymphatic development.[56]

VEGFR-3 is expressed on both early blood and lymphatic vessel endothelium, and plays a pivotal role in the development of both lineages of vasculature.[11,57] In murine embryonic development, from E8.5 to E12.5, VEGFR-3 is expressed by head mesenchyme angioblasts and cardinal vein endothelial cells.[46] At E12.5, VEGFR-3 is detected in both developing venous and lymphatic endothelia, but later its expression becomes restricted primarily to the lymphatics.[46,58–60] Interestingly, under non-homeostatic conditions, VEGFR-3 may not possess sufficient specificity to serve as a lymphatic vascular marker, since VEGFR-3 has been detected in proliferating blood vessel endothelia associated with tumor neovascularization and within wound granulation tissue.[11,61,62]

LYVE-1 is a lymphatic endothelium-specific hyaluronan receptor cloned as a homologue of CD44.[63,64] Although the precise role of LYVE-1 beyond its presumed involvement in hyaluronan metabolism remains unknown, it serves as a key lymphatic-specific cellular marker. LYVE-1 is expressed early during lymphatic competence on blood vascular endothelial cells[48] and on macrophages and hepatic sinusoidal blood capillaries[65] after birth. Apart from these exceptions, LYVE-1 is associated almost exclusively with LECs throughout development.[63] Surprisingly, LYVE-1-deficient mice are without lymphatic vascular abnormality and are phenotypically normal,[66] suggesting that LYVE-1 does not exclusively direct lymphatic development, but that it may exert an influence of more limited scope.[11]

Lymphatic Commitment

Prior to commitment to a lymphatic lineage, mouse embryonic cardinal vein endothelial cells all display competence in expressing VEGFR-3 and LYVE-1.[49] By E9.5–E10.5, an unknown inductive signal derived from the surrounding mesenchyme stimulates a subset of these competent endothelial cells, situated at one pole of the cardinal vein, to express the transcription factor PROX-1.[48,67] It is this expression of PROX-1 that marks lymphatic commitment, and the progenitors of the mature lymphatic system are wholly constituted by this PROX-1-positive subpopulation of venous endothelial cells, arrayed in a polarized fashion along the cardinal vein.[11] Committed LECs ultimately achieve autonomy from the local microenvironment of the cardinal vein and migrate peripherally as they attain higher levels of differentiation through the process of lymphatic specification.[11]

Lymphatic Specification

Between E10.5 and E11.5, committed LECs begin to bud off from the cardinal vein independently of further PROX-1 influence.[11,48,49] As the precursor LECs bud from the cardinal vein in this first ostensibly concerted event of lymphangiogenesis, they alter their gene expression profiles to adopt a lymphatic phenotype. In this stage of lymphatic specification, the expression of additional lymphatic-specific markers including neuropilin-2, secondary lymphoid-organ chemokine, and T1α/podoplanin, is accompanied by the progressive downregulation of blood vascular endothelial markers including CD4, laminin, and type IV collagen.[48,68] VEGFR-3 and LYVE-1 expression continue to be maintained at high levels in the budding LECs.[10,48]

Of note, in PROX-1-deficient mice, immature endothelial cells also proceed to bud from the cardinal vein, but lineage-specific lymphatic specification does not occur.[48,67] This has suggested that PROX-1 may specify lymphatic cell fate by directly reprogramming the transcriptome of embryonic venous endothelial cells.[68] Indeed, the ectopic expression of PROX-1 in differentiated blood vascular endothelial cells has been shown to be sufficient to reprogram these cells to adopt a lymphatic phenotype.[68,69] Together, these studies suggest that the blood vascular phenotype represents the default endothelial differentiation and the aggregate data identify an essential role of PROX-1 in the program specifying LEC fate.[3]

Lymphatic Maturation

The process of LEC maturation into differentiated lymphatic vessels commences at E11.5-E12.5, when budding LECs start to form nascent lymphatic structures known as rudimentary lymph sacs.[48] At E12.5, PROX-1-positive, specified LECs cells begin to sprout from these sacs and spread throughout the embryo,

while remaining Prox1-negative venous endothelial cells lose VEGFR-3 and LYVE-1 expression to adopt a blood vascular endothelial cell phenotype.[48,67] By E14.5, the maturation of LECs into differentiated lymphatic vessels is nearly complete, though the lymphatic vasculature continues to organize through the first few days of postnatal life.[70] Additional lymphatic markers, such as desmoplakin and B-chemokine receptor D6, continue to be expressed in this late stage of terminal differentiation, and are thought to be among the last markers expressed.[11] Shortly after birth, the mature lymphatic network expresses the complete profile of lymphatic markers found in adulthood.[50]

Postnatal Lymphatic Development

The mechanisms governing postnatal secondary lymphangiogenesis, in contrast to embryonic lymphatic development, are less well established. Animal model studies have demonstrated that, in the context of postsurgical lymphatic insufficiency, lymphangiogenesis is a complex processes that is likely regulated by multiple factors including inflammation, interstitial fluid flow, matrix metalloproteinase activity, and VEGF-C/VEGFR-3 signaling.[71–76] These findings have important implications for wound healing, of which lymphangiogenesis is likely an integral feature.[10,62,77]

Secondary lymphangiogenesis has also been studied in the context of tumor biology, with an increasing body of evidence suggesting that growth factors, such as VEGF-C, VEGF-D, hepatocyte growth factor (HGF), and platelet-derived growth factor (PDGF)-BB, play active roles in promoting tumor lymphangiogenesis and lymph node/systemic metastasis. New lymphatics appear to originate solely from pre-existing lymphatic capillaries during cancer-associated lymphangiogenesis,[78] although circulating endothelial progenitor cells have been reported to contribute to lymphatic vessel growth in other settings.[79,80]

Genetic Alterations in Human Lymphatic Disease

Primary hereditary lymphedema can occur as an autosomal dominant condition, termed *Milroy's disease*, where lymphedema is often the only clinically apparent abnormality, or as a condition where lymphedema occurs as a manifestation of a more complex genetic syndrome.[14,56] The clinical heterogeneity of lymphedema syndromes is being confirmed by the emerging molecular genetic studies of these phenotypes.

Milroy's disease is usually characterized by the sole finding of lymphedema of the lower extremity due to hypoplasia of the lymphatic vessels, although other clinical manifestations including hydrocele, large caliber veins, upslanting of the toenail plates, and papillomatosis of the toes have also been reported.[81,82] Although linkage analysis in families with Milroy's disease has suggested genetic heterogeneity and complex inheritance for Milroy's disease, some families clearly have linkage to the VEGFR-3 locus on 5q35.3.[83–85]

Indeed, the only identified molecular causes of Milroy's disease have been VEGFR-3 gene mutations, with 18 mutations described to date, ranging from deletions to missense mutations.[82,83,86–92] The VEGFR-3 receptor, which is activated by VEGF-C and VEGF-D to mediate lymphangiogenesis,[93] is made up of an extracellular ligand-binding domain containing seven immunoglobulin homology regions, and two intracellular tyrosine kinase domains.[94] All VEGFR-3 mutations associated with Milroy's disease localize to one of these two tyrosine kinase domains, illustrating the necessity of these domains in the signal transduction function of VEGFR-3.[82]

It is believed that the VEGFR-3 mutations in Milroy's disease result in defective VEGF-C and VEGF-D signaling due to dominant negative inhibition of VEGFR-3 autophosphorylation.[82] In a study that was designed to establish whether VEGFR-3 missense mutations identified in four primary lymphedema families truly alter VEGFR-3 function, the corresponding mutant receptor cDNAs were overexpressed in heterologous cells to demonstrate that these receptors indeed fail to demonstrate ligand-dependent autophosphorylation in response to VEGF-C.[87] Of interest, humans with heterozygous deletion of the VEGFR-3 gene do not exhibit lymphedema as a phenotype,[95,96] suggesting that lymphedema is not due to haploinsufficiency for VEGFR-3.[83]

Only a minority of patients with primary lymphedema carry VEGFR-3 mutations.[10] Additional genetic loci have also been implicated in families with other lymphedema syndromes. For instance, inactivating mutations in the forkhead box C2 (FOXC2) gene result in the more prevalent hereditary lymphedema–distichiasis syndrome,[97] an autosomal dominant syndrome characterized by pubertal onset of lymphedema and aberrant eyelashes arising from the meibomian glands.[98] Expression of other features of the lymphedema–distichiasis syndrome (cleft lip/palate, congenital heart disease, ptosis, and venous malformations) is variable.[83,99–104] Unlike Milroy's disease, the lymphedema–distichiasis syndrome is characterized by normal or hyperplastic lymphatic vessels, sug-

gesting a participatory role for FOXC2 in the functional integrity of lymphatic vessels, with a less vital developmental role.[11]

Lymphedema–distichiasis has been linked to the FOXC2 locus in chromosome 16q.[97,105–109] in numerous families. FOXC2 is a member of the forkhead family of transcription factors,[110] and plays a critical role in pathways of metabolism and in a wide variety of developmental processes during embryogenesis.[11,83] Nearly all of the more than 30 characterized FOXC2 mutations in lymphedema–distichiasis are predicted to lead to a truncation of the wild-type protein.[107–109] Notably, the lack of correlation between this truncation point and the occurrence of specific phenotypic features suggests that FOXC2 may be a dosage-sensitive gene and that there is a haploinsufficiency effect of FOXC2 for lymphedema–distichiasis.[83]

The rare hypotrichosis–lymphedema–telangiectasia (HLT) syndrome is another condition where lymphedema occurs as a manifestation of a more complex presentation. HLT patients characteristically exhibit childhood hypotrichosis, lymphedema, and telangiectasia or vascular nevi on the palmar surfaces.[11] Mutations in the SRY-related HMG box 18 (SOX18) gene have been identified as playing a role in both recessive and dominant forms of HLT.[111]

The SOX family of transcription factors is characterized by a conserved DNA-binding domain (HMG box), which binds to a specific DNA consensus sequence, resulting in transcriptional modulation of the associated gene. SOX family proteins regulate the specification of cell types and tissue differentiation in diverse developmental processes.[112–115] SOX18 is expressed in a broad range of tissues[111,116,117] and appears to be involved in the regulation of blood vascular development, possibly via the VEGF-VEGFR pathway[118] or interaction with the endothelial transcription factor MEF2C.[119] In addition to its DNA-binding domain, SOX18 encodes a conserved C-terminal domain and a transcriptional *trans*-activation domain.

A study has recently identified affected members of HLT families who express either recessive missense mutations in SOX-18's DNA-binding domain or a spontaneous dominant nonsense mutation in SOX18's *trans*-activation domain.[111] Mutations found in dominant forms of HLT are proposed to generate a SOX18 protein that may act in a dominant-negative fashion to interfere with the functions of proteins important to vascular, hair follicle, and lymphatic development.[112,118,120] The mechanisms by which SOX18 regulates lymphatic development are unclear, but may involve direct modulation of VEGF-C and VEGF-D expression, since the promoters for these genes have been found to contain the specific DNA-binding site for SOX proteins.[111]

Genetic Models of Lymphatic Development and Inheritable Lymphatic Insufficiency

Our understanding of the molecular mechanisms underlying embryonic lymphatic development and disease has paralleled, and been facilitated by, the generation of genetic animal models targeting specific genes proven to be integral to the lymphatic system. These models have shed light on the mechanisms of inheritable lymphatic insufficiency and other pathologic manifestations of lymphatic dysfunction. Studies involving these animal models hold promise for the development of targeted therapies for the treatment of lymphatic disease.

The integral role of VEGFR-3 in the development of the lymphatic system would make it a desirable target for genetic manipulation, but the lymphatic phenotype of VEGFR-3 knockout mice has not been able to be evaluated, as VEGFR-3 deficiency results in death secondary to cardiovascular failure at E9.5, prior to development of the lymphatic system.[47] Meanwhile, mice heterozygous for a null allele at the VEGFR3 locus are phenotypically normal and have normal-appearing lymphatics.[83] However, an existing mouse strain known as the Chy mouse model,[87] which harbors an inactivating VEGFR-3 mutation, is characterized by a phenotype of chylous ascites, hypoplastic cutaneous lymphatic vessels and limb edema that bears resemblance to Milroy's disease in humans.[14,34] The Chy mouse contains a heterozygous I1053F substitution within a highly conserved catalytic domain of VEGFR-3, in close proximity to the locus of VEGFR-3 mutations in Milroy's disease.[121] The described mutation inactivates the receptor tyrosine kinase, a finding consistent with the tyrosine kinase-inactive VEGFR-3s seen in the human disease.[87]

Importantly, in this Chy mouse model that resembles Milroy's disease, virus-mediated delivery of recombinant human VEGF-C promotes lymphangiogenesis and an amelioration of lymphedema that accompanies the generation of new, functional lymphatic vasculature.[34,121] Additionally, crosses between Chy and K14-VEGF-C156S mice, which overexpress the VEGFR3-specific ligand VEGF-C156S,[93] leads to restored lymphatic function in the Chy × K14-VEGF-C156S offspring.[121] These observations indicate that delivery of an excess of ligand can restore normal lymphatic function in the setting of impaired VEGFR-3 signaling,[83] making VEGF-C a promising molecular

tool for the treatment of primary lymphatic insufficiency.

VEGF-C and VEGF-D, originally cloned as ligands for VEGFR-3, were the first factors identified with the ability to induce growth of new lymphatic vessels.[14,53,122–125] VEGF-C is expressed by numerous cell types, including mesenchymal cells in regions where lymphatic vessels sprout from embryonic veins, smooth muscle cells surrounding large arteries, and activated macrophages.[53,126–128] VEGF-C is produced as a pre-proprotein with long C- and N-terminal propeptides flanking the VEGF homology domain, and a succession of proteolytic cleavage steps is needed to generate a fully processed form with high affinity for VEGFR-3.[129] Both VEGF-C and VEGF-D promote, through activation of VEGFR-3, the proliferation of cultured human LECs.[130] VEGF-C and D, through VEGFR-3, are also able to induce lymphangiogenesis *in vivo*, as has been demonstrated in the skin of transgenic mice overexpressing VEGF-C/D.[52,93] It is important to note that, while VEGF-C appears to act primarily as a lymphangiogenic factor, it might also, through potential interaction with VEGFR-2, play a role in angiogensesis.[131]

A transgenic mouse model[130] bearing attributes of human lymphedema[14,34] has been described in which binding of VEGF-C/VEGF-D to their endogenous receptors is reduced. The model expresses a soluble chimeric protein containing the ligand-binding domain of VEGFR-3. Overexpression of this soluble receptor in the skin competes for VEGF-C/VEGF-D binding with endogenous VEGFR-3, thereby inhibiting fetal lymphangiogenesis and leading to regression of developing lymphatic vessels.[1] These mice exhibit a lymphedema-like phenotype of limb edema, increased subcutaneous fat, and dermal fibrosis. Furthermore, there is loss of lymphatic tissue in internal organs and deficient dermal lymphatic vessel development, although there is an escape from this pattern of biological expression over time.

VEGF-C knockout mice demonstrate complete abrogation of early lymphatic vessel formation[128] and underscore the vital role that VEGF-C plays during embryonic lymphangiogenesis.[1] Although PROX-1 expression is not inhibited in these VEGF-C- deficient mice, PROX-1-positive, lymphatically specified endothelial cells are unable to bud from the embryonic cardinal vein to form initial lymph sacs, suggesting that LEC specification and subsequent cell migration are controlled by distinct signaling pathways.[69] This sprouting defect is rescued in the presence of VEGF-C or VEGF-D, but not VEGF-A, indicating specificity for VEGFR-3 and the necessity of

VEGF-3-mediated signaling for initial lymphatic vessel formation.[128] Ultimately, the absence of lymphatic vessels in VEGF-C-deficient mice leads to massive tissue edema and embryonic lethality. Mice heterozygous for VEGF-C develop cutaneous lymphatic hypoplasia and lymphedema, revealing a haploinsufficiency effect of VEGF-C upon normal lymphatic function and development.[69]

In conjunction with the VEGF family, the angiopoietin signaling system appears to modulate both angiogenesis and lymphangiogenesis,[11,131–135] and offers another target for genetic manipulation for the study of lymphatic disease. Angiopoietin 1 (ANG-1) and angiopoietin 2 (ANG-2) exert their effects by binding to TIE-2, a receptor tyrosine kinase expressed on the surface of endothelial cells[131] and whose expression pattern in lymphatic vessels is poorly characterized. In the mouse embryo, ANG-1 is expressed initially in the myocardium and later in a more extensive manner around developing vessels.[136] ANG-2 is expressed in the embryonic dorsal aorta and its branches, and in adult tissues at sites of active vessel remodeling.[137,138]

The collaboration between ANG-1 and ANG-2 and VEGF in angiogenesis is well characterized. ANG-1, an agonist of TIE-2, has a stabilizing effect on the blood vessel network during embryogenesis through maintenance of maximal interactions between endothelial cells and the extracellular matrix.[139] ANG-2, on the other hand, is a context-dependent antagonist or agonist of TIE-2.[136,137,140] In blood vascular development, ANG-2 likely antagonizes TIE-2 to block the constitutive stabilizing action of ANG-1, resulting in less stable pericyte/endothelial cell interactions and reversion of blood vessels to a more plastic state for initiation or regression of angiogenesis.[10,137] Indeed, in rat tumor models, ANG-2 induces angiogenic sprouting in the presence of VEGF and vessel regression in the absence of VEGF.[141,142]

ANG-2 knockout mice, in addition to having defective postnatal angiogenesis, exhibit gross abnormalities of the lymphatic system,[138] thus serving as another model for the study of lymphatic insufficiency. ANG-2-deficient mice display a phenotype characterized by structurally irregular and leaky lymphatic vessels, chylous ascites, lymphedema, and death at 2 weeks postnatally. Mice heterozygous for a null allele at the ANG-2 locus are overtly normal. The molecular targets for ANG-2 during its regulation of lymphatic development remain poorly defined, but it is of interest to note that ANG-1 overexpression in ANG-2-deficient mice rescues the lymphatic, but not blood vascular, phenotype. This suggests that, in the context of lymphatic development, ANG-1 and ANG-2 cooperatively induce lymphangiogenesis as agonists of TIE-2.

Transgenic mouse models have further demonstrated the importance of the angiopoietin signaling system in regulating lymphangiogenesis.[131] Tissues obtained from mice overexpressing ANG-1 are characterized by lymphatic vessel endothelial proliferation, vessel enlargement, and new vessel sprouting, as observed in mice overexpressing VEGF-C. These effects are associated with upregulation of VEGFR-3 and inhibited by systemic blocking of VEGFR-3 signaling, reinforcing the concept that the angiopoietin and VEGF families collaborate during lymphangiogenesis and that ANG-1 therapy, like VEGF-C, may hold promise for the treatment of lymphatic insufficiency.

With the exception of the VEGF family, perhaps no other molecular regulator plays as a pivotal role in lymphatic development as PROX-1, a homeodomain transcription factor that serves as a master regulator of cell fate decisions leading to specific cell lineages in numerous developing organs.[143] Studies of PROX-1 knockout mice have led to the current molecular model for early lymphatic vascular development in which, as aforementioned, PROX-1 induces lymphatic vascular development from preexisting embryonic veins.[48,49,67] Indeed, it appears that PROX-1 specifies LEC fate by directly reprogramming the transcriptome of embryonic venous endothelial cells.[67]

The PROX-1 knockout mouse is unable to develop a lymphatic vascular system, making it an important model for the study of lymphatic insufficiency and disease. Although early budding of immature endothelial cells from the cardinal vein is observed in PROX-1-deficient mice, subsequent budding and sprouting of specified LECs is arrested,[48,67] resulting in severe edema and embryonic lethality at E14.5. Mice heterozygous for a null allele at the PROX-1 locus appear to develop normal lymphatic networks in the cutaneous, digestive, and respiratory systems, but ultimately develop severe edema and chylous ascites, and die within 2 to 3 days after birth, suggesting a haploinsufficiency effect of PROX-1 during the development of the lymphatic system.

As previously discussed, PROX-1 plays a vital role in specifying the lymphatic fate of a subset of embryonic cardinal vein endothelial cells by inducing expression of lymphatic-specific genes in these precursor LECs. The mucin-type transmembrane glycoprotein podoplanin/T1α is one of these key PROX-1-induced genes.[69,144] Podoplanin is expressed by various cell types including type I alveolar cells of the lung, and kidney podocytes.[144,145] In the vascular system, podoplanin expression is first seen at E11.0 in budding PROX-1-positive lymphatic progenitor cells and later becomes specifically restricted to lymphatic endothelium. Ultrastructural analysis has revealed its predominant localization to the luminal plasma membrane of lymphatic vessels.[144]

Podoplanin-knockout mice exhibit cutaneous lymphedema associated with impaired lymphatic transport, dilated lymphatic vessels of the intestine and skin, and neonatal lethality due to respiratory failure.[144,145] Comparative analysis of wild-type, heterozygous, and knockout animals reveals graded hyperplasticity of lymphatic vessels and defective lymphatic capillary plexus formation. Although little is known about podoplanin's specific biological function *in vivo*, it has been proposed that this molecule is responsible for outlining the superficial capillary beds that anastomose with deeper lymphatic networks in subcutaneous tissue.[11,144] *In vitro* studies in cultured endothelial cells indicate that podoplanin may play an important role in LEC migration, adhesion, and tube formation.[144]

Neuropilin-2 is another gene upregulated by PROX-1 in precursor LECs during early lymphatic development. Neuropilins are nonkinase type I transmembrane proteins originally identified as receptors for class III semaphorins, which mediate axon guidance.[146] Neuropilins also modulate angiogenesis and bind to several VEGF family members.[147] These proteins are believed to be non-signaling and are thought to require the presence of a signal-transducing receptor to mediate their effects.[10] By binding with various VEGFs, neuropilins may act as co-receptors for the VEGFRs, augmenting the efficiency of their tyrosine kinase signals.[148] Neuropilin-2, a receptor for VEGF-C,[121,128] is expressed primarily in veins and visceral lymphatic vessels.[1,68,148,149]

Targeted deletion of neuropilin-2 in mice results in absence or severe reduction in small lymphatic vessels during development, whereas arteries, veins, and larger collecting lymphatics are not affected.[149] The precise mechanism of this impairment in lymphatic development has not been elucidated, but the observation that neuropilin-2 can bind VEGF-C has suggested that neuropilin-2 may be involved in VEGFR-3 signaling in lymphatics.[1,121,128,149] The applicability of neuropilin-2-deficient mice for the study of lymphedema is limited by the observation that edema is not observed in neuropilin-2-null mice. It appears that despite the observed reduction in lymphatic vessels in the skin and the internal organs, the intact larger collecting lymphatic vessels may be sufficient for tissue fluid transport.[149]

The aforementioned animal models, particularly those targeting VEGFR-3 signaling, may be taken to approximate Milroy's disease in humans, where lymphedema is often the only clinically apparent abnormality. Interestingly, mice with targeted manipulation of the FOXC2 gene exhibit a more complex phenotype that is astonishingly similar to the lymphedema–distichiasis syndrome in humans, and serve as a valuable model for exploring physiologic events in mesenchymal differentiation associated with lymphatic growth.[150]

As previously described, inactivating mutations in the FOXC2 gene have been identified in human families with the lymphedema–distichiasis syndrome. Both lymphedema and distichiasis (an aberrant row of eyelashes arising from the meibomian glands) are highly penetrant in individuals with these FOXC2 mutations,[150,151] while associated cardiac and skeletal abnormalities, including cleft palate, are less penetrated.[109] FOXC2 knockout mice have provided insight into the relationship between FOXC2 deficiency and this lymphangiodysplastic phenotype. These mice display cardiac and skeletal abnormalities including cleft palate, and die embryologically after E13.5 up to shortly after birth.[150–153] Although lymphatic vessels in FOXC2-null mice appear normal early in development, they later become irregular and dilated, and are associated with abnormally increased recruitment of periendothelial cells.[154] FOXC2-deficient mice furthermore lack lymphatic valves, which results in reflux within lymphatic vessels. These observations suggest that valvular agenesis and abnormal interactions between LECs and periendothelial cells underlie the pathogenesis of lymphedema–distichiasis. Further phenotypic characterization of FOXC2-deficient mice has not been possible because of pre- and perinatal lethality.

Mice with heterozygous deletion of the FOXC2 gene may most closely mimic the human phenotype of lymphedema–distichiasis. Initial studies of these mice had previously reported their phenotype to be overtly normal,[152,153,155,156] but closer examination has revealed that they exhibit generalized lymphatic vessel and lymph node hyperplasia, abnormal reflux through incompetent interlymphangion valves, occasional hindlimb lymphedema, and bilateral distichiasis.[150] Lymphatic vessel hyperplasia and distichiasis are both highly penetrant in the FOXC2 haploinsufficient mice, and are observed in 92% and 100% of heterozygotes, respectively, thereby representing an instance of an engineered heterozygous mouse knockout closely mimicking the phenotype of a dominant human disease.[150]

Mice with SOX18 mutations, like FOXC2 mutants, are another genetic model system that provides insight into the pathogenetic mechanisms of a complex human syndrome. In humans, *de novo* point mutations leading to truncation of the SOX18 transcriptional *trans*-activation domain are implicated in the dominant form of HLT, as previously described. Correspondingly, in mice, spontaneous mutations truncating SOX18 have given rise to strains of mice (Ra and its allelic variants RaJ, Ragl, and RaOP)[120,157] that display phenotypes resembling HLT in humans.

The Ra, RaJ, and Ragl are semi-dominant mutations consisting of single base deletions within the C terminus of SOX18 that result in truncation of the transcriptional *trans*-activation domain of this protein.[120,157] Mice homozygous for Ra, RaJ, and Ragl exhibit severe vascular and hair follicle anomalies, edema, and chylous accumulation that mimic HLT in humans, while heterozygotes display an intermediate phenotype between homozygous mutant and wild-type mice. The RaOp strain[120] similarly contains a single base deletion within SOX18's C terminus of SOX18, but this dominant-negative mutation results in more extensive truncation of the SOX18 gene product. The resulting dominant phenotype is the most severe among the allelic forms of Ra, with heterozygous mice exhibiting a similar phenotype to homozygous strains of the other Ra variants.

Surprisingly, SOX18 knockout mice demonstrate no obvious cardiovascular or lymphatic abnormalities.[158] Taken together, these results suggest that there is functional redundancy between SOX18 and other SOX proteins,[112,118,120] and that the semidominant nature of many of the Ra mutations can be explained by a trans-dominant negative effect mediated by the mutant SOX18 protein rather than haploinsufficiency.[158] Furthermore, the observation that point mutations in the Ra mouse strains and in dominant-negative forms of human HLT both result in a truncation of the transcriptional *trans*-activation domain of SOX18 suggests that the pathogenetic mechanisms leading to the Ra phenotypes and human HLT are similar. Studies of Ra mouse strains, then, may ultimately allow us to penetrate the mechanisms of HLT which is currently so poorly understood.

While many of these genetic models are based on the known importance of a particular molecular regulator to lymphatic development, the comparative transcriptional profiling of cultured human lymphatic and blood vascular endothelial cells has permitted identification of previously uncharacterized lymphangiogenic factors that may serve as potential targets for generating novel animal models of lymphatic disease. The preponderance of genes investigated by microarray analyses are expressed at comparable levels in the two endothelial cell lineages,[159,160] corroborating their close genetic relationship.[1] Human lymphatic and blood vascular endothelial cells do, however, manifest striking differences in the expression of pro-inflammatory cytokines and their receptors, in addition to proteins involved in cytoskeletal and cell–cell interactions, such as the integrins.[10,159,160]

Integrin $\alpha 9$, for instance, is expressed primarily in LECs, while integrin $\alpha 5$ appears to be blood vascular endothelial–cell specific.[159] Integrins are transmembrane $\alpha\beta$ heterodimeric receptors which mediate cell–cell interactions and cell adhesion to extracellular matrix components.[161] Integrin $\alpha 9\beta 1$ is broadly expressed in airway epithelium, skeletal, cardiac, and visceral smooth muscle cells, hepatocytes, and neutrophils,[162,163] and serves as a receptor for multiple ligands, including the inducible endothelial counter-receptor, vascular cell adhesion molecule-1[164] and the extracellular matrix proteins, tenascin C[165] and osteopontin.[166,167] *In vitro*, $\alpha 9\beta 1$ mediates cell adhesion and migration through interaction with these ligands.

Importantly, the integrin $\alpha 9$ subunit appears to be important in the context of lymphatic development. Transgenic PROX-1 overexpression in human umbilical vein endothelial cells (HUVECs), which expectedly induces an upregulation of lymphatic specific markers including VEGFR-3 and podoplanin, also leads to increased integrin $\alpha 9$ expression, suggesting that integrin $\alpha 9$ upregulation is part of a program of LEC differentiation.[168] Notably, this upregulation coincides with an increased HUVEC chemotactic response to VEGF-C, which is abrogated by anti-$\alpha 9\beta 1$ neutralizing antibody. Conversely, knockdown of PROX-1 expression in human LECs leads to reduced integrin $\alpha 9$ expression and a decreased chemotactic response to VEGF-C. These results suggest that integrin $\alpha 9$ may function as a key regulator of lymphangiogenesis acting downstream of PROX-1.

Integrin $\alpha 9$ subunit knockout mice[161] provide the strongest evidence that integrin $\alpha 9\beta 1$ plays a unique and non-redundant *in vivo* role in lymphatic development. Though integrin $\alpha 9$–deficient mice appear normal at birth, they later develop respiratory failure due to bilateral chylothorax, dying between 6 and 12 days of age. These mice further display edema and lymphocytic infiltration in tissues surrounding lymphatic vessels in the chest wall, which is suggestive of a defect in lymphatic function. Taken together, these findings indicate that integrin $\alpha 9\beta 1$ may be required for normal development of the lymphatic system, and that mutation in the integrin $\alpha 9$ subunit gene could represent a cause of congenital chylothorax in humans.[161] Although the mechanisms by which integrin α9β1 mediates lymphatic development require elucidation, interactions between integrin α9β1 and VEGF-C/D[169,170] and/or VEGFR-3[170] may begin to explain the phenotype of α9-deficient mice.

Comparative transcriptional profiling has revealed that hepatocyte growth factor receptor (HGF-R), like integrin α9, is more highly expressed by cultured LECs than by blood endothelial cells.[171] HGF-R, a transmembrane tyrosine kinase[172] whose expression is increased in sites of tumor growth and metastases,[173] serves as a receptor for HGF, a heparin-binding glycoprotein expressed by cells of mesenchymal origin. HGF was originally identified as a mediator of liver generation[174] and also appears to play a major role in tissue repair and promoting tumor invasiveness.[175–178]

The role of HGF/HGF-R signaling during development of the lymphatic system requires elucidation. In mouse embryos, HGF-R appears to be most strongly expressed by maturing LECs of the primary lymph sacs after E14.5, suggesting that HGF-R may play a role during later stages of lymphatic network formation and maturation.[171] Postnatally, while HGF-R is strongly expressed by regenerating lymphatic endothelium during tissue repair and by activated lymphatic vessels in inflamed skin, HGF-R expression is scarcely detected on quiescent lymphatic vessels in normal skin.[171]

Notably, HGF has been found to promote LEC proliferation and migration in vitro.[171] Treating LECs with HGF-R–blocking antibody inhibits this HGF-induced LEC proliferation, whereas treatment with VEGFR-3-blocking antibody does not, suggesting that HGF–HGF-R signaling acts as an alternative lymphangiogenic mechanism that is independent of the VEGF-C/D–VEGFR-3 pathway. The HGF-induced migration of LECs appears to be mediated by integrin α9, since LEC expression of integrin α9 is induced by HGF and the pro-migratory effects of HGF are blocked with integrin α9–specific blocking antibody. Surprisingly, there have been no reports of lymphatic defects in HGF-R- or HGF-knockout mice[179,180]; both mouse strains manifest severe impairment of organ development, leading to early embryonic lethality, which may limit characterization of an abnormal lymphatic phenotype. Mice heterozygous for deletions of HGF-R or HGF appear normal.

A recent study in transgenic HGF-overexpressing mice[171] has demonstrated that HGF is also a potent lymphangiogenic factor in vivo. In comparison to wild-type control mice, HGF-overexpressing mice display an increased number and enlargement of lymphatic vessels in various organ systems, including the skin and gastrointestinal tract. Subcutaneous delivery of exogenous HGF similarly promotes lymphangiogenesis, which is not prevented with VEGFR-3-blocking antibody, indicating that HGF-induced lymphangiogenesis does not require VEGFR-3 signaling. The same study has demonstrated that systemic blockade of HGF-R with anti-HGF-R antibody inhibits cutaneous lymphatic vessel enlargement in a well-characterized mouse model of experimental skin inflammation.[181]

These results identify HGF as a novel lymphangiogenic factor whose mechanism of action is independent of the VEGFR-3 pathway but depends on HGF-R signaling. HGF may thus conceivably have applicability as an alternative strategy to VEGF-C/D for the treatment of lymphatic insufficiency. In addition, HGF-R could serve as a potential target for inhibiting pathologic lymphangiogenesis, such as that which occurs during tumor growth,[171] as will be later discussed. That HGF-R appears to be preferentially expressed by activated, proliferating lymphatic endothelium, but not by quiescent lymphatic vessels in normal skin, has important implications for the potential use of targeted HGF-R-blocking therapeutic strategies.[171]

Ephrin-B2 is another protein not originally known to have importance to the lymphatic system, but which has recently been identified to have an essential role in lymphatic development.[182] Ephrin-B2, a 40-kDa transmembrane belonging to the ephrin ligand family, binds to the EphB2, EphB3, and EphB4 receptor tyrosine kinases and is involved in a range of physiologic systems including tumor metastasis and vascular development.[182–186] In the blood vasculature, ephrin-B2 is expressed in arterial endothelial and smooth muscle cells.[187–189] Although ephrin-B2 can induce forward signaling by binding to its cognate receptors, it can also mediate reverse signaling into the host cell[190] through phosphorylation of tyrosine residues on its cytoplasmic tail by Src-family kinases, or through binding of its cytoplasmic PDZ interaction site with PDZ domain–containing proteins.[191,192] The required in vivo signaling pathways downstream of ephrin reverse signaling are not well understood.

During embryonic vascular development, remodeling of the blood vasculature after initial vascular plexus formation necessitates VEGF and angiopoietin signaling,[138,193] but also requires signaling mediated by ephrin-B2 and its receptors.[187,188,194] Mice with targeted disruption of the ephrin-B2 gene,[187] for instance, initially display normal vasculogenesis, but from E9.0 exhibit markedly defective angiogenesis/remodeling of the primary capillary plexus and die at E11.0. Lymphatic abnormalities have not been reported in ephrin-B2 knockout mice, perhaps in part due to their early embryonic lethality, and, until recently, ephrin-B2's role in lymphatic development had not been specifically examined.

A recent study[182] has reported a knock-in mouse model that expresses a hypomorphic ephrin-B2 gene but lacks its C-terminal PDZ interaction site. Unlike ephrin-B2 knockout mice, this ephrin-B2 mutant strain survives for several weeks postnatally and does not display overt blood vascular abnormalities. Of interest, the mutant mice develop chylothorax and gross defects of the lymphatic system, including hyperplasia and lack of luminal valve formation in the collecting lymphatic vessels. Furthermore, although dermal lymphatics appear normal at birth, these mice fail to remodel their primary lymphatic capillary plexus into a hierarchical vessel network. Notably, these findings are absent in an alternative knock-in mouse model whose expressed ephrin-B2 is incapable of mediating phosphotyrosine signaling, but still has an intact PDZ interaction site. Taken together, these observations indicate that ephrin-B2 is a specific and essential regulator of postnatal lymphatic maturation and development, and that interactions with PDZ domain effectors are required to mediate its functions.[182]

Genetic models have identified many of the molecular mediators that regulate lymphatic development but, to the present, little has been known about the molecular signals that maintain the separation of emerging lymphatics from the blood vasculature. Recently, the hematopoietic signaling protein SLP-76 was identified as an essential regulator of this separation.[195] SLP-76 was originally isolated from T cells as a novel substrate of the T cell antigen receptor (TCR)–stimulated protein tyrosine kinases.[196,197] Although SLP-76 does not have intrinsic enzymatic activity, it possesses have several domains capable of directing intermolecular interactions and modulating TCR-induced signals. Initial studies of SLP-76-deficient mice revealed defective pre-TCR signaling leading to complete block in thymocyte development, marked subcutaneous and intraperitoneal hemorrhage, and frequent perinatal death.[196,198]

A recent study[182] has demonstrated that SLP-76-deficient mice exhibit direct vascular anastomoses between blood and lymphatic vessels, as demonstrated by the identification of chimeric vasculature containing both lymphatic and blood vascular endothelial cells. These blood–lymphatic connections lead to cutaneous hemorrhage in which blood is observed within thin-walled lymphatic vessels, and peritoneal hemorrhage appears chylous. Interestingly, SLP-76 expression is undetectable on either lymphatic or blood venous endothelial cells, but is found in circulating cells within the cardinal vein, suggesting that the effects of SLP-76 are mediated by circulating cells. Finally, the reconstitution of lethally irradiated wild-type mice with SLP-deficient bone marrow is sufficient to confer an identical vascular phenotype to that of SLP-76 knockout mice, indicating that SLP-76 signaling is required exclusively in hematopoietic cells for regulation of lymphatic and blood vascular separation. The downstream effector pathways by which SLP-76 signaling influences lymphatic and blood vascular development require elucidation.

Genetic mouse models have been our primary source for understanding the mechanisms of embryonic lymphatic development, but studies with mice are labor-intensive and are limited in that they are based primarily on a gene-by-gene approach.[199] Recent studies of lymphatic development in the *Xenopus laevis* tadpole[199] and the zebra fish *Danio rerio*[200] have indicated that these model systems will likely be able to be used in large-scale forward screens to more rapidly and efficiently identify gene candidates important to the lymphatic system.

Xenopus laevis is an extensively used laboratory research animal in developmental biology that has the advantage of being a prolific egg-layer and easy to manipulate genetically. Specifically, injection of morpholino antisense oligos into the early embryo allows for the knockdown of specific genes,[201] while injection of DNA or mRNA allows for the rapid functional expression of genes. While adult frogs have lymph hearts,[202,203] and early studies reported the visualization of a lymphatic network in tadpoles via dye injection,[45,204] the molecular identity of the lymphatic system in tadpoles had not previously been characterized.[199]

A recent study[199] has confirmed that the *Xenopus laevis* tadpole possesses a functional lymphatic network that permits its use as a model for testing candidate lymphangiogenic genes. In this species, putative lymphatic vessels staining for PROX-1 are easily distinguished from blood vessels that express mesenchyme-associated serpentine receptor, a known blood vascular marker.[205] The lymphatic identity of these vessels has been confirmed by structural analyses demonstrating their lack of a basement membrane, and functional studies showing that these vessels drain lymph. Developmentally, the lymphatic vasculature appears to arise both from venous endothelial cell transdifferentiation, as seen in mice, as well from lymphangioblasts as has been observed in avian species.[206]

Importantly, in *Xenopus laevis*, morpholino-mediated knockdown of PROX-1 expression impairs the commitment of precursor endothelial cells to a lymphatic lineage, thereby resulting in a hypoplastic lymphatic vasculature and lymphedema. Of interest, knockdown of VEGF-C also induces a lymphedematous phenotype but, unlike tadpoles subjected to PROX-1 knockdown, the lymphatic defects in VEGF-C knockdown animals arise as a consequence of impaired LEC sprouting and migration, rather than by failure to commit to a lymphatic lineage. These findings parallel what is observed in PROX-1- and VEGF-C-deficient mice, suggesting that the mechanisms governing lymphatic development in the *Xenopus laevis* tadpole are similar to that in mammalian systems. This species thus appears to be a promising animal model for screening candidate genes important to lymphatic disease and development.

The zebrafish *Danio rerio* is another model organism used in developmental biology that was only recently identified as having a functional lymphatic system.[200] During zebrafish embryogenesis, independent cell clusters are noted to give rise to a putative lymphatic trunk which is morphologically distinct from the main blood vessels, does not contain blood cells, and which appears capable of draining lymph from the interstitium for delivery to the venous system. This trunk subsequently sprouts to give rise to a lymphatic tree. The molecular identity of the initial cell clusters giving rise to the primary trunk and the mechanisms governing subsequent lymphatic sprouting require elucidation. Lymphatic development in the zebrafish does, however, appear to depend on VEGF-C and VEGFR-3 signaling. Injection of antisense morpholinos directed against VEGF-C leads to agenesis of the primary lymphatic trunk, as does injection of mRNA encoding for a soluble form of VEGFR-3 that binds zebrafish VEGF-C. These observations suggest that evolutionarily conserved mechanisms underlie zebrafish lymphatic development,[200] and delineate the potential use of this model system in screening for genes and pathways critical for lymphatic development and function.

New Models for the Study of Acquired Lymphatic Insufficiency

While genetic animal models have identified key regulators of embryonic lymphatic development and provided insight into the mechanisms of primary lymphatic insufficiency, the secondary tissue responses to chronic lymph stasis and the interplay between local inflammation, matrix remodeling, and lymphangiogenesis in secondary lymphatic insufficiency are less well understood. Even as we have become aware of the array of gene programs activated in LECs in response to lymphangiogenic stimuli,[207,208] it has become increasingly apparent that numerous other physiologic processes are likely to participate in the response to lymph stasis.[209,210] The use of sophisticated molecular tools to study experimental animal models of secondary lymphatic insufficiency, which simulate the regional, acquired lymph stagnation that can arise after trauma, surgery, and cancer therapeutics in humans, has begun to elucidate the molecular basis of this response to lymph stasis.

A recent study[211] of a mouse tail model of acquired lymphatic insufficiency has correlated the timing of tissue swelling with the evolution of specific secondary tissue responses to lymph stasis. To create the model, a circumferential incision is made through the dermis at the mouse tail base to sever dermal lymphatic vessels, after which the incisional edges are pushed apart with a cauterizing iron to disturb deeper lymphatics and leave a 2- to 3-mm gap to delay wound healing. Using this technique, sustained, reproducible edema is created for a period of at least 30 days, with tail swelling that occurs immediately after disruption of the lymphatics and peaks at 15 days. The early responses observed after lymphatic disruption includes macrophage infiltration and degradation of dermal/hypodermal collagen. Thereafter, LEC proliferation leading to hyperplastic lymphatic vessels is seen, in addition to increased hypodermal lipid deposition.

Notably, in this model of postsurgical lymphatic insufficiency, the degree of lymphatic hyperplasia is greatest when tissue swelling is most severe, with an observed decline in LEC proliferation as edema volume subsides, indicating that edema may be a driving force for lymphangiogenesis. Sustained swelling also appears to upregulate VEGF-C expression, but interestingly, LEC proliferation occurs before VEGF-C upregulation, suggesting that initial lymphangiogenesis in this model may occur via mechanisms independent of the VEGF-C/VEGFR-3 pathway and that VEGF-C supports lymphatic hyperplasia only at later stages. Interestingly, the hyperplastic lymphatic vessels in this model are poorly functional, suggesting that secondary responses to lymph stasis, including VEGF-C upregulation, may paradoxically exacerbate underlying edema.

A study has reported a similar mouse tail model of acquired lymphatic insufficiency that, in combination with large-scale transcriptional profiling experiments, has extended our insights into the pathophysiology of secondary lymphedema and the tissue responses engendered by sustained lymph stagnation.[71] This study reports the establishment of postsurgical lymphedema via circumferential skin incision close to the mouse tail base, after which major lymphatic trunks are identified through vital dye injection and cauterized. Sustained lymphedema is thereby observed with an approximately twofold increase in tail volume over base line at postsurgical day 7 and persistence of swelling until time of sacrifice at day 14. Whole-body lymphoscintigraphy of the lymphedematous mice demonstrates significant dermal backflow in the affected tails, with absent flow beyond the tail base, simulating the changes seen in analogous imaging of human acquired lymphedema.

Histologic assessment of postmortem specimens in this model reveals dramatic inflammatory and architectural changes in the skin that simulate the histopathologic characteristics of human acquired lymphedema, with a notable increase in cellularity and an increase in the number of observed fibroblasts, histiocytes, and neutrophils compared to control animals.[71] Other changes include hyperkeratosis and spongiosis of the epidermis, elongated dermal papillae, and expansion of the tissue between the bone and the epidermis. LYVE-1 staining reveals striking increases in both the number and size of cutaneous lymphatic vessels, corroborating another report[211] of lymphatic vessel hyperplasia induced by lymph stagnation. *In vivo* imaging of luciferase-expressing immunocytes injected into the distal tail demonstrates delayed clearance of these immunocompetent cells, confirming prior reports that impaired immune trafficking is a characteristic feature of lymph stasis.[17,212]

In this mouse model that closely simulates human acquired lymphedema, large-scale transcriptional profiling of lymphedematous skin samples has revealed the differential upregulation of genes related to acute inflammation (e.g., calgranulin B, tenascin C, and cyclophilin B, among many others) as well those involved in immune response, complement activation, wound healing, and oxidative stress response.[71] These findings confirm that lymph stasis is likely to engender a pathologic response more complex than that of impaired fluid homeostasis alone. Furthermore, because the reported gene expression signature of this model is derived from whole skin, it reflects the contributions of a heterogenous cellular population and thereby provides important insights into the whole-tissue response to lymphatic vascular insufficiency.

A salient finding in this mouse tail lymphedema model is the histologic and gene expression evidence that inflammation is a dominant feature of the response to lymph stasis. Indeed, the gene expression signature of this model resembles that of other inflammatory conditions,[71] such as multiple sclerosis,[213] psoriasis,[214]

and atherosclerosis.[215,216] It is unclear whether the observed tissue inflammation is solely a reflection of impaired immunocyte clearance, or if deranged immune trafficking actually induces an inflammatory response due to inefficient clearance of immunogenic antigens.[71,209,212] It is notable that this inflammatory response to lymph stasis has been conjectured to be an important trigger for postnatal lymphangiogenesis[72,73] and may in part be responsible for the striking lymphatic vessel hyperplasia observed in this animal model.[209]

In establishing inflammatory and immune mechanisms as key components of the tissue response to lymph stasis, this model has identified broad biological processes that may be targeted to reverse the pathologic tissue responses seen in lymphedema. Such a strategy for developing treatments for lymphatic insufficiency holds promise in light of other systemic inflammatory disease states for which effective therapies already exist.[71] Such therapies will ideally diminish the sequelae of prolonged lymph stasis, including the soft tissue fibrosis and adipose deposition that characterize the late disease.[15,71]

Although inflammation has been conjectured to trigger lymphangiogenesis, and various lymphangiogenic factors, such as VEGF-C, VEGF-D, and HGF, have been identified, the dominant physiologic driving forces that initiate and coordinate postnatal lymphangiogenesis remain to be elucidated. Interestingly, it has been proposed that, inasmuch as the lymphatic system serves to maintain interstitial fluid homeostasis as one of its primary functions,[217] interstitial fluid flow itself may be an important regulator of lymphangiogenesis,[74] much as oxygen concentration is a key regulator of blood angiogenesis.[218] However, until recently, our understanding of the morphologic, spatial, and temporal aspects of lymphangiogenesis in relation to interstitial fluid flow has been limited by the lack of an adequate animal model with which to study the spatial organization of LECs relative to lymph flow.

A study[74] has reported a new model of secondary lymphangiogenesis in regenerating mouse tail skin that has permitted the study of LEC organization and lymphatic vessel growth in relation to interstitial fluid flow. To create this model, a 2-mm-wide circumferential band of dermal tissue is removed from the mid-tail without disturbing the tail core (major blood vessels, tendon, muscle and bone), thereby preserving blood flow to the tail but completely interrupting the lymphatic network and interstitial fluid flow out of the tail tip. A collagen scaffold is subsequently inserted into this prepared gap so as to bridge the distal and proximal portions of the tail skin and to restore the proximally routed flow of lymph. This scaffold serves the dual function of both accommodating fluid transport and acting as a matrix on which LECs can organize to form bridging lymphatic vessels.

Prior to this surgical manipulation, *in vivo* fluorescence microlymphangiography (FM) of the normal mouse tail revealed a discrete hexagonal network of lymphatic vessels in the skin.[74] After surgery, when these lymphatics were no longer present, lymph was observed on FM to traverse the scaffold in a diffuse manner, without distinct channeling. Twenty-five days after surgery, FM demonstrated that lymph had concentrated within crude, yet discrete, fluid channels that developed to bridge the distal and proximal lymphatic networks. Concurrent immunohistochemical staining revealed single LECs migrating along these fluid channels in a distal to proximal orientation in the direction of lymph flow. By day 60, FM showed that the fluid channels were remodeled to exhibit the hexagonal patterning of a normal lymphatic network. Immunohistochemical staining demonstrated that these fluid channels were fully colocalized with LECs as seen in normal skin.

This murine model of skin regeneration thus identifies lymphangiogenesis as a process in which interstitial fluid channeling and LEC migration along these channels precedes the ultimate organization of the fluid channels into endothelium-lined vessels, with lymphatic network formation proceeding distally to proximally, in the direction of lymph flow.[74] The mechanism by which lymph flow directs lymphangiogenesis may be related to actions of collagen-degrading proteases and VEGF-C. For instance, *in situ* zymography in this model demonstrates evidence of matrix metalloproteinase activity at the distal, and not proximal, edge of the regenerating region, prior to the onset of fluid channeling. This observation suggests a mechanism of fluid channel formation in which proteases, transported by interstitial fluid flow, preferentially degrade matrix in a downstream direction, leading to fluid channel formation and directed cell migration in the direction of flow. Immunohistochemical studies reveal that VEGF-C expression is likewise more pronounced at the distal than proximal edge of the regenerating region prior to LEC infiltration, indicating that growth factor transport, expression and subsequent LEC mitogenesis may also be influenced by interstitial fluid flow.

Subsequent studies[75,76] in this mouse model of skin regeneration have provided additional insight into the molecular regulation of secondary lymphangiogenesis. Endogenous VEGF-C expression has been shown to be greatest immediately after surgical intervention, when lymphatic cells are seen migrating into the regenerating region, with a subsequent decline in VEGF-C expression as LECs coalesce and organize into functional lymphatic vessels.[75] VEGF-C signaling may thus be most important in early stages of lymphangiogenesis when LECs are migrating and proliferating, rather than at later stages of LEC organization and maturation. This is supported by the observation that delivery of exogenous VEGF-C to the regenerating region augments early lymphatic cell proliferation, without increasing the rate of functional lymphangiogenesis nor the density of regenerating lymphatic vessels compared to controls.[219]

This model has additionally been used to study the regulation of secondary lymphangiogenesis by VEGFR-3 and VEGFR-2, both of which bind VEGF-C.[76] After surgical disruption of tail lymphatics, systemic administration of neutralizing antibodies to either VEGFR-3 or VEGFR-2 prevents LEC migration into the regenerating region, resulting in complete inhibition of lymphangiogenesis.[76] This is consistent with VEGFR-3's known importance to lymphangiogenesis and recent reports that secondary lymphangiogenesis requires VEGFR-2 signaling.[207,220,221] Interestingly, blocking either VEGFR-3 or VEGFR-2 at a later timepoint, after LECs have been allowed to migrate into the regenerating region, prevents further LEC migration, but allows existing LECs in the regenerating region to organize into functional lymphatic vessels. These observations indicate that LEC migration and proliferation, but not LEC organization, require the cooperative signaling of both VEGFR-3 and VEGFR-2. VEGFR-3 appears to be redundant with VEGFR-2 for LEC organization into functional capillaries, as the simultaneous neutralization of both VEGFR-3 and VEGFR-2 is needed to inhibit LEC organization in this model.

Studies of Therapeutic Lymphangiogenesis in Models of Acquired Lymphatic Insufficiency

As previously discussed, the amelioration of primary lymphedema with growth factor therapy has been demonstrated in the Chy mouse model[121] of heritable lymphatic insufficiency, whereby virus-mediated delivery of excess VEGF-C restores normal lymphatic function even in the setting of impaired VEGFR-3 signaling. Animal studies have similarly demonstrated the potential for growth factor–mediated therapy in secondary forms of lymphatic insufficiency, thus enhancing the prospects for growth factor–mediated therapies in a larger segment of the disease population than is represented by Milroy's disease alone.[1,14,34]

The first study[18] of successful growth factor–mediated therapeutic lymphangiogenesis in an animal model of acquired lymphatic insufficiency used a modification of the previously described approach in the rabbit ear.[41] To establish this model, a 2-cm circumferential strip of skin, subcutaneous tissue, and perichondrium is excised from the ear base, with sparing of the neurovascular bundle and chondrium. After identification of lymphatics via intradermal vital dye injection, major lymphatic trunk segments are resected. Sustained lymphedema is thereby achieved, with close simulation of the human disease. Radionuclide lymphoscintigraphy, for instance, demonstrates dermal backflow in the affected ears, an imaging attribute of insufficient lymphatic transport. Light microscopic assessment of postmortem specimens furthermore reveals profound architectural changes, including a markedly thickened epidermis, a highly cellular dermis, and an overall increase in thickness of the tissues. Remarkably, in this model, a single dose of locally administered human recombinant VEGF-C has been demonstrated to be sufficient to induce lymphangiogenesis, improve dynamic lymphatic function, and reverse the tissue hypercellularity that characterizes the untreated lymphedematous state.[18]

The potential for therapeutic lymphangiogenesis in acquired lymphedema has been demonstrated in various additional animal models of acquired lymphatic insufficiency,[222–224] further enhancing prospects for future therapy of the human disease. In a study of a postsurgical mouse tail lymphedema model established via techniques previously described,[71] exogenous administration of human recombinant VEGF-C was reported to significantly reduce tail edema, restore normal histologic patterning, and ameliorate the impaired immune trafficking that characterizes the untreated disease.[222] Therapeutic lymphangiogenes in the mouse tail also appears to be feasible using gene transfer to deliver VEGF-C. Local transfer of naked plasmid DNA encoding human VEGF-C (phVEGF-C) has been reported to augment new lymphatic vessel growth and ameliorate secondary lymphedema.[223] The same study has reported successful therapeutic lymphangiogenesis with phVEGF-C in the rabbit ear.

Although the aforementioned studies of therapeutic lymphangiogenesis have been based on the known importance of VEGF-C/VEGFR-3 signaling to lymphatic development, the recent identification of HGF as a novel lymphangiogenic factor[171] whose mechanism of action is independent of the VEGFR-3 pathway but depends on HGF-R signaling offers promise

that HGF may have applicability as an alternative strategy to VEGF-C/D for the treatment of lymphatic insufficiency. Indeed, a study has recently demonstrated that local transfer of plasmid DNA encoding human HGF in a rat tail model of postsurgical lymphatic insufficiency results in an amelioration of lymphedema via promotion of lymphangiogenesis.[224] The applicability of other newly characterized lymphangiogenic factors, including platelet-derived growth factor (PDGF)-BB,[225] for the treatment of secondary lymphatic insufficiency remains unknown.

Optimism over the promise of growth factor therapy for the treatment of secondary lymphedema should be balanced with an understanding that postnatal lymphangiogenesis is a complex process regulated by multiple factors. Indeed, conflicting reports have arisen regarding the effects of growth factor therapies designed to induce lymphangiogenesis. In the mouse tail model of skin regeneration described previously,[74] for instance, delivery of exogenous VEGF-C to the regenerating region augments early LEC proliferation, but does not ultimately induce an increase in functional lymphatic vasculature.[219] This observation is supported by other studies[75,76] in this model, suggesting that VEGF-C/VEGFR-3 signaling, while necessary for LEC migration and proliferation, may not serve as important a role in the organizational evolution of lymphangiogenesis. These observations have important implications inasmuch as VEGF-C might only be potentially useful in augmenting lymphangiogenesis in areas where lymphatics are absent or are present in suboptimal densities, and not in cases where lymphatics are intact but poorly functional.[76]

It is, furthermore, important to recognize that, as growth factor–mediated tumor lymphangiogenesis in various animal models has been associated with enhanced lymph node metastasis,[78,225–229] the risk of enhanced growth and spread of tumors during growth factor therapy needs careful evaluation.[10,18]

Animals Models for the Study of Tumor Lymphangiogenesis

Malignant cells have historically been thought to enter the lymphatic circulation in a passive manner, invading local lymphatics at random with tumor expansion.[11] Recent animal studies have suggested a more active natural history of metastatic transformation through tumor-mediated lymphangiogenesis, with VEGF-C/VEGFR-3 signaling playing a central role in the process.[11,78,226,227,230,231] Indeed, there is ample evidence for the expression of VEGF-C in human tumors. VEGF-C has been detected in breast, colon, gastric, lung, squamous cell, and thyroid tumors in addition to mesotheliomas and neuroblastomas.[232–243] Furthermore, a correlation between VEGF-C expression and rate of metastasis to lymph nodes has been found in breast, colorectal, gastric, lung, prostate, and thyroid cancer.[232–238,240,242,244]

Experimental murine models have supported the role of VEGF-C in promoting tumor lymphangiogenesis and metastasis.[1] In one such model system,[226] transgenic (RipVEGF-C) mice that express VEGF-C specifically in pancreatic beta cells and that, accordingly, develop a lymphatic network around the beta cells, are crossed with a second transgenic (Rip1Tag2) strain that is known to develop nonmetastatic pancreatic beta cell tumors. These double-transgenic mice exhibit VEGF-C-induced lymphangiogenesis around the beta cell tumors, in addition to metastatic spread of tumor cells to pancreatic and regional lymph nodes. The increased density of peritumoral lymphatics seen in this double-transgenic strain supports the notion that VEGF-C-mediated increases in peritumoral lymphatic vessels render them more accessible to tumor cells.[1,10]

The functional importance of VEGF-C-mediated lymphangiogenesis for tumor progression has further been assessed in orthotopic tumor models,[227,229,245,246] in which cancer cells overexpressing VEGF-C are implanted into immunosuppressed mice. In one such human breast carcinoma model system,[227] VEGF-C overexpression was noted to increase both peritumoral and intratumoral lymphatic vessel density, without affecting intratumoral blood vessels, and to enhance tumor metastasis to regional lymph nodes and lung. The degree of tumor metastasis appears to correlate with intratumoral lymphatic vessel density as well as with the depth of lymphatic vessel invasion into the tumors. Other comparable tumor models have corroborated the importance of VEGF-C-mediated lymphangiogenesis in facilitating tumor spread.[229,245,246] VEGF-D has similarly been shown to promote metastasis via induction of tumor lymphangiogenesis in a mouse 293EBNA tumor model.[228]

Studies have demonstrated that adenoviral expression of a soluble VEGFR-3, which competes for VEGF-C/D binding with endogenous receptors, can inhibit VEGF-C-mediated tumor lymphangiogenesis.[78,229] Indeed, in a nonorthotopic murine tumor model in which human lung cancer cells are implanted subcutaneously,[78] lymph node metastases are inhibited by soluble VEGFR-3 expression. Interestingly, systemic metastasis of tumor cells continues to occur at a frequency similar to that of control animals that do not express soluble VEGFR-3. This observation suggests that, while VEGF-C/D-mediated lymphangiogenesis contributes to the metastatic process in this model, factors unrelated to VEGF-C/D-mediated lymphangiogenesis, such as tumor cell transport through the bloodstream, may also influence metastatic spread.[78]

Even in the context of tumor lymphangiogenesis alone, it has become evident that VEGF-C and -D are not the sole determinants of lymphatic vessel growth. A recent study in a squamous cell carcinoma model[247] has established that VEGF-A, the predominant tumor angiogenesis factor in the majority of forms of human cancer,[248] is also a tumor lymphangiogenic factor. In this model, chemical skin carcinogenesis is used to establish squamous cell tumors in transgenic mice that overexpress VEGF-A specifically in the skin. While VEGF-A overexpression expectedly augments tumor angiogenesis, it also results in proliferation of peritumoral lymphatic vessels compared to non-VEGF-A-expressing controls, likely through interaction with VEGFR-2 present on tumor-associated lymphatic vessels. This increase in tumor lymphangiogenesis is accompanied by an increase in metastasis to lymph nodes. Interestingly, even prior to metastasizing, VEGF-A-overexpressing primary tumors induce sentinel lymph node lymphangiogenesis, suggesting an anticipatory component of lymphangiogenesis that facilitates cancer spread.[11,247] A similar mechanism of increased metastatic potential has recently been proposed in VEGF-C-expressing tumors.[249]

Further highlighting the complexity of the processes regulating tumor lymphangiogenesis has been the identification of PDGF-BB, a known angiogenic factor, as a novel lymphangiogenic factor.[225] PDGF-BB is member of the PDGF protein family that binds to the PDGFR-α and -β tyrosine kinase receptors.[225,250] PDGF-BB, when implanted into the mouse cornea, induces growth of both blood and lymphatic vessels that express PDGFR-α and –β receptors.[225] Notably, the administration of soluble VEGFR-3 and VEGFR-3 does not inhibit lymphangiogenesis in this model, indicating that PDGF-BB-induced lymphangiogenesis is not dependent on VEGF-C/-D/VEGFR-3 signaling. PDGF-BB has furthermore been demonstrated, in a murine fibrosarcoma model, to induce intratumoral lymphangiogenesis and lymph node metastasis. Blocking PDGF receptor activation with STI571, a potent PDGFR inhibitor, results in inhibition of tumor lymphangiogenesis, indicating that PDGF receptor activation is necessary for PDGF-BB induced lymphangiogenesis.

There are a number of other lymphangiogenic factors and regulators, including HGF[171] and α4β1,[251] whose abilities to influence tumor lymphangiogenesis are currently under investigation. *In vivo* studies have demonstrated that HGF can promote tumor invasiveness by activating proteases that mediate extracellular matrix/basement membrane dissolution.[171,175,177] However, the finding that HGF-R is highly expressed on tumor-associated lymphatic vessels,[171] together with the observed correlation between HGF expression in human tumors and lymph node metastasis,[252] suggests that HGF, in addition to exerting its direct effects on tumor cells, may contribute to tumor spread by promoting lymphangiogenesis.[171] The integrin $\alpha4\beta1$, with its capacity to regulate lymphangiogenesis *in vivo*,[251] may similarly be important for tumor lymphangiogenesis. Indeed, integrin $\alpha9\beta1$ is expressed on proliferating tumoral lymphatics, but not on resting lymphatics, invoking a potential role for this integrin in the metastatic process. To test this hypothesis, studies in a variety of spontaneous orthotopic and nonorthotopic mouse tumor models are currently investigating the effects of integrin $\alpha4\beta1$ blockade on tumor lymphangiogenesis and metastasis.[251]

Animal tumor models, such as those described, provide important insights into the mechanisms governing tumor lymphangiogenesis, and support the role of this process in facilitating metastasis. Delineation of the *in vivo* mediators of tumor lymphangiogenesis has identified potential targets for the therapeutic blockade of lymphatic cancer spread, as well for markers predicting disease progression. Expression profiles of VEGF-C, VEGF-D, and VEGFR-3 appear to have prognostic value in relation to tumor progression/metastasis and patient survival, but further studies are needed to validate their utility for precancerous and cancerous staging. The prognostic value of other tumor lymphangiogenic factors including PDGF-BB and HGF in predicting clinical outcomes warrants evaluation.

Conclusions

The development of animal model systems for the study of the lymphatic system has resulted in an explosion of information regarding the mechanisms governing lymphatic development and the diseases associated with lymphatic dysfunction. Studies in these animal models have led to a new molecular model of embryonic lymphatic vascular development, and have provided insight into the pathophysiology of both inherited and acquired lymphatic insufficiency. It has become apparent, however, that the importance of the lymphatic system to human disease extends, beyond its role in lymphedema, to many other diverse pathologic processes, including, very notably, inflammation and tumor lymphangiogenesis. The translation of these advances into therapies for human diseases associated with lymphatic dysfunction will require the continued study of the lymphatic system through robust animal disease models that simulate their human counterparts.

Conflicts of Interest

The authors declare no conflicts of interest.

References

1. SHIN, W.S. & S.G. ROCKSON. 2006. Lymphangiogenesis: recapitulation of angiogenesis in health and disease. *In* New Frontiers in Angiogenesis. R. Forough, Ed.: 159–202. Springer. Dordrecht.
2. SZUBA, A. *et al*. 2003. The third circulation: radionuclide lymphoscintigraphy in the evaluation of lymphedema. J. Nucl. Med. **44:** 43–57.
3. HONG, Y.K. & M. DETMAR. 2003. Prox1, master regulator of the lymphatic vasculature phenotype. Cell. Tissue. Res. **314:** 85–92.
4. LEAK, L.V. & J.F. BURKE. 1966. Fine structure of the lymphatic capillary and the adjoining connective tissue area. Am. J. Anat. **118:** 785–809.
5. LEAK, L.V. 1970. Electron microscopic observations on lymphatic capillaries and the structural components of the connective tissue-lymph interface. Microvasc. Res. **2:** 361–391.
6. CASLEY-SMITH, J.R. 1980. The fine structure and functioning of tissue channels and lymphatics. Lymphology **13:** 177–183.
7. BARSKY, S.H. *et al*. 1983. Use of anti-basement membrane antibodies to distinguish blood vessel capillaries from lymphatic capillaries. Am. J. Surg. Pathol. **7:** 667–677.
8. EZAKI, T. *et al*. 1990. A new approach for identification of rat lymphatic capillaries using a monoclonal antibody. Arch. Histol. Cytol. **53**(Suppl): 77–86.
9. OH, S.J. *et al*. 1997. VEGF and VEGF-C: specific induction of angiogenesis and lymphangiogenesis in the differentiated avian chorioallantoic membrane. Dev. Biol. **188:** 96–109.
10. LOHELA, M. *et al*. 2003. Lymphangiogenic growth factors, receptors and therapies. Thromb. Haemost. **90:** 167–184.
11. NAKAMURA, K. & S.G. ROCKSON. 2007. Biomarkers of lymphatic function and disease: state of the art and future directions. Mol. Diagn. Ther. **11:** 227–238.
12. OLSZEWSKI, W. 1991. Lymphology and the lymphatic system. *In* Lymph Stasis: Pathophysiology, Diagnosis and Treatment. W. Olszewski, Ed.: 4–12. CRC Press. Boca Raton, FL.
13. SZUBA, A. & S.G. ROCKSON. 1998. Lymphedema: classification, diagnosis and therapy. Vasc. Med. **3:** 145–156.
14. SHIN, W.S., A. SZUBA & S.G. ROCKSON. 2003. Animal models for the study of lymphatic insufficiency. Lymphat. Res. Biol. **1:** 159–169.

15. ROCKSON, S.G. 2001. Lymphedema. Am. J. Med. **110:** 288–295.

16. PILLER, N.B. 1980. Lymphoedema, macrophages and benzopyrones. Lymphology **13:** 109–119.

17. PILLER, N.B. 1990. Macrophage and tissue changes in the developmental phases of secondary lymphoedema and during conservative therapy with benzopyrone. Arch. Histol. Cytol. **53:** 209–218.

18. SZUBA, A. *et al.* 2002. Therapeutic lymphangiogenesis with human recombinant VEGF-C. FASEB J. **16:** 1985–1987.

19. ROCKSON, S. 2000. Primary Lymphedema. *In* Current Therapy in Vascular Surgery, 4th ed. C. Ernst & J. Stanley, Eds.: 915–918. Mosby. Philadelphia.

20. KEISER, P.B. & T.B. NUTMAN. 2002. Update on lymphatic filarial infections. Curr. Infect. Dis. Rep. **4:** 65–69.

21. TAYLOR, M.J. 2002. Wolbachia endosymbiotic bacteria of filarial nematodes. A new insight into disease pathogenesis and control. Arch. Med. Res. **33:** 422–424.

22. VELANOVICH, V. & W. SZYMANSKI. 1999. Quality of life of breast cancer patients with lymphedema. Am. J. Surg. **177:** 184–187; discussion 188.

23. TOBIN, M.B. *et al.* 1993. The psychological morbidity of breast cancer-related arm swelling: psychological morbidity of lymphoedema. Cancer **72:** 3248–3252.

24. MAUNSELL, E., J. BRISSON & L. DESCHENES. 1993. Arm problems and psychological distress after surgery for breast cancer. Can. J. Surg. **36:** 315–320.

25. PASSIK, S. *et al.* 1995. Predictors of psychological distress, sexual dysfunction and physical functioning among women with upper extremity lymphedema related to breast cancer. Psycho-Oncology **4:** 255–263.

26. ROCKSON, S.G. 2002. Lymphedema after surgery for cancer: the role of patient support groups in patient therapy. Dis. Manage. Health Outcomes **10:** 345–347.

27. ROCKSON, S.G. *et al.* 1998. American Cancer Society Lymphedema Workshop. Workgroup III: Diagnosis and management of lymphedema. Cancer **83:** 2882–2885.

28. OLSZEWSKI, W. 1968. Experimental lymphedema in dogs. J. Cardiovasc. Surg. (Torino) **9:** 178–183.

29. DANESE, C., M. GEORGALAS-BERTAKIS & L. MORALES. 1968. A model of chronic postsurgical lymphedema in dogs' limbs. Surgery **64:** 814–820.

30. HAN, L., T. CHANG & W. HWANG. 1985. Experimental model of chronic limb lymphedema and determination of lymphatic and venous pressures in normal and lymphedematous limbs. Ann. Plastic Surg. **15:** 303–312.

31. SEGERSTROM, K. *et al.* 1992. Factors that influence the incidence of brachial edema after treatment of breast cancer. Scand. J. Plast. Reconstr. Surg. Hand. Surg. **26:** 223–227.

32. ERICKSON, V. *et al.* 2001. Arm edema in breast cancer patients. J. Nat. Cancer. Inst. **93:** 96–111.

33. BEAULAC, S.M. *et al.* 2002. Lymphedema and quality of life in survivors of early-stage breast cancer. Arch. Surg. **137:** 1253–1257.

34. ROCKSON, S.G. 2002. Preclinical models of lymphatic disease: the potential for growth factor and gene therapy. Ann. N.Y. Acad. Sci. **979:** 64–75; discussion 76–79.

35. WANG, G.Y. & S.Z. ZHONG. 1985. A model of experimental lymphedema in rats' limbs. Microsurgery **6:** 204–210.

36. KANTER, M.A., S.A. SLAVIN & W. KAPLAN. 1990. An experimental model for chronic lymphedema. Plast. Reconstr. Surg. **85:** 573–580.

37. LEE-DONALDSON, L. *et al.* Refinement of a rodent model of peripheral lymphedema. Lymphology **32:** 111–117.

38. GRAY, H. 1939. Studies of the regeneration of lymphatic vessels. J. Anat. **74:** 309.

39. GOFFRINI, P. & P. BOBBIO. 1964. [The lymph circulation of the upper extremity following the radical operation of mammary cancer and its relations to the secondary edema of the arm.]. Chirurg **35:** 145–148.

40. MALEK, R. 1972. Lympho-venous anastomoses. *In* Handbuch der Allgemein Pathologie. F. Buchner, E. Letterer & S. Roulet, Eds. Springer. Berlin.

41. HUANG, G.K. & Y.P. HSIN. 1983. An experimental model for lymphedema in rabbit ear. Microsurgery **4:** 236–242.

42. SLAVIN, S.A. *et al.* 1999. Return of lymphatic function after flap transfer for acute lymphedema. Ann. Surg. **229:** 421–427.

43. ASELLI, G. 1627. De Lactibus Sive Lacteis Venis. J.B. Bidellius. Milan.

44. SABIN, F. 1904. On the development of the superficial lymphatics in the skin of the pig. Am. J. Anat. **3:** 183–195.

45. SABIN, F. 1902. On the origin of the lymphatic system from the veins and the development of the lymph hearts and thoracic duct in the pig. Am. J. Anat. **1:** 367–391.

46. KAIPAINEN, A. *et al.* 1995. Expression of the fms-like tyrosine kinase 4 gene becomes restricted to lymphatic endothelium during development. Proc. Natl. Acad. Sci. USA **92:** 3566–3570.

47. DUMONT, D.J. *et al.* 1998. Cardiovascular failure in mouse embryos deficient in VEGF receptor-3. Science **282:** 946–949.

48. WIGLE, J.T. & G. OLIVER 1999. Prox1 function is required for the development of the murine lymphatic system. Cell **98:** 769–778.

49. OLIVER, G. & M. DETMAR. 2002. The rediscovery of the lymphatic system: old and new insights into the development and biological function of the lymphatic vasculature. Genes. Dev. **16:** 773–783.

50. OLIVER, G. 2004. Lymphatic vasculature development. Nat. Rev. Immunol. **4:** 35–45.

51. VEIKKOLA, T. *et al.* 2000. Regulation of angiogenesis via vascular endothelial growth factor receptors. Cancer Res. **60:** 203–212.

52. JELTSCH, M. *et al.* 1997. Hyperplasia of lymphatic vessels in VEGF-C transgenic mice. Science **276:** 1423–1425.

53. JOUKOV, V. *et al.* 1996. A novel vascular endothelial growth factor, VEGF-C, is a ligand for the Flt4 (VEGFR-3) and KDR (VEGFR-2) receptor tyrosine kinases. EMBO J. **15:** 1751.

54. KAIPAINEN, A. *et al.* 1993. The related FLT4, FLT1, and KDR receptor tyrosine kinases show distinct expression patterns in human fetal endothelial cells. J. Exp. Med. **178:** 2077–2088.

55. ENHOLM, B. *et al*. 2001. Adenoviral expression of vascular endothelial growth factor-C induces lymphangiogenesis in the skin. Circ. Res. **88:** 623–629.

56. FERRELL, R. 2002. Research perspectives in inherited lymphatic disease. Ann. N.Y. Acad. Sci. **979:** 39–51.

57. JUSSILA, L. & K. ALITALO. 2002. Vascular growth factors and lymphangiogenesis. Physiol. Rev. **82:** 673–700.

58. PAJUSOLA, K. *et al*. 1992. FLT4 receptor tyrosine kinase contains seven immunoglobulin-like loops and is expressed in multiple human tissues and cell lines. Cancer Res. **52:** 5738–5743.

59. APRELIKOVA, O. *et al*. 1992. FLT4, a novel class III receptor tyrosine kinase in chromosome 5q33-qter. Cancer Res. **52:** 746–748.

60. GALLAND, F. *et al*. 1993. The FLT4 gene encodes a transmembrane tyrosine kinase related to the vascular endothelial growth factor receptor. Oncogene **8:** 1233–1240.

61. PARTANEN, T.A. *et al*. 2000. VEGF-C and VEGF-D expression in neuroendocrine cells and their receptor, VEGFR-3, in fenestrated blood vessels in human tissues. FASEB J. **14:** 2087–2096.

62. PAAVONEN, K. *et al*. 2000. Vascular endothelial growth factor receptor-3 in lymphangiogenesis in wound healing. Am. J. Pathol. **156:** 1499–1504.

63. BANERJI, S. *et al*. 1999. LYVE-1, a new homologue of the CD44 glycoprotein, is a lymph-specific receptor for hyaluronan. J. Cell. Biol. **144:** 789–801.

64. PREVO, R. *et al*. 2001. Mouse LYVE-1 is an endocytic receptor for hyaluronan in lymphatic endothelium. J. Biol. Chem. **276:** 19420–19430.

65. MOUTA CARREIRA, C. *et al*. 2001. LYVE-1 is not restricted to the lymph vessels: expression in normal liver blood sinusoids and down-regulation in human liver cancer and cirrhosis. Cancer Res. **61:** 8079–8084.

66. GALE, N.W. *et al*. 2007. Normal lymphatic development and function in mice deficient for the lymphatic hyaluronan receptor LYVE-1. Mol. Cell. Biol. **27:** 595–604.

67. WIGLE, J.T. *et al*. 2002. An essential role for Prox1 in the induction of the lymphatic endothelial cell phenotype. EMBO J. **21:** 1505–1513.

68. HONG, Y.K., J.W. SHIN & M. DETMAR. 2004. Development of the lymphatic vascular system: a mystery unravels. Dev. Dyn. **231:** 462–473.

69. HONG, Y., HARVEY, *et al*. 2002. Prox1 is a master control gene in the program specifying lymphatic endothelial cell fate. Dev. Dyn. **225:** 351–357.

70. CUENI, L.N. & M. DETMAR. 2006. New insights into the molecular control of the lymphatic vascular system and its role in disease. J. Invest. Dermatol. **126:** 2167–2177.

71. TABIBIAZAR, R. *et al*. 2006. Inflammatory manifestations of experimental lymphatic insufficiency. PLoS Med. **3:** e254.

72. RISTIMAKI, A. *et al*. 1998. Proinflammatory cytokines regulate expression of the lymphatic endothelial mitogen vascular endothelial growth factor-C. J. Biol. Chem. **273:** 8413–8418.

73. MOUTA, C. & M. HEROULT. 2003. Inflammatory triggers of lymphangiogenesis. Lymphat. Res. Biol. **1:** 201–218.

74. BOARDMAN, K.C. & M.A. SWARTZ. 2003. Interstitial flow as a guide for lymphangiogenesis. Circ. Res. **92:** 801–808.

75. RUTKOWSKI, J.M., K.C. BOARDMAN & M.A. SWARTZ. 2006. Characterization of lymphangiogenesis in a model of adult skin regeneration. Am. J. Physiol. Heart Circ. Physiol. **291:** H1402–H1410.

76. GOLDMAN, J. *et al*. 2007. Cooperative and redundant roles of VEGFR-2 and VEGFR-3 signaling in adult lymphangiogenesis. FASEB J. **21:** 1003–1012.

77. CLARK, E. & E. CLARK. 1932. Observations on the new growth of lymphatic vessels as seen in transparent chambers introduced into the rabbit's ear. Am. J. Anat. **51:** 43–87.

78. HE, Y. *et al*. 2002. Suppression of tumor lymphangiogenesis and lymph node metastasis by blocking vascular endothelial growth factor receptor 3 signaling. J. Natl. Cancer Inst. **94:** 819–825.

79. SHI, Q. *et al*. 1998. Evidence for circulating bone marrow-derived endothelial cells. Blood **92:** 362–367.

80. KERJASCHKI, D. *et al*. 2006. Lymphatic endothelial progenitor cells contribute to de novo lymphangiogenesis in human renal transplants. Nat. Med. **12:** 230–234.

81. BRICE, G. *et al*. 2005. Milroy disease and the VEGFR-3 mutation phenotype. J. Med. Genet. **42:** 98–102.

82. BUTLER, M.G. *et al*. 2007. A novel VEGFR3 mutation causes Milroy disease. Am. J. Med. Genet. Part A **143A:** 1212–1217.

83. FERRELL, R.E. *et al*. 1998. Hereditary lymphedema: evidence for linkage and genetic heterogeneity. Hum. Mol. Genet. **7:** 2073–2078.

84. EVANS, A.L. *et al*. 1999. Mapping of primary congenital lymphedema to the 5q35.3 region. Am. J. Hum. Genet. **64:** 547–555.

85. HOLBERG, J.H. *et al*. 2001. Segregation analysis and a genome wide linkage search confirm genetic heterogeneity and suggest oliogogenic inheritance in some Milroy congenital primary lymphedema families. Am. J. Med. Genet. **98:** 303–312.

86. IRRTHUM, A. *et al*. 2000. Congenital hereditary lymphedema caused by a mutation that inactivates VEGFR3 tyrosine kinase. Am. J. Hum. Genet. **67:** 295–301.

87. KARKKAINEN, M.J. *et al*. 2000. Missense mutations interfere with VEGFR-3 signaling in primary lymphoedema. Nat. Genet. **25:** 153–159.

88. EVANS, A.L. *et al*. 2003. Identification of eight novel *VEGFR-3* mutations in families with primary congenital lymphodema. J. Med. Genet. **40:** 697–703.

89. DANIEL-SPIEGEL, E. *et al*. 2005. Hydrops fetalis: an unusual prenatal presentation of hereditary congenital lymphedema. Prenat. Diagn. **25:** 1015–1018.

90. MIZUNO, S. *et al*. 2005. Clinical variability in a Japanese hereditary lymphedema type I family with an FLT4 mutation. Congen. Anom. **4:** 59–61.

91. GHALAMKARPOUR, A. *et al*. 2006. Hereditary lymphedema type I associated with VEGFR3 mutation: the first de novo case and atypical presentations. Clin. Genet. **70:** 330–335.

92. SPIEGEL, R. *et al*. 2006. Wide clinical spectrum in a family with hereditary lymphedema type I due to a novel

missense mutation in VEG FR3. J. Hum. Genet. **51:** 846–850.

93. VEIKKOLA, T. *et al*. 2001. Signaling via vascular endothelial growth factor receptor-3 is sufficient for lymphangiogenesis in transgenic mice. EMBO J. **20:** 1223–1231.

94. ILJIN, K. *et al*. 2001. VEGFR3 gene structure, regulatory region, and sequence polymorphisms. FASEB J. **15:** 1028–1036.

95. BARBER, J.C. 1996. Unbalanced translocation in a mother and her son in one of two 5;10 translocation families. Am. J. Med. Genet. **62:** 84–90.

96. GROEN, S.E. *et al*. 1998. Repeated unbalanced offspring due to a familial translocation involving chromosomes 5 and 6. Am. J. Med. Genet. **80:** 448–453.

97. FANG, J. *et al*. 2000. Mutations in FOXC2 (MFH-1), a forkhead family transcription factor, are responsible for the hereditary lymphedema-distichiasis syndrome. Am. J. Hum. Genet. **67:** 1382–1388.

98. FALLS, H.F. & E.D. KERTESZ. 1964. A new syndrome combining pterygium colli with fevelopmental anomalies of the eyelids and lymphatics of the lower extremities. Trans. Am. Ophthalmol. Soc. **62:** 248–275.

99. DALE, R.F. 1987. Primary lymphoedema when found with distichiasis is of the type defined as bilateral hyperplasia by lymphography. J. Med. Genet. **24:** 170–171.

100. CHYNN, K.Y. 1967. Congenital spinal extradural cyst in two siblings. Am. J. Roentgenol. Radium. Ther. Nucl. Med. **101:** 204–215.

101. PAP, Z. *et al*. 1980. Syndrome of lymphoedema and distichiasis. Hum. Genet. **53:** 309–310.

102. CORBETT, C.R. *et al*. 1982. Congenital heart disease in patients with primary lymphedemas. Lymphology **15:** 85–90.

103. GOLDSTEIN, S. *et al*. 1985. Distichiasis, congenital heart defects and mixed peripheral vascular anomalies. Am. J. Med. Genet. **20:** 283–294.

104. BARTLEY, G.B. & I.T. JACKSON. 1989. Distichiasis and cleft palate. Plast. Reconstr. Surg. **84:** 129–132.

105. MANGION, J. *et al*. 1999. A gene for lymphedema-distichiasis maps to 16q24.3. Am. J. Hum. Genet. **65:** 427–432.

106. BELL, R. *et al*. 2000. Reduction of the genetic interval for lyphoedema-distichiasis to below 2 Mb. J. Med. Genet. **37:** 725.

107. BELL, R. *et al*. 2001. Analysis of lymphoedema-distichiasis families for FOXC2 mutations reveals small insertions and deletions throughout the gene. Hum. Genet. **108:** 546–551.

108. FINEGOLD, D.N. *et al*. 2001. Truncating mutations in FOXC2 cause multiple lymphedema syndromes. Hum. Mol. Genet. **10:** 1185–1189.

109. ERICKSON, R.P. *et al*. 2001. Clinical heterogeneity in lymphoedema-distichiasis with FOXC2 truncating mutations. J. Med. Genet. **38:** 761–766.

110. WEIGEL, D. *et al*. 1989. The homeotic gene fork head encodes a nuclear protein and is expressed in the terminal regions of the Drosophila embryo. Cell **57:** 645–658.

111. IRRTHUM, A. *et al*. 2003. Mutations in the transcription factor gene *SOX18* underlie recessive and dominant forms of

112. MATSUI, T. *et al*. Redundant roles of Sox17 and Sox18 in postnatal angiogenesis in mice. J. Cell Sci. **119:** 3513–3526.

113. PEVNY, L.H. & R. LOVELL-BADGE. 1997. Sox genes find their feet. Curr. Opin. Genet. Dev. **7:** 338–344.

114. WEGNER, M. 1999. From head to toes: the multiple facets of Sox proteins. Nucleic Acids Res. **27:** 1409–1420.

115. BOWLES, J., G. SCHEPERS & P. KOOPMAN. 2000. Phylogeny of the SOX family of developmental transcription factors based on sequence and structural indicators. Dev. Biol. **227:** 239–255.

116. PENNISI, D. *et al*. 2000. Structure, mapping, and expression of human *SOX18*. Mamm. Genome **11:** 1147–1149.

117. HOSKING, B.M., WYETH, *et al*. Cloning and functional analysis of the *Sry*-related HMG box gene, *Sox18*. Gene **262:** 239–247.

118. DOWNES, M. & P. KOOPMAN. 2001. *SOX18* and the transcriptional regulation of blood vessel development. Trends Cardiovasc. Med. **11:** 318–324.

119. HOSKING, B.M. *et al*. 2001. SOX18 directly interacts with MEF2C in endothelial cells. Biochem. Biophys. Res. Commun. **287:** 493–500.

120. JAMES, K. *et al*. 2003. Sox18 mutations in the ragged mouse alleles ragged-like and opassum. Genesis **36:** 1–6.

121. KARKKAINEN, M.J. *et al*. 2001. A model for gene therapy of human hereditary lymphedema. Proc. Natl. Acad. Sci. USA **98:** 12677–12682.

122. LEE, J. *et al*. 1996. Vascular endothelial growth factor-related protein: a ligand and specific activator of the tyrosine kinase receptor Flt4. Proc. Natl. Acad. Sci. USA **93:** 1988–1992.

123. ORLANDINI, M. *et al*. 1996. Identification of a c-fos-induced gene that is related to the platelet-derived growth factor/vascular endothelial growth factor family. Proc. Natl. Acad. Sci. USA **93:** 11675–11680.

124. YAMADA, Y. *et al*. 1997. Molecular cloning of a novel vascular endothelial growth factor, VEGF-D. Genomics **42:** 483–488.

125. ACHEN, M.G. *et al*. 1998. Vascular endothelial growth factor D (VEGF-D) is a ligand for the tyrosine kinases VEGF receptor 2 (Flk1) and VEGF receptor 3 (Flt4). Proc. Natl. Acad. Sci. USA **95:** 548–553.

126. KUKK, E. *et al*. 1996. VEGF-C receptor binding and pattern of expression with VEGFR-3 suggests a role in lymphatic vascular development. Development **122:** 3829–3837.

127. EICHMANN, A. *et al*. 1998. Avian VEGF-C: cloning, embryonic expression pattern and stimulation of the differentiation of VEGFR2-expressing endothelial cell precursors. Development **125:** 743–752.

128. KARKKAINEN, M.J. *et al*. 2004. Vascular endothelial growth factor C is required for sprouting of the first lymphatic vessels from embryonic veins. Nat. Immunol. **5:** 74–80.

129. JOUKOV, V. *et al*. 1997. Proteolytic processing regulates receptor specificity and activity of VEGF-C. EMBO J. **16:** 3898–3911.

130. MAKINEN, T. *et al*. 2001. Inhibition of lymphangiogenesis with resulting lymphedema in transgenic mice expressing soluble VEGF receptor-3. Nat. Med. **7:** 199–205.

131. CAO, Y. *et al*. 1998. Vascular endothelial growth factor C induces angiogenesis in vivo. Proc. Natl. Acad. Sci. USA **95:** 14389–14394.

132. CARMELIET, P. 2003. Angiogenesis in health and disease. Nat. Med. **9:** 653–660.

133. ERIKSSON, U. & K. ALITALO. 1999. Structure, expression and receptor-binding properties of novel vascular endothelial growth factors. Curr. Top. Microbiol. Immunol. **237:** 41–57.

134. THURSTON, G. 2003. Role of angiopoietins and Tie receptor tyrosine kinases in angiogenesis and lymphangiogenesis. Cell. Tissue. Res. **314:** 61–68.

135. PETERS, K.G. *et al*. 2004. Functional significance of Tie2 signaling in the adult vasculature. Recent Prog. Horm. Res. **59:** 51–71.

136. DAVIS, S. *et al*. 1996. Isolation of angiopoietin-1, a ligand for the TIE2 receptor, by secretion-trap expression cloning. Cell **87:** 1161–1169.

137. MAISONPIERRE, P.C. *et al*. 1997. Angiopoietin-2, a natural antagonist for Tie2 that disrupts in vivo angiogenesis. Science **277:** 55–60.

138. GALE, N.W. *et al*. 2002. Angiopoietin-2 is required for postnatal angiogenesis and lymphatic patterning, and only the latter role is rescued by Angiopoietin-1. Dev. Cell. **3:** 411–423.

139. SURI, C. *et al*. 1996. Requisite role of angiopoietin-1, a ligand for the TIE2 receptor, during embryonic angiogenesis. Cell **87:** 1171–1180.

140. TEICHERT-KULISZEWSKA, K. *et al*. 2001. Biological action of angiopoietin-2 in a fibrin matrix model of angiogenesis is associated with activation of Tie2. Cardiovasc. Res. **49:** 659–670.

141. HOLASH, J. & MAISONPIERRE, *et al*. 1999. Vessel cooption, regression, and growth in tumors mediated by angiopoietins and VEGF. Science **284:** 1994–1998.

142. HOLASH, J., S.J. WIEGAND & G.D. YANCOPOULOS. 1999. New model of tumor angiogenesis: dynamic balance between vessel regression and growth mediated by angiopoietins and VEGF. Oncogene **18:** 5356–5362.

143. OLIVER, G. *et al*. 1993. Prox 1, a prospero-related homeobox gene expressed during mouse development. Mech. Dev. **44:** 3–16.

144. SCHACHT, V. *et al*. 2003. T1alpha/podoplanin deficiency disrupts normal lymphatic vasculature formation and causes lymphedema. EMBO J. **22:** 3546–3556.

145. RAMIREZ, M.I. *et al*. 2003. T1alpha, a lung type I cell differentiation gene, is required for normal lung cell proliferation and alveolus formation at birth. Dev. Biol. **256:** 61–72.

146. NEUFELD, G. *et al*. 2002. The neuropilins: multifunctional semaphorin and VEGF receptors that modulate axon guidance and angiogenesis. Trends Cardiovasc. Med. **12:** 13–19.

147. MATSUMOTO, T. & L. CLAESSON-WELSH. 2001. VEGF receptor signal transduction. Sci. STKE **112:** RE21.

148. GIRAUDO, E. *et al*. 1998. Tumor necrosis factor-alpha regulates expression of vascular endothelial growth factor receptor-2 and of its co-receptor neuropilin-1 in human vascular endothelial cells. J. Biol. Chem. **273:** 22128–22135.

149. YUAN, L. 2002. Abnormal lymphatic vessel development in neuropilin 2 mutant mice. Development **129:** 4797–4806.

150. KRIEDERMAN, B.M. *et al*. 2003. FOXC2 haploinsufficient mice are a model for human autosomal dominant lymphedema-distichiasis syndrome. Hum. Mol. Genet. **12:** 1179–1185.

151. BRICE, G. *et al*. 2002. Analysis of the phenotypic abnormalities in lymphoedema-distichiasis syndrome in 74 patients with FOXC2 mutations or linkage to 16q24. J. Med. Genet. **39:** 478–483.

152. IIDA, K. *et al*. 1997. Essential roles of the winged helix transcription factor MFH-1 in aortic arch patterning and skeletogenesis. Development **124:** 4627–4638.

153. WINNIER, G.E., L. HARGETT & B.L. HOGAN. 1997. The winged helix transcription factor MFH1 is required for proliferation and patterning of paraxial mesoderm in the mouse embryo. Genes. Dev. **11:** 926–940.

154. PETROVA, T.V. *et al*. 2004. Defective valves and abnormal mural cell recruitment underlie lymphatic vascular failure in lymphedema distichiasis. Nat. Med. **10:** 974–981.

155. WINNIER, G.E. *et al*. 1999. Roles for the winged helix transcription factors MF1 and MFH1 in cardiovascular development revealed by nonallelic noncomplementation of null alleles. Dev. Biol **213:** 418–431.

156. SMITH, R.S. *et al*. 2000. Haploinsufficiency of the transcription factors *FOXC1* and *FOXC2* results in aberrant ocular development. Hum. Mol. Genet. **9:** 1021–1032.

157. PENNISI, D. *et al*. 2000. Mutations in Sox18 underlie cardiovascular and hair follicle defects in ragged mice. Nat. Genet. **24:** 434–437.

158. PENNISI, D. *et al*. 2000. Mice null for Sox18 are viable and display a mild coat defect. Mol. Cell. Biol. **20:** 9331–9336.

159. PETROVA, T. *et al*. 2002. Lymphatic endothelial reprogramming of vascular endothelial cells by the Prox-1 homeobox transcription factor. EMBO J. **21:** 4593–4599.

160. HIRAKAWA, S. *et al*. 2003. Identification of vascular lineage-specific genes by transcriptional profiling of isolated blood vascular and lymphatic endothelial cells. Am. J. Pathol. **162:** 575–586.

161. HUANG, X.Z. *et al*. 2000. Fatal bilateral chylothorax in mice lacking the integrin alpha9beta1. Mol. Cell. Biol. **20:** 5208–5215.

162. CHEN, C. *et al*. 2006. The integrin α9β1 contributes to granulopoiesis by enhancing granulocyte colony-stimulating factor receptor signaling. Immunity **25:** 895–906.

163. PALMER, E.L. *et al*. 1993. Sequence and tissue distribution of the integrin α9 subunit, a novel partner of β1 that is widely distributed in epithelia and muscle. J. Cell. Biol. **123:** 1289–1297.

164. TAOOKA, Y. *et al*. 1999. The integrin α9β1 mediates adhesion to activated endothelial cells and transendothelial neutrophil migration through interaction with vascular cell adhesion molecule-1. J. Cell. Biol. **145:** 413–420.

165. YOKOSAKI, Y. *et al*. 1994. The integrin α9β1 mediates cell attachment to a non-RGD site in the third fibronectin

type III repeat of tenascin. J. Biol. Chem. **269:** 26691–26696.

166. SMITH, L.L. *et al.* 1996. Osteopontin N-terminal domain contains a cryptic adhesive sequence recognized by alpha9beta1 integrin. J. Biol. Chem. **371:** 28485–28491.

167. YOKOSAKI, Y. *et al.* 1999. The integrin α9β1 binds to a novel recognition sequence (SVVYGLR) in the thrombin-cleaved amino-terminal fragment of osteopontin. J. Biol. Chem. **274:** 36328–36334.

168. MISHIMA, K. *et al.* 2007. Prox1 induces lymphatic endothelial differentiation via integrin α9 and other signaling cascades. Mol. Biol. Cell. **18:** 1421–1429.

169. VLAHAKIS, N.E. *et al.* 2005. The lymphangiogenic vascular endothelial growth factors VEGF-C and –D are ligands for the integrin α9β1. J. Biol. Chem. **280:** 4544–4552.

170. WANG, J.F., X.F. ZHANG & J.E. GROOPMAN. 2001. Stimulation of beta 1 integrin induces tyrosine phosphorylation of vascular endothelial growth factor receptor-3 and modulates cell migration. J. Biol. Chem. **276:** 41950–41957.

171. KAJIYA, K. *et al.* 2005. Hepatocyte growth factor promotes lymphatic vessel formation and function. EMBO J. **24:** 2885–2895.

172. GIORDANO, S. *et al.* 1989. Biosynthesis of the protein encoded by the c-met proto-oncogene. Oncogene **4:** 1383–1388.

173. DANILKOVITCH-MIAGKOVA, A. & B. ZBAR. 2002. Dysregulation of Met receptor tyrosine kinase activity in invasive tumors. J. Clin. Invest. **109:** 863–867.

174. NAKAMURA, T. *et al.* 1989. Molecular cloning and expression of human hepatocyte growth factor. Nature **342:** 440–443.

175. ROSEN, E.M. *et al.* 1994. Scatter factor modulates the metastatic phenotype of the EMT6 mouse mammary tumor. Int. J. Cancer **57:** 706–714.

176. SILVAGNO, F. *et al.* 1995. *In vivo* activation of met tyrosine kinase by heterodimeric hepatocyte growth factor molecule promotes angiogenesis. Arterioscler. Thromb. Vasc. Biol. **15:** 1857–1865.

177. JEFFERS, M., S. RONG & G.F. VANDE WOUDE. 1996. Enhanced tumorigenicity and invasion-metastasis by hepatocyte growth factor/scatter factor-met signalling in human cells concomitant with induction of the urokinase proteolysis network. Mol. Cell. Biol. **16:** 1115–1125.

178. MATSUMOTO, K. & T. NAKAMURA. 2001. Hepatocyte growth factor: renotropic role and potential therapeutics for renal diseases. Kidney Int. **59:** 2023–2038.

179. UEHARA, Y. *et al.* 1995. Placental defect and embryonic lethality in mice lacking hepatocyte growth factor/scatter factor. Nature **373:** 702–705.

180. BLADT, F. *et al.* 1995. Essential role for the c-met receptor in the migration of myogenic precursor cells into the limb bud. Nature **376:** 768–771.

181. KUNSTFELD, R. *et al.* 2004. Induction of cutaneous delayed-type hypersensitivity reactions in VEGF-A transgenic mice results in chronic skin inflammation associated with persistent lymphatic hyperplasia. Blood **104:** 1048–1057.

182. MAKINEN, T. *et al.* 2005. PDZ interaction site in ephrinB2 is required for the remodeling of lymphatic vasculature. Genes Dev. **19:** 397–410.

183. OIKE, Y. *et al.* 2002. Regulation of vasculogenesis and angiogenesis by EphB/ephrin-B2 signaling between endothelial cells and surrounding mesenchymal cells. Blood **100:** 1326–1333.

184. TAKAHASHI, T. *et al.* 2001. Temporally compartmentalized expression of ephrin-B2 during renal glomerular development. J. Am. Soc. Nephrol. **12:** 2673–2682.

185. MARTINEZ, A. & E. SORIANO. 2005. Functions of ephrin/Eph interactions in the development of the nervous system: Emphasis on the hippocampal system. Brain Res. Rev. **49:** 211–226.

186. MEYER, S. *et al.* 2005. Ephrin-B2 overexpression enhances integrin-mediated ECM-attachment and migration of B16 melanoma cells. Int. J. Oncol. **27:** 1197–1206.

187. WANG, H.U., Z.F. CHEN, D.J. ANDERSON. 1998. Molecular distinction and angiogenic interaction between embryonic arteries and veins revealed by ephrin-B2 and its receptor Eph-B4. Cell **93:** 741–753.

188. ADAMS, R.H. *et al.* 1999. Roles of ephrinB ligands and EphB receptors in cardiovascular development: demarcation of arterial/venous domains, vascular morphogenesis, and sprouting angiogenesis. Genes Dev. **13:** 295–306.

189. SHIN, D. *et al.* 2001. Expression of ephrinB2 identifies a stable genetic difference between arterial and venous vascular smooth muscle as well as endothelial cells, and marks subsets of microvessels at sites of adult neovascularization. Dev. Biol. **230:** 139–150.

190. KULLANDER, K. & R. KLEIN. 2002. Mechanisms and functions of Eph and ephrin signalling. Nat. Rev. Mol. Cell. Biol. **3:** 475–486.

191. LU, Q. *et al.* 2001. Ephrin-B reverse signaling is mediated by a novel PDZ-RGS protein and selectively inhibits G protein-coupled chemoattraction. Cell **105:** 69–79.

192. PALMER, A., ZIMMER, *et al.* 2002. EphrinB phosphorylation and reverse signaling: regulation by Src kinases and PTP-BL phosphatase. Mol. Cell. **9:** 725–737.

193. THURSTON, G. 2002. Complementary actions of VEGF and angiopoietin-1 on blood vessel growth and leakage. J. Anat. **200:** 575–580.

194. ADAMS, R.H. 2002. Vascular patterning by Eph receptor tyrosine kinases and ephrins. Semin. Cell Dev. Biol. **13:** 55–60.

195. ABTAHIAN, F. *et al.* 2003. Regulation of blood and lymphatic vascular separation by signaling proteins SLP-76 and Syk. Science **299:** 247–251.

196. CLEMENTS, J.L. *et al.* 1999. Fetal hemorrhage and platelet dysfunction in SLP-76-deficient mice. J. Clin. Invest. **103:** 19–25.

197. JACKMAN, J.K. *et al.* 1995. Molecular cloning of SLP-76, a 76 kDa tyrosine phosphoprotein associated with Grb2 in T cells. J. Biol. Chem. **270:** 7029–7032.

198. PIVNIOUK, V. *et al.* 1998. Impaired viability and profound block in thymocyte development in mice lacking the adaptor protein SLP-76. Cell **94:** 229–238.

199. Ny, A., M. Koch, *et al*. 2005. A genetic *Xenopus laevis* tadpole model to study lymphangiogenesis. Nat. Med. **11:** 998–1004.

200. Küchler, A.M., Gjini, *et al*. 2006. Development of the zebrafish lymphatic system requires Vegfc signaling. Curr. Biol. **16:** 1244–1248.

202. Olszewski, W. 1993. Regulation of water balance between blood and lymph in the frog, *Rana pipiens*. Lymphology **26:** 54–155.

201. Nutt, S.L. *et al*. 2001. Comparison of morpholino based translational inhibition during the development of *Xenopus laevis* and *Xenopus tropicalis*. Genesis **30:** 110–113.

203. Liu, Z.Y. & J.R. Casley-Smith. 1989. The fine structure of the amphibian lymph sac. Lymphology **22:** 31–35.

204. Hoyer, M. 1905. Untersuchungen ueber das Lymphgefaessystem der Froschlarven. Bull. Acad. Cracov. Teill **II:** 451–464.

205. Devic, E. *et al*. 1996. Expression of a new G protein-coupled receptor X-msr is associated with an endothelial lineage in *Xenopus laevis*. Mech. Dev. **59:** 129–140.

206. Wilting, J. *et al*. 2000. An avian model for studies of embryonic lymphangiogenesis. Lymphology **33:** 81–94.

207. Hong, Y.K. *et al*. 2004. VEGF-A promotes tissue repair-associated lymphatic vessel formation via VEGFR-2 and the alpha1beta1 and alpha2beta1 integrins. FASEB J. **18:** 1111–1113.

208. Tammela, T., T.V. Petrova & K. Alitalo. Molecular lymphangiogenesis: new players. Trends Cell. Biol. **15:** 434–441.

209. Schneider *et al*. 2006. A new mouse model to study acquired lymphedema. PLoS Med. **3:** e264.

210. Harvey, N.L. *et al*. 2005. Lymphatic vascular defects promoted by Prox1 haploinsufficiency cause adult-onset obesity. Nat. Genet. **37:** 1072–1081.

211. Rutkowski, J.M. *et al*. 2006. Secondary lymphedema in the mouse tail: lymphatic hyperplasia, VEGF-C upregulation, and the protective role of MMP-9. Microvasc. Res. **72:** 161–171.

212. Olszewski, W.L. *et al*. 1990. Immune cells in peripheral lymph and skin of patients with obstructive lymphedema. Lymphology **23:** 23–33.

213. Lock, C. *et al*. 2002. Gene-microarray analysis of multiple sclerosis lesions yields new targets validated in autoimmune encephalomyelitis. Nat. Med. **8:** 500–508.

214. Terui, T. 2000. Inflammatory and immune reactions associated with stratum corneum and neutrophils in sterile pustular dermatoses. Tohoku J. Exp. Med. **190:** 239–248.

215. Tabibiazar, R. *et al*. 2005. Signature patterns of gene expression in mouse atherosclerosis and their correlation to human coronary disease. Physiol. Genomics **22:** 213–226.

216. Tabibiazar, R. *et al*. 2005. Mouse strain-specific differences in vascular wall gene expression and their relationship to vascular disease. Arterioscler. Thromb. Vasc. Biol. **25:** 302–308.

217. Swartz, M.A. 2001. The physiology of the lymphatic system. Adv. Drug. Deliv. Rev. **50:** 3–20.

218. Bunn, H.F. & R.O. Poyton. 1996. Oxygen sensing and molecular adaptation to hypoxia. Physiol. Rev. **76:** 839–885.

219. Goldman, J. *et al*. 2005. Overexpression of VEGF-C causes transient lymphatic hyperplasia but not increased lymphangiogenesis in regenerating skin. Circ. Res. **96:** 1193–1199.

220. Maruyama, K. *et al*. 2005. Inflammation-induced lymphangiogenesis in the cornea arises from CD11b-positive macrophages. J. Clin. Invest. **115:** 2363–2372.

221. Shibuya, M. & L. Claesson-Welsh. 2006. Signal transduction by VEGF receptors in regulation of angiogenesis and lymphangiogenesis. Exp. Cell. Res. **312:** 549–560.

222. Cheung, L. *et al*. 2006. An experimental model for the study of lymphedema and its response to therapeutic lymphangiogenesis. BioDrugs **20:** 363–70.

223. Yoon, Y.S. *et al*. 2003. VEGF-C gene therapy augments postnatal lymphangiogenesis and ameliorates secondary lymphedema. J. Clin. Invest. **111:** 717–725.

224. Saito, Y. *et al*. 2006. Transfection of human hepatocyte growth factor gene ameliorates secondary lymphedema via promotion of lymphangiogenesis. Circulation **114:** 1177–1184.

225. Cao, R. *et al*. 2004. PDGF-BB induces intratumoral lymphangiogenesis and promotes lymphatic metastasis. Cancer Cell **6:** 333–345.

226. Mandriota, S.J. *et al*. 2001. Vascular endothelial growth factor-C-mediated lymphangiogenesis promotes tumour metastasis. EMBO J. **20:** 672–682.

227. Skobe, M. *et al*. 2001. Induction of tumor lymphangiogenesis by VEGF-C promotes breast cancer metastasis. Nat. Med. **7:** 192–198.

228. Stacker, S.A. *et al*. VEGF-D promotes the metastatic spread of tumor cells via the lymphatics. Nat. Med. **7:** 186–191.

229. Karpanen, T. *et al*. 2001. Vascular endothelial growth factor C promotes tumor lymphangiogenesis and intralymphatic tumor growth. Cancer Res. **61:** 1786–1790.

230. Mattila, M.M., Ruohola, *et al*. 2002. VEGF-C induced lymphangiogenesis is associated with lymph node metastasis in orthotopic MCF-7 tumors. Int. J. Cancer **98:** 946–9451.

231. Cassella, M. & M. Skobe. 2002. Lymphatic vessel activation in cancer. Ann. N.Y. Acad. Sci. **979:** 120–130.

232. Kurebayashi, J. *et al*. 1999. Expression of vascular endothelial growth factor (VEGF) family members in breast cancer. Jpn. J. Cancer Res. **90:** 977–981.

233. Salven, P. *et al*. 1998. Vascular endothelial growth factors VEGF-B and VEGF-C are expressed in human tumors. Am. J. Pathol. **153:** 103–108.

234. Akagi, K. *et al*. 2000. Vascular endothelial growth factor-C (VEGF-C) expression in human colorectal cancer tissues. Br. J. Cancer **83:** 887–891.

235. Andre, T. *et al*. 2000. Vegf, Vegf-B, Vegf-C and their receptors KDR, FLT-1 and FLT-4 during the neoplastic progression of human colonic mucosa. Int. J. Cancer **86:** 174–181.

236. Niki, T. *et al*. 2000. Expression of vascular endothelial growth factors A, B, C, and D and their relationships to

lymph node status in lung adenocarcinoma. Clin. Cancer Res. **6:** 2431–2439.

237. OHTA, Y. *et al*. 2000. Increased vascular endothelial growth factor and vascular endothelial growth factor-c and decreased nm23 expression associated with microdissemination in the lymph nodes in stage I non-small cell lung cancer. J. Thorac. Cardiovasc. Surg. **119:** 804–813.

238. BUNONE, G., VIGNERI, *et al*. 1999. Expression of angiogenesis stimulators and inhibitors in human thyroid tumors and correlation with clinical pathological features. Am. J. Pathol. **155:** 1967–1976.

239. OHTA, Y. *et al*. 1999. VEGF and VEGF type C play an important role in angiogenesis and lymphangiogenesis in human malignant mesothelioma tumours. Br. J. Cancer **81:** 54–61.

240. FELLMER, P.T. *et al*. 1999. Vascular endothelial growth factor-C gene expression in papillary and follicular thyroid carcinomas. Surgery **126:** 1056–1061; discussion 1061–1062.

241. SHUSHANOV, S. *et al*. 2000. VEGFc and VEGFR3 expression in human thyroid pathologies. Int. J. Cancer **86:** 47–52.

242. YONEMURA, Y. *et al*. 1999. Role of vascular endothelial growth factor C expression in the development of lymph node metastasis in gastric cancer. Clin. Cancer Res. **5:** 1823–1829.

243. EGGERT, A. *et al*. 2000. High-level expression of angiogenic factors is associated with advanced tumor stage in human neuroblastomas. Clin. Cancer Res. **6:** 1900–1908.

244. TSURUSAKI, T. *et al*. 1999. Vascular endothelial growth factor-C expression in human prostatic carcinoma and its relationship to lymph node metastasis. Br. J. Cancer **80:** 309–313.

245. SKOBE, M. *et al*. 2001. Concurrent induction of lymphangiogenesis, angiogenesis, and macrophage recruitment by vascular endothelial growth factor-C in melanoma. Am. J. Pathol. **159:** 893–903.

246. KAWAKAMI, M. *et al*. 2005. Vascular endothelial growth factor C promotes lymph node metastasis in a rectal cancer orthotopic model. Surg. Today **35:** 131–138.

247. HIRAKAWA, S. *et al*. 2005. VEGF-A induces tumor and sentinel lymph node lymphangiogenesis and promotes lymphatic metastasis. J. Exp. Med. **201:** 1089–1099.

248. FERRARA, N., H.P. GERBER & J. LECOUTER. 2003. The biology of VEGF and its receptors. Nat. Med. **9:** 669–676.

249. HIRAKAWA, S. *et al*. 2007. VEGF-C induced lymphangiogenesis in sentinel lymph nodes promotes tumor metastasis to distant sites. Blood **109:** 1010–1017.

250. HAMMACHER, A. *et al*. 1989. Isoform-specific induction of actin reorganization by platelet-derived growth factor suggests that the functionally active receptor is a dimer. EMBO J. **8:** 2489–2495.

251. GARMY-SUSINI, B. *et al*. 2007. Methods to study lymphatic vessel integrins. Methods Enzymol. **426:** 415–438.

252. BIRCHMEIER, C. *et al*. Met, metastasis, motility and more. Nat. Rev. Mol. Cell. Biol. **4:** 915–925.

Lymphatic Vasculature Development

Current Concepts

GUILLERMO OLIVER AND R. SATHISH SRINIVASAN

Department of Genetics and Tumor Cell Biology,
St. Jude Children's Research Hospital, Memphis, Tennessee, USA

The recent identification of lymphatic endothelial cell–specific markers and the generation of mouse models harboring mutations in many genes that encode those markers have enabled us to begin to elucidate the developmental steps that go toward establishing the lymphatic vasculature network. Understanding these basic mechanisms of lymphatic development will improve our ability to diagnose, prevent, and ideally treat lymphatic disorders.

Key words: lymphangiogenesis; Prox1; Foxc2; ephrinB2; podoplanin

Introduction

The functions of the lymphatic vasculature are to return extravasated tissue fluid back to the blood circulation, to enhance immune surveillance, and to absorb lipids from the intestine. This system also provides a means for metastatic spread of tumor cells. Absence of functional lymphatic vasculature results in lymphedema, a disabling disease with limited treatment options.[1] Despite these important functional roles, our understanding of the genes and mechanisms regulating the development of the lymphatic vasculature is still rudimentary. However, thanks to a score of recent new findings the field is quickly moving forward, and as our understanding improves, more questions arise. Here we review some of the current views and highlight many of those yet unanswered questions.

Specification of Lymphatic Endothelial Cell Type

From an embryologic perspective, the development of a specialized cell type is called differentiation. The stepwise differentiation of a specialized lymphatic endothelial cell (LEC) type is initiated by the commitment of a progenitor cell toward an LEC fate. At least in mammals and at the molecular level, the earliest indication that this process has started is the expression of the lymphatic vessel endothelial hyaluronan receptor 1 (Lyve-1)[2] in some of the endothelial cells (ECs) lining the anterior cardinal vein.[1] Early Lyve-1 expression could indicate that venous endothelial progenitor cells are now competent (committed) to respond to a specific lymphatic inductive signal. However, competence is not a passive state, but rather an actively acquired condition, and *Lyve-1*-null embryos lack any obvious lymphatic phenotype.[3] These findings indicate that Lyve-1 is not the competence factor; the identity of such a factor remains unknown.

Following this initial step, polarized expression of the homeobox protein Prox1 is detected in the anterior cardinal vein in a subpopulation of "competent" Lyve-1-expressing venous ECs.[4] As development progresses, the number of Prox1-expressing cells in the anterior cardinal vein increases, and its expression is detected in more caudally located embryonic veins (e.g., the posterior cardinal vein and the peri-mesonephric veins).[4] This expression is in agreement with earlier detailed anatomic studies[5] indicating that the mammalian lymphatic vasculature arises from the anterior cardinal vein (at the junction of the jugular and primitive ulnar veins) and later continues in the posterior cardinal and iliac veins. The mechanisms responsible for initiating and maintaining the precise temporal and spatial expression of Prox1 in the veins is not yet known. Nevertheless, it is worth mentioning that concomitant with the timing of Prox1 expression in the veins at embryonic day (E)9.5-E13.5, major venous remodeling is also taking place.[5] Could this remodeling process cell autonomously activate Prox1 expression in the veins?

After the initial phase of Prox1 expression in the veins, Prox1-expressing LECs appear to migrate from

Address for correspondence: Guillermo Oliver, Department of Genetics and Tumor Cell Biology, St. Jude Children's Research Hospital Memphis, TN 38105.

guillermo.oliver@stjude.org

the veins and form the primitive lymph sacs, which are scattered along the anteroposterior embryonic axis.[4] Functional analysis of *Prox1*-null embryos revealed that Prox1 activity is crucial not only for the formation of the entire lymphatic vasculature (*Prox1*-null embryos lack a lymphatic vasculature[4]), but also for specification of the LEC phenotype in committed venous EC progenitors.[6] In *Prox1*-null embryos, blood endothelial cells (BECs) fail to acquire an LEC phenotype. Therefore, it was proposed that BECs are the default phenotype upon which the LEC phenotype is specified after initiation of Prox1 expression.[1,6] Therefore, migration of Prox1-expressing LECs from the cardinal vein can be considered the first indication that an LEC phenotype has been specified. During the process of LEC specification and migration, the expression of other lymphatic markers (e.g., neuropilin-2 and podoplanin) is detected in the migrating Prox1-expressing LECs; expression of some BEC markers (e.g., CD34 and laminin) concurrently decreases.[1]

The analysis of *Prox1*-null embryos also suggested that signals from the forming LECs maintain Prox1 expression in the veins. During the generation of the *Prox1*-mutant strain, a *LacZ* reporter cassette was inserted in frame downstream of the *Prox1* endogenous ATG translation initiation site.[4] Therefore, detection of LacZ expression indicated normal *Prox1* promoter activity. Interestingly, despite the fact that no LECs were present in *Prox1*$^{-/-}$ embryos, the *Prox1* promoter (as indicated by *LacZ* expression) remained active in the mutant ECs until approximately E11.5; later on LacZ expression was no longer detected.[4,6] This result suggested that budding LECs provide a regulatory feedback signal to the veins to maintain Prox1 expression and, therefore, LEC specification. Such feedback loop regulation could be an important mechanism to tightly control the number of LECs specified by Prox1 activity in the veins and the number and timing of LECs migrating from the veins. This regulation could also influence the ultimate location of the forming primitive lymph sacs. The positioning could be closely related to van der Putte's[5] observation in 1975 which indicated that the lymphatic primordia (lymph sacs) arise at almost every place where large veins merge. These *in vivo* results demonstrated that Prox1 activity in default BECs was required to specify the LEC phenotype. Furthermore, results from *in vitro* experiments using BECs maintained in culture supported these initial results by showing that forced expression of Prox1 was sufficient to upregulate LEC markers and downregulate BEC markers.[7,8] Therefore, it was proposed that Prox1 is a master control gene of LEC differentiation.[9] In this context, Prox1 could function as a binary switch

turning off the BEC program and turning on the LEC program.

Recently, a similar mechanism was described for cell fate determination in the pituitary gland.[10] The generation of somatotrope, lactotrope, and thyrotrope cell types in the pituitary depends on the function of the homeodomain transcription factor Prop1 and Wnt activity.[10] Prop1 associates with β-catenin and forms two distinct transcription factor complexes: one that represses the lineage-inhibiting transcription factor Hesx1 and one that activates the lineage-determining transcription factor Pit1.[10] During LEC specification, Prox1 may also form such distinct complexes and therefore have the capacity to simultaneously repress the BEC program and activate the LEC-specification pathway (FIG. 1).

What signaling mechanisms regulate LEC migration from the veins? Recently, the vascular endothelial growth factor receptor-3 (*Vegfr-3*) ligand Vegf-c was functionally inactivated in mice.[11] *Vegf-c*-null embryos also lacked a lymphatic vasculature, because Vegf-c activity is essential for the migration of the Prox1-expressing LECs from the embryonic veins.[11] Importantly, LEC specification still occurs in *Vegf-c*-null embryos, as indicated by normal *Prox1* expression in venous LEC progenitors.[11] Interestingly, *Vegf-c*-heterozygous mice exhibit lymphedema and accumulation of chyle (milky fluid) in the peritoneal cavity.[11] In addition, missense mutations in VEGFR-3 were identified in patients with hereditary lymphedema.[12] During early embryonic development, Vegfr-3 is expressed at comparable levels in BECs and LECs; during later development, its expression is downregulated in BECs but remains in LECs.[6,13,14] This receptor–ligand signaling system appears to be at least one of the mechanisms required to promote (and maybe also guide) the polarized budding of Prox1-expressing LECs, thereby ensuring the precise location of the primary lymph sacs.[1,11] However, *Vegf-c* appears to be uniformly expressed by the mesenchyme surrounding the cardinal veins; therefore, it is not clear how this uniform expression provides directional guidance for LEC migration.

Although we do not yet know much about the migration of LECs, we speculate that it shares some similarities with the migration of BECs during angiogenesis, a process about which much more information is available.[15,16] For LECs to migrate, veins most likely need to compromise their smooth muscle/pericyte coverage. This could be accomplished by the temporal and spatial downregulation of molecules required to maintain the interaction between ECs and smooth muscle cells (SMCs) (e.g., Tie2/Ang1; PDGF/PDGFR).[15] In

A

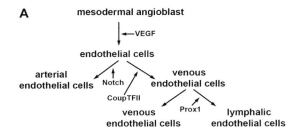

B

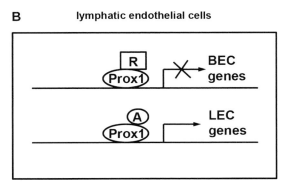

FIGURE 1. (A) Schematic representation of various steps involved in blood and lymphatic vasculature development. The first step in vascular development is the differentiation of ECs from primitive mesodermal cells known as angioblasts. VEGF signaling is known to be required for this process.[16] Subsequently, Notch signaling maintains arterial identity of ECs and expression of CoupTFII directs the venous specification by downregulating the Notch pathway.[16] Finally, expression of Prox1 in a subpopulation of venous ECs results in the specification of LECs. **(B)** Schematic representation of Prox1 activity as a "binary switch" in LEC specification. In the absence of Prox1 expression, BEC-specific genes are on and LEC-specific genes are off. Upon Prox1 expression, Prox1 can form two distinct transcriptional complexes: a transcriptional repressor complex (R) that switches off BEC-specific genes or a transcriptional activator complex (A) that switches on the LEC-specific genes.

addition, the extracellular matrix also must be degraded so that migrating LECs can reach the surrounding mesenchyme.[15]

As development progresses, expression of additional LEC markers starts to be detected in the migrating Prox1-expressing LECs. One of those is the mucin-type transmembrane glycoprotein T1α/Podoplanin.[17] *T1α/Podoplanin*[−/−] pups die soon after birth on account of respiratory failure and they exhibit severe lymphedema resulting from defects in lymphatic vascular patterning and function.[17] In addition, *T1α/Podoplanin*[−/−] pups appear to have defects in establishing proper connections between the

deeper and superficial lymphatics.[17] These defects have been attributed to disruption of the suggested roles of *T1α/Podoplanin* in regulating the migration of LECs and lumen formation.[17]

Neuropilin 2 (*Nrp2*), which is a receptor for class III semaphorins, can also interact with VEGFR-2 and VEGFR-3[18,19] and may mediate their signaling pathways. In the EC compartment, *Nrp2* is expressed predominantly in the veins at around E10.0. Subsequent to the formation of the lymph sacs at E13.0, *Nrp2* expression is restricted to LECs.[19] Functional inactivation of *Nrp2* results in reduced LEC proliferation and fewer lymphatic capillaries, but no obvious defects in deeper lymphatic vessels or lymphatic function were identified.[19] The exact mechanism of *Nrp2* action has not yet been determined. The expression of another neuropilin, *Nrp1*, is restricted to arteries.[19] However, although *Nrp2*[−/−] pups survive until adulthood and *Nrp1*[−/−] embryos survive until around E13.5, *Nrp1;Nrp2* double-null embryos have severe vascular defects and die around E8.5.[20] Hence, *Nrp1* might compensate for the loss of *Nrp2*, which would explain the mild lymphatic phenotype observed in *Nrp2*[−/−] mice.

Lymphovenous Separation

An important step during developmental lymphangiogenesis is the separation of the forming lymphatics from the forming blood vasculature. The signaling molecules *Slp-76, Syk, and PLCg2* are key players in lymphovascular separation.[21] *Slp-76* is expressed in T cells, platelets, macrophages, mast cells, neutrophils, natural killer cells, and developing B cells, but it is absent in ECs.[21] Syk is expressed by all hematopoietic lineages and by ECs.[21,22] Finally, *PLCg2* is a predominantly B cell–specific signaling molecule.[23] Functional inactivation of any of these genes in mice results in a blood-filled lymphatics phenotype.[21] In addition, *Slp-76*-null embryos also have defective immune receptor signaling and arteriovenous shunting.[21] Recently, Slp-76 expression was also detected in a subpopulation of circulating endothelial progenitors (CEPs).[24] Injection of GFP-labeled *Syk*[−/−] or *Slp-76*[−/−] embryonic stem (ES) cells into wild-type blastocysts revealed the presence of blood-filled lymphatics in restricted regions of the chimeric embryos derived from the mutant cells.[24] On the basis of these results, it was suggested that Slp-76-expressing CEPs cell-autonomously contribute to the developing lymphatic vasculature.[24] However, lineage-tracing experiments have so far failed to identify any hematopoietic contribution to the EC compartment.[25,26] It is nevertheless intriguing that chimeric embryos generated using $Syk^{-/-}$ ES cells appeared to have a more drastic blood-filled lymphatic phenotype than those generated using Slp-$76^{-/-}$ ES cells.[24] Whether this difference can be attributed to Syk expression in ECs cells is not yet known.

In Slp-$76^{-/-}$ embryos, the lymph sacs are chimeric with respect to Lyve-1 expression.[21,24] Although constant in lymphatic capillaries, Lyve-1 expression is normally downregulated in SMC-covered collecting lymphatic vessels.[27] In Slp-76-null embryos, SMCs may be abnormally recruited, and this defect leads to the blood-filled lymph sac phenotype.

Different degrees of defective lymphovenous separation have also been reported in other mutant mouse strains in which genes as diverse as fasting-induced adipose factor ($Fiaf$ or $ANGPTL4^{28}$), $ephrinB2$,[27,29] or $Foxc2^{30}$ have been functionally inactivated. Blood-filled intestinal lymphatics were observed in $Fiaf$-null pups, and downregulation of $Prox1$ expression in postnatal intestinal lymphatics was suggested as a cause for this phenotype.[28] Similar to Slp-$76^{-/-}$ embryos, $Fiaf^{-/-}$ mice also exhibit abnormal arteriovenous shunting.[28]

Lymphatic Maturation and Remodeling

Following LEC specification and migration and concomitant with lymph sac formation and the lymphovenous separation, LECs sprout from the sacs to give rise to the entire lymphatic network. Almost no information is available about this complex process that assembles individual sprouting LECs to form lymphatic capillaries and vessels and eventually form a lumen for lymph transport. Recent work in zebrafish proposed that during angiogenesis, pinocytosis (the fusion of large vacuoles) is at least partially responsible for forming blood vessel lumens.[31] Could it be that a similar mechanism forms lumens in the lymphatic vasculature? For the network of lymphatic vessels and capillaries to form, LEC differentiation and lymphatic vasculature remodeling must take place. Expression of additional gene products is first detected in the forming lymphatics during later stages of embryonic and postnatal lymphangiogenesis.[1,32]

EphrinB2 is one of the transmembrane ligands of the Eph group of receptors and is expressed in arterial ECs, SMCs covering the ECs, and LECs of the collecting lymphatic vessels.[27,33,34] EphB4, the receptor for ephrinB2, is expressed by venous ECs and LECs, but not by arterial ECs.[27,35] EphrinB2-EphB4

bidirectional reciprocal signaling is thought to maintain proper arteriovenous separation.[35] Deletion of the cytoplasmic PDZ-interaction domain of ephrinB2 results in postnatal lethality at 3 weeks of age. The mutant pups exhibit a range of lymphatic defects, such as defective dermal lymphatic remodeling, absence of lymphatic valves, abnormal partial accumulation of SMCs by lymphatic capillaries, and chylothorax.[27] On the basis of these results, a cell-autonomous role of $ephrinB2$ during postnatal lymphatic remodeling was proposed.[27] However, the recent conditional deletion of $ephrinB2$ in SMCs resulted in early postnatal death, and the mutant pups exhibited severe blood and lymphatic vasculature defects.[29] Abnormal migration of SMCs to the lymphatic vessels and blood-filled lymphatics were observed in the mutant pups.[29] Because of the early postnatal death, whether the mutant pups developed any of the lymphatic remodeling defects described above for the standard mutants is unknown. Conditional deletion of $ephrinB2$ in LECs might provide additional information about the role(s) of this molecule.

Forkhead transcription factor $Foxc2$ is expressed by LECs and by the paraxial presomitic mesoderm and developing somites.[36] Mutations in FOXC2 have been associated with familial lymphedema–distichiasis.[37] Like $ephrinB2$-mutant mice, $Foxc2^{-/-}$ mice have an abnormal accumulation of pericytes in the lymphatic vasculature, lymphatic valve agenesis, and lymphatic dysfunction.[30] The occasional presence of red blood cells in the lymphatic vasculature has also been reported.[30] In addition, in association with $Foxc1$, $Foxc2$ regulates the expression of $Vegf$-c and Notch target genes including $ephrinB2$.[38] On the basis of this observation and the similarities in the lymphatic vascular phenotypes identified in $Foxc2^{-/-}$ and $ephrinB2$-mutant mice (i.e., valve defects and the abnormal recruitment of SMCs), we propose that $Foxc2$'s direct regulation of $ephrinB2$ could regulate the recruitment of SMCs by the lymphatic vasculature. Obviously, a conditional deletion of $Foxc2$ in LECs or pericytes could provide valuable information about its mechanism(s) of action. Finally, functionally relevant cooperation between $Foxc2$ and $Vegfr$-3 in the establishment of a pericyte-free lymphatic capillary network has been proposed.[30]

The Tie2 ligand Ang2 is involved in postnatal lymphatic remodeling and maturation.[39] $Ang2^{-/-}$ neonatal mice are edematous, have chylous ascites, and have distorted leaky lymphatic vessels.[39] In contrast to $ephrinB2$- and $Foxc2$-mutant mice that have abnormal accumulations of SMCs, $Ang2^{-/-}$ mice lack proper SMC coverage, a phenotype that may be responsible

TABLE 1. Different types of lymphatic phenotypes identified in various mouse mutant models together with the affected gene

Phenotype	Gene/genotype	Relevant expression	References
No LEC specification/lack of lymphatic vasculature	*Prox1$^{-/-}$*	LECs	4, 6
No migration of LECs from the veins/lack of lymphatic vasculature	*Vegf-c$^{-/-}$*	Mesenchyme	11
Blood-filled lymph sacs	*Slp-76$^{-/-}$*, *Syk$^{-/-}$*, *PLCg2$^{-/-}$*	Hematopoietic cells	21, 24
Sprouting of LECs from lymph sacs appears defective	*Nrp2$^{-/-}$*	LECs	19
	T1α/Podoplanin$^{-/-}$	LECs	17
Abnormal SMC accumulation in the lymphatic vasculature	*ephrinB2*	LECs of collecting lymphatics, SMCs	27, 29
	Foxc2$^{-/-}$	LECs, SMCs	30
Abnormal lymphatic valve development	*ephrinB2* (mutated in the PDZ interaction domain)	LECs of the collecting lymphatics, SMCs	27
	Foxc2$^{-/-}$	LECs, SMCs	30
Abnormal postnatal lymphatic remodeling	*ephrinB2* (mutated in the PDZ interaction domain)	LECs of the collecting lymphatics, SMCs	27
	Ang2$^{-/-}$	Arterial SMCs and ECs undergoing remodeling	39
Blood-filled intestinal lymphatics	*Fiaf$^{-/-}$*	Enterocytes of the small intestine	28
Chylothorax	*ephrinB2* (mutated in the PDZ interaction domain)	LECs of the collecting lymphatics, SMCs	27
	Elk3 (Net)$^{-/-}$	ECs	40
	Integrin α9$^{-/-}$	SMCs	41
Chylous ascites	*Prox1$^{+/-}$*	LECs	42
	Vegf-c$^{+/-}$	Mesenchyme	11
	Ang2$^{-/-}$	Arterial SMCs and EC undergoing remodeling	39
	Vefgr3$^{+/-}$ (*Chy* mice with a dominant negative mutation in VEGFR3	BECs and LECs	43
	Sox18$^{Ra/Ra}$ (dominant ragged mutation)	ECs	44, 45
	p110α (defective in its capacity to interact with Ras)	Ubiquitous	46

for the leaky lymphatic vessels.[39] Interestingly, *Ang1* can substitute for *Ang2* in regulating proper lymphatic remodeling.[39]

TABLE 1 lists the different types of lymphatic phenotypes identified in various mouse mutant models together with the affected gene.

Acknowledgments

We thank the present and past members of our laboratory for their ideas and extensive discussions. This work was supported by funds from the National Institutes of Health (R01-HL073402) to G.O., Cancer Center Support Grant CA-21765, and the American Lebanese Syrian Associated Charities (ALSAC). R.S.S is supported by a postdoctoral fellowship award from the Lymphatic Research Foundation (LRF).

Conflicts of Interest

The authors declare no conflicts of interest.

References

1. OLIVER, G. 2004. Lymphatic vasculature development. Nat. Rev. Immunol. **4:** 35–45.
2. BANERJI, S. *et al.* 1999. LYVE-1, a new homologue of the CD44 glycoprotein, is a lymph-specific receptor for hyaluronan. J. Cell Biol. **144:** 789–801.
3. GALE, N.W. *et al.* 2007. Normal lymphatic development and function in mice deficient for the lymphatic hyaluronan receptor LYVE-1. Mol. Cell Biol. **27:** 595–604.
4. WIGLE, J.T. & G. OLIVER. 1999. Prox1 function is required for the development of the murine lymphatic system. Cell **98:** 769–778.
5. VAN DER PUTTE, S.C. 1975. The early development of the lymphatic system in mouse embryos. Acta Morphol. Neerl. Scand. **13:** 245–286.
6. WIGLE, J.T. *et al.* 2002. An essential role for Prox1 in the induction of the lymphatic endothelial cell phenotype. EMBO J. **21:** 1505–1513.
7. HONG, Y.K. *et al.* 2002. Prox1 is a master control gene in the program specifying lymphatic endothelial cell fate. Dev. Dyn. **225:** 351–357.
8. PETROVA, T.V. *et al.* 2002. Lymphatic endothelial reprogramming of vascular endothelial cells by the Prox-1 homeobox transcription factor. EMBO J. **21:** 4593–4599.
9. HONG, Y.K. & M. DETMAR. 2003. Prox1, master regulator of the lymphatic vasculature phenotype. Cell Tissue Res. **314:** 85–92.
10. OLSON, L.E. *et al.* 2006. Homeodomain-mediated beta-catenin-dependent switching events dictate cell-lineage determination. Cell **125:** 593–605.
11. KARKKAINEN, M.J. *et al.* 2004. Vascular endothelial growth factor C is required for sprouting of the first lymphatic vessels from embryonic veins. Nat. Immunol. **5:** 74–80.
12. KARKKAINEN, M.J. *et al.* 2000. Missense mutations interfere with VEGFR-3 signalling in primary lymphoedema. Nat. Genet. **25:** 153–159.
13. KAIPAINEN, A. *et al.* 1995. Expression of the fms-like tyrosine kinase 4 gene becomes restricted to lymphatic endothelium during development. Proc. Natl. Acad. Sci. USA **92:** 3566–3570.
14. DUMONT, D.J. *et al.* 1998. Cardiovascular failure in mouse embryos deficient in VEGF receptor-3. Science **282:** 946–949.
15. CARMELIET, P. 2000. Mechanisms of angiogenesis and arteriogenesis. Nat. Med. **6:** 389–395.
16. ADAMS, R.H. & K. ALITALO. 2007. Molecular regulation of angiogenesis and lymphangiogenesis. Nat. Rev. Mol. Cell Biol. **8:** 464–478.
17. SCHACHT, V. *et al.* 2003. T1alpha/podoplanin deficiency disrupts normal lymphatic vasculature formation and causes lymphedema. EMBO J. **22:** 3546–3556.
18. NEUFELD, G. *et al.* 2002. The neuropilins: multifunctional semaphorin and VEGF receptors that modulate axon guidance and angiogenesis. Trends Cardiovasc. Med. **12:** 13–19.
19. YUAN, L. *et al.* 2002. Abnormal lymphatic vessel development in neuropilin 2 mutant mice. Development **129:** 4797–4806.
20. TAKASHIMA, S. *et al.* 2002. Targeting of both mouse neuropilin-1 and neuropilin-2 genes severely impairs developmental yolk sac and embryonic angiogenesis. Proc. Natl. Acad. Sci. USA **99:** 3657–3662.
21. ABTAHIAN, F. *et al.* 2003. Regulation of blood and lymphatic vascular separation by signaling proteins SLP-76 and Syk. Science **299:** 247–251.
22. YANAGI, S. *et al.* 2001. Syk expression and novel function in a wide variety of tissues. Biochem. Biophys. Res. Commun. **288:** 495–498.
23. WANG, D. *et al.* 2000. Phospholipase C gamma2 is essential in the functions of B cell and several Fc receptors. Immunity **13:** 25–35.
24. SEBZDA, E. *et al.* 2006. Syk and Slp-76 mutant mice reveal a cell-autonomous hematopoietic cell contribution to vascular development. Dev. Cell. **11:** 349–361.
25. STADTFELD, M. & T. GRAF. 2005. Assessing the role of hematopoietic plasticity for endothelial and hepatocyte development by non-invasive lineage tracing. Development **132:** 203–213.
26. SAMOKHVALOV, I.M., N.I. SAMOKHVALOVA & S. NISHIKAWA. 2007. Cell tracing shows the contribution of the yolk sac to adult haematopoiesis. Nature **446:** 1056–1061.
27. MAKINEN, T. *et al.* 2005. PDZ interaction site in ephrinB2 is required for the remodeling of lymphatic vasculature. Genes Dev. **19:** 397–410.
28. BACKHED, F. *et al.* 2007. Postnatal lymphatic partitioning from the blood vasculature in the small intestine requires fasting-induced adipose factor. Proc. Natl. Acad. Sci. USA **104:** 606–611.
29. FOO, S.S. *et al.* 2006. Ephrin-B2 controls cell motility and adhesion during blood-vessel-wall assembly. Cell **124:** 161–173.
30. PETROVA, T.V. *et al.* 2004. Defective valves and abnormal mural cell recruitment underlie lymphatic vascular failure in lymphedema distichiasis. Nat. Med. **10:** 974–981.
31. KAMEI, M. *et al.* 2006. Endothelial tubes assemble from intracellular vacuoles in vivo. Nature **442:** 453–456.
32. OLIVER, G. & K. ALITALO. 2005. The lymphatic vasculature: recent progress and paradigms. Annu. Rev. Cell Dev. Biol. **21:** 457–483.
33. ADAMS, R.H. *et al.* 1999. Roles of ephrinB ligands and EphB receptors in cardiovascular development: demarcation of arterial/venous domains, vascular morphogenesis, and sprouting angiogenesis. Genes Dev. **13:** 295–306.
34. GALE, N.W. *et al.* 2001. Ephrin-B2 selectively marks arterial vessels and neovascularization sites in the adult, with expression in both endothelial and smooth-muscle cells. Dev. Biol. **230:** 151–160.
35. WANG, H.U., Z.F. CHEN & D.J. ANDERSON. 1998. Molecular distinction and angiogenic interaction between embryonic arteries and veins revealed by ephrin-B2 and its receptor Eph-B4. Cell **93:** 741–753.
36. KUME, T. *et al.* 2001. The murine winged helix transcription factors, Foxc1 and Foxc2, are both required for cardiovascular development and somitogenesis. Genes Dev. **15:** 2470–2482.

37. FANG, J. *et al.* 2000. Mutations in FOXC2 (MFH-1), a forkhead family transcription factor, are responsible for the hereditary lymphedema-distichiasis syndrome. Am. J. Hum. Genet. **67:** 1382–1388.

38. SEO, S. *et al.* 2006. The forkhead transcription factors, Foxc1 and Foxc2, are required for arterial specification and lymphatic sprouting during vascular development. Dev. Biol. **294:** 458–470.

39. GALE, N.W. *et al.* 2002. Angiopoietin-2 is required for post-natal angiogenesis and lymphatic patterning, and only the latter role is rescued by Angiopoietin-1. Dev. Cell. **3:** 411–423.

40. AYADI, A. *et al.* 2001. Net-targeted mutant mice develop a vascular phenotype and up-regulate egr-1. Embo J. **20:** 5139–5152.

41. HUANG, X.Z. *et al.* 2000. Fatal bilateral chylothorax in mice lacking the integrin alpha9beta1. Mol. Cell Biol. **20:** 5208–5215.

42. HARVEY, N.L. *et al.* 2005. Lymphatic vascular defects promoted by Prox1 haploinsufficiency cause adult-onset obesity. Nat. Genet. **37:** 1072–1081.

43. KARKKAINEN, M.J. *et al.* 2001. A model for gene therapy of human hereditary lymphedema. Proc. Natl. Acad. Sci. USA **98:** 12677–12682.

44. PENNISI, D. *et al.* 2000. Mutations in Sox18 underlie cardiovascular and hair follicle defects in ragged mice. Nat. Genet. **24:** 434–437.

45. PENNISI, D. *et al.* 2000. Mice null for sox18 are viable and display a mild coat defect. Mol. Cell Biol. **20:** 9331–9336.

46. GUPTA, S. *et al.* 2007. Binding of Ras to Phosphoinositide 3-kinase p110alpha is required for Ras-driven tumorigenesis in mice. Cell **129:** 957–968.

The Link between Lymphatic Function and Adipose Biology

Natasha L. Harvey

Division of Haematology, The Hanson Institute, IMVS, Adelaide, South Australia, Australia

Despite observations of a link between lymphatic vessels and lipids that date as far back as 300 BC, a link between lymphatic vessels and adipose tissue has only recently been recognized. This review will summarize documented evidence that supports a close relationship between lymphatic vessels and adipose tissue biology. Lymphatic vessels mediate lipid absorption and transport, share an intimate spatial association with adipose tissue, and regulate the traffic of immune cells that rely on specialized adipose tissue depots as a reservoir of energy deployed to fight infection. Important links between inflammation and adipose tissue biology will also be discussed in this article, as will recent evidence connecting lymphatic vascular dysfunction with the onset of obesity. There seems little doubt that future research in this topical field will ensure that the link between lymphatic vascular function and adipose tissue is firmly established.

Key words: lymphatic; lymphangiogenesis; obesity; adipogenesis; inflammation; lymph node; lymph

Introduction

The first documented observation that a connection exists between lymphatic vessels and lipids dates as far back as *ca.* 300 BC, when the Ancient Greeks documented distinct mesenteric vessels filled with milk in suckling young.[1,2] In the 17th century, almost 2000 years later, Gaspar Aselli was credited with being the first to recognize the role of these vessels in lipid absorption and transport; he described the mesenteric lymphatic vessels in the gut of a dog that had consumed a lipid-rich meal as "white veins."[3] In extensive work that followed Aselli's initial recognition of the absorptive nature of these vessels, it was established that the "milky veins" he described in fact made up a vascular network distinct from the blood vasculature: the lymphatic vessels.[1] Work from many 17th century anatomists demonstrated that the lymphatic vessels of the mesentery drained their lipid-rich contents progressively via the cisterna chyli and thoracic duct to the bloodstream, at the junction of the thoracic duct with the great veins of the neck.[1] Were it not for the "illumination" of lymphatic vessels by ingested lipid-rich lymph, the discovery of these vessels

would almost certainly have occurred much later in the course of history. While the role and mechanism of lymphatic vascular lipid transport have been extensively established since Gaspar Aselli's initial studies, a link between lymphatic vascular function and adipose tissue biology has only recently been recognized. This article will reflect on both historic and recent data that underpin a close relationship between lymphatic vessels and adipose tissue, and will consider the possibility that lymphatic vascular dysfunction may underlie a proportion of cases of human obesity.

Lymph Nodes, Lymphatic Vessels, and Adipose Tissue Depots

A close relationship exists between lymph nodes and adipose tissue; indeed, lymph nodes, the organizing centers of immune surveillance and response, are always found surrounded by adipose tissue. By extrapolation then, it follows that lymphatic vessels are also in close physical association with adipose tissue. Subcutaneous adipose tissue lies in close proximity to the dermal lymphatic vasculature (Fig. 1), while visceral adipose tissue surrounds the collecting lymphatic vessels of the mesentery, cisterna chyli and thoracic duct, as well as the efferent and afferent lymphatic vessels of intra-abdominal lymph nodes. The efferent and afferent lymphatic vessels of the superficial lymph nodes are also encapsulated by adipose tissue. Extensive work by Pond and colleagues has revealed that

Address for correspondence: Natasha L. Harvey, Florey Research Fellow, Division of Haematology, The Hanson Institute, IMVS, P.O. Box 14, Rundle Mall, Adelaide, South Australia, 5000, Australia. Voice: +61-8-8222-3569; fax: +61-8-8222-3139.

natasha.harvey@imvs.sa.gov.au

Ann. N.Y. Acad. Sci. 1131: 82–88 (2008). © 2008 New York Academy of Sciences.
doi: 10.1196/annals.1413.007

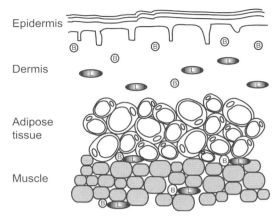

FIGURE 1. Lymphatic capillaries in the dermis are located in close proximity to the subcutaneous adipose tissue layer. L: lymphatic vessels, B: blood vessels.

the adipose tissue surrounding lymph nodes serves as a reservoir of energy that is deployed to power local immune responses.[4–7] Their work has demonstrated that the adipocytes within lymph node fat pads that are most closely apposed to the lymph node (perinodal adipocytes) respond to local immune challenge by increasing their rate of lipolysis compared to adipocytes that are more distantly located from the node.[5] Increased lipolysis liberates fatty acids and other lipid-derived mediators that fuel the ensuing immune response. Prolonged inflammation has been shown to propagate energy release by stimulating lipolysis in adipocytes closer to the periphery of the lymph node fat pad, as well as in adipose depots further afield from the site of challenge.[8] These observations suggest that the scope and magnitude of adipocyte lipolysis is progressively increased to meet the requirements of sustained immune cell activation.

Chronic lymph node stimulation has also been shown to result in the expansion of lymph node–associated adipose tissue in a rat model.[9] In this model, chronic lymph node stimulation resulted in increased adipose tissue mass due to an increase in both the size and number of adipocytes within the stimulated fat pad. Adipose tissue expansion is presumably an immune system insurance strategy that ensures that sufficient energy is always in reserve to power the fight against infectious agents. Taken together, these data link inflammation with adipose tissue metabolism and may help to explain the changes in adipose tissue biology that accompany chronic human inflammatory disorders, including HIV-associated adipose redistribution syndrome,[10] Crohn's disease, and obesity.

An intriguing observation that provides further direct evidence of an intimate lymph node–adipose tissue relationship is that, in at least one mouse model devoid of lymph nodes, the absence of nodes results in a failure of the associated lymph node fat pads to develop.[11] This observation suggests that the establishment of adipose tissue around lymph nodes is directly dependent on as yet unidentified lymph node– or immune cell–derived signals that are liberated during lymph node growth and maturation.

The Link between Lymphatic Vessels and Adipose Tissue in Human Disease

Lymphatic vessels are a critical conduit of interstitial fluid transport and immune cell traffic. Lymphatic vascular insufficiency due to developmental lymphatic vascular abnormalities, injury, obstruction or infection results in the accumulation of interstitial fluid and protein in affected tissues—a situation known as lymphedema. There are a number of characterized primary, or inherited, lymphedema syndromes,[12,13] several of which can be ascribed to inherited mutations in genes important for the growth and development of the lymphatic vasculature. Thus far, inactivating mutations have been described in the vascular endothelial growth factor receptor-3 (*VEGFR-3*) signaling pathway in patients suffering from Milroy's disease,[14,15] and in the transcription factors *FOXC2*[16–18] and *SOX18*[19] in lymphedema–distichiasis and hypotrichosis–lymphedema–telangiectasia syndromes, respectively. By far the most prevalent form of lymphedema is secondary, or acquired lymphedema, which arises as a result of lymphatic vascular injury or infection. Secondary lymphedema is estimated to occur in up to 20% of breast cancer patients after the surgical resection of axillary lymph nodes.[20] These patients experience painful and disabling edema of the affected arm for which very little effective treatment is currently available. Secondary lymphedema of the lower limbs has also been documented to occur in 10–25% of patients after surgery and radiotherapy for the treatment of gynecologic cancer.[21] The most predominant incidence of secondary lymphedema, however, is in the tropical world, where it is estimated that more than 120 million people suffer lymphedema as a result of lymphatic filariasis.[22] Whether primary or secondary in origin, if lymphedema is not resolved, the affected tissue manifests changes that include chronic inflammation, fibrosis and, most pertinent to this review, adipose tissue accumulation.[12,13] In fact, as early as the 19th century, German dermatologist Paul Unna proposed that the stagnation of tissue because of lymphatic or venous interruption resulted in fat accumulation.[23]

One report of a cutaneous lymphatic malformation that resulted in secondary late-onset adipose tissue hypertrophy exists in the literature.[24]

Lipedema is a syndrome found predominantly in postpubertal women and is characterized by bilateral, symmetrical enlargement of the legs as a result of adipose tissue accumulation, with sparing of the feet.[12] While the term lipedema suggests the existence of edematous tissue due to lymphatic vascular insufficiency, this disease is being progressively considered to be primarily a lipodystrophy syndrome. This conclusion is being drawn because a number of groups of patients with lipedema have shown normal, or only slightly reduced lymphatic function.[25,26] The diminished lymphatic function recorded in patients with lipedema could perhaps occur secondarily to adipose tissue accumulation in the legs by virtue of an obstructive effect on lymphatic flow. A role for lymphatic vessels in the etiology of this disease cannot yet be ruled out, however, as some investigators have shown functional alterations in lymph flow in lipedema patients,[27] and microlymphatic aneurysms of the lymphatic capillaries have been described in the skin of lipedema patients.[28] Further investigations should reveal whether lipedema is primarily a lymphatic vascular disease, or a lipodystrophy syndrome that interferes with normal lymphatic vascular function.

Adipose tissue accumulation has also been described in mouse models of lymphatic vascular dysfunction. The *Chy* mouse, a naturally occurring mouse model of lymphedema due to heterozygous inactivating mutations in VEGFR-3, exhibits lymphedema as a result of hypoplastic cutaneous lymphatic vessels.[29] *Chy* mice display adipose tissue accumulation predominantly in the edematous subcutaneous adipose layer that lies in close physical proximity to the dysfunctional hypoplastic lymphatic vessels of the dermis.[29] While the mechanism of adipose tissue accumulation due to lymphatic vascular dysfunction has not yet been precisely determined in *Chy* mice, or in human lymphedema patients, insights into a potential mechanism were recently discovered in mice haploinsufficient for *Prox1*, a gene encoding a homeobox transcription factor crucial for lymphatic vascular development.[30]

A Mouse Model of Adult Onset Obesity Caused by Lymphatic Vascular Disruption

Prox1 was the first gene identified to be critical for the specification of lymphatic endothelial cell fate.[31,32] Targeted inactivation of *Prox1* in the mouse results in embryonic lethality at approximately embryonic day (E) 15, by which stage *Prox1*-nullizygous embryos display pronounced edema due to the complete absence of lymphatic vessels.[31] While many *Prox1* heterozygous mice die soon after birth, displaying phenotypes characteristic of lymphatic vascular dysfunction, such as peritoneal and/or thoracic chylous ascites, a striking feature of surviving *Prox1* heterozygous mice is adult-onset obesity.[30] Our extensive characterization of food consumption, energy expenditure, mediators of appetite and satiety control and lipid metabolism, failed to reveal any changes in any of these parameters that could account for the onset of obesity in *Prox1* heterozygous mice. A consistent correlate was, however, observed between the degree of lymphatic vascular disorganization and dysfunction and the magnitude of adipose tissue accumulation in *Prox1*[+/−] mice. The lymphatic vessels that were most severely affected in *Prox1*[+/−] animals were those of the viscera, particularly the intestine, mesentery, and thoracic duct.[30]

In addition to the abnormal patterning and dilation of lymphatic vessels in the mesentery of *Prox1*[+/−] mice, feeding of the mice with a fluorescent lipid illuminated regions of mesenteric lipid leakage. These were most obvious in close proximity to areas abundant in disorganized lymphatic vessels. How could lymphatic vascular disorganization and leakage be linked with adipose tissue accumulation? A search of the literature revealed early work in which both total lymph and the lipid-rich chylomicron fraction of lymph, had been shown to promote the differentiation of adipocyte precursors isolated from the stromal-vascular fraction of embryonic rabbit adipose tissue.[33] We therefore hypothesized that the lymph leaking from ruptured lymphatic vessels of *Prox1*[+/−] mice could potentially contain an adipogenic stimulus, and we tested this hypothesis by culturing mouse 3T3-L1 pre-adipocytes with lymph collected from newborn *Prox1*[+/−] pups. Our data demonstrated that lymph was a potent stimulant of 3T3-L1 adipogenic differentiation and that lymph was able to synergize with insulin to even more potently promote adipogenic differentiation in this model. No effect on 3T3-L1 differentiation was observed when safflower oil was added to culture media as a source of exogenous lipid, indicating that the lipid accumulation we observed in 3T3-L1 adipocytes cultured with lymph was indicative of true adipogenic differentiation and that this effect was mediated by an unidentified factor/factors contained in lymph. Quantitation of the size and number of adipocytes in the fat pads of *Prox1*[+/−] mice and their wild-type littermates revealed a two-step mechanism of increased adipose tissue mass, the first step of which was adipocyte hypertrophy, and the second (in general restricted to the most obese $Prox1^{+/-}$ mice), the promotion of adipogenic differentiation to generate additional adipocytes in which to store lipid. These data provided further evidence that a true adipogenic stimulus indeed resides within lymph. We further demonstrated that increased mesenteric adipose tissue accumulation was evident in $Prox1^{+/-}$ mice even prior to a significant increase in total body weight, and that adipose tissue accumulation was associated with an increased number of lymphatic vascular endothelial HA receptor (LYVE-1)–positive macrophages in the mesentery.[30] Final confirmation of our hypothesis that obesity was caused by lymphatic vascular rupture and resultant lymph leakage was cemented when we inactivated *Prox1* specifically in the vasculature and demonstrated that these mice developed adult-onset obesity.[30] Generation of this mouse model proved to us that lymphatic vascular defects caused by *Prox1* haploinsufficiency were sufficient to result in obesity. What is the identity of the adipogenic factor contained in lymph? No doubt future studies will reveal the answer to this intriguing question.

Inflammation and Obesity

Low-grade inflammation is increasingly recognized as being linked with, and contributing to, obesity and obesity-associated metabolic complications such as insulin resistance, type 2 diabetes, and cardiovascular disease.[34] The adipose tissue of obese mice and humans produces proinflammatory cytokines, chemokines, and peptides including TNF-α,[35] TGF-β,[36] interleukin 6,[37] monocyte chemoattractant protein-1,[38] and leptin[39]— all of which are able to recruit and stimulate cells of the immune system. Many of these proinflammatory mediators also have documented angiogenic activity,[40,41] or are able to indirectly stimulate angiogenesis via promoting the production of angiogenic growth factors from adipocytes[42] or macrophages. Macrophages, increased in number in the adipose tissue of both obese mice and humans,[43,44] have recently been shown to produce lymphangiogenic growth factors including vascular endothelial growth factors A, C and D (VEGF-A, -C, -D) in response to inflammatory stimuli.[45–47] Macrophages have also been demonstrated to promote lymphangiogenesis in mouse models of inflammatory disease[46,47] and to stimulate angiogenesis in epididymal adipose tissue.[48] It is thereby plausible that obesity-stimulated inflammation could result in the promotion of both angiogenesis and lymphangiogenesis within, and in close vicinity

to, adipose tissue. Increased lymphangiogenesis and lymphatic vascular hyperplasia have previously been associated with inflammatory conditions, including psoriasis[49] and chronic airway inflammation.[50]

Could inflammation contribute to adipose tissue accumulation in $Prox1^{+/-}$ mice? On the basis of the data discussed above, we reasoned that the influx of inflammatory cells that we observed in the liver and adipose tissue of obese $Prox1^{+/-}$ mice could indeed potentially contribute to further adipose tissue accumulation in this mouse model. Two likely mechanisms of inflammation-stimulated adipose tissue accumulation could be envisioned: (*1*) that chronic inflammation could promote an increase in adipose tissue mass in order to fulfill the energy requirements of sustained immune cell activation via direct immune cell–adipocyte signaling events; and (*2*) that chronic inflammation could promote increased adipogenesis by stimulating neo-lymphangiogenesis in the mesentery of $Prox1^{+/-}$ mice, thereby exacerbating the release of lymph-derived adipogenic stimuli (FIG. 2). This mechanism would likely involve the direct transmission of lymph-derived signal(s) to local adipocytes, although the involvement of cells of the immune system in this process cannot be ruled out. Multiple lines of evidence suggested to us that the initial trigger of increased adipose tissue mass in $Prox1^{+/-}$ mice was ruptured leaky lymphatic vessels, as defects in lymphatic vessels were obvious at early stages of embryonic lymphatic development, and an increase in mesenteric adipose tissue mass was obvious in $Prox1^{+/-}$ mice compared to their wild-type counterparts even prior to a noticeable increase in overall body weight. It seems entirely plausible, though, and even likely, that once established, inflammation could propagate pro-adipogenic stimuli even further in $Prox1^{+/-}$ mice, resulting in a cyclical exacerbation of adipose tissue accumulation.

Adipose Tissue as a Source of Lymphangiogenic Factors

As discussed above, adipose tissue has been demonstrated to liberate numerous signals that have an impact on vascular development. The requirement for macrophages in adipose tissue angiogenesis, important to sustain adipose tissue expansion, has recently been demonstrated by Cho and colleagues.[48] This work illustrated that macrophages are recruited into the hypoxic tip region of epididymal adipose tissue via signals mediated by stromal cell–derived factor 1, VEGF, and matrix metalloproteinases (MMP), and then act to promote angiogenesis dependent on MMP and VEGF

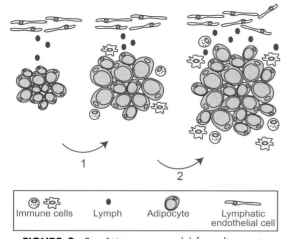

FIGURE 2. *Prox1*[+/−] mouse model for adipose tissue expansion initiated by lymphatic vascular dysfunction. In this model, it is proposed that the leakage of lymph from ruptured lymphatic vessels results in a biphasic increase in adipose tissue mass due to the exposure of adipose tissue to lymph-derived adipogenic signals. The initial phase of increased adipose tissue mass is proposed to occur via adipocyte hypertrophy (1), while the second phase is facilitated via the promotion of adipocyte differentiation from precursor adipocytes (2). The identity of the factor(s) within lymph that promote adipose tissue accumulation have not yet been identified. In addition to the direct adipogenic activity contained within lymph, it is proposed that adipose tissue accumulation is exacerbated by the inflammation that accompanies lymph leakage. Chronic inflammation has been closely linked with increased adipose tissue accumulation, obesity, neo-angiogenesis, and neo-lymphangiogenesis.

signals. While no lymphatic vessels were observed growing into the tip region of epididymal adipose tissue during the timeframe of this study, it is possible that lymphangiogenesis might secondarily follow establishment of the blood vascular network, primary construction of which is critical to sustain the metabolic demands of tissue growth and expansion. Indeed, studies from a number of laboratories have demonstrated that mouse adipose tissue can be ablated by targeted destruction of the vasculature,[51–53] confirming the importance of maintaining an adequate supply of oxygen and nutrients for adipose tissue integrity. Investigation of the ocular adipose tissue in human patients suffering the inflammatory conditions of orbital mucormyosis and panendophthalmitis suggest that lymphangiogenesis can be promoted within adipose tissue.[54] The studies of this group revealed that lymphatic vessels were present within ocular adipose granulation tissue, while

absent from normal healthy controls,[54] suggesting that inflammation is able to promote lymphangiogenesis within adipose tissue.

An interesting, recently described factor produced by white adipocytes, among other tissues, is fasting-induced adipose factor (Fiaf).[55] Fiaf seems to play a role in lymphatic partitioning from the blood vasculature, as the majority of *Fiaf*-deficient mice die within 3 weeks of birth, and exhibit abnormal lymphatic–venous connections in the intestinal and mesenteric lymphatic vasculature. Analysis of embryonic and postnatal lymphatic development in *Fiaf*[−/−] mice demonstrated that the importance of Fiaf for lymphatic vascular development and function appears to be restricted to vessels of the intestine and mesentery. This effect could potentially be due to localized production of *Fiaf* in this anatomic region, or to distinct Fiaf-dependent developmental events that occur selectively in the mesenteric and intestinal lymphatics during a specified timeframe. The mechanism by which Fiaf acts to effect the separation of blood and lymphatic vascular networks in the mesentery remains, thus far, uncharacterized.

Questions to Be Answered in Order to Establish the Link between Lymphatic Function and Adipose Biology

The data summarized in this review document some of the many observations made by scientists and clinicians that link lymphatic vascular function with adipose tissue biology. Lymphatic vessels, lymph nodes, and cells of the immune system all interact extensively with adipose tissue. Many questions remain to be answered before we will fully understand the complex interplay between the cells that make up these two biological systems, both of which are of fundamental importance to human health. Some of the most topical in the current research climate include: What is the identity of the adipogenic stimulus/stimuli in lymph? Are dysfunctional lymphatics responsible for human obesity? Do other models of lymphatic dysfunction result in obesity? Are mutations/SNPs in *Prox1* associated with human obesity? Which signaling mechanisms are used in adipocyte–immune cell communications? If answers to any of the foregoing questions implicate lymphatic vascular dysfunction in the etiology of obesity in the human population, how might lymphatic vessels be targeted in order to treat obesity and obesity-associated metabolic complications? In view of the epidemic of obesity in the Western world, the answers to these questions could have a potentially large impact on the generation of new therapeutics for the treatment of obesity-associated health complications. New-generation therapeutics targeted to ablate inflammation-stimulated adipogenesis might also provide a treatment option for patients who suffer from lymphedema that is more efficient than those currently available. The next chapter in this exciting research field will no doubt report answers to some of these questions, thereby further confirming the link between lymphatic vessels and adipose tissue biology.

Acknowledgment

N.L.H. is supported by a Royal Adelaide Hospital Florey Fellowship.

Conflicts of Interest

The author declares no conflicts of interest.

References

1. LORD, R.S. 1968. The white veins: conceptual difficulties in the history of the lymphatics. Med. Hist. **12:** 174–184.
2. YOFFEY, J.M. & F.C. COURTICE. 1970. Lymphatics, Lymph and the Lymphomyeloid Complex. Academic Press. London–New York.
3. ASELLIUS, G. 1627. De lactibus sive lacteis venis. Mediolani. Milan.
4. POND, C.M. & C.A. MATTACKS. 1995. Interactions between adipose tissue around lymph nodes and lymphoid cells in vitro. J. Lipid. Res. **36:** 2219–2231.
5. POND, C.M. & C.A. MATTACKS. 1998. In vivo evidence for the involvement of the adipose tissue surrounding lymph nodes in immune responses. Immunol. Lett. **63:** 159–167.
6. POND, C.M. & C.A. MATTACKS. 2003. The source of fatty acids incorporated into proliferating lymphoid cells in immune-stimulated lymph nodes. Br. J. Nutr. **89:** 375–383.
7. MATTACKS, C.A., D. SADLER & C.M. POND. 2004. Site-specific differences in fatty acid composition of dendritic cells and associated adipose tissue in popliteal depot, mesentery, and omentum and their modulation by chronic inflammation and dietary lipids. Lymphat. Res. Biol. **2:** 107–129.
8. POND, C.M. & C.A. MATTACKS. 2002. The activation of the adipose tissue associated with lymph nodes during the early stages of an immune response. Cytokine **17:** 131–139.
9. MATTACKS, C.A., D. SADLER & C.M. POND. 2003. The cellular structure and lipid/protein composition of adipose tissue surrounding chronically stimulated lymph nodes in rats. J. Anat. **202:** 551–561.
10. POND, C.M. 2003. Paracrine relationships between adipose and lymphoid tissues: implications for the mechanism of HIV-associated adipose redistribution syndrome. Trends Immunol. **24:** 13–18.
11. EBERL, G. et al. 2004. An essential function for the nuclear receptor RORgamma(t) in the generation of fetal lymphoid tissue inducer cells. Nat. Immunol. **5:** 64–73.
12. ROCKSON, S.G. 2000. Lymphedema. Curr. Treat. Options Cardiovasc. Med. **2:** 237–242.
13. WITTE, M.H. et al. 2001. Lymphangiogenesis and lymphangiodysplasia: from molecular to clinical lymphology. Microsc. Res. Tech. **55:** 122–145.
14. KARKKAINEN, M.J. et al. 2000. Missense mutations interfere with VEGFR-3 signalling in primary lymphoedema. Nat. Genet. **25:** 153–159.
15. IRRTHUM, A. et al. 2000. Congenital hereditary lymphedema caused by a mutation that inactivates VEGFR3 tyrosine kinase. Am. J. Hum. Genet. **67:** 295–301.
16. FANG, J. et al. 2000. Mutations in FOXC2 (MFH-1), a forkhead family transcription factor, are responsible for the hereditary lymphedema-distichiasis syndrome. Am. J. Hum. Genet. **67:** 1382–1388.
17. FINEGOLD, D.N. et al. 2001. Truncating mutations in FOXC2 cause multiple lymphedema syndromes. Hum. Mol. Genet. **10:** 1185–1189.
18. BELL, R. et al. 2001. Analysis of lymphoedema-distichiasis families for FOXC2 mutations reveals small insertions and deletions throughout the gene. Hum. Genet. **108:** 546–551.
19. IRRTHUM, A. et al. 2003. Mutations in the transcription factor gene SOX18 underlie recessive and dominant forms of hypotrichosis-lymphedema-telangiectasia. Am. J. Hum. Genet. **72:** 1470–1478.
20. PETREK, J.A., P.I. PRESSMAN & R.A. SMITH. 2000. Lymphedema: current issues in research and management. CA Cancer J. Clin. **50:** 292–307; quiz 308–311.
21. BEESLEY, V. et al. 2007. Lymphedema after gynecological cancer treatment : prevalence, correlates, and supportive care needs. Cancer **109:** 2607–14.
22. World Health Organization. 2000. Lymphatic filariasis. www.who.int.
23. RYAN, T.J. 1995. Lymphatics and adipose tissue. Clin. Dermatol. **13:** 493–498.
24. TAVAKKOLIZADEH, A., K.Q. WOLFE & L. KANGESU. 2001. Cutaneous lymphatic malformation with secondary fat hypertrophy. Br. J. Plast. Surg. **54:** 367–369.
25. HARWOOD, C.A. et al. 1996. Lymphatic and venous function in lipoedema. Br. J. Dermatol. **134:** 1–6.
26. BRAUTIGAM, P. et al. 1998. Analysis of lymphatic drainage in various forms of leg edema using two compartment lymphoscintigraphy. Lymphology **31:** 43–55.
27. BILANCINI, S. et al. 1995. Functional lymphatic alterations in patients suffering from lipedema. Angiology **46:** 333–339.
28. AMANN-VESTI, B.R., U.K. FRANZECK & A. BOLLINGER. 2001. Microlymphatic aneurysms in patients with lipedema. Lymphology **34:** 170–175.
29. KARKKAINEN, M.J. et al. 2001. A model for gene therapy of human hereditary lymphedema. Proc. Natl. Acad. Sci. USA **98:** 12677–12682.
30. HARVEY, N.L. et al. 2005. Lymphatic vascular defects promoted by Prox1 haploinsufficiency cause adult-onset obesity. Nat. Genet. **37:** 1072–1081.

31. WIGLE, J.T. & G. OLIVER. 1999. Prox1 function is required for the development of the murine lymphatic system. Cell **98:** 769–778.

32. WIGLE, J.T. *et al.* 2002. An essential role for Prox1 in the induction of the lymphatic endothelial cell phenotype. EMBO J. **21:** 1505–1513.

33. NOUGUES, J., Y. REYNE & J.P. DULOR. 1988. Differentiation of rabbit adipocyte precursors in primary culture. Int. J. Obes. **12:** 321–333.

34. HOTAMISLIGIL, G.S. 2006. Inflammation and metabolic disorders. Nature **444:** 860–867.

35. HOTAMISLIGIL, G.S., N.S. SHARGILL & B.M. SPIEGELMAN. 1993. Adipose expression of tumor necrosis factor-alpha: direct role in obesity-linked insulin resistance. Science **259:** 87–91.

36. SAMAD, F. *et al.* 1997. Elevated expression of transforming growth factor-beta in adipose tissue from obese mice. Mol. Med. **3:** 37–48.

37. FRIED, S.K., D.A. BUNKIN & A.S. GREENBERG. 1998. Omental and subcutaneous adipose tissues of obese subjects release interleukin-6: depot difference and regulation by glucocorticoid. J. Clin. Endocrinol. Metab. **83:** 847–850.

38. SARTIPY, P. & D.J. LOSKUTOFF. 2003. Monocyte chemoattractant protein 1 in obesity and insulin resistance. Proc. Natl. Acad. Sci. USA **100:** 7265–7270.

39. HAMILTON, B.S. *et al.* 1995. Increased obese mRNA expression in omental fat cells from massively obese humans. Nat. Med. **1:** 953–956.

40. LEIBOVICH, S.J. *et al.* 1987. Macrophage-induced angiogenesis is mediated by tumour necrosis factor-alpha. Nature **329:** 630–632.

41. SIERRA-HONIGMANN, M.R. *et al.* 1998. Biological action of leptin as an angiogenic factor. Science **281:** 1683–1686.

42. REGA, G. *et al.* 2007. Vascular endothelial growth factor is induced by the inflammatory cytokines interleukin-6 and oncostatin m in human adipose tissue in vitro and in murine adipose tissue in vivo. Arterioscler. Thromb. Vasc. Biol. **27:** 1587–1595.

43. WEISBERG, S.P. *et al.* 2003. Obesity is associated with macrophage accumulation in adipose tissue. J. Clin. Invest. **112:** 1796–1808.

44. XU, H. *et al.* 2003. Chronic inflammation in fat plays a crucial role in the development of obesity-related insulin resistance. J. Clin. Invest. **112:** 1821–1830.

45. BERSE, B. *et al.* 1992. Vascular permeability factor (vascular endothelial growth factor) gene is expressed differentially in normal tissues, macrophages, and tumors. Mol. Biol. Cell. **3:** 211–220.

46. SCHOPPMANN, S.F. *et al.* 2002. Tumor-associated macrophages express lymphatic endothelial growth factors and are related to peritumoral lymphangiogenesis. Am. J. Pathol. **161:** 947–956.

47. CURSIEFEN, C. *et al.* 2004. VEGF-A stimulates lymphangiogenesis and hemangiogenesis in inflammatory neovascularization via macrophage recruitment. J. Clin. Invest. **113:** 1040–1050.

48. CHO, C.H. *et al.* 2007. Angiogenic role of LYVE-1-positive macrophages in adipose tissue. Circ. Res. **100:** e47–e57.

49. KUNSTFELD, R. *et al.* 2004. Induction of cutaneous delayed-type hypersensitivity reactions in VEGF-A transgenic mice results in chronic skin inflammation associated with persistent lymphatic hyperplasia. Blood **104:** 1048–1057.

50. BALUK, P. *et al.* 2005. Pathogenesis of persistent lymphatic vessel hyperplasia in chronic airway inflammation. J. Clin. Invest. **115:** 247–257.

51. RUPNICK, M.A. *et al.* 2002. Adipose tissue mass can be regulated through the vasculature. Proc. Natl. Acad. Sci. USA **99:** 10730–10735.

52. KOLONIN, M.G. *et al.* 2004. Reversal of obesity by targeted ablation of adipose tissue. Nat. Med. **10:** 625–632.

53. BRAKENHIELM, E. *et al.* 2004. Angiogenesis inhibitor, TNP-470, prevents diet-induced and genetic obesity in mice. Circ. Res. **94:** 1579–1588.

54. FOGT, F. *et al.* 2004. Observation of lymphatic vessels in orbital fat of patients with inflammatory conditions: a form fruste of lymphangiogenesis? Int. J. Mol. Med. **13:** 681–683.

55. BACKHED, F. *et al.* 2007. Postnatal lymphatic partitioning from the blood vasculature in the small intestine requires fasting-induced adipose factor. Proc. Natl. Acad. Sci. USA **104:** 606–611.

Molecular Regulation of Lymphatic Contractility

MARIAPPAN MUTHUCHAMY AND DAVID ZAWIEJA

Department of Systems Biology and Translational Medicine, College of Medicine,
Cardiovascular Research Institute Division of Lymphatic Biology,
Texas A&M Health Science Center, College Station, Texas, USA

The lymphatic system plays critical roles in body fluid and macromolecular homeostasis, lipid absorption, immune function, and metastasis. To accomplish these tasks, the lymphatics must move lymph and its contents from the interstitial space through the lymph vessels and nodes and into the great veins. Contrary to popular belief, lymph does not passively "drain" down this pathway, because the net pressure gradients oppose flow. Instead, the lymphatics must act as both the conduits that direct and regulate lymph flow and the pumps that generate the lymph flow. Thus, to regulate lymph transport and function, both lymphatic pumping and flow resistance must be controlled. Both of these processes occur via regulation of lymphatic muscle contractions, which are classically thought to occur via the interaction of cell calcium with regulatory and contractile proteins. However, our knowledge of this regulation of lymphatic contractile function is far from complete. In this chapter we review our understanding of the important molecular mechanisms, the calcium regulation, and the contractile/regulatory proteins that control lymphatic contractions. A better understanding of these mechanisms could provide the basis for the development of better diagnostic and treatment modalities for lymphatic dysfunction. While progress has been made in our understanding of the molecular biology of lymphangiogenesis as a result of the development of potential lymphangiogenic therapeutic targets, there are currently no therapeutic agents that specifically modulate lymphatic pump function and lymph flow via lymphatic muscle. However, their development will not be possible until the molecular basis of lymphatic contractility is more fully understood.

Key words: lymph flow; lymph pump; lymphatic muscle; calcium regulation; regulatory proteins

General Overview of the Lymphatic System

The lymphatic system consists of a network of lymphatic vessels and interconnected lymph nodes distributed throughout most of the body. The lymph nodes are complex structures that allow the interaction of elements of the lymph and blood associated with the immune response to occur in complicated, orchestrated patterns. The network of lymphatic vascular structures is necessary for the controlled transport of immune cells, antigens, lipids, macromolecules, fluid, and particulate matter in the form of lymph. The transportation of lymph along the lymphatic network is directed from the parenchymal interstitial spaces to the nodes and between the nodes, and eventually the lymph exits the lymphatic system, emptying into the blood in the great veins of the neck. Thus the lymphatic transport system differs from the other vascular transport system, the blood, in that the lymph transport is essentially unidirectional, as opposed to the true "circulation" of blood around the heart–arterial–capillary–venous heart circuit. Interstitial tissue fluid enters the lymphatic system through the initial lymphatics, the blind-ended tubes that are made up of endothelial cells, usually without smooth muscle (SM) cells. This lymph formation is thought to occur between adjacent endothelial cells.[1] This fluid becomes lymph and carries dissolved and suspended substances, including large particles. The removal of large macromolecules (like extravasated plasma proteins) and particulate matter from the interstitial space is a unique quality and critical function of the lymphatic system, given that the blood vessels are generally not well suited for that task. If these macromolecules were left in the interstitium, they would lead to increases in tissue oncotic pressures, resulting in an imbalance of the transvascular

Address for correspondence: David Zawieja, Department of Systems Biology and Translational Medicine, College of Medicine, Cardiovascular Research Institute Division of Lymphatic Biology, Texas A&M Health Science Center, College Station, TX 77843-1114. Voice: +979/845-7465. dcz@tamu.edu exchange and edema formation.[2,3] Since the prevailing steady-state pressure in the interstitial fluid and the initial lymphatics typically oppose lymph formation, the exact mechanisms that drive the fluid uptake into the initial lymphatics to form lymph is not precisely understood and may be different in different tissues. However, the predominant mechanism driving lymph formation appears to be the development of transient fluid pressure gradients between the interstitium, the initial lymphatic, and downstream collecting lymphatics favorable to lymph formation that can occur during variations in the local interstitial fluid pressures on account of tissue movement and/or compression.[4–7] The initial lymphatics are uniquely tethered to the surrounding interstitial matrix through their characteristic "anchoring filaments."[8–11] Compression of the initial lymphatic vessels by local tissue forces, closes the inter-endothelial junctions or primary lymphatic valves[12–14] and propels lymph into the collecting lymphatics. Since the lymphatic transport is *not* circuitous and the prevailing steady-state pressures *do not* support lymph formation or lymph flow down the lymphatic network, luminal unidirectional valves must prevent the backflow of lymph.[15–22] These important structures are present throughout the lymphatic network beyond the initial lymphatics. The lymphatic valves appear to depend primarily on some unique properties of the lymphatic endothelium and their underlying matrix,[15,17,20,21,23,24] but there is also some evidence for potential involvement of specialized lymphatic muscle.[25] The function of these unique valves is critical for the function of the lymphatic transport system. Recently published genetic studies of lymphedema and distichiasis have shown the impact of lymphatic valvular failure and lymphedema resulting from a mutation of the FOXC2 gene in mice.[26,27]

The result of the development of oscillatory tissue-pressure gradients and normal valve structure/function is the formation and net propulsion of lymph from the interstitium to the collecting lymphatics. Once lymph has moved into the collecting lymphatic system, it must then be propelled further along the lymphatic network toward the first set of lymph nodes. Again the prevailing pressure gradients along the lymphatic network do not normally allow the passive movement of fluid down a standing pressure gradient.[28] Thus, lymph usually does not "drain" down the lymphatic network passively, as is commonly described. Instead, lymph must be carefully pumped, in a controlled fashion, along the lymphatic network from the early collecting lymphatics, through the nodes, and to the great veins of the neck.[16,29–36] To accomplish this movement of lymph requires the input of energy

to overcome the prevailing lymph pressure gradients. This is particularly true in the case of humans, who spend much of their time upright, generating significant gravitational hydrostatic pressures that must be overcome to move lymph from collecting vessels anywhere below the upper thorax/neck to the outflow of the thoracic duct into the veins of the neck.[36,37] The collecting and transport lymphatic vessels have SM cell layers invested in their outer walls to generate and control the movement of lymph along the lymphatic network, even against significant opposing pressure gradients. The functional units within the muscular lymphatic vessels, called lymphangions, are arranged in series and separated by highly competent valves.[32,38]

The lymphatic system must impart energy to the lymph via pumping mechanisms (extrinsic and intrinsic) to propel it along the lymphatic network. Several motive forces, when combined with functional lymphatic valves, move lymph centrally. These forces can be divided into two categories on the basis of their energy source: (1) the "intrinsic" or "active" lymph pump generates forces by spontaneous contractions of lymphatic muscle and (2) the "extrinsic" or "passive" lymph pump combines all other forces outside the lymphatic that can compress the lymphatics and move fluid centrally. The active pumps are essential for lymph flow in most lymphatic beds. However, mammals do not possess the specialized lymph hearts that lower vertebrate species use to pump lymph. Instead, the lymphatic vessels themselves must act as both the pumps that generate the lymph flow and the conduits that direct and regulate flow. Thus, in mammals, including humans, without the intrinsic phasic contractile activity of the lymphatic muscle cells (i.e., the intrinsic lymph pump), lymph flow, the hallmark of lymphatic function, is not possible.

Lymphatic Contractility and Lymph Transport

Many studies conducted on lymphatic function have focused on the regulation of lymphatic diameter and thus its resistance to control lymph flow. While such studies are important, it is clear that many lymphatic beds also possess a very different type of contractile activity, the phasic contractile activity of the lymph pump. In general, it appears that lymphatic vessels that are not encased in a parenchymal tissue that undergoes regular periodic oscillations in tissue pressure, themselves demonstrate regular, strong, fast phasic contractions that drive lymph flow through the lymphatic network. Thus, to regulate overall lymph transport and function, both lymphatic pumping and flow resistance must be controlled. Both of these processes occur via the regulation of lymphatic muscle contractions. The lymphatic vessels must function as both regulated conduits and regulated pumps, and thus they have functional characteristics of both blood vessels and hearts.[39–42] Tonic contraction/relaxation of the lymph vessel via local, neural, and humoral factors can modulate the flow by altering the outflow resistance of the lymphatic bed. It is well documented that neural and humoral agents, such as α-adrenergic agonists, prostaglandins, bradykinin, substance P, and others, can affect the tone of lymphatics and thus alter flow resistance and lymphatic function.

In the intrinsic lymph pump, flow through a lymphatic bed is generated by coordinated contractions of the lymphatic muscle cells.[32,43–47] The brisk contraction of these cells leads to a rapid reduction of the lymphatic diameter, an increase in the local lymph pressure, and ejection of lymph downstream. The contractions are initiated by pacemaker activity located within the muscle layer of the lymphatic wall.[48–55] Clearly, the electrophysiological properties of lymphatic muscle are key cellular regulators of the lymphatic contractile function. Lymphatic muscle electrical properties have been measured in cow, sheep, and guinea pig mesenteric vessels using different electrophysiological methods. These techniques have been used to study the resting membrane potential, pacemaker activity, and action potentials that are associated with changes in lymphatic muscle contractile activity. The nature of the resting membrane potential, the pacemaker activity, and the ions and channels involved in determining these is outside of the scope of this paper, but is the focus of a recent review.[56] The depolarization and action potential spreads from cell to cell along the lymphatics, eliciting a propagated contraction both upstream and downstream. The action potential produces increases in intracellular calcium that lead to contraction of the actin/myosin filaments within the muscle cells.[57–59]

The phasic contractile cycle of these vessels can be divided into periods of lymphatic systole and diastole using a cardiac cycle analogy.[34] Contraction frequency, ejection fraction, stroke volume, and lymph pump flow can be determined to evaluate lymphatic pumping function by means of this cardiac pump analogy.[34,60] The phasic contractions can be modulated in an inotropic (i.e., mediated by changes in the strength of contraction) and/or chronotropic (i.e., mediated by changes in the contraction frequency) fashion by transmural pressure, lymph flow/shear, neural input, and humoral influences.[16,30,31,61–75] An increase in stretch

of the vessel wall is the classic activating pathway of the lymph pump. Stretch of the lymphatics will produce increases in the lymphatic contraction frequency and initial rises in the strength of the phasic contraction,[16,30,38] followed by a plateau and decrease in the phasic contraction strength. However, we have shown in various rat lymphatics that increased stretch leads to decreases in lymphatic tonic contraction strength.[41] Recently we have demonstrated that in fact the lymphatic muscles possess both functional and molecular characteristics of blood vessels and the heart.[40] Vascular and cardiac muscle contractions are both critically regulated by calcium. However, the contractile/regulatory proteins and calcium dynamics involved in cardiac and vascular muscle contractions are dramatically different from each other. The question of how lymphatic muscles can regulate both tonic and phasic lymphatic contractions, presumably via the same factor, calcium, remains to be determined. Additionally the molecular regulation of tonic and phasic contractions in lymphatic muscle is another understudied and poorly understood principle that needs to be understood if we are to develop a better understanding of the physiological principles that drive and regulate lymphatic function in order to treat lymphatic diseases that are associated with lymphatic muscle dysfunction. The rest of this review will focus on what is known about the molecular mechanisms, specifically the role of intracellular calcium and the contractile and regulatory proteins that control the lymphatic muscle activity necessary for the generation and regulation of the main function of the lymphatic system —lymph transport.

The Regulation of Lymphatic Muscle Contraction by Calcium

While intracellular calcium has been implicated as a critical molecule, necessary for the phasic lymphatic contractions, the exact roles it plays in the regulation of lymphatic contractions, both tonic and phasic, remain to be determined. First we will describe what we know about the parallel roles of calcium in the regulation of contraction of tonic vascular smooth muscle and phasic striated muscle, and then we will discuss the state of our current understanding of the role of calcium in the regulation of lymphatic muscle.

Increases in the level of $[Ca^{2+}]i$ is thought to be the primary mechanism that initiates blood vascular SM contraction. Numerous mechanisms can alter the sensitivity of the blood vascular contractile elements to calcium, resulting in a calcium-independent modulation of vascular SM contraction.[76] However, it is thought that the principal regulator of vascular contractions is indeed $[Ca^{2+}]i$. In vascular SM, the simplified view is that Ca^{2+} binds to calmodulin, and the Ca^2–calmodulin complex then activates the myosin light chain kinase (MLCK). MLCK then phosphorylates the regulatory myosin light chain (RMLC). RMLC phosphorylation activates myosin ATPase and the interaction of myosin with actin and initiates the smooth muscle contraction. Reductions in $[Ca^{2+}]i$ inactivate MLCK, and RMLC is then dephosphorylated by myosin light chain phosphatase (MLCP). This deactivates the myosin ATPase activity and relaxes the vascular muscle. Thus, in the regulation of SM contraction, the changes in $[Ca^{2+}]i$ through various channels, stores, and exchangers alters the balance of phosphorylation and dephosphorylation of the RMLC. In addition it is clear that there are numerous mechanisms that can modulate the "calcium-sensitivity" of vascular SM, resulting in the modulation of force generation at any given calcium level. These topics have been the focus of numerous studies and reviews.[77–84] What roles the calcium-dependent mechanisms seen in blood vessels may play in controlling lymphatic tonic and phasic contractions are not well understood.

In striated muscle, the membrane depolarization associated with action potentials results in the rapid rise in $[Ca^{2+}]i$. The mechanisms by which this occurs in cardiac and skeletal muscle differ, but both use calcium-induced calcium release (CICR) from sarcoplasmic reticulum (SR) calcium stores upon opening of the ryanodine receptor (RyR) channels in the SR. Whereas in skeletal muscle, the SR CICR results in essentially all the $[Ca^{2+}]i$ needed for the contraction, in cardiac muscle, calcium from both SR CICR and calcium influx through membrane potential–sensitive calcium channels contribute significantly to the phasic contraction. Excitation–contraction coupling in striated muscle is then due to calcium binding to troponin C (TnC). Through interactions of troponin I (TnI), troponin T (TnT), and tropomyosin (TM), the myosin heads bind to actin, and through the hydrolysis of adenosine triphosphate (ATP) results in striated muscle contraction. Clearly, a tremendous number of studies have evaluated these mechanisms and these are reviewed elsewhere.[85–94] Given the phasic nature of the spontaneous lymphatic contractions, what role, if any, that the striated muscle calcium-regulation plays in lymphatic muscle contraction is unclear.

While we have known for a few decades that calcium is required for either the electrical activity or excitation–contraction coupling that

drives phasic lymphatic contractions,[53,57,58,95–103] relatively little is known about the specific molecular mechanisms by which calcium regulates the lymphatic muscle contraction. However, we know that phasic lymphatic contractions require extracellular calcium and that they are inhibited by dihydropyridines, especially L-type calcium channel blockers.[49,50,57,95,100,104] Other calcium-dependent currents involved in the lymphatic electrical activity associated with the phasic contractions include a T-type calcium current and a calcium-activated Cl^- current.[53,104] Work by several groups has shown that calcium, specifically calcium released from SR stores, appears to be involved with the pacemaking mechanisms that drive the lymphatic pressure responsible for phasic contractions.[53,96,99,100,105] While the exact store and membrane channels responsible for this are not clear, they likely involve calcium-activated Cl^- currents and quantal calcium release from SR stores. It is known that guinea pig pacemaker activity was inhibited by the intracellular calcium chelator BAPTA-AM and increased by agonists that stimulate calcium release from internal stores. In guinea pig lymphatics, release of calcium from IP3-sensitive calcium stores appears to be more important than release from ryanodine-sensitive calcium stores.[106] The work of Atchison et al., studying the effect of intracellular calcium store modulation on the actively contracting isolated bovine mesenteric lymph vessels, has produced interesting results.[96] SR calcium store modulators, caffeine, ryanodine, and cyclopiazonic acid (CPA) all inhibited lymphatic pumping when transmural pressure was fixed, implicating SR calcium stores (including a ryanodine-sensitive store) in the phasic contractile activity. However, when these investigators evaluated the effect of these agents on the relationship between lymph pump activity and transmural pressure, the magnitude of inhibition by caffeine and CPA was greater than that produced by ryanodine. This could be interpreted that stores other than the ryanodine-sensitive stores (i.e., IP3-sensitive stores) are also involved and/or that the caffeine and CPA both deplete all stores more thoroughly than does ryanodine. In rat mesenteric lymphatics, we have shown that the lymphatic muscle exhibits calcium changes that are very different from those seen in typical vascular SM.[59] Associated with each phasic contraction is a spike in lymphatic calcium that precedes the mechanical event. Other investigators[58,99] have shown similar changes in calcium associated with lymphatic pumping in other vessels. Shirasawa and Benoit have measured isometric force and calcium simultaneously in rat thoracic duct preparations in response to stretch.[58] Their results suggest that stretch induces lymphatic pump activity by increasing the pacemaker activity and the calcium sensitivity of contractile apparatus, but it did not increase the levels of calcium within the thoracic duct. This is in contradiction to the results of our studies in rat mesenteric lymphatics, where we show clear elevations in both basal and peak calcium levels as pressure in the lymphatics is increased. The reason for this difference is not known, although we speculate that it may be due to tissue differences since we already know that the rat thoracic duct phasic contractile activity is relatively insensitive to increases in transmural pressure when compared to rat mesenteric lymphatics.[41]

We know the critical importance of both extra and intracellular sources of calcium to the contraction of lymphatic muscle. Muscle calcium has always played a crucial role in contraction regulation and excitation–contraction coupling. However, there is much more work to do in evaluating the stores, channels, and mechanisms involved in the regulation of lymphatic muscle calcium and learning how these influence the tonic and phasic lymphatic contractile activity. To better understand the role of intracellular calcium modulation in the regulation of lymphatic muscle contraction, we must determine the mechanisms that regulate calcium inside lymphatic muscle and thus play important roles in the lymphatic contractile regulatory apparatus. This information is badly needed to better understand the mechanisms that drive the lymphatic pump, thus determining lymphatic function.

Lymphatic Muscle Contractile and Regulatory Protein Isoforms and Tonic/Phasic Contractions

Until recently it was unclear what types of contractile and regulatory proteins the lymphatic muscle contractile apparatus was composed of or what the lymphatic muscle contractile cycle entailed. We now know that lymphatic muscle contains both striated and smooth muscle contractile elements.[40] In addition (unpublished observations), we have seen that lymphatic muscles express some striated muscle regulatory components, such as TnC and TnT. How these different mechanisms are employed simultaneously within lymphatic muscle to control tonic and phasic contractions is not yet clear. We describe what little is known about lymphatic muscle contractile regulation below, starting with a review of the processes regulating other smooth

and striated muscle contraction and how lymphatics may utilize those processes.

The contractile apparatus of SM is generally similar to that in striated muscle, although it is not highly organized into a discrete sarcomeric structure. This is principally due to the variation of the content of myosin and actin and other filament-organizing proteins in the two muscle types. The molecular components of the contractile apparatus have been characterized in skeletal and cardiac muscle as well as various types of SM.[107–109] Biochemical and molecular analyses show that SM expresses at least four muscle-specific and two non-muscle-specific myosin heavy chains (MHCs), and four types of myosin light chains (MLCs) during growth, development, and disease.[110] The SM MHC variants are SM1A and SM1B and SM2A and SM2B, in which SM1 and SM2 vary in their carboxy-terminal ends and the SM-B isoform has a specific 7–amino acid insert that resides in surface loop 1, adjacent to the ATP-binding domain of MHC, of the myosin head. Several *in vitro* studies have demonstrated that the SM-B (SM1/SM2) isoform expressed in the SM types with faster contractile properties, such as bladder, has nearly two-fold higher ATPase activity. In addition, the rate of MgATP binding to and MgADP release from the active site is increased in the SM-B isoform. Furthermore, *in vitro* biochemical studies have shown that preparations containing the SM-B isoform have an increase in the velocity of shortening when compared to SM-A preparations.[111–113] The recent mouse model developed by Babu *et al*. lacking the SM-B MHC isoform exhibits the abnormal contractile function of aorta and bladder tissues.[114] Our recent data show the presence of significant amounts of SM-B MHCs in mesenteric lymphatic muscle that correlates well with its rapid phasic contractile nature.

The SM MLC variants are SM-specific MLC_{20} and non-muscle MLC_{20}, and MLC_{17} A and B isoforms.[107] It is well established that Ca^{2+}-dependent MLC phosphorylation plays a major role in the regulation of SM contraction.[115] Although myosin ATPase activity is highly correlated with MLC phosphorylation, the Ca^{2+}-regulated thin-filament system also plays an important modulatory role in SM contraction.[83,116–119] SM contains TM and caldesmon (CaD) at a ratio of 1 CaD: 2 TM: 14 actin monomers,[120,121] in addition to smaller amounts of filamin, calponin, and calmodulin.[122] CaD acts as an inhibitor of actin activation of myosin ATPase activity, and many characteristics of this inhibition are similar to the role of TnI in striated muscle.[123] In SM, the phosphorylation of CaD by p21-activated kinase modulates the CaD-inhibition of myosin ATPase activity, thereby controlling the Ca^{2+} sensitivity of SM contraction.[124] The phosphorylation of CaD is also regulated by PKC and extracellular-regulated kinase 1 (ERK1); however, different results have been reported on the inhibitory function of CaD on myosin ATPase activity.[125] Similar to the roles of CaD, the role of calponin is also controversial in SM contraction. There are indications that calponin may directly inhibit the actin-activated myosin ATPase activity.[126] On the other hand, biochemical studies have shown that calponin acts as a signaling molecule that facilitates PKC and ERK redistribution, which then phosphorylate CaD, and activate contraction.[127,128] In addition, both structural and biochemical studies support a role for SM TM in the cooperativity of SM activation.[129] It has also been reported that TM is necessary for the full inhibition of actin-activated myosin ATPase activity by CaD.[130] SM-specific TM isoforms also potentiate actin-activated myosin ATPase activity.[131–133] Recent studies suggest an involvement of TM movement in the on/off switching of SM contraction in an as-yet-undefined manner.[134]

In striated muscle, calcium binding to TnC triggers an allosteric effect on TnI and TnT, allowing TM to flex.[135] This causes TM to move away from its steric hindrance position, allowing the attachment of myosin heads to actin. The transition from a weakly bound crossbridge to a strongly bound state locks TM in an open position; this produces a cooperative effect of promoting further crossbridge attachment.[136,137] After the generation of a power stroke, the actomyosin interaction shifts to a rigor state. Subsequent ATP binding then promotes the detachment process. Although the precise mechanism by which the TM–Tn complex regulates muscle contraction is still unknown, various models have been proposed (e.g., steric hindrance model, biochemical blocking model,[138,139] and a "continuum" model[140]). Current models for thin-filament activation and the crossbridge cycle show that TM exists in blocked, closed (off), and open (on) states.[136,137,141,142] In addition to Ca^{2+} binding to TnC, full activation of the thin filament requires strongly bound crossbridges.[137] Preliminary evidence from our lab (unpublished observations) indicates that cardiac troponin-C (cTn-C) is present in lymphatic muscle; however, the role, if any, this key determinant of thin-filament activation plays in lymphatic muscle contraction is unknown.

Despite the preponderance of information on both smooth and striated muscle regulatory mechanisms, very little is known about lymphatic muscle contraction. In fact, our recent studies provide the first evidence that lymphatic muscle contractile apparatus consists of both striated and SM contractile elements.[40] In addition, preliminary data demonstrate that lymphatic muscles express striated muscle regulatory components, such as cTn-C and cardiac troponin-T (cTn-T) (unpublished observations). Other than this, the only other reviewed paper that describes the regulatory mechanisms that may be involved in the regulation of lymphatic muscle contraction is that published by Hosaka *et al.*,[143] which evaluated the participation of the Rho-Rho kinase pathway in the regulation of lymphatic contractile activity. This study employed the Rho-kinase inhibitor, Y-27632, and the phosphatase inhibitor, okadaic acid, on lymph pump activity and myogenic, pressure- and agonist-induced tone in isolated rat lymphatics. Y-27632 produced a reduction of basal and activated tonic contraction with a cessation of the lymph pump activity at high doses. Okadaic acid produced an increase in lymphatic tonic constriction, but reduced the frequency of the phasic lymphatic contractions. Additionally, the effects of Y-27632 were in part reversed by the pretreatment with okadaic acid. These findings indicate that Rho kinase and myosin phosphatase contribute to the regulation of lymphatic tonic contractile activity and may be involved in the development of activity associated with the phasic contractions. While there have been a few other published manuscripts that discuss the presence of contractile or regulatory proteins in some lymphatic tissues, to date these are the only two published papers[40,143] describing the specific muscle contractile regulatory pathways that have been carefully studied in lymphatic muscle models. Given the scarcity of mechanistic evaluation of lymphatic contractile function, it is unfortunately impossible at this point to define the mechanisms that are responsible for the combined regulation of lymphatic tonic and phasic contractile activity. Given the critical dependence of lymphatic function (i.e., the generation and regulation of lymph flow) on lymphatic muscle contraction, clearly this is an area that needs a great deal of additional study.

Conclusions

Thus, although numerous studies have provided information on lymphatic pump activity, the basic molecular regulation of lymphatic muscle contraction is not well understood. Studies of the regulation of lymphatic contractile function are much less numerous than those in blood vessels, and our understanding of the molecular processes that regulate the tonic and phasic lymphatic muscle contractions is far from complete. What is sorely needed is a better understanding of the nature of the regulatory mechanisms that modulate force generation in lymphatic muscle during tonic and phasic contractions. This would provide valuable insight into lymphatic muscle contraction, and a better understanding of its mechanisms could provide the basis for better diagnosis and treatment of lymphatic dysfunction. Dysfunctional lymph flow can result in a wide range of clinical disorders, including lymphedema, altered lymphocyte circulation, depressed immune function, and impaired lipid metabolism, among others. Worldwide there are hundreds of millions patients afflicted with lymphedema, primarily due to the debilitating parasitic infection lymphatic filariasis. In the United States, secondary lymphedema is the most commonly diagnosed lymphatic disease, resulting from mastectomy, congestive heart failure, or reconstructive surgery. There are millions of patients in the U.S. suffering from secondary lymphedema after surgical node resection. In addition, there are likely other diseases in which lymphatic dysfunction plays roles that have not yet been defined. Currently, there are very few efficacious therapies, and a complete lack of medical treatment options for lymphatic dysfunction.[144–147] In large part, this is because of our poor understanding of even the most basic of lymphatic functions—the generation of lymph flow. So, while progress has recently been made in our understanding of lymphangiogenesis, currently there are no therapeutic agents that specifically modulate lymphatic muscle function. However, their development will not be possible until the molecular basis of lymphatic contractility is more fully understood.

Acknowledgments

This study was supported by Grants HL080526 and KO2HL086650 from the NIH (to M.M.) and HL75199 and HL07688 (to D.Z.). We would like to thank the following people for their help in the studies that support this work—Anatoliy A. Gashev, Michael J. Davis, and Pierre-Yves von der Weid.

Conflicts of Interest

The authors declare no conflicts of interest.

References

1. CASLEY-SMITH, J.R. 1972. The role of the endothelial intercellular junctions in the functioning of the initial lymphatics. Angiologica **9:** 106–131.
2. TAYLOR, A.E. 1990. The lymphatic edema safety factor: the role of edema dependent lymphatic factors (EDLF). Lymphology **23:** 111–123.
3. YOFFEY, J.M. & F.C. COURTICE. 1970. Lymphatics, Lymph and the Lymphomyeloid Complex. Academic Press. London and New York.
4. MORIONDO, A., S. MUKENGE & D. NEGRINI. 2005. Transmural pressure in rat initial subpleural lymphatics during spontaneous or mechanical ventilation. Am. J. Physiol. Heart Circ. Physiol. **289:** H263–H269.
5. NEGRINI, D., A. MORIONDO & S. MUKENGE. 2004. Transmural pressure during cardiogenic oscillations in rodent diaphragmatic lymphatic vessels. Lymphat. Res. Biol. **2:** 69–81.
6. NEGRINI, D. & M.D. FABBRO. 1999. Subatmospheric pressure in the rabbit pleural lymphatic network. J. Physiol. **520**(Pt 3): 761–769.
7. NEGRINI, D., S.T. BALLARD & J.N. BENOIT. 1994. Contribution of lymphatic myogenic activity and respiratory movements to pleural lymph flow. J. Appl. Physiol. **76:** 2267–2274.
8. LEAK, L. & J. BURKE. 1968. Ultrastructural studies on the lymphatic anchoring filaments. J. Cell Biol. **36:** 129–149.
9. COLLIN, H. 1969. The ultrastructure of conjunctival lymphatic anchoring filaments. Exp. Eye Res. **8:** 102–105.
10. LEAK, L.V. 1976. The structure of lymphatic capillaries in lymph formation. Fed. Proc. **35:** 1863–1871.
11. BOCK, P. 1978. Histochemical staining of lymphatic anchoring filaments. Histochemistry **58:** 343–345.
12. LYNCH, P.M., F.A. DELANO & G.W. SCHMID-SCHONBEIN. 2007. The primary valves in the initial lymphatics during inflammation. Lymphat. Res. Biol. **5:** 3–10.
13. MENDOZA, E. & G.W. SCHMID-SCHONBEIN. 2003. A model for mechanics of primary lymphatic valves. J. Biomech. Eng. **125:** 407–414.
14. TRZEWIK, J., S.K. MALLIPATTU, G.M. ARTMANN, *et al.* 2001. Evidence for a second valve system in lymphatics: endothelial microvalves. FASEB J. **15:** 1711–1717.
15. ALBERTINE, K., L. FOX & C. O'MORCHOE. 1982. The morphology of canine lymphatics valves. Anat. Rec. **202:** 453–461.
16. MCHALE, N.G. & I.C. RODDIE. 1976. The effect of transmural pressure on pumping activity in isolated bovine lymphatic vessels. J. Physiol. **261:** 255–269.
17. GNEPP, D. 1976. The bicuspid nature of the valves of the peripheral collecting lymphatic vessels of the dog. Lymphology **9:** 75–77.
18. ZWEIFACH, B. & J. PRATHER. 1975. Micromanipulation of pressure in terminal lymphatics in the mesentery. Am. J. Physiol. **228:** 1326–1335.
19. PAPP, M., S. VIRAGH & G. UNGVARY. 1975. Structure of cutaneous lymphatics propelling lymph. Acta Med. Acad. Sci. Hung. **32:** 311–320.
20. TAKADA, M. 1971. The ultrastructure of lymphatic valves in rabbits and mice. Am. J. Anat. **32:** 207–217.
21. LAUWERYNS, J.M. 1971. Stereomicroscope funnel-like architecture of pulmonary lymphatic valves. Lymphology **4:** 125–132.
22. EISENHOFFER, J., A. KAGAL, T. KLEIN, *et al.* 1995. Importance of valves and lymphangion contractions in determining pressure gradients in isolated lymphatics exposed to elevations in outflow pressure. Microvasc. Res. **49:** 97–110.

23. MAZZONI, M.C., T.C. SKALAK & G.W. SCHMID-SCHONBEIN. 1987. Structure of lymphatic valves in the spinotrapezius muscle of the rat. Blood Vessels **24:** 304–312.

24. GNEPP, D. & F. GREEN. 1980. Scanning electron microscopic study of canine lymphatic vessels and their valves. Lymphology **13:** 91–99.

25. PETRENKO, V.M. 2007. [Conceptions on the structural organization of the active lymph flow between the neighboring lymphangions]. Morfologiia **132:** 87–92.

26. KRIEDERMAN, B.M., T.L. MYLOYDE, M.H. WITTE, *et al.* 2003. Foxc2 haploinsufficient mice are a model for human autosomal dominant lymphedema-distichiasis syndrome. Hum. Mol. Genet. **12:** 1179–1185.

27. PETROVA, T.V., T. KARPANEN, C. NORRMEN, *et al.* 2004. Defective valves and abnormal mural cell recruitment underlie lymphatic vascular failure in lymphedema distichiasis. Nat. Med. **10:** 974–981.

28. SCHMID-SCHONBEIN, G.W. 1990. Microlymphatics and lymph flow. Physiol. Rev. **70:** 987–1028.

29. SZEGVARI, M., A. LAKOS, F. SZONTAGH, *et al.* 1963. Spontaneous contractions of lymphatic vessels in man. Lancet **7294:** 1329.

30. HARGENS, A.R. & B.W. ZWEIFACH. 1977. Contractile stimuli in collecting lymph vessels. Am. J. Physiol. **233:** H57–H65.

31. MISLIN, H. 1961. [Experimental detection of autochthonous automatism of lymph vessels]. Experentia **17:** 29–30.

32. MISLIN, H. 1976. Active contractility of the lymphangion and coordination of lymphangion chains. Experientia **32:** 820–822.

33. SMITH, R. 1949. Lymphatic contractility: a possible intrinsic mechanism of lymphatic vessels for the transport of lymph. J. Exp. Med. **90:** 497–509.

34. GRANGER, H.J., S. KOVALCHECK, B.W. ZWEIFACH, *et al.* 1977. Quantitative analysis of active lymphatic pumping. Proceedings of the VII Summer Computer Simulation Conference: 562–565. Simulation Council.

35. ORLOV, R.S. & R.P. BORISOVA. 1974. Spontaneous and evoked contractile activity of lymphatic vessel smooth muscle [Russian]. Dokl. Akad. Nauk. SSSR **215:** 1013–1015.

36. OLSZEWSKI, W.L. & A. ENGESET. 1980. Intrinsic contractility of prenodal lymph vessels and lymph flow in human leg. Am. J. Physiol. Heart Circ. Physiol. **239:** H775–H783.

37. OLSZEWSKI, W.L., A. ENGESET & J. SOKOLOWSKI. 1977. Lymph flow and protein in the normal male leg during lying, getting up, and walking. Lymphology **10:** 178–183.

38. MISLIN, H. 1966. Structural and functional relations of the mesenteric lymph vessels. New Trends in Basic Lymphology; Proceedings of a Symposium held at Charleroi (Belgium), July 11–13, 1966. Experientia Suppl. **14:** 87–96.

39. GASHEV, A.A. & D.C. ZAWIEJA. 2001. Comparison of the active lymph pumps of the rat thoracic duct and mesenteric lymphatics. Presented at the 7th World Congress for Microcirculation, Sidney, Australian and New Zealand Microcirculation Society, P1–19.

40. MUTHUCHAMY, M., A. GASHEV, N. BOSWELL, *et al.* 2003. Molecular and functional analyses of the contractile apparatus in lymphatic muscle. FASEB J. **17:** 920–922.

41. GASHEV, A.A., M.J. DAVIS, M.D. DELP, *et al.* 2004. Regional variations of contractile activity in isolated rat lymphatics. Microcirculation **11:** 477–492.

42. QUICK, C.M., A.M. VENUGOPAL, A.A. GASHEV, *et al.* 2007. Intrinsic pump-conduit behavior of lymphangions. Am. J. Physiol. Regul. Integr. Comp. Physiol. **292:** R1510–R1518.

43. MCHALE, N.G. & I.C. RODDIE. 1975. Pumping activity in isolated segments of bovine mesenteric lymphatics. J. Physiol. **244:** 70P–72P.

44. ALLEN, J.M. & N.G. MCHALE. 1986. Neuromuscular transmission in bovine mesenteric lymphatics. Microvasc. Res. **31:** 77–83.

45. ZAWIEJA, D.C., K.L. DAVIS, R. SCHUSTER, *et al.* 1993. Distribution, propagation, and coordination of contractile activity in lymphatics. Am. J. Physiol. **264:** H1283–H1291.

46. MCHALE, N.G. & M.K. MEHARG. 1992. Co-ordination of pumping in isolated bovine lymphatic vessels. J. Physiol. **450:** 503–512.

47. CROWE, M.J., P.Y. VON DER WEID, J.A. BROCK, *et al.* 1997. Co-ordination of contractile activity in guinea-pig mesenteric lymphatics. J. Physiol. **500:** 235–244.

48. ORLOV, R.S., R.P. BORIGORA & E.S. MUNDRIKO. 1976. Investigation of contractile and electrical activity of smooth muscle of lymphatic vessels. *In* Physiology of Smooth Muscle. E. Bulbring and M.F. Shuba, Eds.: 147–152. Raven Press. New York.

49. OHHASHI, T. & T. AZUMA. 1979. Electrical activity and ultrastructure of bovine mesenteric lymphatics. Lymphology **12:** 4–6.

50. LOBOV, G.I. & R.S. ORLOV. 1983. [Electrical and contractile activity of the lymphangions of the mesenteric lymphatic vessels]. Fiziol. Zh. SSSR Im. I. M. Sechenova. **69:** 1614–1620.

51. WARD, S.M., K.M. SANDERS, K.D. THORNBURY, *et al.* 1991. Spontaneous electrical activity in isolated bovine lymphatics recorded by intracellular microelectrodes. J. Physiol. **438:** 168.

52. MCCLOSKEY, K.D., H.M. TOLAND, M.A. HOLLYWOOD, *et al.* 1999. Hyperpolarisation-activated inward current in isolated sheep mesenteric lymphatic smooth muscle. J. Physiol. **521**(Pt 1): 201–211.

53. TOLAND, H.M., K.D. MCCLOSKEY, K.D. THORNBURY, *et al.* 2000. Ca^{2+}-activated Cl- current in sheep lymphatic smooth muscle. Am. J. Physiol. Cell Physiol. **279:** C1327–C1335.

54. VAN HELDEN, D.F. 1993. Pacemaker potentials in lymphatic smooth muscle of the guinea-pig mesentery. J. Physiol. **471:** 465–479.

55. VON DER WEID, P.Y., M.J. CROWE & D.F. VAN HELDEN. 1996. Endothelium-dependent modulation of pacemaking in lymphatic vessels of the guinea-pig mesentery. J. Physiol. **493:** 563–575.

56. VON DER WEID, P. 2001. Review article: lymphatic vessel pumping and inflammation—the role of spontaneous constrictions and underlying electrical pacemaker potentials. Aliment. Pharmacol. Ther. **15:** 1115–1129.

57. McHale, N.G. & J.M. Allen. 1983. The effect of external Ca^{2+} concentration on the contractility of bovine mesenteric lymphatics. Microvasc. Res. **26:** 182–192.

58. Shirasawa, Y. & J.N. Benoit. 2003. Stretch-induced calcium sensitization of rat lymphatic smooth muscle. Am. J. Physiol. Heart Circ. Physiol. **285:** H2573–H2577.

59. Zawieja, D.C., E. Kossman & J. Pullin. 1999. Dynamics of the microlymphatic system. J. Progr. Appl. Microcircul. **23:** 100–109.

60. Benoit, J.N., D.C. Zawieja, A.H. Goodman, et al. 1989. Characterization of intact mesenteric lymphatic pump and its responsiveness to acute edemagenic stress. Am. J. Physiol. **257:** H2059–H2069.

61. Gasheva, O.Y., D.C. Zawieja & A.A. Gashev. 2006. Contraction-initiated NO-dependent lymphatic relaxation: A self-regulatory mechanism in rat thoracic duct. J. Physiol. **575:** 821–832.

62. Amerini, S., M. Ziche, S.T. Greiner, et al. 2004. Effects of substance P on mesenteric lymphatic contractility in the rat. Lymphat. Res. Biol. **2:** 2–10.

63. Gashev, A.A., M.J. Davis & D.C. Zawieja. 2002. Inhibition of the active lymph pump by flow in rat mesenteric lymphatics and thoracic duct. J. Physiol. **540:** 1023–1037.

64. Yokoyama, S. & J.N. Benoit. 1996. Effects of bradykinin on lymphatic pumping in rat mesentery. Am. J. Physiol. **270:** G752–G756.

65. Ferguson, M.K. & V.J. Defilippi. 1994. Nitric oxide and endothelium-dependent relaxation in tracheobronchial lymph vessels. Microvasc. Res. **47:** 308–317.

66. Zawieja, D.C., S.T. Greiner, K.L. Davis, et al. 1991. Reactive oxygen metabolites inhibit spontaneous lymphatic contractions. Am. J. Physiol. **260:** H1935–H1943.

67. Schmid-Schonbein, G.W. 1990. Mechanisms causing initial lymphatics to expand and compress to promote lymph flow. Arch. Histol. Cytol. **53**(Suppl): 107–114.

68. Ferguson, M.K., H.K. Shahinian & F. Michelassi. 1988. Lymphatic smooth muscle responses to leukotrienes, histamine and platelet activating factor. J. Surg. Res. **44:** 172–177.

69. Elias, R.M., M.G. Johnston, A. Hayashi, et al. 1987. Decreased lymphatic pumping after intravenous endotoxin administration in sheep. Am. J. Physiol. **253:** H1349–H1357.

70. Hogan, R. & J. Unthank. 1986. Mechanical control of initial lymphatic contractile behavior in bat's wing. Am. J. Physiol. **251:** H357–H363.

71. McHale, N.G. & I.C. Roddie. 1983. The effects of catecholamines on pumping activity in isolated bovine mesenteric lymphatics. J. Physiol. **338:** 527–536.

72. Johnston, M.G. & J.L. Gordon. 1981. Regulation of lymphatic contractility by arachidonate metabolites. Nature **293:** 294–297.

73. McHale, N.G., I.C. Roddie & K.D. Thornbury. 1980. Nervous modulation of spontaneous contractions in bovine mesenteric lymphatics. J. Physiol. **309:** 461–472.

74. Ohhashi, T., Y. Kawai & T. Azuma. 1978. The response of lymphatic smooth muscles to vasoactive substances. Pflugers Arch. **375:** 183–188.

75. Orlov, R.S. & T.A. Lobacheva. 1977. Intravascular pressure and spontaneous lymph vessels contractions [Russian]. Bull. Exp. Biol. Med. **83:** 392–394.

76. Pfitzer, G. 2001. Signal transduction in smooth muscle: Invited Review. Regulation of myosin phosphorylation in smooth muscle. J. Appl. Physiol. **91:** 497–503.

77. Hirano, K. 2007. Current topics in the regulatory mechanism underlying the Ca^{2+} sensitization of the contractile apparatus in vascular smooth muscle. J. Pharmacol. Sci. **104:** 109–115.

78. Akata, T. 2007. Cellular and molecular mechanisms regulating vascular tone. Part **2:** regulatory mechanisms modulating Ca^{2+} mobilization and/or myofilament Ca^{2+} sensitivity in vascular smooth muscle cells. J. Anesth. **21:** 232–242.

79. Hirano, K., M. Hirano & H. Kanaide. 2004. Regulation of myosin phosphorylation and myofilament Ca^{2+} sensitivity in vascular smooth muscle. J. Smooth Muscle. Res. **40:** 219–236.

80. Iino, M. 2002. Molecular basis and physiological functions of dynamic Ca^{2+} signalling in smooth muscle cells. Novartis Found. Symp. **246:** 142–146; discussion, 147–53, 221–227.

81. Hill, M.A., H. Zou, S.J. Potocnik, et al. 2001. Invited review: arteriolar smooth muscle mechanotransduction: Ca(2+) signaling pathways underlying myogenic reactivity. J. Appl. Physiol. **91:** 973–983.

82. Savineau, J.P. & R. Marthan. 1997. Modulation of the calcium sensitivity of the smooth muscle contractile apparatus: molecular mechanisms, pharmacological and pathophysiological implications. Fund. Clin. Pharmacol. **11:** 289–299.

83. Walsh, M.P. 1994. Regulation of vascular smooth muscle tone. Can. J. Physiol. Pharmacol. **72:** 919–936.

84. Himpens, B. 1992. Modulation of the Ca(2+)-sensitivity in phasic and tonic smooth muscle. Verh. K. Acad. Geneeskd. Belg. **54:** 217–251.

85. Moss, R.L. 1992. Ca^{2+} regulation of mechanical properties of striated muscle: mechanistic studies using extraction and replacement of regulatory proteins. Circ. Res. **70:** 865–884.

86. Dibb, K.M., H.K. Graham, L.A. Venetucci, et al. 2007. Analysis of cellular calcium fluxes in cardiac muscle to understand calcium homeostasis in the heart. Cell Calcium **42:** 503–512.

87. Williams, A.J. 1997. The functions of two species of calcium channel in cardiac muscle excitation-contraction coupling. Eur. Heart J. **18**(Suppl A): A27–A35.

88. Wier, W.G., J.R. Lopez-Lopez, P.S. Shacklock, et al. 1995. Calcium signalling in cardiac muscle cells. Ciba Found. Symp. **188:** 146–160; discussion, 160–164.

89. McDonald, T.F., S. Pelzer, W. Trautwein, et al. 1994. Regulation and modulation of calcium channels in cardiac, skeletal, and smooth muscle cells. Physiol. Rev. **74:** 365–507.

90. Lucchesi, B.R. 1989. Role of calcium on excitation-contraction coupling in cardiac and vascular smooth muscle. Circulation **80:** IV1–IV13.

91. Shibata, S. 1989. Role of calcium in the cardiac and vascular smooth muscle contraction–overview. Microcirc. Endothelium Lymphatics **5:** 3–11.

92. BERCHTOLD, M.W., H. BRINKMEIER & M. MUNTENER. 2000. Calcium ion in skeletal muscle: its crucial role for muscle function, plasticity, and disease. Physiol. Rev. **80:** 1215–1265.

93. SCHNEIDER, M.F. 1994. Control of calcium release in functioning skeletal muscle fibers. Annu. Rev. Physiol. **56:** 463–484.

94. CATTERALL, W.A. 1991. Excitation-contraction coupling in vertebrate skeletal muscle: a tale of two calcium channels. Cell **64:** 871–874.

95. ATCHISON, D.J. & M.G. JOHNSTON. 1997. Role of extra- and intracellular Ca^{2+} in the lymphatic myogenic response. Am. J. Physiol. **272:** R326–R333.

96. ATCHISON, D.J., H. RODELA & M.G. JOHNSTON. 1998. Intracellular calcium stores modulation in lymph vessels depends on wall stretch. Can. J. Physiol. Pharmacol. **76:** 367–372.

97. AZUMA, T., T. OHNASHI & M. SAKAGUCHI. 1977. Electrical activity of lymphatic smooth muscles. Proc. Soc. Exp. Biol. Med. **155:** 270–273.

98. MCHALE, N.G., J.M. ALLEN & H.L. IGGULDEN. 1987. Mechanism of alpha-adrenergic excitation in bovine lymphatic smooth muscle. Am. J. Physiol. **252:** H873–H878.

99. FERRUSI, I., J. ZHAO, D. VAN HELDEN, et al. 2004. Cyclopiazonic acid decreases spontaneous transient depolarizations in guinea pig mesenteric lymphatic vessels in endothelium-dependent and -independent manners. Am. J. Physiol. Heart Circ. Physiol. **286:** H2287–H2295.

100. COTTON, K.D., M.A. HOLLYWOOD, N.G. MCHALE, et al. 1997. Outward currents in smooth muscle cells isolated from sheep mesenteric lymphatics. J. Physiol. (Lond.). **503:** 1–11.

101. ORLOV, R.S. & G.I. LOBOV. 1984. Ionic mechanisms of the electrical activity of the smooth-muscle cells of the lymphatic vessels [Russian]. Fiziol. Zh. SSSR Imeni I. M. Sechenova **70:** 712–721.

102. TAKESHITA, T., M. KAWAHARA, K. FUJII, et al. 1989. The effects of vasoactive drugs on halothane inhibition of contractions of rat mesenteric lymphatics. Lymphology **22:** 194–198.

103. MIZUNO, R., G. DORNYEI, A. KOLLER, et al. 1997. Myogenic responses of isolated lymphatics: modulation by endothelium. Microcirculation **4:** 413–420.

104. HOLLYWOOD, M.A., K.D. COTTON, K.D. THORBURY, et al. 1997. Isolated sheep mesenteric lymphatic smooth muscles possess both T- and L-type calcium currents [abstract]. J. Physiol. **501:** 109P–110P.

105. VON DER WEID, P.Y. & D.F. VAN HELDEN. 1996. Beta-adrenoceptor-mediated hyperpolarization in lymphatic smooth muscle of guinea pig mesentery. Am. J. Physiol. **270:** H1687–H1695.

106. ZHAO, J. & D.F. VAN HELDEN. 2003. Et-1-associated vasomotion and vasospasm in lymphatic vessels of the guinea-pig mesentery. Br. J. Pharmacol. **140:** 1399–1413.

107. EDDINGER, T.J. 1998. Myosin heavy chain isoforms and dynamic contractile properties: skeletal versus smooth muscle. Comp. Biochem. Physiol. B Biochem. Mol. Biol. **119:** 425–434.

108. SCHWARTZ, K., D. DE LA BASTIE, P. BOUVERET, et al. 1986. Alpha-skeletal muscle actin mRNA's accumulate in hypertrophied adult rat hearts. Circ. Res. **59:** 551–555.

109. OWENS, G.K. 1995. Regulation of differentiation of vascular smooth muscle cells. Physiol. Rev. **75:** 487–517.

110. BABU, G.J., D.M. WARSHAW & M. PERIASAMY. 2000. Smooth muscle myosin heavy chain isoforms and their role in muscle physiology. Microsc. Res. Tech. **50:** 532–540.

111. LAUZON, A.M., M.J. TYSKA, A.S. ROVNER, et al. 1998. A 7-amino-acid insert in the heavy chain nucleotide binding loop alters the kinetics of smooth muscle myosin in the laser trap. J. Muscle Res. Cell Motil. **19:** 825–837.

112. KHROMOV, A.S., A.V. SOMLYO & A.P. SOMLYO. 1996. Nucleotide binding by actomyosin as a determinant of relaxation kinetics of rabbit phasic and tonic smooth muscle. J. Physiol. **492:** 669–673.

113. SWEENEY, H.L., S.S. ROSENFELD, F. BROWN, et al. 1998. Kinetic tuning of myosin via a flexible loop adjacent to the nucleotide binding pocket. J. Biol. Chem. **273:** 6262–6270.

114. BABU, G.J., E. LOUKIANOV, T. LOUKIANOVA, et al. 2001. Loss of SM-b myosin affects muscle shortening velocity and maximal force development. Nat. Cell Biol. **3:** 1025–1029.

115. KAMM, K.E. & J.T. STULL. 1985. The function of myosin and myosin light chain kinase phosphorylation in smooth muscle. Annu. Rev. Pharmacol. Toxicol. **25:** 593–620.

116. MARSTON, S.B. & P.A. HUBER. 1996. Caldesmon. *In* Biochemistry of Smooth Muscle Contraction. M. Barany, Ed.: 77–90. Academic Press. San Diego, CA.

117. SMILLIE, L. 1996. Tropomyosin. *In* Biochemistry for Smooth Muscle Contraction. M. Barany, Ed.: 63–76. Academic Press. San Diego, CA.

118. GIMONA, A. & J.V. SMALL. 1996. Calponin. *In* Biochemistry of Smooth Muscle Contraction. M. Barany, Ed.: 91–103. Academic Press. San Diego, CA.

119. WALSH, M.P. 1994. Calmodulin and the regulation of smooth muscle contraction. Mol. Cell Biochem. **135:** 21–41.

120. MARSTON, S. 1990. Stoichiometry and stability of caldesmon in native thin filaments from sheep aorta smooth muscle. Biochem. J. **272:** 305–310.

121. LEHMAN, W., D. DENAULT & S. MARSTON. 1993. The caldesmon content of vertebrate smooth muscle. Biochim. Biophys. Acta **1203:** 53–59.

122. MARSTON, S.B. & C.W. SMITH. 1984. Purification and properties of Ca^{2+}-regulated thin filaments and f-actin from sheep aorta smooth muscle. J. Muscle Res. Cell Motil. **5:** 559–575.

123. MARSTON, S.B. & C.S. REDWOOD. 1993. The essential role of tropomyosin in cooperative regulation of smooth muscle thin filament activity by caldesmon. J. Biol. Chem. **268:** 12317–12320.

124. FOSTER, D.B., L.H. SHEN, J. KELLY, et al. 2000. Phosphorylation of caldesmon by p21-activated kinase: implications for the Ca(2+) sensitivity of smooth muscle contraction. J. Biol. Chem. **275:** 1959–1965.

125. MORGAN, K.G. & S.S. GANGOPADHYAY. 2001. Signal transduction in smooth muscle: Invited review. Cross-bridge regulation by thin filament-associated proteins. J. Appl. Physiol. **91:** 953–962.

126. WINDER, S.J. & M.P. WALSH. 1990. Smooth muscle calponin: inhibition of actomyosin MgATPase and regulation by phosphorylation. J. Biol. Chem. **265:** 10148–10155.

127. LEINWEBER, B.D., P.C. LEAVIS, Z. GRABAREK, *et al.* 1999. Extracellular regulated kinase (ERK) interaction with actin and the calponin homology (ch) domain of actin-binding proteins. Biochem. J. **344**(Pt 1): 117–123.

128. MENICE, C.B., J. HULVERSHORN, L.P. ADAM, *et al.* 1997. Calponin and mitogen-activated protein kinase signaling in differentiated vascular smooth muscle. J. Biol. Chem. **272:** 25157–25161.

129. LEHRER, S.S., N.L. GOLITSINA & M.A. GEEVES. 1997. Actin-tropomyosin activation of myosin subfragment 1 ATPase and thin filament cooperativity: the role of tropomyosin flexibility and end-to-end interactions. Biochemistry **36:** 13449–13454.

130. SMITH, C.W., K. PRITCHARD & S.B. MARSTON. 1987. The mechanism of Ca^{2+} regulation of vascular smooth muscle thin filaments by caldesmon and calmodulin. J. Biol. Chem. **262:** 116–122.

131. SMALL, J.V. & A. SOBIESZEK. 1980. The contractile apparatus of smooth muscle. Int. Rev. Cytol. **64:** 241–306.

132. CHACKO, S. & E. EISENBERG. 1990. Cooperativity of actin-activated ATPase of gizzard heavy meromyosin in the presence of gizzard tropomyosin. J. Biol. Chem. **265:** 2105–2110.

133. MARSTON, J.B. 1983. The regulation of smooth muscle contractile proteins. Prog. Biophys. Mol. Biol. **41:** 1–41.

134. GRACEFFA, P. 2000. Phosphorylation of smooth muscle myosin heads regulates the head-induced movement of tropomyosin. J. Biol. Chem. **275:** 17143–17148.

135. SOLARO, R.J. & H.M. RARICK. 1998. Troponin and tropomyosin: proteins that switch on and tune in the activity of cardiac myofilaments. Circ. Res. **83:** 471–480.

136. GEEVES, M.A. & S.S. LEHRER. 1994. Dynamics of the muscle thin filament regulatory switch: the size of the cooperative unit. Biophys. J. **67:** 273–282.

137. SWARTZ, D.R., R.L. MOSS & M.L. GREASER. 1996. Calcium alone does not fully activate the thin filament for s1 binding to rigor myofibrils. Biophys. J. **71:** 1891–1904.

138. HILL, T.L., E. EISENBERG & L. GREENE. 1980. Theoretical model for the cooperative equilibrium binding of myosin subfragment 1 to the actin-troponin-tropomyosin complex. Proc. Natl. Acad. Sci. USA **77:** 3186–3190.

139. GREENE, L.E. & E. EISENBERG. 1980. Cooperative binding of myosin subfragment-1 to the actin-troponin-tropomyosin complex. Proc. Natl. Acad. Sci. USA **77:** 2616–2620.

140. EL-SALEH, S.C., K.D. WARBER & J.D. POTTER. 1986. The role of tropomyosin-troponin in the regulation of skeletal muscle contraction. J. Muscle Res. Cell. Motil. **7:** 387–404.

141. VIBERT, P., R. CRAIG & W. LEHMAN. 1997. Steric-model for activation of muscle thin filaments. J. Mol. Biol. **266:** 8–14.

142. MCKILLOP, D.F. & M.A. GEEVES. 1993. Regulation of the interaction between actin and myosin subfragment 1: evidence for three states of the thin filament. Biophys. J. **65:** 693–701.

143. HOSAKA, K., R. MIZUNO & T. OHHASHI. 2003. Rho-rho kinase pathway is involved in the regulation of myogenic tone and pump activity in isolated lymph vessels. Am. J. Physiol. Heart Circ. Physiol. **284:** H2015–H2025.

144. ROCKSON, S.G. 2001. Lymphedema. Am. J. Med. **110:** 288–295.

145. ROCKSON, S.G. 2000. Lymphedema. Curr. Treat Opt. Cardiov. Med. **2:** 237–242.

146. SZUBA, A. & S.G. ROCKSON. 1998. Lymphedema: classification, diagnosis and therapy. Vasc. Med. **3:** 145–156.

147. SZUBA, A. & S.G. ROCKSON. 1997. Lymphedema: anatomy, physiology and pathogenesis. Vasc. Med. **2:** 321–326.

Lymphatic Vessels: Pressure- and Flow-dependent Regulatory Reactions

ANATOLIY A. GASHEV

Department of Systems Biology and Translational Medicine,
Cardiovascular Research Institute, Division of Lymphatic Biology,
College of Medicine, Texas A&M Health Science Center, Temple, Texas, USA

This chapter discusses the current status of knowledge on basic pressure/stretch– and flow/shear–dependent regulatory reactions of contracting lymphangions in lymphatic vessels. The balance of passive and active lymph pumps is discussed; the data on pressure measurements in different lymphatic nets of several species are provided. The characteristics of typical changes in lymphatic contractility to the increases in transmural pressures are described and discussed, as is the role of pre-existing distension in initiation of lymphatic contractions. The up-to-date status of knowledge on flow-dependent adaptive reactions of contracting lymphangions is provided with detailed characterization of the influences of imposed and contraction-generated flows on lymphatic contractility. The interaction between extrinsic and intrinsic flows in regulation of lymphatic contractility is considered.

Key words: lymphatic vessels; lymphangion; lymph pressure; lymph flow; lymph pump

Intrinsic and Extrinsic Lymph Pumps

Several driving forces support lymph flow. Traditionally these forces are divided into two groups by their relation to the spontaneous contractility of lymphangions, sections of lymphatic vessels between adjacent valves.[1–3] The terms *active* or *intrinsic* lymph pump describe the lymph driving force that is generated by the active spontaneous contractions of lymphangions. The terms *passive* or *extrinsic* lymph pump encompass the influences of all other forces that are not connected with active contractions of lymphatic muscle cells in lymphatic vessel wall and that may support to greater or lesser degree lymph flow in different regions of body.

Intrinsic Lymph Pump

There is no single pump in the lymphatic system as there is in the cardiovascular system, where the energy of the heart's contractions is enough to move blood through all the circulation. Because the lymphatic capillary net have no direct connections with blood capillaries, the energy of cardiac contractions cannot directly create a pressure gradient in the lymphatic

system that is sufficient to propel lymph centripetally all way toward the central veins. The lymphatic system possesses its own numerous intrinsic pumps—lymphangions, the contraction energy of which is absolutely necessary for lymph flow in the majority of mammals including humans. Since the driving force created by one such small pump is not enough to propel lymph all way down through the lymphatic system, lymphatic vessels are organized in chains of these pumps. There is no evidence that in any regional lymphatic network contraction of lymphatic vessel occurs simultaneously along all of its length from capillaries toward the local "output" of this regional net. On the contrary, numerous reports demonstrated the propagation of peristaltic-like contractile waves along the lymphatic vessels.[4,5] For some of regional lymphatic nets (like those in the lower limbs), the presence of an interrupted fluid column in lymphatic vessels was demonstrated during the normal contractile activity of lymphangions[6–8]; in such situations adjacent lymphangions contract in counter-phase fashion.[8] Consequently, each lymphangion can be principally described as a short-distance local pump whose primarily "task" is to drive a bolus of lymph only to fill the next or few next ones. However, all together the chains of such pumps are capable of maintaining effective long-distance transport of lymph. During the active contractions of lymphangions, the lymphatic muscle cells create an increase of intralymphatic pressure and form a local positive pressure gradient to propel lymph

Address for correspondence: Anatoliy A. Gashev, Department of Systems Biology and Translational Medicine, Cardiovascular Research Institute, Division of Lymphatic Biology, College of Medicine, Texas A&M Health Science Center, 702 SW H.K. Dodgen Loop, Temple, TX 76504. Voice: +1-254-742-7147; fax: +1-254-742-7145.

gashev@tamu.edu

Ann. N.Y. Acad. Sci. 1131: 100–109 (2008). © 2008 New York Academy of Sciences.
doi: 10.1196/annals.1413.009

centripetally. As a result of such intrinsic spontaneous pumping activity of lymphangions, the positive (supportive to lymph flow) pressure gradient occurs near the downstream front of the propagating contracting zone in lymphatic vessels (no matter what, this zone includes one or several lymphangions contracting at the same time). At the upstream edge of the contracting zone, the negative (opposing lymph flow) pressure gradient always exists between contracting lymphangions and upstream relaxing lymphangions. This gradient generates the short-lasting local reversed flow and subsequent valve closure at the input edge of the contracting zone in lymphatic vessels.[4,9]

The majority of the contractile force during the lymphangion contractions is used to produce an increase of intraluminal pressure and propel lymph. But some part of the energy generated by lymphatic contraction will be used to cause deformation of collagen and elastic fibers in the lymphatic wall. During diastole these fibers will release their remaining tension and lead the lymphatic wall to expand. Recently[10,11] the pressure inside single isolated bovine mesenteric lymphangions has been measured under conditions when diastolic pressure inside lymphangions was set to $0\,cmH_2O$. The pressure tracings obtained in this study demonstrated the development of negative fluctuations of pressure inside lymphangions up to $-5\,cmH_2O$. These negative intraluminal pressures were coupled with contractions of lymphatic wall. It was concluded that at low levels of lymphangion filling during the relatively low lymph formation states, such negative pressure waves may produce suction in lymphangions, so energy of the active lymphatic pump will be used not only for emptying the lymphangion, but also for its subsequent filling.

Extrinsic Lymph Pump

The unique feature of the lymph dynamics is that the driving forces generated by the active lymphatic pump are greatly influenced by the action of different extralymphatic forces. These forces can have a sometimes greater impact on lymph flow than the active pump does, and the sum vector of these forces is not always favorable to lymph flow. The action of variable extralymphatic forces complicates pressure and flow patterns in the lymphatic net and may vary dramatically, even in the same region of body depending on the level of activity of tissues and organs surrounding lymphatics.

The term "passive" or "extrinsic" lymph pump combines all extralymphatic forces which influence lymph flow and when these forces are favorable to lymph flow. The origin of these forces is not connected

with active contractions of muscle cells in lymphatic vessel wall. But the use of the term "passive" is correct only in referring to the active contractions of lymphangions and is not completely correct from the point of view that all of these forces are generated by different active processes that are unrelated to these lymphangions' contractions. Extrinsic lymph driving forces include: driving force of lymph formation (historically also called *vis a tergo*), influences of cardiac and arterial pulsations, contractions of skeletal muscles adjacent to the lymphatic vessels, central venous pressure fluctuations, gastrointestinal peristalsis, and respiration. All of these forces may produce hydrostatic gradients in the lymphatic network which could effectively propel lymph, even sometimes in the absence of lymphatic vessel contractions. Some of these extrinsic lymph driving forces, such as the influences of lymph formation and central venous pressure fluctuations, affect lymph flow in the whole body; whereas influences of others are more or less localized.

Lymph flow directly depends on the intensity of lymph formation. In general, lymphatic capillaries have their own limits to capacity. Therefore, the faster the volume of recently formed lymph rises inside the lymphatic capillary net, the faster excessive lymph volumes will move downstream into collecting lymphatic vessels. Other extralymphatic forces may influence the emptying of lymphatic capillaries as well as the pumping too. In the gastrointestinal lymphatic net, intestinal peristalsis has a great impact on lymph formation and transport. An increase in intraluminal pressure in the gut directly increases the rates of lymph formation and pressure in intestinal initial lymphatics, with subsequent increase in lymph flow through the net of mesenteric collecting lymphatics.[12] In the thoracic cavity, cardiogenic and respiratory tissue motions have been shown to have a direct influence on lymphatic pressure by the generation of rhythmic expansions and compressions of the lymphatic wall.[13-15] Many lymphatic vessels are affected by contractions of adjacent or surrounding skeletal muscles, which cause compression of the lymphatic wall and create an additional force to empty lymphatic vessels. It has been demonstrated that local values of lymphatic pressure and levels of lymph flow directly correlate with intensity of skeletal muscle activity in legs.[16]

Thus, the values of intraluminal pressure and its gradients in lymphatic networks depend on contractile activity of lymphangions and actions of extrinsic forces. Pressure measurements in the different parts of the lymphatic system demonstrated the presence of pressure fluctuations. Primarily pressure oscillations are connected with the contraction of lymphangions.

Although an increment of pressure of $1-1.5\,\mathrm{cmH_2O}$ has been seen to be sufficient to open a closed valve, lymphatic vessels generate pressure peaks in about $5-10\,\mathrm{cmH_2O}$ higher than baseline diastolic intralymphatic pressure.[10,17–19]

Because of the tremendous difficulties in measuring pressures in different parts of the lymphatic system, only a few reports are able to present a comparatively complete range of pressures along the lymphatic system or along some of its comparatively long parts. Until now the most inclusive representation of pressure patterns has been made by Szabo and Magyar.[20] These authors published the results of their systematic measurements of pressure during cannulation of different lymphatic vessels in major lymphatic trunks in dogs. The authors demonstrated that mean intralymphatic pressure in the thoracic duct was $5.11\,\mathrm{mmHg}$ ($= 6.947\,\mathrm{cmH_2O}$) and in right lymphatic trunk $2.13\,\mathrm{mmHg}$, whereas in the jugular vein it was measured as $5.83\,\mathrm{mmHg}$. In more peripheral lymphatic vessels the mean intralymphatic pressures were as follows: left jugular trunk, $0.85\,\mathrm{mmHg}$; efferent lymphatic trunk of heart, $2.91\,\mathrm{mmHg}$; bronchomedial trunk, $2.09\,\mathrm{mmHg}$; hepatic trunk, $3.43\,\mathrm{mmHg}$; intestinal trunk, $3.60\,\mathrm{mmHg}$; left lumbar trunk, $2.72\,\mathrm{mmHg}$; and femoral lymph vessel, $0.51\,\mathrm{mmHg}$. This study was the first in which detailed evidence was presented that there is no constant positive pressure gradient along the lymphatic system. The data of this study confirmed that the action of so-called *vis a tergo* (or force of lymph formation) is limited, probably by the initial capillary segments of the lymphatic net and by the actual levels of lymph formation there. The majority of time the lymphangions in collecting lymphatic vessels had to produce an extra force to overcome the negative pressure gradient between the segments. The reasons for the existence of such pressure differences are the action of gravitational forces, wide variability of influences of skeletal muscle contractions, and the presence of competent valves along the lymphatic vessels.[20]

In bipedal humans the negative pressure gradient along the lymphatic net may be even much higher than in animals. Olszewski and Engeset[16,21,22] performed lymphatic cannulations using T-shape tubes to measure lateral pressures. At rest, the mean lateral systolic pressure in the leg lymphatic vessels with free lymph flow (measurements of side pressure) was $13.5 \pm 8.01\,\mathrm{mmHg}$, lymph pulse amplitude was $8.8 \pm 4.6\,\mathrm{mmHg}$, and contraction frequency was $2.42 \pm 1.88/\mathrm{min}$. All these values rose sequentially from a horizontal resting position, to a horizontal position with flexing of the foot, to an upright resting position, to upright position with rising toes. In the last case

these values were: peak pressure, $23.8 \pm 6.15\,\mathrm{mmHg}$; lymph pulse amplitude, $9.67 \pm 4.08\,\mathrm{mmHg}$; and contraction frequency, $5.5 \pm 1.04/\mathrm{min}$. The authors calculated mean lymph flow by measuring the movement of a minute air bubble introduced into tubing inserted into both ends of the leg lymph vessel. It was $0.25 \pm 0.04\,\mathrm{mL/h}$ ($0.004\,\mathrm{mL/min}$) in the horizontal position at the rest, and $0.76 \pm 0.26\,\mathrm{mL/h}$ ($0.012\,\mathrm{mL/min}$) in the upright position rising on toes. Lymph flow was only observed during the pulse waves, and there was no flow in the period between the pulses despite massaging. Several other groups measured lymphatic pressures in human legs and these data generally gave the same order of magnitudes of intralymphatic pressure in this region of the body. In 1991, Krylov *et al.*[23] published measurements of lymphatic pressure before lymphography when they catheterized leg lymphatic vessels and immediately measured the end-lymphatic pressure. The values of these pressures measured on the dorsal surface of the foot were low—$0.86-1.1\,\mathrm{mmHg}$. Then, during the catheterization of more downstream lymphatic vessels in the leg, pressure waves have been recorded about $2-20\,\mathrm{mmHg}$ in amplitude from basal levels of $8-17\,\mathrm{mmHg}$. The authors visibly correlated the elevations of lymph pressure with the contractions of the lymphatic vessel. Zaugg-Vesti and others[24] measured lymphatic capillary pressure in healthy volunteers using the servo-null technique in the distal forefoot proximal to the base of the first and second toe. Mean lymph capillary pressure was $7.9 \pm 3.4\,\mathrm{mmHg}$ and pressure fluctuations of more than $3\,\mathrm{mmHg}$ were found. In another study[25] this group described the influences of postural changes on cutaneous lymph capillary pressure at the dorsum of the foot. Mean lymph capillary pressure was $9.9 \pm 3.0\,\mathrm{mmHg}$ in the sitting and $3.9 \pm 4.2\,\mathrm{mmHg}$ in the supine position.

In general the comparatively high values of resting intralymphatic pressures and peak pressure fluctuations described above reflect much higher outflow resistance for leg lymphatic net in humans and the physiological demand for local lymphangions to develop much stronger contractions than are seen in animals. At the same time the data on pressure measurements in the human thoracic duct are much more limited and some of them represent the tracings of only end-lymphatic pressure,[26] with pressure fluctuations of about $5-10\,\mathrm{mmHg}$ in amplitude from a basal level near $30\,\mathrm{mmHg}$. But even in those pressure measurements, which could be considered as measurements of side pressure,[27] most of the time thoracic duct pressure fluctuated in average between 14 and $22\,\mathrm{mmHg}$. These levels are much higher than basal resting pressures in the leg lymphatic net. Only during inspiration the intralymphatic pressure in the thoracic duct may change to slightly negative or slightly positive values,[27] which could temporarily make net pressure gradient favorable to lymph flow. However, such a short temporal pattern of pressure changes in the thoracic duct may influence lymph flow only locally in the thoracic cavity, but not in the legs, in which lymphatic vessels extend to almost the height of the human body from the thoracic duct and often possess interrupted fluid columns.[6-8]

Modulation of Lymphatic Vessel Contractility by Pressure/Stretch

Transmural pressure is an important physical factor of lymph dynamics and influences the contractile activity of lymphangions causing inotropic (changes in the strength of contraction) and chronotropic (changes in the contraction frequency) effects. Transmural pressure is defined as a pressure gradient across the vessel wall, and depends on intralymphatic as well as extralymphatic forces. In collecting lymphatics, two main forces may produce increases in intravascular (intraluminal) pressure and cause the lymphangion's filling and distension of the lymphatic wall. These are the driving force of lymph formation and pressure pulses generated by contractions of upstream lymphangions. The influences of several extralymphatic forces on the lymphatic wall may help to expand lymphatics, but in other situations may lead to vessel compression. Since the work of Florey,[28,29] Smith,[30] and Horstmann[31,32] has appeared, it has been postulated that the generation and distribution of lymphatic contractions depend exclusively on mechanical stimuli. The traditional paradigm postulates that distension of the lymphatic wall activates the lymphatic contraction, which generates pressure pulse sufficient to propel lymph to the next lymphatic segment. Later, in numerous studies performed both *in vivo* and *in vitro*,[4,33-40] it was shown that increases in transmural pressure caused positive inotropic and chronotropic effects in lymphatic vessels. Lymphatic vessels from different tissues and species reach the maximums of their pumping at the different values of intravascular pressure. These values found for many tissues are comparatively low and vary in different species and different regions between 3 and 15 cmH$_2$O. Further increase of intraluminal pressure causes overdistension of the lymphatic wall and diminishes pumping.

Of particular note, McHale and Roddie[4] demonstrated that isolated bovine mesenteric lymphatic segments which contained 5–7 lymphangions were able to increase frequency of contractions and stroke volumes during the increase of transmural pressure from 1 to 4 cmH$_2$O. These lymphatic vessels reached their maximums of pumping at transmural pressure about 4–5 cmH$_2$O. Further increase in transmural pressure led to decrease in stroke volume. The frequency of lymphatic contractions continued to rise, but this positive chronotropic effect of increase in transmural pressure did not compensate for the negative inotropic influences of continuing distension and flow felt at transmural pressures of 6 cmH$_2$O and higher. Principally the same patterns of lymphatic contractile behavior in response to increased transmural pressure were observed by Ohhashi *et al.*[35] Isolated one- or two-lymphangion segments of bovine mesenteric lymphatic vessels with outer diameter 0.5–3 mm had pumping maximums between 5 and 10 cmH$_2$O of intraluminal end-diastolic pressure.

A typical bell-shaped curve pressure/pumping relationship was shown for different regions and for different species. But it is important to mention that for smaller lymphatic vessels located more peripherally, the maximums of lymphatic pumping occur at higher values of transmural pressure. For example, the greatest pumping ability in sheep prenodal popliteal lymphatic vessels was observed at values of transmural pressure near 18–26 cmH$_2$O and greater than 50% pumping was seen between 12 and 43 cmH$_2$O.[40] Recent studies demonstrated the same tendency in the bovine mesenteric lymphatic net.[41] Bovine prenodal mesenteric lymphatic vessels are able to increase contractility during the increases in transmural pressure and to reach the maximums in their pumping at transmural pressures of 6–9 cmH$_2$O. During the further increases in transmural pressure, the contractility and pumping in these vessels were not significantly depressed up to 15 cmH$_2$O, whereas pumping in bovine postnodal mesenteric lymphatic vessels is typically depressed at transmural pressures higher than 10 cmH$_2$O. These data may indicate that more peripheral lymphatic vessels must develop much higher pressures to overcome greater outflow resistance in their particular location and the conditions for their optimal pumping are shifted toward the greater values of intraluminal/transmural pressures accordingly.

Recently new evidence was obtained to demonstrate the regional variations in pressure-induced changes in lymphatic contractility. Studies were performed on lymphatic vessels taken from four different regions of one species, the rat.[42] The local differences in pressure sensitivities and pumping ability were determined for the thoracic duct, cervical, mesenteric, and femoral lymphatic vessels. All lymphatics investigated were able to increase pumping during moderate increases in transmural pressure up to some pumping maximum. The largest pump productivity was observed at 3 cmH_2O transmural pressure for all lymphatics except mesenteric lymphatics, where maximum pumping occurred at a pressure of 5 cmH_2O. Moreover, detailed analysis demonstrated that all these lymphatics had a range of transmural pressures over which there were no significant differences in pumping. Experimental data demonstrate that these ranges of pressure were: 2–4 cmH_2O for the thoracic duct, 2–8 cmH_2O for cervical lymphatic vessels, 2–7 cmH_2O for mesenteric lymphatic vessels, and 2–9 cmH_2O for femoral lymphatic vessels. These data reveal that all selected lymphatics have their optimal pumping conditions at comparatively low levels of transmural pressure comparable to typical *in situ* lymph pressures,[38] and these pressure levels have a tendency to be higher in more peripheral lymphatic vessels. The highest pumping (at the optimal pressure levels) of 6–8 volumes/min was demonstrated for mesenteric lymphatic vessels and the lowest pumping (near 2 volumes/min) was seen in the thoracic duct.

Because of the importance of pressure stimuli for lymphatic contractility, the idea that distension stimuli are mandatory to generate lymphatic contractions has dominated the literature for many decades. But several studies reported that lymphatic vessels could contract in a coordinated fashion without distension stimuli.[4,19,33,43] Moreover, experiments performed on lymphatics from different tissues and species showed a high percentage of cases in which the contractile wave propagates in retrograde direction along the vessel.[3,5,38,44,45] At low or normal levels of lymph formation, in many tissues lymphangions, at the end of contractions, are often empty or close to empty.[8] Because of the presence of highly competent valves in lymphatics, the stretch-dependent activation of several upstream lymphangions in such situations is very unlikely. Particularly, Mislin and Rathenow noted[3] that the contractile wave could propagate in the retrograde direction through several lymphangions unconnected with the increase of the local transmural pressure. Contractions of upstream lymphangions could be activated after the contraction of a downstream lymphangion by the retrograde propagation of electrical excitation. More recent studies[10,11,46] demonstrated for 80% of lymphangions poor or no correlation between experimentally generated fluctuations of their intraluminal pressure and lymphatic contractions. Moreover, it was also shown[10,11,46] that isolated

bovine and rat mesenteric lymphatics can have a stable long-lasting spontaneous contractility at 0 cmH_2O intraluminal pressure, and in the absence of radial and axial distension. These data lead to the reasonable conclusion that the distension of the lymphatic wall by intraluminal/transmural pressure is an important factor in modulating the contractile activity in lymphatic vessels, but is not a mandatory factor for pacemaking.

Modulation of the Lymphatic Vessel Contractility by Intrinsic and Extrinsic Flows

As discussed above, the lymph flow generates a complicated combination of influences of active and passive lymph driving forces/pumps. Because the peaks of the actions of passive lymph pumps often are not synchronized with intrinsic contractile activity of lymphangions, flow profiles in lymphatic nets are extremely variable and bidirectional. Only the presence of valves in lymphatic vessels prevents the existence of extended periods of backflow and supports net unidirectional lymph flow. On other hand the presence of lymphatic valves additionally complicates the lymph flow profile. During the lymphangion systole, the pressure difference between contracting and relaxing lymphangions causes the temporal lymph backflow and leads to closure of downstream valve(s).[4,9,35] The bulb-like shape of valve sinuses (parts of lymphangions immediately downstream to the valve), which exist in majority of lymphatic vessels, is a structural factor that could promote the formation of local turbulent flows during the lymphatic valve closure.[9] The unique shape of lymphatic valves is another structural factor which by itself could complicate the lymph flow profile locally during the lymphangion's diastolic filling. The space between two valve leaflets is much narrower than the lymphatic lumen in non-valve areas of lymphangions. During lymphatic filling, the narrow valve-containing section between lymphangions could be a factor that causes local temporal accelerations and turbulences in lymph flow. Moreover, although lymphatic valves more commonly consist of two leaflets, valves having from one to five leaflets have also been described in the literature.[47] Our own yet unpublished observations demonstrated that in rat mesenteric lymphatic vessels the adjacent lymphatic valves could be oriented by 90° to each other, which may be additional factor in forming not only local turbulence near the valves, but also extended turbulence along the lymphatic vessel. Such variability in lymphatic valve structure and position predetermines even greater unpredictability in lymph flow profiles in lymphatic vessels.

Because of the numerous limitations in the methods of quantitative investigations of lymph flow, only recently have first reliable data been obtained on lymph flow velocities and values of shear stress in lymphatic vessels.[48–50] A high-speed video system was used to capture multiple contraction cycles in rat mesenteric lymphatic preparations *in situ*. The images were analyzed to determine fluid velocity, volume flow rate, wall shear stress, and retrograde flow. It was determined that lymphocyte density and flux varied from 326 to 35,500 cells/μL and 206 to 2030 cells/min, respectively. Lymphatics contracted phasically, with a mean diameter of 91 ± 9.0 μm and amplitudes of 39%, and lymph velocity in them varied with the phasic contractions in both direction and magnitude with an average of 0.87 ± 0.18 and peaks of 2.2–9.0 mm/s. The velocity was $\sim 180°$ out of phase with the lymphatic contractile cycle. The average lymph flow was 13.95 ± 5.27 μL/h with transient periods of reversed flow. This resulted in an average shear of 0.64 ± 0.14 with peaks of 4–12 dynes/cm^2. These studies confirmed that shear rate in mesenteric lymphatics is low, but had large variations in magnitude compared to that seen in blood vessels. It is important to note that in this work the authors measuring the velocity fluctuations in contracting mesenteric lymphatics and. more importantly, approximating the wall shear stresses that occur *in situ* validated that the velocities experimentally induced in the isolated vessels experiments were physiologically relevant.

Historically researchers started to investigate the influences of different lymph flows/velocities on lymphatic pumping, creating the different levels of shear in experiments with steady increases in lymph flow rates. In particular, Benoit *et al.*[38] increased lymph flow in rat mesenteric lymphatics by elevating lymph formation and, accordingly, lymph flow rates as a result of plasma dilution. They found the increased parameters of active lymph contractility in mesenteric lymphatics during the periods of increased lymph flow. But they also mentioned that pressure in the lymphatic network became less pulsatile in high lymph flow states. Of course such kinds of experiments with increased lymph formation *in situ* give important information on lymphatic contractile behavior in situations similar to those that occur during different phases of tissue and organ activities. However, it is very difficult to separate and study the effects of increased flow from well-known effects of increased transmural pressure in such states.

An additional indication of how lymphatic vessels behave in response to the increases of flow in them was given in experiments on isolated bovine mesenteric lymphangions which were exposed to the elevations of axial pressure gradient.[39,51] In these studies the expected increase in total flow through the lymphangion during the increases of input pressure was demonstrated. But it was also found that lymphangions contract only during the elevations of pressure gradient up to $+3–+5$ cmH$_2$O. Further increase of flow by rising axial pressure gradient caused a complete inhibition of active pumping in lymphangions and led to the periods of uninterrupted flow through them. It was hypothesized that lymphangions do not actively contract when axial flow gradient in them exceeds 3–5 cmH$_2$O, which is enough to move lymph through the lymphatic segment without the need for active fluid propulsion. In these experiments axial positive flow gradients (direction from upstream end to downstream end) have been created by increases of input pressure with unchanged output pressure. This led to the increases in transmural pressure in lymphatic segments at the same time as flow was increased. But an even higher level of contractile inhibition during the periods of increased flow in these lymphangions can be expected if stimulatory effects of increased transmural pressure will be eliminated by equal lowering of output pressure.

Another set of studies on influences of imposed flow on contractile activity of rat isolated lymphatic vessel was presented recently by Gashev *et al.*,[42,52] who performed studies on isolated and perfused lymphatic vessels from four different regions of body from the same species—the rat—which allowed comparisons of the imposed flow-induced modulations of lymphatic contractility in different regional lymphatic beds. In all of these studies, the inflow pressures were changed simultaneously with changes of outflow pressure made on the same level but in the opposite direction to maintain the mean transmural pressure equal during the periods of increased imposed flow. Parameters of lymphatic contractions were evaluated immediately after imposed flow was set or changed and were monitored during the subsequent 5 min with an analysis of time-dependent changes in 1-min intervals. In some series of experiments, investigators changed the direction of imposed flow in isolated mesenteric lymphatic segments to retrograde.

By means of these experimental approaches the potent imposed flow-dependent inhibition of the active lymph pump has been found in mesenteric lymphatics as well in the thoracic duct[52] and later in femoral and cervical lymphatic vessels.[42] Imposed flow gradient caused reductions in the frequency and amplitude of lymphatic contractions. As a result of these negative chronotropic and inotropic effects, the active pumping of lymphatics was greatly diminished. However, it is difficult to conclude that such imposed flow-dependent inhibition of the active lymph pump decreases the total lymph flow *in vivo*. Because total lymph flow is the sum of passive and active flows, it is likely that the increase in imposed (or passive) flow could overwhelm any decreases in active lymph flow. A potentially important overriding factor would be an enhanced rate of lymph formation. At high levels of lymph formation, passive lymph flow could become a greater driving force to move lymph than the active lymph pump. Imposed flow-dependent inhibition of the active lymph pump in such situations could be a reasonable physiological mechanism to save metabolic energy by temporarily decreasing or stopping contractions during the time when the lymphatic vessel does not need it. An additional outcome of the inhibition of the lymph pump under these circumstances would be a reduction in lymph outflow resistance. This reduction in outflow resistance is the result of the net increase in average lymphatic diameter that occurs when contractions are inhibited. For example, complete cessation of the mesenteric active lymph pump (at zero imposed flow and 5 cmH$_2$O transmural pressure gradient) would result in a net increase in the time-averaged diameter by about 23%, thus theoretically reducing resistance by approximately 56%.[52] This reduction in the outflow resistance could ease the removal of fluid from the affected compartment that is producing the high lymph flows and facilitate the resolution of edema.

Imposed flow-induced inhibition of the lymph pump followed two temporal patterns.[52] The first is the rapidly developing inhibition of contraction frequency. Upon imposition of flow, the contraction frequency immediately decreased and then partially recovered over time during continued flow. This effect was dependent on the magnitude of imposed flow, but did not depend on the direction of flow. The effect also depended upon the rate of change in the direction of flow. The second pattern was a slowly developing reduction of the amplitude of the lymphatic contractions, which increased over time during continued flow. The inhibition of contraction amplitude was dependent on the direction of the imposed flow, but independent of the magnitude of flow.

Therefore, the chronotropic and inotropic imposed flow-induced inhibitory responses appear somewhat different. In the first minute of the initiation of imposed flow, the lymphatic response to imposed flow occurs primarily through a rapid inhibition of the contraction frequency. *In vivo*, short periods of increased flow occur very often on account of the contractions of upstream lymphangions. It is possible that a fast chronotropic response of lymphatics is an important short-term regulatory reaction to rapid but short-lasting periods of increased flow. At high rates of lymph formation, which can be present in the mesenteric lymphatic bed *in vivo*, long periods of increased flow may occur. The slow inotropic effect, which develops in lymphatics in minutes, could be an important long-term regulatory reaction to slow but long-lasting periods of increased flow. This slowly developing flow-induced inhibition of lymphatic contractility could conserve energy in lymphatics when there are sufficient passive forces to move lymph without the active lymph pump and decrease local outflow resistance.

Additionally, imposed flow-induced inhibition of contraction frequency does not depend on direction of flow, whereas imposed flow-induced inhibition of amplitude of lymphatic contractions does. Long-term inhibition of lymphatic contractility is stronger during orthograde flow, which occurs in lymphatics more often in comparison with long periods of retrograde flow, which rarely occur because of the competency of the lymphatic valves. This evidence allows proposing that the mechanisms of the rapidly developing chronotropic and the slowly developing inotropic lymphatic responses to flow could be different.

As mentioned above, lymphangions work in a short-distance pump fashion and the force of their phasic contractions generates flow and shear in the lymphatics by themselves. Recently it was demonstrated[53] that the flow generated during phasic contractions in rat thoracic duct itself plays an important self-regulatory role in the lymphatic contractile cycle in a shear-dependent manner. In this study the thoracic duct was chosen as a vessel sensitive to an imposed flow and for its variable contractile behavior.[27,42,54–58] In many cases contractions may occur in one part of the thoracic duct, but do not propagate between different segments.[27,42,54–58] Contractile waves often do not propagate along this vessel, and many times the phasic contractions develop locally while adjacent parts of the duct are not contracting.[59] This feature of the thoracic duct was used to design the experiments to evaluate the importance of flow and shear generated by lymphatic phasic contractions in the regulation of the lymphatic contractile cycle. Two types of segments of thoracic duct were taken into this study: phasically active segments and phasically non-active segments. Close attention was paid to maintain the input and output pressures at the same level, thereby excluding any imposed flow. Thus in the phasically active lymphatic segments, flow and shear occurred only as a result of their inherent contractions. In phasically non-active segments flow and shear did not take place. As a result experimental conditions that allowed investigating the influences of flow and shear, generated solely by the phasic lymphatic pump, on the contractile function without any extra imposed flow were used. Gasheva *et al.*[53] found that lymphatic resting tone in phasically active segments of the thoracic duct was 2–2.6 times lower than in phasically non-active segments. The only difference between the experimental conditions of the active and non-active segments was the existence of actively generated flow in the contracting lymphatic segments. Hence in this study a relaxation in the thoracic duct that was connected exclusively with its spontaneous phasic pumping activity, was discovered. Investigating the possible mechanisms of this relaxation, it was demonstrated that blockade of NO-synthase by L-NAME completely abolished this difference in lymphatic tone between the phasically active and non-active segments. Therefore this evidence indicates that the contraction-generated reduction of lymphatic tone in the thoracic duct *is* mediated only by NO. Basing their conclusions on this fact, these investigators determined what happens with active lymphatic pumping when the mechanism of the contraction-generated reduction of tone is completely blocked. They found that the reduction of tone in lymphatic segments generated by the phasic contractions improves their diastolic filling (enhanced lusitropy, lowering half-relaxation time, indicates the speed of diastolic filling), makes lymphatic contractions stronger (enhanced inotropy, higher contraction amplitudes), and propels more fluid forward during each contraction (elevated ejection fraction), while decreasing contraction frequency (reduced chronotropy). After NO-synthase blockade, the lymphatic segment must contract more often (higher contraction frequency) to maintain the minute productivity (fractional pump flow) appropriate to the existing level of preload (transmural pressure). It was concluded that the reduction in lymphatic tone due to the flow/shear generated by phasic contractions is a regulatory mechanism that maintains lymphatic pumping in an energy-saving efficient mode (stronger, but fewer contractions per minute).

Importantly, the flow-mediated relaxation can exist in any phasically contracting lymphatic. But, the lymph flow profile is a complicated and variable sum of different forces, not only the result of phasic lymphatic contractions. When discussing the flow conditions in a single lymphangion, it is reasonable to divide the flow pattern into two components: *intrinsic flow* (meaning the flow that is a result of the contractions of that lymphangion) and *extrinsic flow* (meaning the flow that is a result of all influences from outside that single lymphangion, predominantly influences from upstream of that single lymphangion, but some influences

from downstream as well). From the experiments with imposed flow it was known that as the imposed flow was increased, the degree of inhibition of lymphatic pumping increased.[11,42,52] In the thoracic duct it was observed[42] that during periods of high imposed flow (a transaxial gradient of $5\,cmH_2O$) normalized diastolic diameter increased, resulting in a 57% reduction in resting tone (in comparison to the absence of imposed flow at the same transmural pressure level). On the other hand at this level of imposed flow, the spontaneous contractions of the thoracic duct were almost completely abolished.[42] This leads to the conclusion that *in situ*, where the extrinsic flow varies dramatically and is dependent on many factors, the lymphangions are constantly operating under a combination of intrinsic and extrinsic flows. When extrinsic flow is not enough to move lymph downstream, the maintenance of low lymphatic tone by the extrinsic flow (demonstrated in Refs. 11, 42, and 52) is supported or completely replaced by the reduction in lymphatic tone mediated by intrinsic flow during the pumping-effective phasic contractions (demonstrated in Gasheva *et al.*[53]). When extrinsic flow is high enough to propel lymph by itself, spontaneous contractions may be inhibited to save energy and only the lowering of tone by the extrinsic flow will exist.

Currently available data support the idea that contractile activity of the transporting lymphatic vessels *in situ* constantly adjusts to the local "need" to propel variable volumes of lymph by a continuous interplay between the influences of extrinsic and intrinsic flows. At low levels of inflow in the transporting lymphatics, the influences of intrinsic flow will dominate and NO release due to the phasic contractions will maintain the effective energy-saving lymphatic pumping patterns. As soon as the levels of lymph formation and, accordingly, inflow are increased in the transporting lymphatics, the influences of extrinsic flow will dominate, leading to a massive release of NO which will temporarily inhibit the intrinsic contractility of transporting lymphatics. In future studies, direct measurements of the NO concentrations in lymphatics will be able to provide more detailed information on this matter. However, current findings strongly support the idea that flow/shear–dependent self-regulatory mechanisms in the lymphangions continuously adjust lymphatic tone and phasic contractions to the physiologically variable preloads and outflow resistances.

In conclusion, it is important to mention that knowledge of basic principles of physiological adaptive reactions of lymphatic vessels to the changes in main physical factors of their environment (pressure and flow) is far from complete. It was thought for many decades that the adaptive reactions of lymphatic vessels to the increases in their volume loads are simple and could be explained by simple "bell-shaped" pressure-dependent regulation. The recent findings described above demonstrated that the contractions of lymphangions are constantly influenced by a complicated interplay of different factors of lymph dynamics: pressure/stretch, generated by active and passive lymph pumps, and subsequent combination of intrinsic and extrinsic flows/shears. Moreover, lymphangions react differently to changes in these factors, depending on species, and also on their location in the whole body and inside the regional lymphatic nets. However, because of the natural limitations and difficulties scientists face in attempting to investigate lymphatic contractility, the picture of all possible adaptive contractile events in lymphatic vessels is far from complete. To facilitate the first steps in investigating the basic regulation in lymphatic contractility, researchers artificially separated the studies on pressure and flow effects on it, while *in vivo* lymphangions exist in conditions of simultaneously changing both pressure and flow. The next step in discovering basic principles of lymph flow must be focused on a combined evaluation of interaction between pressure/stretch– and flow/shear–dependent adaptive reactions in lymphangions. A reasonable question may be posed here: Do the flow-dependent regulatory mechanisms act similarly or differently at different levels of lymphatic wall distension by intraluminal diastolic lymphatic pressure because it is more likely that the increases of extrinsic flow will be followed by the increases in the lymphangions' filling pressure? Obtaining the answers to such questions of principle concerning the functioning of lymphatic vessels will not only promote our ability to create a commonly accepted theory of the physiology of lymph flow, but also will help us to precisely target the clinical efforts that must be made toward the correction of the main "driving engine" of lymph flow altered during lymphatic diseases—partially or completely incompetent lymphatic muscle cells unable to generate sufficient flow in lymphatic nets.

Conflicts of Interest

The author declares no conflicts of interest.

References

1. MISLIN, H. 1961. Experimental detection of autochthonous automatism of lymph vessels [in German]. Experientia **17:** 29–30.
2. MISLIN, H. 1966. Structural and functional relations of the mesenteric lymph vessels. *In* New Trends in Basic Lymphology; Proceedings of a Symposium held at Charleroi (Belgium); July 11–13, 1966. Experientia **Suppl. 14:** 87–96.
3. MISLIN, H. & D. RATHENOW. 1962. Eksperimentelle Untersuchungen uber die bewegungskoordination der Lymphangione [in German]. Rev. Suisse Zool. **69:** 334–344.
4. MCHALE, N.G. & I.C. RODDIE. 1976. The effect of transmural pressure on pumping activity in isolated bovine lymphatic vessels. J. Physiol. **261:** 255–269.
5. ZAWIEJA, D.C. *et al.* 1993. Distribution, propagation, and coordination of contractile activity in lymphatics. Am. J. Physiol. **264:** H1283–h1291.
6. ARMENIO, S. *et al.* 1981. Spontaneous contractility in the human lymph vessels. Lymphology **14:** 173–178.
7. ARMENIO, S. *et al.* 1981. Spontaneous contractility of the lymphatic vessels in man [in Spanish]. Angilogia **33:** 325–327.
8. GASHEV, A.A. *et al.* 1990. The mechanisms of lymphangion interaction in the process of the lymph movement [in Russian]. Fiziol Zh SSSR Im IM **76:** 1489–1508.
9. GASHEV, A.A. 1991. The mechanism of the formation of a reverse fluid filling in the lymphangions [in Russian]. Fiziol Zh SSSR Im IM Sechenova **77:** 63–69.
10. GASHEV, A.A., R.S. ORLOV & D.C. ZAWIEJA. 2001. Contractions of the lymphangion under low filling conditions and the absense of stretching stimuli: the possibility of the sucking effect [in Russian]. Ross Fizio Zh Im IM Sechenova **87:** 97–109.
11. GASHEV, A.A. 2002. Physiologic aspects of lymphatic contractile function: current perspectives. Ann. N.Y. Acad. Sci. **979:** 178–187; discussion 188–96.
12. LEE, J. 1986. Distension pressure on subserosal and mesenteric lymph pressures of rat jejunum. Am. J. Physiol. **251:** G611–G614.
13. NEGRINI, D., S.T. BALLARD & J.N. BENOIT. 1994. Contribution of lymphatic myogenic activity and respiratory movements to pleural lymph flow. J. Appl. Physiol. **76:** 2267–2274.
14. NEGRINI, D., A. MORIONDO & S. MUKENGE. 2004. Transmural pressure during cardiogenic oscillations in rodent diaphragmatic lymphatic vessels. Lymphat. Res. Biol. **2:** 69–81.
15. MORIONDO, A., S. MUKENGE & D. NEGRINI. 2005. Transmural pressure in rat initial subpleural lymphatics during spontaneous or mechanical ventilation. Am. J. Physiol. Heart. Circ. Physiol. **289:** H263–H269.
16. OLSZEWSKI, W.L. & A. ENGESET. 1980. Intrinsic contractility of prenodal lymph vessels and lymph flow in human leg. Am. J. Physiol. **239:** H775–H783.
17. ZWEIFACH, B. & J. PRATHER. 1975. Micromanipulation of pressure in terminal lymphatics in the mesentery. Am. J. Physiol. **228:** 1326–1335.
18. HARGENS, A.R. & B.W. ZWEIFACH. 1976. Transport between blood and peripheral lymph in intestine. Microvasc. Res. **11:** 89–101.
19. HARGENS, A.R. & B.W. ZWEIFACH. 1977. Contractile stimuli in collecting lymph vessels. Am. J. Physiol. **233:** H57–H65.
20. SZABO, G. & Z. MAGYAR. 1967. Pressure measurements in various parts of the lymphatic system. Acta Med. Acad. Sci. Hung. **23:** 237–241.

21. OLSZEWSKI, W.L. & A. ENGESET. 1979. Lymphatic contractions. N. Engl. J. Med. **300**: 316.

22. OLSZEWSKI, W.L. & A. ENGESET. 1979. Intrinsic contractility of leg lymphatics in man. Preliminary communication. Lymphology **12**: 81–84.

23. KRYLOV, V.S. *et al.* 1991. Endolymphatic pressure in the evaluation of the status of peripheral lymph flow in the extremities [in Russian]. Khirurgiia **6**: 63–69.

24. ZAUGG-VESTI, B. *et al.* 1993. Lymphatic capillary pressure in patients with primary lymphedema. Microvasc. Res. **46**: 128–134.

25. FRANZECK, U.K. *et al.* 1996. Effect of postural changes on human lymphatic capillary pressure of the skin. J. Physiol. **494**: 595–600.

26. TILNEY, N.L. & J.E. MURRAY. 1968. Chronic thoracic duct fistula: operative technic and physiological effects in man. Ann. Surg. **167**: 1–8.

27. KINNAERT, P. 1973. Pressure measurements in the cervical portion of the thoracic duct in man. Br. J. Surg. **60**: 558–561.

28. FLOREY, H. 1927. Observations on the contractility of lacteals. Part I. J. Physiol. **62**: 267–272.

29. FLOREY, H. 1927. Observations on the contractility of lacteals. Part II. J. Physiol. **63**: 1–18.

30. SMITH, R. 1949. Lymphatic contractility: a possible intrinsic mechanism of lymphatic vessels for the transport of lymph. J. Exp. Med. **90**: 497–509.

31. HORSTMANN, E. 1952. Uber die funktinelle Struktur der mesenterialen Lymphgefasse [in German]. Morphol. Jahrb. **91**: 483–510.

32. HORSTMANN, E. 1959. Beobachtungen zur Motorik der Lymphgefasse [in German]. Pflugers Arch. **269**: 511–519.

33. MCHALE, N.G. & I.C. RODDIE. 1975. Pumping activity in isolated segments of bovine mesenteric lymphatics. J. Physiol. **244**: 70P–72P.

34. ORLOV, R.S. & T.A. LOBACHEVA. 1977. Intravascular pressure and spontaneous lymph vessels contractions [in Russian]. Bull. Exp. Biol. Med. **83**: 392–394.

35. OHHASHI, T., T. AZUMA & M. SAKAGUCHI. 1980. Active and passive mechanical characteristics of bovine mesenteric lymphatics. Am. J. Physiol. **239**: H88–H95.

36. HAYASHI, A. *et al.* 1987. Increased intrinsic pumping of intestinal lymphatics following hemorrhage in anesthetized sheep. Circ. Res. **60**: 265–272.

37. REDDY, N.P. & N.C. STAUB. 1981. Intrinsic propulsive activity of thoracic duct perfused in anesthetized dogs. Microvasc. Res. **21**: 183–192.

38. BENOIT, J.N. *et al.* 1989. Characterization of intact mesenteric lymphatic pump and its responsiveness to acute edemagenic stress. Am. J. Physiol. **257**: H2059–H2069.

39. GASHEV, A.A. 1989. The pump function of the lymphangion and the effect on it of different hydrostatic conditions [in Russian]. Fiziol Zh SSSR Im IM Sechenova **75**: 1737–1743.

40. EISENHOFFER, J., S. LEE & M.G. JOHNSTON. 1994. Pressure-flow relationships in isolated sheep prenodal lymphatic vessels. Am. J. Physiol. **267**: H938–H943.

41. GASHEV, A.A. *et al.* 2007. Characteristics of the active lymph pump in bovine prenodal mesenteric lymphatics. Lymph. Res. Biol. **5**: 71–79.

42. GASHEV, A.A. *et al.* 2004. Regional variations of contractile activity in isolated rat lymphatics. Microcirculation **11**: 477–492.

43. MAWHINNEY, H.J. & I.C. RODDIE. 1973. Spontaneous activity in isolated bovine mesenteric lymphatics. J. Physiol. **229**: 339–348.

44. MCHALE, N.G. & M.K. MEHARG. 1992. Co-ordination of pumping in isolated bovine lymphatic vessels. J. Physiol. **450**: 503–512.

45. CROWE, M.J. *et al.* 1997. Co-ordination of contractile activity in guinea-pig mesenteric lymphatics. J. Physiol. **500**: 235–244.

46. GASHEV, A.A. & D.C. ZAWIEJA. 1999. Lymphatic contractions: the role of distension mechanisms. FASEB J. **13**: A11.

47. MAZZONI, M.C., T.C. SKALAK & G.W. SCHMID-SCHONBEIN. 1987. Structure of lymphatic valves in the spinotrapezius muscle of the rat. Blood Vessels **24**: 304–312.

48. DIXON, J.B. *et al.* 2007. Image correlation algorithm for measuring lymphocyte velocity and diameter changes in contracting microlymphatics. Ann. Biomed. Eng. **35**: 387–396.

49. DIXON, J.B. *et al.* 2006. Lymph flow, shear stress, and lymphocyte velocity in rat mesenteric prenodal lymphatics. Microcirculation **13**: 597–610.

50. DIXON, J.B. *et al.* 2005. Measuring microlymphatic flow using fast video microscopy. J. Biomed. Opt. **10**: 064016.

51. GASHEV, A.A. 1989. Pumping function of lymphangion depending on various hydrostatic gradients [in Russian]. Doklady Akad. Nauk SSSR **308**: 1261–1264.

52. GASHEV, A.A., M.J. DAVIS & D.C. ZAWIEJA. 2002. Inhibition of the active lymph pump by flow in rat mesenteric lymphatics and thoracic duct. J. Physiol. **540**: 1023–1037.

53. GASHEVA, O.Y., D.C. ZAWIEJA & A.A. GASHEV. 2006. Contraction-initiated NO-dependent lymphatic relaxation: a self-regulatory mechanism in rat thoracic duct. J. Physiol. **575**: 821–832.

54. NUSBAUM, M. *et al.* 1964. Roentgenographic and direct visualization of thoracic duct. Arch. Surg. **88**: 127–135.

55. BROWSE, N.L., R.S. LORD & A. TAYLOR. 1971. Pressure waves and gradients in the canine thoracic duct. J. Physiol. **213**: 507–524.

56. CAMPBELL, T. & T. HEATH. 1973. Intrinsic contractility of lymphatics in sheep and in dogs. Quart. J. Exp. Physiol. **58**: 207–217.

57. ORLOV, R.S., R.P. BORISOVA & E.S. MANDRYKO. 1975. The contractile and electrical activity of the smooth muscles of the major lymph vessels [in Russian]. Fiziol Zh SSSR Im IM Sechenova **61**: 1045–1053.

58. ORLOV, R.S. & R.P. BORISOVA. 1982. Contractions of the lymphatic vessels, their regulation and functional role [in Russian]. Vestn Akad Med Nauk SSSR (7): 74–83.

59. MUTHUCHAMY, M. *et al.* 2003. Molecular and functional analyses of the contractile apparatus in lymphatic muscle. FASEB J. **17**: 920–922.

Contractility Patterns of Human Leg Lymphatics in Various Stages of Obstructive Lymphedema

WALDEMAR L. OLSZEWSKI

Department of Surgical Research and Transplantation, Medical Research Center, Polish Academy of Sciences, Warsaw, Poland

Central Clinical Hospital, MIA, Warsaw, Poland

Norwegian Radium Hospital, Oslo, Norway

In healthy human leg, lymphatics contract spontaneously, rhythmically propelling lymph. The pressures generated by lymphatic contractions constitute the main force for lymph flow. This mechanism is of utmost importance during night rest, anesthesia, and immobilization, as well as in those with damaged peripheral motor neurons. Under physiological conditions, limb muscular activity and position only slightly change lymph flow. In obstructive lymphedema, lymphatic wall muscular fibers become damaged and the spontaneous contractility becomes ineffective in lymph transport because of low generated pressures and lymphatic valve insufficiency. The lymph-propelling task is taken over by leg muscle contractions. Measuring intralymphatic pressures and flow gives some insight into the mechanism of lymph flow in healthy limbs and loss of this function in lymphedema. This knowledge will be useful in the derivation of rational treatments for lymphedema.

Key words: lymphatics; contractility; lymph; flow; pressure

Under normal conditions, the main propelling forces for lymph flow are the rhythmic contractions of lymphangions (segments of lymphatics between two unidirectional valves), which generate lymph (L) pressures high enough to move the intralymphatic fluid centripetally. All other factors, such as muscular contractions, respiratory movements, and arterial pulsations, are secondary to the spontaneous contractions of lymphatics.[1] Spontaneous lymph flow is essential for prevention of edema, both under physiological conditions and in cases of mechanical obstruction of lymphatics called lymphedema. Lymphedema is a swelling of tissue caused by excess of tissue fluid and lymph resulting from the lack of proper centripetal drainage due to damage to the lymphatic collector wall. The damage may be of different degrees, depending on the etiology. The posterysipelas changes bring about total obstruction of lymphatic trunks. Bacterial staphylococcal dermatitis causes alterations in the vessel wall partly affecting muscular fibers. The posttraumatic lymphedema is characterized by local scars, whereas after excision of lymph nodes in cancer, a gradual destruction of intima and media are seen first in the groin lymphatics and progress distally (the so-called die-back). The common denominator for all these changes is impaired spontaneous lymph flow, leading to secondary degenerative changes in the limb tissues.

The aim of this article is to present the representative L pressure and flow recordings obtained from lymphedematous lower limbs in various stages of the disease. Furthermore, the effects of manual massage and pneumatic compression on L pressure and flow were studied and documentation will be provided.

Material and Methods

Studies were carried out on a group of 21 patients, aged 25 to 60 years, with postinflammatory obstructive lymphedema of the legs in stages II, III, and IV. All patients had a soft tissue inflammatory episode in the past and edema developed 1 to 15 months thereafter. We classified the advancement of lymphedema according to the level edema reached from the foot to the groin and intensity of skin changes: stage I,

Address for correspondence: Waldemar L. Olszewski, Department of Surgical Research and Transplantology, Medical Research Center, Polish Academy of Sciences, 5 Pawinskiego Str., 02 106 Warsaw, Poland. Voice: +48-22-6685316; fax: +48-22-66855334.

wlo@cmdik.pan.pl

Ann. N.Y. Acad. Sci. 1131: 110–118 (2008). © 2008 New York Academy of Sciences.
doi: 10.1196/annals.1413.010

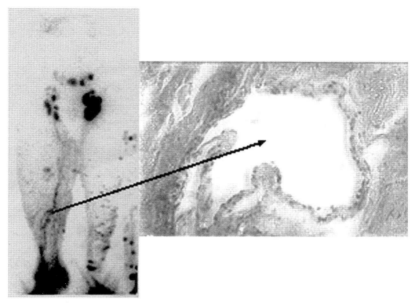

FIGURE 1. Lymphoscintigram and histologic cross-section of a lymphatic in stage II of obstructive lymphedema showing dilated lymphatics and remnants of inguinal lymph nodes. A dilated calf lymphatic specimen was obtained during cannulation for pressure recordings (stained with hematoxylin–eosin; ×200). Lymph pressure and flow pattern in this patient is shown on FIGURE 2.

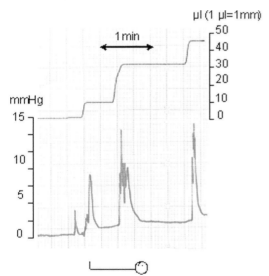

FIGURE 2. Pressure (*lower curve*) and flow (*upper curve*) pattern in calf lymphatics in stage II obstructive lymphedema of the leg with patient in a horizontal position. Irregular pulse waves of the cannulated lymphangion with various amplitudes are shown. Small peaks superimposed upon high peaks are the pulse waves produced by the interconnecting neighboring lymphangions. There is no correlation between contractions and pulse waves of individual interconnected lymphangions. Relatively high flow for this stage of lymphedema is seen.

edema limited to the foot, no skin changes; stage II, edema of the foot and lower part of the calf, pitting, no hyperkeratosis; stage III, edema reaching knee level, hard, hyperkeratosis of toes; and stage IV, edema of the whole extremity, hard, hyperkeratosis of foot and calf skin. All patients had lymphoscintigraphy performed to outline the superficial and deep lymphatics and inguinal lymph nodes.

For measuring L pressure and flow, a subcutaneous lymph vessel of the leg was cannulated, against the direction of lymph flow, according to the techniques described previously.[2,3] The cannulated lymph vessel drained lymph from the skin, subcutaneous tissue, and perimuscular fascia of the foot and anterior aspect of the lower leg. Intralymphatic pressures were recorded by an electronic micromanometer. To measure flow we used a low-flow flowmeter of own construction, with an accuracy of $0.1\,\mu L$ and range between 0.1 and $6\,\mu L/$.[4]

The pressure and flow measurements were carried out with the patient at rest and during contractions of leg muscles as well as during manual massage and pneumatic compression of foot and lower part of the calf.

The study was approved by the Institute's Ethical Committee. Patients were acquainted with the study protocol and signed consent forms.

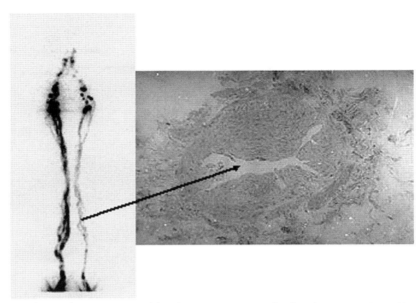

FIGURE 3. Lymphoscintigram and histologic cross-section of a lymphatic in patient with stage III obstructive lymphedema. Only a weak outline of left limb lymphatics and remnants of inguinal lymph nodes can be seen. A partly obliterated calf lymphatic specimen was obtained during cannulation for pressure recordings (stained with hematoxylin–eosin; ×200). Lymph pressure and flow pattern in this patient is shown on FIGURE 4.

Results

Lymph Pressure and Flow in Various Stages of Lymphedema

Stage II. Active contractility of cannulated lymphangion was observed in all cases, although pulsation was irregular compared with normal lymphatic rhythm (FIGS. 1 and 2).[1] There was lack of correlation between the pulse amplitude and lymph stroke volume. In two cases high systolic pressures propelled little lymph (FIGS. 3 and 4).

Stage III. Lymphatics were actively contracting, although the generated pressures were not high enough to propel lymph (FIGS. 5 and 6). Changing from horizontal to upright position brought about increase in lymph input into lymphatics, which was followed by increase in pulse amplitude and lymph flow (FIG. 6).

Stage IV. Only rudimentary pulse waves could be seen and there was no spontaneous lymph flow (FIGS. 7 and 8).

Lymph Pressure and Flow in Various Conditions

Lymphatic valve insufficiency. Insufficiency of valves brought about lymph backflow during diastole (FIG. 9). Calf muscular contractions increased pulse frequency

and amplitude, but lymph flow did not rise significantly (FIG. 9).

The effect of calf muscular contractions. In stage I/II calf muscle contractions increased pulse frequency and lymph flow (FIG. 10, left panel). In the more advanced stage there were no spontaneous pulses, but calf muscle contraction generated pressures propelling lymph (FIG. 10, left panel). In stage IV, with sclerotic changes in skin and subcutaneous tissue, muscular contractions did not have any effect of pulse wave and flow (FIG. 11).

The effect of foot manual massage. Manual massage in lymphedema stage II and III produced pressure waves propelling lymph. Discontinuation of massage was followed by an immediate stop in lymph flow (FIG. 12).

The effects of foot bandaging. Gradual increase in bandage pressure raised L pressure to a mean of 50 mmHg with a series of low-amplitude pulses. Calf muscular contraction increased pressure to even higher levels (above 100 mmHg). The most effective was the bandage pressure of 40 mmHg. Higher pressures did not have any additional effect (FIG. 13).

The effect of sequential pneumatic compression of the foot and lower part of the leg. In stage II and early stage III of lymphedema, the effect of pneumatic compression was similar to that of manual massage. In some cases

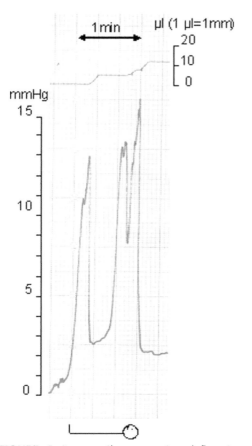

FIGURE 4. Pressure (*lower curve*) and flow (*upper curve*) in calf lymphatics in advanced stage III of obstructive lymphedema of lower limb with patient in a horizontal position. High systolic pressures but low flow are seen (compare data with those in Fig. 1). Lack of correlation between the pulse amplitude and stroke volume may be an individual trait or the result of decreased contracting and lymph-propelling capacity against proximal resistance (partial obliterative changes in lymphatic collectors).

it evoked spontaneous lymphatic pulse waves (Fig. 14). In stage IV, cannulation of lymphatics was usually unsuccessful. In some cases blind segments deprived of contractility were cannulated.

Discussion

Under normal conditions, the L pressures in the prenodal lymphatics are similar irrespective of the drained areas of the extremity. They mainly depend on the intrinsic contractility of lymphatics,[5] but not much on active movements of the muscles of the drained extremity. The first studies of spontaneous rhythmic contractility of prenodal lymphatics in a normal man, with intravascular pressure and lymph flow recordings were performed by Olszewski and Engeset.[1,5,6] To study the efficiency of intrinsic contractions of the lymphatics on lymph propulsion in human legs, the end and lateral pressures, as well as lymph flow, were measured in subcutaneous lymph vessels in the normal leg. The mean lateral systolic pressure in the lymphatics with free lymph flow was 13.5 mmHg during rest. The mean pulse amplitude was 8.8 mmHg, and the pulse frequency was 2.42 per minute. The pressure in the periods between the pulse waves ranged between zero and several mmHg.

In obstructive lymphedema, L pressure and flow in legs depend on factors affecting the anatomy of the lymphatic vessel wall. A great diversity of pressure patterns was found in the individuals studied with respect to the shape of the pulse curve, the levels of diastolic and systolic pressure, and the time to reach maximum lymph lateral pressure. These differences may depend on the size of the vessel intercommunicating branches and valve competency. Lymphatics of draining segments of skin, subcutaneous tissue, and perifascial and perivascular connective tissue act independently, determined by the local metabolism and lymph production. This being so, some lymphatics may be obstructed, some escape the damage, and some still contract rhythmically.[7]

Degenerative changes in the lymphatic wall include hyperplasia of endothelial cells, thickening of the subendothelial layer of collagen fibers, a decrease in number of muscular fibers, and eventually total occlusion of the lumen.[8] The pathologic factors that damage lymphatic endothelium and the muscular fibers are usually microorganisms transported from the site of skin penetration to the lymph node and accumulation of activated lymphocytes and chemical tissue products.

In the initial stages after mechanical or inflammatory damage to lymphatics, deceleration of lymph flow develops. Overloading of lymphatics with an excess of continuously produced lymph brings about dilatation of vessels with subsequent insufficiency of unidirectional valves. The stretched lymphatics contract continuously; however, they do not generate effective pressures sufficient to propel lymph. A "to and fro" movement of lymph can be observed. In late stages muscular fibers undergo atrophy and no more spontaneous pressure waves are generated.

In healthy limbs with normal lymphatics, active movements of the foot and leg do not significantly increase the lymph mean systolic pressures and flow.

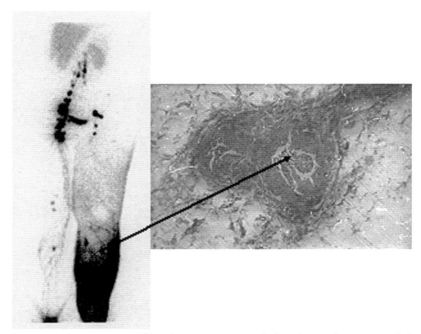

FIGURE 5. Lymphoscintigram and histologic cross-section of a lymphatic in late stage III of obstructive lymphedema. There is no outline of left limb lymphatics, but dermal backflow of tracer in the calf and remnants of inguinal lymph nodes can been seen. An almost totally obliterated calf lymphatic vessel was obtained during cannulation for pressure recordings (stained with hematoxylin–eosin; ×200). Lymph pressure and flow pattern in this patient are shown in FIGURE 6.

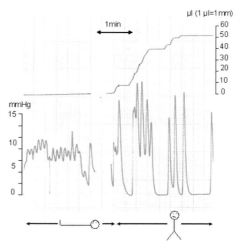

FIGURE 6. Pressure (*lower curve*) and flow (*upper curve*) in calf lymphatics in obstructive lymphedema of lower limb stage III in a patient changing from a horizontal to an upright position. A totally chaotic pulse wave pattern is evident. With the patient in a horizontal position, high pulse frequency, intralymphatic mean pressure around 7 mmHg, but no lymph flow (*upper curve*) are seen. After the patient changed to a standing position there was a sudden increase in pulse amplitude with systolic pressure reaching 18 mmHg. Lymphatic contractions generate pressures propelling lymph (*upper curve*). Stretching of the lymphatic wall by excess of lymph after a standing position stimulated contractions. These pressure recordings illustrate how high is the contracting capacity of human leg lymphatics in overcoming proximal flow resistance (fibrotic changes of proximal collecting lymphatics).

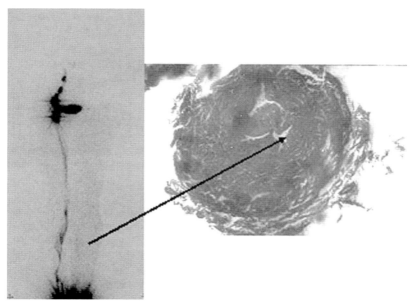

FIGURE 7. Lymphoscintigram and histologic cross-section of a lymphatic in stage IV of obstructive lymphedema. There is a lack of outline of left limb lymphatics, and inguinal lymph nodes are shown. A totally obliterated calf lymphatic specimen was obtained during cannulation for pressure recordings (stained with hematoxylin–eosin; ×200). Lymph pressure and flow pattern in this patient are shown on FIGURE 8.

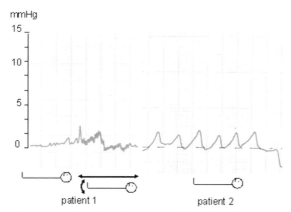

FIGURE 8. Pressure (*lower curve*) and flow (*upper curve*) in calf lymphatics in obstructive lymphedema of lower limb in advanced stage IV with patient in a horizontal position. Contractions of a lymphatic collector wall do not generate pressures high enough to propel lymph, nor do leg muscular contractions create effective intralymphatic pressures for flow (patient 1). Pressure waves of 2 mmHg are too low to be effective (*dashed horizontal line*) (patient 2).

However, an increase in pulse frequency may be observed. The pressure waves produced by the muscular contractions of the limb have an amplitude of 1–3 mmHg and are too weak to produce any lymph flow. To the contrary, in lymphedema, leg muscle contractions either stimulate lymphatic contractions, subsequently increasing lymph flow, or generate intralymphatic pressures, propelling lymph.

Under normal conditions change from horizontal to upright position does not affect L pressure. This could be explained by lack of lymph in lymphatics, and thus no hydrostatic pressure. In lymphedema, lymphatics are filled up with excess of lymph, and although the hydrostatic pressure in a lying position is low, it increases immediately after an upright position is resumed.

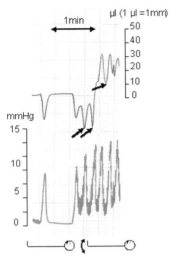

FIGURE 9. Pressure (*lower curve*) and flow (*upper curve*) in calf lymphatics in obstructive lymphedema of lower limb stage III in a patient with lymphatic valve insufficiency; patient in a horizontal position. In the *left panel*, at rest lymphatic systolic pressure propels lymph, but during the diastolic phase, lymph flows in retrograde fashion (*arrow*). In the *right panel*, calf muscular contractions increase pulse frequency and systolic pressure. In addition small pressure waves are produced by muscle contractions superimposed upon spontaneous waves. Flow rises, but during diastole there is a backflow (*arrows*). Forward-flow during pulse wave is followed by backflow in the diastolic phase (*upper curve*).

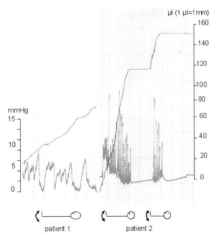

FIGURE 10. The influence of calf muscle contractions on lymph pressure and flow. Intralymphatic pressure and flow recordings during contractions of calf muscles are shown for two different patients with obstructive lymphedema in early stage I/II. In patient 1, fast movements generated low pressures waves superimposed upon high spontaneous waves. There was continuous lymph flow. In patient 2, there was a lack of spontaneous lymphatic pulse and flow at rest, but fast leg muscle movements propel lymph.

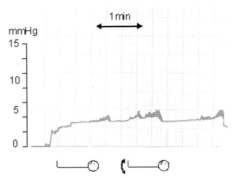

FIGURE 11. Obstructive lymphedema of lower limb stage III/IV, with patient in a horizontal position. There was a lack of spontaneous pulses, low mean pressure, and lack of pressure waves produced by leg muscle movements and no lymph flow.

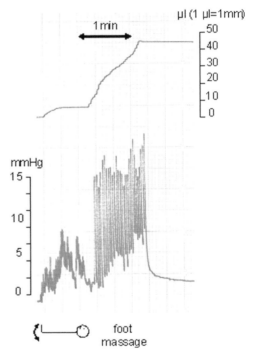

FIGURE 12. The effects of massage on lymph pressure and flow. Intralymphatic pressure and flow recordings in a patient with obstructive lymphedema stage I/II undergoing foot massage. Rhythmic leg movements produce minor waves superimposed upon three spontaneous peaks and there is a low lymph flow, whereas manual foot massage brings about significant increase in flow.

Massaging the top of the lymphedematous foot performed at a rate of 30 compressions per minute increased pulse frequency to almost the same level as did active movements of the foot. Flow was observed

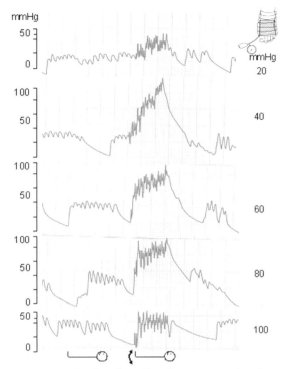

FIGURE 13. The effect of bandaging on intralymphatic pressures in a patient with obstructive lymphedema of lower limb stage II/III. A sphygmomanometer cuff has been placed around the foot and wrapped by an elastic bandage. The intra-cuff pressures were gradually increased and end pressures were simultaneously measured in a cannulated calf lymphatic. Raising cuff pressures caused increase of mean lymphatic pressure to 40–50 mmHg. There was no further increase in intralymphatic pressures even at external pressure of 80–100 mmHg. Contractions of calf muscles increased spontaneous pressures to 110 mmHg at cuff pressure of 40 mmHg. Higher cuff pressures during calf muscle contractions did not have any additional effect on lymphatic pressure. The lymphangion's contracting force reached peak values.

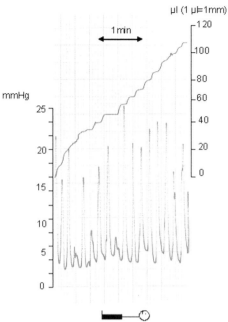

FIGURE 14. The effect of pump compression of the foot on calf lymph pressures and flow in obstructive lymphedema stage I/II. Rhythmic sleeve inflations to 60 mmHg evoked high-frequency spontaneous contractions of the cannulated calf lymphatic, producing systolic pressures of 50 mmHg. Foot tissue fluid and lymph were squeezed and subsequently moved proximally through the patent lymphatics. Filling of lymphangions caused their contractions. Each contraction propelled lymph. This pressure and flow pattern can be seen only in early stages of lymphedema with patent lymphatics. In advanced stages there is no flow through lymphatics and recordings are as shown in FIGURE 12.

as long as massage was continued. There was no flow after its discontinuation.

Bandaging of the lymphedematous foot raised calf intralymphatic pressures and evoked spontaneous vessel contractions. The optimum bandage pressure was around 40 mmHg. Higher pressures did not add to L pressures. This was presumably due to emptying tissue space and distal lymphatics of lymph.

Sequential intermittent pneumatic compression brought about the same effect as massaging. However, this effect was short-lasting, most likely due to faster evacuation of tissue fluid from peripheral tissues compared with manual massage.

Differences among lymph parameters in normal and lymphedematous legs are presented in TABLE 1.[9]

Taken together, lymphatic wall destruction by pathologic factors in lower limbs results in loss of contractile properties and subsequently stagnation of tissue fluid and lymph flow. Depending on the site of the pathologic process, lymphatics can be damaged in a circumscribed area or all superficial and deep trunks become unable to carry lymph. This being so, a variety of lymphatic pulse and flow patterns can be seen. Importantly, limb contracting muscles take over the lymph propelling function which is not seen under normal conditions. Bandaging of the limb with pressures around 40 mmHg and muscle contractions effectively propel lymph, as does manual massage and pneumatic compression. This can be seen as long as lymphatics are still partially patent. In stage IV of lymphedema there is no more lymphatic transport. In this case

TABLE 1. Differences in lymph pressure and flow in health and obstructive lymphedema

	Normal	Lymphedema
Spontaneous contractions	Present	Irregular, superimposed pressure waves from various lymphangions; low amplitude
Hydrostatic pressure component	Absent	Present
Lymphatic pulse	Approximately 6/min at rest, up to 18/min during walking	Irregular, low mean pressure, ineffective for propelling lymph; in advanced stages no pulse, no flow
Ability of leg muscle contractions to propel lymph	Absent	Present
Ability of elastic support to improve flow	Absent	Present
Ability of massage to improve flow	Absent	Present
Ability of pneumatic compression to improve flow	Absent	Present

measuring tissue fluid pressure remains as the only available method for rational setting of applied pressures by elastic support, massage, and pneumatic compression.

Conflicts of Interest

The author declares no conflicts of interest.

References

1. OLSZEWSKI, W.L. & A. ENGESET. 1980. Intrinsic contractility of prenodal lymph vessels and lymph flow in human leg. Am. J. Physiol. **239:** H775–H783.
2. ENGESET, A., B. HARGER, A. NESHEIM & A. KOLBESTVEDT. 1973. Studies on human peripheral lymph. I. Sampling method. Lymphology **6:** 1–5.
3. OLSZEWSKI, W.L. 1977. Collection and physiological measurements of lymph and interstitial fluid in man. Lymphology **10:** 137–145.
4. SORENSEN, O., A. ENGESET, W.L. OLSZEWSKI & T. LINDMO. 1982. High-sensitivity optical flowmeter. Lymphology **15:** 29–31.
5. OLSZEWSKI, W.L. & A. ENGESET. 1979. Lymphatic contractility. N. Engl. J. Med. **300:** 316.
6. OLSZEWSKI, W.L. & A. ENGESET. 1979. Intrinsic contractility of leg lymphatics in man: preliminary communication. Lymphology **12:** 81–84.
7. VENUGOPAL, A.M., R.H. STEWART, G.A. LAINE, *et al.* 2007. Lymphangion coordination minimally affects mean flow in lymphatic vessels. Am. J. Physiol. Heart Circ. Physiol. **293:** H1183–H1189.
8. OLSZEWSKI, W.L. 1977. Pathophysiological and clinical observations of obstructive lymphedema of limbs. *In* Lymphedema. L. Clodius, Ed. Thieme Verlag. Stuttgart.
9. OLSZEWSKI, W.L. 2002. Contractility patterns of normal and pathologically changed human lymphatics. Ann. N.Y. Acad. Sci. **979:** 52–63.

Cell Traffic and the Lymphatic Endothelium

LOUISE A. JOHNSON AND DAVID G. JACKSON

Weatherall Institute of Molecular Medicine, MRC Human Immunology Unit, John Radcliffe Hospital, Oxford OX3 9DS, United Kingdom

The principal immune function of the afferent lymphatics is to bear antigen and leukocytes from peripheral tissues to the draining lymph nodes. Recent research has shown that passage of leukocytes into the afferent lymphatic capillaries is far from an indolent process; rather it is carefully orchestrated by an array of adhesion molecules, as well as by chemokines and their receptors. Here we review the current knowledge of leukocyte trans-lymphatic endothelial migration and its role in the development of an immune response.

Key words: ICAM-1; VCAM-1; dendritic cell; CCL21; CCR7; transmigration; inflammation; LYVE-1

Introduction

The early history of the research into the lymphatics is contemporary with the founding of modern medicine. The ancient Greeks first recorded the observation of vessels containing colorless fluid: Hippocrates spoke circa 400 BC of "white blood." However, the important immunologic functions performed by the lymphatics were not recognized until the 20th century, with the characterization of lymph nodes and understanding of antigen presentation. The pioneering works of Gowans and later Morris are credited with paving the way for our modern understanding (reviewed by Young[1]). The presence of cells (lymphocytes) within the lymphatics had been noted, although it was initially believed that these were newly formed cells. Gowans demonstrated that the cannulation of the thoracic duct in rats resulted in a progressive loss of lymphocytes in the thoracic duct lymph, but that this loss could be reversed by returning the collected cells to the animal by intravenous injection. Furthermore, by labeling these cells with radioactive tracking agents and later recovering them from the thoracic duct, he also showed that lymph-borne cells recirculate through both the blood and efferent lymphatic systems.

Clearly the efferent lymphatics traced in these seminal studies of lymphocyte recirculation are only part of the larger lymphatic network, which consists of both afferent and efferent vessels. It is the afferent lymphatics, which begin as blind-ended capillaries and possess the characteristic loose overlapping endothelial junctions, that control admission of cellular traffic, while the efferent lymphatics act largely as conduits. This functional dichotomy is reflected in the different cellular composition of afferent and efferent lymph as determined from earlier analyses in domestic animals.[1] Whereas efferent lymph is composed almost exclusively of lymphocytes, indicating the predominant contribution from recirculating blood lymphocytes, afferent lymph comprises a mixture of 85–90% lymphocytes, and 10–15% monocytes and dendritic cells (DCs), reflecting the leukocyte populations that predominate in the tissues. This chapter will therefore focus on the afferent lymphatics and on current knowledge of the processes regulating leukocyte transmigration during the development of the immune response.

Lymphatic Trafficking in the Context of the Immune Response

The acquired immune response is initiated within the lymph nodes, wherein a unique environment has evolved to optimize encounters between naïve T and B cells with foreign antigen, leading to lymphocyte activation and clonal expansion. Naïve lymphocytes migrate from the blood vasculature through specialized lymph node postcapillary venules lined by distinctive high endothelial cells (high endothelial venules [HEVs]) to enter the paracortical regions of these organs. Unlike activated effector or memory T cells, naïve T cells are largely excluded from entering peripheral tissue and thus rely upon antigen from those tissues to be delivered to them in the lymph node, a service provided by DCs, which migrate from the

Address for correspondence: Prof. David G. Jackson, MRC Human Immunology Unit, Weatherall Institute of Molecular Medicine, John Radcliffe Hospital, Headington, OXFORD OX3 9DS, UK. Voice: +44(0)1865-222313.

djackson@hammer.imm.ox.ac.uk

Ann. N.Y. Acad. Sci. 1131: 119–133 (2008). © 2008 New York Academy of Sciences.
doi: 10.1196/annals.1413.011

periphery via the afferent lymphatics.[2] Although other cell types, such as macrophages or activated parenchymal cells, may serve as antigen-presenting cells and activate memory T cells, DCs are the only class of antigen-presenting cells that can activate naïve T cells, stimulating clonal expansion and initiation of the immune response.[3,4] Immature DCs reside mainly within epithelial and connective tissue, poised to acquire antigens that might infect peripheral tissue. As stated above, most DCs do not circulate and hence cannot enter the lymph nodes directly from the blood; rather they are inherent in the tissues, from which they traffic to lymph nodes via the afferent lymphatics.

It has been assumed that passage of leukocytes into the afferent lymphatics is an indolent process. However, transmission electron microscopy studies have shown that DCs must undergo a dramatic change in shape in order to pass through the initial interendothelial junctions.[5,6] Moreover, the discovery that entry and migration of DCs to the lymphatics and trafficking to the lymph nodes is dependent upon chemokines has provided compelling evidence that these processes are highly regulated. In the next sections we will discuss lymphatic endothelial transmigration by reviewing first what is known about the role of chemokines in the process and then move on to the translymphatic migration steps by including a summary of candidate endothelial–leukocyte and endothelial junction adhesion molecules from the blood vasculature. We conclude by discussing a hypothetical model of lymphatic transmigration in the light of recent revelations about lymphatic endothelial ultrastructure from confocal microscopy.

Lymphatic Trafficking of Dendritic Cells and Its Regulation by Chemokines

Under normal physiological conditions, immature and semi-mature DCs migrate constitutively from tissues to lymph nodes via afferent lymphatics, permitting both normal immune surveillance and the induction of peripheral tolerance.[7,8] Subsequent to infection or inflammation, the numbers of migrating cells increase dramatically, as the result of inflammation-induced maturation events that promote increased DC motility and responsiveness to lymph tropic chemoattractants and convert these cells from phagocytes to professional antigen-presenting cells (reviewed by Steinman *et al.*[9]).

The number of factors that induce maturation of DCs is large. However, the best known examples include microbial cell wall products, such as bacterial lipopolysaccharides (LPSs), which bind and activate

toll-like pattern recognition receptors (TLRs), and classical inflammatory cytokines, such as IL-1 and TNFα, as well as engagement with T cells via CD40L. Maturation results in several phenotypic changes that enhance uptake of antigen through formation of MHC–peptide complexes and facilitate activation of naïve T cells through DC expression of co-stimulatory molecules, such as CD83 and CD86. Furthermore, mature DCs downregulate expression of chemokine receptors CCR1, CCR2, CCR5, and CXCR1 required for pro-inflammatory chemotaxis in the tissues and upregulate chemokine receptors required for efficient migration from the skin to the lymph node. The key receptors and their lymph tropic ligands are outlined below.

CCL21 and CCR7

The importance of CCR7 and by implication its major ligands CCL21 and CCL19 in attracting mature skin DCs to the afferent lymphatics and migration to the draining lymph nodes was underlined by seminal studies with CCR7-deficient mice.[10] These mice exhibited delayed immune responses to foreign antigen and profound morphologic alterations in all secondary lymphoid organs. More particularly, they lacked contact sensitivity and delayed hypersensitivity responses, failing to mobilize epidermal DCs to the draining lymph nodes as assessed by FITC skin painting. Compelling evidence to show that CCR7 regulates the translymphatic migration step was subsequently provided by Ohl *et al.*, who demonstrated that the absence of this chemokine receptor in CCR7$^{-/-}$ mice caused retention of DCs within the dermis and their failure to accumulate in dermal lymphatics, during both constitutive and inflammatory trafficking.[11]

Additional studies have used the *plt* mouse (paucity of lymph node T cells), which carries a naturally occurring compound mutation simultaneously deleting the tandemly linked genes for CCL21[12] and CCL19—the two known ligands for CCR7.[13] As in the CCR7$^{-/-}$ mice, both naïve T cells and DCs failed to accumulate in lymph nodes of *plt* mice, implicating either or both chemokines in lymphatic transmigration or subsequent lymphatic trafficking. Interestingly, there are two genes for CCL21 in the mouse genome, the products of which differ only by a single amino acid: CCL21Leu is thought to be expressed on peripheral lymphatic endothelia, whereas CCL21Ser is found in lymph node.[14] Although *plt* mice retain expression of CCL21Leu in the periphery, this does not appear to be sufficient for DCs to mobilize to the lymph node, at least after contact sensitization with FITC. The involvement of lymphatic endothelial cells (LECs) in generating CCL21 for transmigration was implied from studies that visualized MHC class II[+] antigen-presenting cells entering CCL21[+] lymphatic vessels and demonstrated anti-CCL21 antibodies could block *in vivo* migration of ^{51}Cr-labeled skin-derived DCs to the draining lymph nodes.[15] Direct visualization of CCL21 production in lymphatic endothelium by a combination of immunohistochemical studies and *in situ* RNA hybridization was also reported by Martín-Fontecha *et al.*, as was the apparent increase in CCL21 in response to the proinflammatory cytokines TNFα or IL-1α, which promote DC transmigration.[16] More recent studies with *in vitro* cultured primary dermal LECs have confirmed the production of CCL21, but not CCL19, by these cells, albeit at low levels.[17,18] However, in contrast to the findings of Martín-Fontecha *et al.*,[16] secretion was reduced rather than increased after stimulation with TNFα or IL-1. Further studies will be required to determine whether this is a genuine phenomenon or an artifact of *ex vivo* culture conditions.

CCR7 and its ligands are also important in regulating lymphatic trafficking of T cells. Previously thought to function solely on naïve and central memory T cells that recirculate directly between blood and secondary lymphoid organs (via HEV) and to stimulate their motility therein,[19] CCR7 recently has been found on a subset of T cells within peripheral tissues that resemble central memory cells but which can further exit the tissues through the afferent lymphatics.[20,21] Thus CCR7 appears to control access of a subset of memory T cells to lymph nodes via both the blood and lymphatic routes and lymphatic endothelial-derived CCL21 could regulate the immune response. Very recent research has also implicated sphingosine 1-phosphate receptor-1 in regulating T cell entry into afferent lymphatics, suggesting that increased levels of the lysophospholipid sphingosine 1-phosphate present in inflamed tissue may induce T cell retention and suppress T cell exit via the lymphatics.[22] Thus CCR7 and CCL21 promote constitutive T cell entry into afferent lymph, while sphingosine 1-phosphate acts as an antagonist during an immunologic challenge, to hold T cells in the periphery until inflammation resolves.

CXCL12 and CXCR4

Deletion of neither CCR7 nor its ligands CCL19 and CCL21 (in the *plt* mouse) fully impairs DC migration to lymph nodes, implying that the process is complex and that additional chemokine receptor pairs are likely to be involved. One such candidate is the chemokine CXCL12, recently shown to be present in murine dermal lymphatics by fluorescent immunohistochemistry, and its receptor CXCR4, shown to be expressed in cutaneous MHC class II[+] DCs. Treatment with a pharmacologic CXCR4 antagonist 4-F-benzoyl-TN14003 was found to impair lymph node migration of both dermal DCs and epidermal (Langerhans cell) DCs by up to 50% during the sensitization phase of contact hypersensitivity in mice.[23] The comparative contributions of CCR7/CCL21 and CXCR4/CXCL12 to lymphatic migration in different contexts have yet to be addressed. However, Langerhans cells displayed a preference for chemotaxis towards CXCL12 rather than CCL21. In addition, the effects of the two chemokines were not additive. Hence these chemokine–receptor pairs may function independently of one another.

Other Chemokines and Receptors

Besides CCL21 and CXCL12, a number of additional chemokines can be considered as candidates in controlling lymphatic transmigration of DCs or other leukocyte populations, on the basis of their transcriptional upregulation in primary human dermal LECs in response to TNFα[24] or their expression in adjuvant-induced murine lymphangiomas.[25] These include monocyte, lymphocyte, and neutrophil chemokines, such as CX$_3$CL1, CCL2, CCL5, CCL20, CXCL2, CXCL5, CXCL9, CXCL10, CXCL13, and XCL1. Although confirmation of the chemotactic activity of these agents must await further experimentation, it is likely that the lymphatic endothelium will be found to play a much more active role in coordinating leukocyte transmigration than previously envisaged.

Finally, the unusual orphan chemokine receptor D6, which binds inflammatory CC (β) chemokines, is itself expressed by lymphatic endothelium in various different tissues.[26] However, unlike the receptors described above, D6 does not promote signaling via G-proteins or chemotaxis, but instead acts as an endocytic receptor for chemokine clearance during the resolution of inflammation. In D6-deficient mice, clearance of inflammatory chemokines is impaired and the animals develop a skin condition resembling psoriasis.[27]

Chemokines and Remote Control of the Lymph Node by Peripheral Lymphatics

Growing evidence suggests that peripheral tissues may exert "remote control" over the downstream draining lymph node by chemokine secretion and cell trafficking through the afferent lymphatics. Palframan *et al.* followed the path of CCL2 secreted from unspecified cells within the dermis to the luminal surface of HEVs by reticular network fibers, triggering

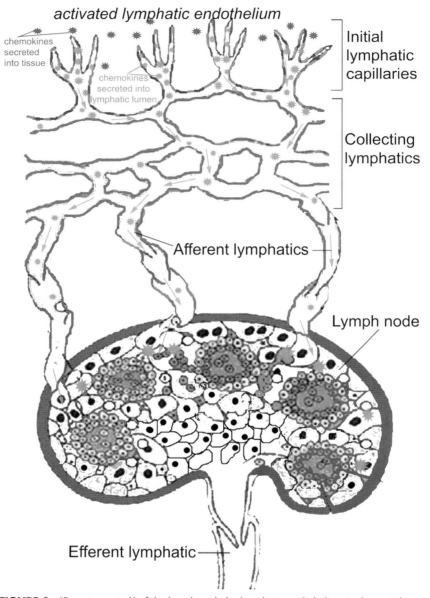

FIGURE 1. "Remote control" of the lymph node by lymphatic endothelium in the periphery.

integrin-dependent arrest of rolling monocytes from the blood circulation and thus enhanced recruitment to the lymph node.[28] As primary dermal LECs secrete CCL2 almost entirely from the luminal surface alone[17] (L.A. Johnson and D.G. Jackson, unpublished observations), it is highly likely that lymphatic endothelium in the periphery is at least partly responsible for this increase in monocyte recruitment downstream in the lymph node (Fig. 1). Similarly, CCL3, CXCL8, and CXCL9, chemotactic for macrophages, neutrophils and monocytes, respectively, are also transported from

their sites of production within inflamed tissue via the lymphatics and then captured on HEVs, subcapsular sinuses, and reticular fibers to contribute to the chemotaxis of a range of leukocytes in the lymph node.[29]

Entry of Innate Immune Cells to the Afferent Lymphatics

In addition to lymphocytes (T, B, and NK cells), monocytes, macrophages and DCs, and neutrophils are also present in afferent lymph, where they constitute 0–5% of cells.[30] Neutrophils play a crucial role in innate immune defense and are rapidly recruited from the blood to inflammatory sites in the periphery to fight microbial infection. Within the tissues they phagocytose pathogens, which they kill by oxidative burst, before they also die and in turn are phagocytosed by resident macrophages. It has thus been assumed that neutrophils live and die in the tissues and do not enter lymph. However, recent studies by Abadie *et al.*[31] and Maletto *et al.*[32] have shown that neutrophils are indeed present in the lymph and can carry antigen from the periphery to the draining lymph node.

In the first of these studies, Abadie and co-workers investigated the early innate immune response to a *Mycobacterium bovis* bacillus Calmette–Guérin (BCG) vaccine administered intradermally in mice.[31] Using a green fluorescent recombinant BCG, they found that, surprisingly, DCs were minimally associated with BCG, both at the vaccination site and in the draining lymph node. Instead, neutrophils were the main early BCG host and, once infected in the periphery, could be seen inside dermal lymphatic capillaries and accumulating in the draining lymph node. Thus neutrophils can phagocytose BCG, but the oxidative burst, essential for killing, is inhibited. This suggests that neutrophils may play an important role in mycobacterial antigen-presentation. However, as they are not efficient at degrading BCG, they could protect live bacilli and impair or delay the induction of the adaptive immune response. Alternatively, they may serve to polarize the T helper response towards a T_H2 profile to limit a potentially harmful pro-inflammatory cytokine storm. It is difficult to say whether the neutrophils are acting in the host's best interest or are simply hijacked by the pathogen.

Similarly, Maletto *et al.* found that after subcutaneous exposure of mice to the antigen ovalbumin (OVA) in Freund's complete adjuvant, neutrophils constituted the majority of antigen-bearing leukocytes in the draining lymph node of mice.[32] After recruitment to the lymph node, neutrophils secreted T_H1 cytokines, principally TNFα, rather than T_H2 cytokines, such as IL-5, which were secreted only when neutrophils were depleted. Hence, neutrophils not only can enter the lymphatics and transport antigen, but they also can provide help in sustaining the immune response and polarize the T cell response.

The molecular mechanisms by which neutrophils gain access to the afferent lymphatic capillaries remain to be elucidated, although LECs have been shown to express neutrophil chemokines such as CXCL2 and CXCL5.[24]

Lymphatic Adhesion/Transmigration and the Role of Leukocyte–Endothelial Cell Adhesion Molecules (CAMs)

Current knowledge of the mechanisms for leukocyte trans-lymphatic migration is extremely sparse. However, our own recent work on DC transmigration has already provided evidence that key events in the process are mediated by the same adhesion molecules that regulate blood vascular transmigration. Our initial studies with cultured human and murine dermal LECs prepared by LYVE-1 antibody immunomagnetic bead selection showed that resting cells express low levels of both ICAM-1 and ICAM-2, but not VCAM-1.[24] However, after stimulation with TNFα, both ICAM-1 and VCAM-1 were both massively upregulated (40- to 80-fold), similar to published findings for blood vascular endothelium. To investigate the involvement of these molecules in DC adhesion/transmigration, we performed conventional static binding assays as well as reverse transmigration assays using 5-chloromethylfluorescein diacetate (CMFDA)-labeled, LPS-activated DCs generated from monocyte-derived macrophages. For the latter transmigration assays, LECs were plated as monolayers on the underside of 3-μm pore membranes and migration was measured in the basolateral to luminal direction, mirroring the situation *in vivo*. We found that monoclonal antibodies, which blocked ICAM-1 and VCAM-1 binding to their cognate integrin ligands LFA-1 (αLβ2) and VLA-4 (α4β1), respectively, inhibited adhesion of DCs to TNFα-activated dermal LECs, as well as blocking transmigration across LEC monolayers. As would be expected, DC transmigration was enhanced dramatically by TNFα treatment of the LEC monolayer, and only the TNFα-stimulated component was inhibited by CAM-blocking antibodies. Thus we concluded that ICAM-1 and VCAM-1 mediate leukocyte adhesion steps that are prerequisite for activation-induced transmigration, similar to the adhesion molecule–mediated firm adhesion that precedes leukocyte diapedesis across blood vascular endothelium.[33–40] An interesting feature of these assays was the relatively slow rate at which the DCs transmigrated by comparison with blood vessel endothelium (requiring hours rather than minutes) and the delay (1–3 h) between initial adhesion and the onset of transmigration. Furthermore, the early stages of adhesion were insensitive to blockade by ICAM-1 and VCAM-1 antibodies. Possible explanations for these phenomena include pre-binding to other receptors, such as ICAM-2, successive rounds of transmigration events by DCs or a lag period due to a requirement for chemokine-induced activation of DC integrins.[24,41,42] Consistent with this latter possibility, we have found that pretreatment of DCs with the chemokine CCL21 greatly enhances their rate of transmigration in the *in vitro* assay and that the effect is ablated fully by pertussis toxin, an inhibitor of chemokine receptor coupled Gi proteins (L.A. Johnson and D.G. Jackson, unpublished observations).

In addition to the *in vitro* data outlined above, we also confirmed the involvement of ICAM-1 and VCAM-1 in DC translymphatic migration *in vivo*. First, in an oxazolone-induced skin hypersensitivity model that approximates a delayed-type hypersensitivity (DTH) response, both ICAM-1 and VCAM-1 expression was clearly detectable on small lymphatics within the inflamed skin areas, and administration of CAM-blocking antibodies impaired migration of resident skin Langerhans cells as well as exogenously added bone marrow–derived DCs to the draining lymph nodes. Moreover, the blocked DCs could be seen to accumulate around the antibody-blocked lymphatics, many of which resembled classical blind-ended initial capillaries. Second, using F5 T cell receptor transgenic mice vaccinated intradermally with an influenza virus nucleoprotein epitope construct (MVA.HIVA.NP), we showed that similar ICAM-1/VCAM-1 antibody blockade of DC entry to skin lymphatics was sufficient to impair development of the primary T cell response (D. Teoh *et al.*, manuscript in preparation). Thus ICAM-1 and VCAM-1 are involved in a mechanism of lymphatic entry that is a key control point for regulating DC migration to lymph nodes in inflammation and immunity (FIG. 2) and hence ICAM-1 deficient mice exhibit a defect in lymph node recruitment of DCs.[43,44] Clearly the control exerted by the lymphatic endothelium represents a potential target for immunosuppressive therapy.

CAM-Mediated Transmigration: Paracellular or Transcellular?

The studies outlined above demonstrate participation of ICAM-1 and VCAM-1 in leukocyte translymphatic migration. Although adhesion to these molecules is clearly not the whole explanation, we can envisage that it plays a pivotal role in determining the next steps in translocation from the basolateral surface to the vessel lumen. In transmigration of blood vessel endothelium, for example, binding via ICAM-1 permits lateral locomotion of adherent leukocytes toward intercellular junctions,[45] where subsequent interactions with the homophilic adhesion molecules CD31 (platelet-endothelial cell adhesion molecule-1

[PECAM-1]) and CD99 as well as tight junction molecules, such as junctional adhesion molecule-1 (JAM-1) and the adherens junction molecule VE-cadherin, promote diapedesis.[46–52] Moreover, engagement of ICAM-1 and VCAM-1 can also direct leukocytes to migrate by a second, transcellular pathway—directly through the endothelial cell body. The first description of this unusual process came from ultrastructural studies of lymphocyte extravasation in lymph nodes capillaries, carried out in the 1960s.[53] These findings were later corroborated by serial electron microscopic sectioning of neutrophil transmigration across inflamed venules.[54] More recent studies have shed light not only on the nature of transcellular migration, but also on the ways in which ICAM-1 and VCAM-1 mediate migration by both transcellular and paracellular pathways. Most notably, during leukocyte transmigration, ICAM-1 and VCAM-1 have been observed to cluster at the apical surface of endothelial cells with the actin-binding proteins, moesin and ezrin, at the tips of microvilli and microspikes, forming arrangements initially termed "docking structures."[55,56] These structures, which have subsequently been reinterpreted as "migratory cups," appear to surround leukocytes and induce redistribution of integrins in the direction of transmigration, thereby providing directional guidance for passage thorough the endothelium.[57,58] We have made similar observations in our own studies of transmigration across authentic LYVE-1-immunobead-selected human dermal LEC, in which we observed transmigrating DCs to be surrounded by a ring-like arrangement of ICAM-1, although in the case of LECs, both ICAM-1 and VCAM-1 appeared to be present on both surfaces, whereas they are more polarized in blood endothelial cells (BECs).

Precisely how transmigrating leukocytes make the choice between transcellular or paracellular routes is not quite clear. Presumably the leukocyte searches out the site most permissive for penetration according to its own activation program and that of the target vessel. Although the paracellular route is the more common in the blood vasculature, the transcellular route may still be quite significant, with between 11 and 30% of transmigrating leukocytes (lymphocytes, monocytes, and neutrophils) demonstrating transcellular migration in one reported *in vitro* study.[59] Transcellular migration appears to involve apical-basolateral transit of the leukocyte through transcellular pores formed from either caveolae or vesicles of the vesiculovacuolar organelle system in concert with actin, α-actinin, and the intermediate filament protein vimentin.[55–60] In experiments with HDMVECs (human dermal microvascular endothelial cells) and HLMVECs (human lung

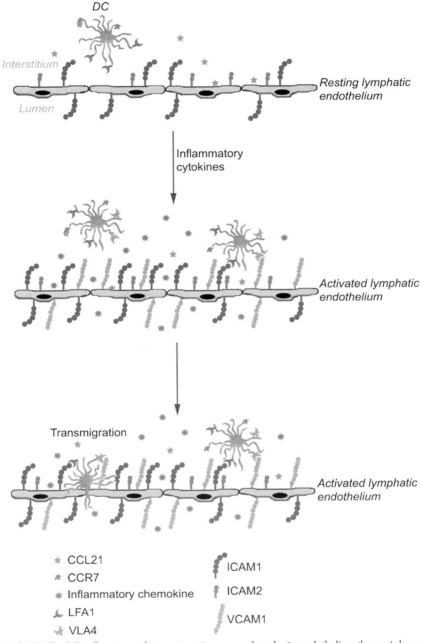

FIGURE 2. DC adhesion and transmigration across lymphatic endothelium the periphery.

microvascular endothelial cells), finger-like projections of the transmigrating lymphocyte surface membrane termed "invadosomes," enriched for LFA-1, were seen to extend into complementary "podoprints" enriched for ICAM-1 in EC early during transcellular pore formation and transcellular migration.[59]

It is not yet known whether leukocytes utilize the transcellular pathway for entry or exit of lymphatic vessels *in vivo*. However, in the studies described above, there is little doubt that the commercial blood vascular endothelial preparations analyzed were in fact mixed blood and lymphatic vascular preparations. Indeed this was explicitly demonstrated for the HDMVECs, in one case by means of LYVE-1 immunostaining, which revealed that T lymphoblasts transmigrated with similar efficiency through both lymphatic and blood vascular cell types.[56] The extent to which transcellular migration across lymphatics occurs may well be determined by the nature of the vascular bed and the identity of the transmigrating cells.

E-selectin

E-selectin (also known as CD62E or ELAM-1 [endothelial leukocyte adhesion molecule-1]) is exclusively expressed on blood and lymphatic endothelia, following stimulation with pro-inflammatory cytokines, although the majority of studies to date have focused on its role in the blood vasculature. This 100-kDa glycoprotein is the first adhesion molecule to be upregulated on blood vascular endothelium, typically 1 to 2 h after the onset of antigen challenge,[61] and can bind complex sialylated carbohydrate groups related to the Lewis-X or Lewis-A family found on various surface proteins of granulocytes, monocytes, and certain memory T cells (reviewed in von Andrian and Mackey[62]). E-selectin is also responsible for the initial attachment of neutrophils[63] and eosinophils[64] to inflamed blood endothelia. Blood endothelial-expressed E-selectin interacts with its ligand P-selectin glycoprotein ligand-1 (PSGL-1) for initial capture of a neutrophil, followed by an interaction with E-selectin ligand-1 (ESL-1), which converts initial tethers into steady slow rolling.[65] E-selectin binding to neutrophil CD44 controls rolling velocity and polarization of selectin ligands on the neutrophil. However, E-selectin-mediated tethering of any leukocyte is reversible and may be followed by either the release of the leukocyte and its return to circulation or a firmer interaction, mediated by VCAM-1 and ICAMs.

It is curious therefore, that E-selectin is transiently upregulated on the surface of inflamed lymphatic endothelium,[24] where it has been calculated that a leukocyte only has to contend with a shear stress of approximately 0.08 dynes/cm^2 (as opposed to 1–4 dynes/cm^2 in blood capillaries[66,67]). It may be that E-selectin is involved in initial capture of leukocytes onto lymphatic endothelium under low shear stress. However, as selectin engagement can trigger signals in both the selectin-expressing cell and the ligand-expressing cell, it may also be that the primary role of lymphatic E-selectin is in this domain. For example, binding of isolated human neutrophils to E-selectin has been shown to activate integrins via the p38 mitogen-activated protein kinase (MAPK)-dependent pathway.[68] Therefore, binding to LEC-derived E-selectin, either membrane-

bound or in soluble form, could be a necessary prerequisite for VCAM-1- and ICAM-1-mediated adhesion and transmigration.

Junctional and Adhesion Molecules Involved in Transendothelial Cell Migration

In addition to VCAM-1 and ICAM-1, various other adhesion and junctional molecules are expressed by both LECs and BEC. Their roles in leukocyte adhesion and transendothelial migration are summarized in TABLE 1.

Specialized Lymphatic Endothelial Receptors with Possible Roles in Transmigration

Despite the commonalities between ECs and BECs, both in terms of molecules expressed and the fact that they are morphologically indistinguishable, they are nevertheless specialized and adapted for their distinct functions. Therefore it is not surprising that certain molecules are distinct to either LECs or BECs, where they constitute not only immunomarkers but perhaps also a "molecular address" to direct homing of specific leukocyte populations. Lymphatic endothelial–specific receptors and their possible roles in regulating leukocyte adhesion and transmigration are discussed below.

CLEVER-1

Common lymphatic endothelial and vascular endothelial receptor-1 (CLEVER-1, also known as stabilin-1 and FEEL-1) is a large multidomain molecule comprising multiple epidermal growth factor–like repeats, tandem fasciclin-like domains, and a single membrane-proximal Link module, a lectin-like domain. Curiously, unlike other members of the Link family, CLEVER-1 cannot bind the matrix glycosaminoglycans hyaluronan (HA) and chondroitin sulfate[85] and is almost entirely intracellular in lymph node HEVs, lymphatic sinus endothelium, and cultured endothelial cells. CLEVER-1 has been shown to cycle rapidly between the plasma membrane and early endosomes, suggesting that its primary function is that of a rapidly recycling scavenger receptor.[85] A role for CLEVER-1 in transmigration has been suggested by Salmi and co-workers, who found that it is expressed on dermal lymphatic vessels and in inflamed blood vascular endothelium. They demonstrated CLEVER-1-dependent transmigration of peripheral blood mononuclear cells across LECs and HUVECs, as well as homing of leukocytes across lymph node HEVs.[86] However, the ligand of CLEVER-1 in transmigration remains unidentified, and the mechanism by which it mediates

TABLE 1. Molecules involved in leukocyte transendothelial cell migration

Adhesion molecule	Alternative Names	Ligands	Expression	Role in adhesion/transmigration?	Refs.
Junctional adhesion molecule-A (JAM-A)	JAM-1	Homophilic interactions in *cis* and *trans*; LFA-1; JAM-C	In tight junctions of blood vascular and lymphatic endothelium; DCs	In the blood vasculature, aids monocyte and neutrophil recruitment and plays a role in transmigration at sites of injury. Also expressed in DCs and is implicated in regulating motility, reducing lymph node homing and response to contact hypersensitivity.	48, 69, 70
Junctional adhesion molecule-B (JAM-B)	JAM-2	VLA-4; JAM-C	In tight junctions of blood and lymphatic endothelium	Implicated in regulating leukocyte transmigration across HEV	71, 72
Junctional adhesion molecule-C (JAM-C)	JAM-3; VE-JAM	Homophilic interactions; MAC-1; JAM-A; JAM-B	In tight junctions of endothelial cells of lymphoid organs and peripheral tissues	Minimal role in transendothelial migration across blood vasculature	73, 74
Intracellular adhesion molecule-2 (ICAM-2)		LFA-1; DC-SIGN; MAC1, other(s) unknown	Blood and lymphatic endothelium	Constitutively expressed for lymphocyte recirculation through lymph nodes and peripheral tissues. Mediates migration of immature murine DCs across non-stimulated endothelium through unknown ligand(s).	24, 42, 75, 76
CD31	Platelet endothelial cell adhesion molecule-1, PECAM-1	Homophilic interactions; αvβ3 integrin; CD38	Hematopoietic cells and both blood and lymphatic endothelium, especially in intercellular adherens junctions	Required for monocyte transmigration across both resting and pro-inflammatory cytokine-stimulated HUVECs and neutrophil transmigration across stimulated HUVECs, most likely through homophilic interactions	77–82
CD99		Homophilic interactions	Most hematopoietic cells and both blood and lymphatic endothelium, particularly in intercellular adherens junctions	Required for transmigration of monocytes and neutrophils across blood vascular endothelium	47, 83
VE-cadherin	Cadherin-5	Homophilic interactions	Expressed within adherens junctions of blood vascular and lymphatic endothelium	Contributes to stability of intercellular adherens junctions and vascular integrity but does not directly participate in leukocyte transmigration.	50, 73
Endothelial cell-selective adhesion molecule (ESAM)		Unknown	Tight junctions of endothelial cells and on activated platelets	Participates in extravasation of neutrophils.	84

these distinct transmigratory events is as yet unclear, particularly as it is predominantly found intracellularly and not always observed on lymphatic vessels.[85,86]

Mannose Receptor

Another possible candidate for regulating leukocyte migration into afferent lymphatics is mannose receptor (MR), a 180-kDa transmembrane protein that contains lectin-like carbohydrate recognition domains.[87] MR is expressed on tissue macrophages and immature DCs and has a well-established role both in the endocytic clearance of certain glycoproteins and in macrophage phagocytosis of micro-organisms. MR is also expressed in lymphatic endothelium and in lymph node sinuses, where, using frozen-section adhesion assays, MR has been shown to bind leukocyte-expressed L-selectin, possibly via the lectin domain of L-selectin binding to a carbohydrate epitope on MR.[88] Thus lymphatic endothelial–expressed MR may assist in adhesion of L-selectin[+] lymphocytes prior to transmigration into afferent lymphatics.

Podoplanin

Podoplanin is one of the most highly expressed genes in lymphatic endothelium[89,90] and, although it is also found on a variety of normal cells (such as mesothelia, neuronal cells, osteocytes, and epithelia), it is not expressed on blood vascular endothelium.[91] Podoplanin contributes to the active recruitment of mononuclear leukocytes by LECs through its predominant expression on the basolateral surface, where it forms a complex with CCL21 that is subsequently shed into the stroma (shown by double-labeling immuno-electron microscopy of human renal tissue), thus establishing a perilymphovascular CCL21 gradient.[92] The relatively high-affinity interaction between podoplanin and CCL21 (K_d 89.7 nM) is dependent upon glycosylation of podoplanin.

Podoplanin expression is upregulated in various human tumors, for example, in squamous cell carcinoma, where its expression is frequently restricted to the invasive front.[93,94] The key roles that podoplanin plays in cell adhesion, migration, and vessel formation have been indicated by studies on podoplanin[−/−] mice, which die immediately after birth on account of respiratory failure, precipitated by impaired formation of alveolar airspaces, which was associated with a reduction in the number of differentiated type I alveolar epithelial cells in the lung.[95] Neonates also exhibited cutaneous lymphedema, dilated cutaneous intestinal lymphatic vessels, and impaired

lymphatic transport.[96] Furthermore, overexpression of podoplanin in vascular endothelial cells has been shown to promote the formation of elongated cell extensions and to increase cell adhesion, migration and tube formation.[96] The mechanisms through which podoplanin performs these functions appear to be dependent upon interactions between a juxtamembrane cluster of basic amino acids in the short cytoplasmic tail of podoplanin and members of the ezrin, radixin, moesin (ERM) family.[93] ERM proteins in turn bind to the cortical actin cytoskeleton and participate in a RhoA GTPase signal transduction pathway that regulates cell motility and adhesion. As podoplanin expression is concentrated at actin-rich microvilli and plasma membrane projections and can stimulate actin rearrangement, it is possible that this small membrane mucin may also play a role in regulating transcellular transmigration of cells across a monolayer of podoplanin-expressing cells, such as lymphatic endothelia or lung alveolar epithelia.

LYVE-1

A highly glycosylated integral membrane glycoprotein and member of the Link superfamily of HA binding receptors, LYVE-1 is another abundant surface component of LECs.[97,98] The similarity in amino acid sequence with the inflammatory homing receptor CD44—a leukocyte molecule that binds HA on blood vascular endothelium—originally suggested that LYVE-1 might play an analogous role in lymphatic homing. For example, it was reasoned that LYVE-1 on lymphatic endothelium might support adhesion of CD44-positive leukocytes through mutual binding to HA synthesized either by the leukocyte or by other cells within the tissue matrix.[99,100] Despite the fact that such tripartite binding can be demonstrated *in vitro* using recombinant receptor ectodomains and purified HA *in vitro*, there is little evidence to date suggesting that such interactions occur widely *in vivo*. For example, mice lacking the gene for LYVE-1 display apparently normal trafficking of DCs from skin to lymph nodes.[101] Although studies of leukocyte adhesion/transmigration across primary LEC monolayers have only been undertaken recently, preliminary analyses have failed to show any disruption to either process by exogenously added soluble LYVE-1 (L.A. Johnson and D.G. Jackson, unpublished observations). Furthermore, exposure of LECs to factors, such as the inflammatory cytokine TNFα, that stimulate trafficking of DCs to lymph nodes induces downregulation of LYVE-1—a finding that is difficult to reconcile with an obvious role in adhesion/transmigration.[102] Despite such considerations, the recent confocal microscopic observation that LYVE-1 is concentrated within the cleft-like openings of overlapping interendothelial junctions in initial lymphatics and only sparse within adjacent tight junction structures[70] is still consistent with a role for the receptor in maintaining some aspect of junctional permeability. Further analysis will be required in order to resolve the issue.

Finally, an important complication in all the above experiments is that the HA-binding activity of LYVE-1 is functionally masked in LECs both *in vivo* and *in vitro*. Intriguingly, the silencing mechanism appears to involve receptor sialylation, a phenomenon shared with CD44, where the process is reversible and serves as an important regulator of biological activity.[103] The significance of this feature is still not clear. It remains to be seen what factors trigger HA binding to LYVE-1 *in vivo*, how binding could mediate cell trafficking, and whether LYVE-1 might bind additional ligands in order to promote adhesion/transmigration.

Broader Responses of Lymphatic Endothelium to Inflammation

In addition to the induction of adhesion molecules and chemokines, lymphatic endothelium responds to inflammation by upregulating a wide range of other transcripts, such as the metalloproteinases MMP-19 and ADAMTS3, the Toll-like receptors TLR-1 and TLR2, and TNF superfamily members such as CD137 (TNFRSF9) and lymphotoxin-β (TNFSF3).[24] It is tempting to speculate that metalloproteinases could potentially facilitate extravasation of leukocytes from the blood by degrading basement membrane structures on interendothelial tight junctions of the blood capillaries juxtaposed to the initial lymphatics. They may also contribute to the establishment of chemokine gradients by releasing immobilized chemokines from their heparan sulfate binding sites within the extracellular matrix or bound to the lymphatic endothelium itself and, through cleavage of the chemokine, activate chemotactic activity.

The increase in transcript abundance of CD137 (4-1BB) is surprising as this activation-induced glycoprotein has not been detected on endothelial cells before. Ligation of this receptor is reported to interrupt T cell apoptotic programs associated with activation-induced cell death,[104] which suggests that LECs too are capable of protecting themselves from apoptotic signals during inflammation. Thus LECs can continue to exert effective control over leukocyte entry into the afferent lymphatics while dynamically influencing the chemotactic propensity of neighboring leukocytes.

Conclusions: A Hypothetical Mechanism for Translymphatic Migration

The limited picture of leukocyte translymphatic migration that we have described here has been based almost entirely on the results of *in vitro* transmigration assays and the effects of antibody blockade on murine T cell and DTH responses. Surprisingly we still know little about the actual sites in lymphatic capillaries at which leukocyte entry occurs. As stated earlier in this chapter, migration has been assumed to proceed through loose interendothelial gaps in initial blind-ended capillaries, essentially in the same way as fluids and dissolved macromolecules are taken into lymph. Yet the organization of these gaps has remained unclear. Just recently, however, this situation has changed dramatically as the result of detailed confocal and electron microscopic analyses of these very structures in the small lymphatic capillaries of the mouse trachea.

In their seminal studies, Baluk *et al.*[70] have shown that the endothelial cells which form overlapping junctions in initial dermal lymphatic capillaries have an oak-leaf shape, and that the gaps form at the points where the finger-like borders interdigitate. More specifically, the borders are tethered on either side by button-like structures containing a mixture of both tight and adherens junctional components including JAMs, ESAM, occludins, claudins, ZO-1, and VE-cadherin. In contrast, the gaps beneath the tips of borders are lined exclusively by CD31 and LYVE-1. These features are interesting for a number of reasons. First, similar proteins when present in blood vessel endothelial junctions are not just structural components, but engage in dynamic interactions between leukocytes and endothelial cells. For example JAMs, including JAM-A and JAM-C, translocate from tight junctions to the luminal face of blood vessel endothelium in response to inflammation, and molecules at the latter location support leukocyte adhesion through binding to LFA-1 and Mac-1, respectively.[46,105] Furthermore, deletion of the gene for JAM-A was recently shown to increase trafficking of DCs within the afferent lymphatics.[69] The button-like structures could therefore be dynamic entities that act partially as gatekeepers for leukocyte entry. Although Baluk *et al.*[70] were not sufficiently fortunate to obtain images of leukocytes in the act of transmigration, their measurements indicated that button/gap structures were the likely points of leukocyte entry.

On the basis of the foregoing data we envisage a sequence of events in which immigrant leukocytes form adhesion complexes with "activated" initial lymphatics via ICAM-1 and VCAM-1, perhaps using these interactions for lateral locomotion toward gaps adjacent to the button-like structures. Given that these gaps are small (approximately 2 μm) compared to the diameter of a leukocyte, we propose that leukocytes must induce transient gap opening, perhaps through engaging JAMs or ESAM within the buttons. Analogous loosening of junctional VE–cadherin contacts is thought to occur during paracellular transmigration of blood vessel endothelium as a result of interactions between JAM-A, CD31, and catenins,[46] as well as by activation of the small GTPase Rho by ESAM.[84] This speculative model is thus a variant of paracellular transmigration, but occurring specifically at points in the initial lymphatic vessel where the "wall" is not completely sealed but where permanent gaps persist. In contrast to the blood vasculature. in which the doors are firmly closed, in the lymphatics they may be always half open. Whether transcellular migration can also occur in the vicinity of these gaps, or whether it is reserved for more distal lymphatics in which the endothelial junctions more closely resemble the zipper-like rearrangement of blood vessels must await further experimentation.

Thus many new questions are generated by these considerations. Do buttons undergo reorganization and do the gaps enlarge in response to inflammation? Do JAMs play a role in leukocyte adhesion/transmigration and do they redistribute from buttons to gaps in the process? Why is LYVE-1 segregated within gaps, yet apparently dispensible for lymphatic trafficking? Where are ICAM-1 and VCAM-1 expressed in relation to the gaps and buttons? Answers to these important questions should soon be forthcoming.

Conflicts of Interest

The authors declare no conflicts of interest.

References

1. YOUNG, A.J. 1999. The physiology of lymphocyte migration through the single lymph node *in vivo*. Semin. Immunol. **11:** 73–83.
2. AUSTYN, J.M. 1989. Migration patterns of dendritic leukocytes. Res. Immunol. **140:** 898–902.
3. ZINKERNAGEL, R.M. 1996. Immunology taught by viruses. Science **271:** 173–178.
4. RANDOLPH, G.J. 2001. Dendritic cell migration to lymph nodes: cytokines, chemokines, and lipid mediators. Semin. Immunol. **13:** 267–274.
5. STOITZNER, P. *et al.* 2002. A close-up view of migrating Langerhans cells in the skin. J. Invest. Dermatol. **118:** 117–125.
6. STOITZNER, P. *et al.* 2003. Visualization and characterization of migratory Langerhans cells in murine skin and lymph nodes by antibodies against Langerin/CD207. J. Invest. Dermatol. **120:** 266–274.
7. HUANG, F.-P. *et al.* 2000. A discrete subpopulation of dendritic cells transports apoptotic intestinal epithelial cells to T cell areas of mesenteric lymph nodes. J. Exp. Med. **191:** 435–443.
8. SCHEINECKER, C. *et al.* 2002. Constitutive presentation of a natural tissue autoantigen exclusively by dendritic cells in the draining lymph node. J. Exp. Med. **196:** 1079–1090.
9. STEINMAN, R.M., D. HAWIGER & M.C. NUSSENZWEIG. 2003. Tolerogenic dendritic cells. Annu. Rev. Immunol. **21:** 685–711.
10. FÖRSTER, R. *et al.* 1999. CCR7 coordinates the primary immune response by establishing functional microenvironments in secondary lymphoid organs. Cell **99:** 23–33.
11. OHL, L. *et al.* 2004. CCR7 govern skin dendritic cell migration under inflammatory and steady-state conditions. Immunity **21:** 279–288.
12. GUNN, M.D. *et al.* 1999. Mice lacking expression of secondary lymphoid organ chemokine have defects in lymphocyte homing and dendritic cell localization. J. Exp. Med. **189:** 451–460.
13. LUTHER, S.A. *et al.* 2000. Coexpression of the chemokines ELC and SLC by T zone stromal cells and deletion of the ELC gene in the plt/plt mouse. Proc. Natl. Acad. Sci. USA **97:** 12694–12699.
14. VASSILEVA, G. *et al.* 1999. The reduced expression of 6Ckine in the plt mouse results from the deletion of one of two 6Ckine genes. J. Exp. Med. **190:** 1183–1188.
15. SAEKI, H. *et al.* 1999. Cutting edge: secondary lymphoid-tissue chemokine (SLC) and CC chemokine receptor 7 (CCR7) participate in the emigration pathway of mature dendritic cells from the skin to regional lymph nodes. J. Immunol. **162:** 2472–2475.
16. MARTÍN-FONTECHA, A. *et al.* 2003. Regulation of dendritic cell migration to the draining lymph node: impact on T lymphocyte traffic and priming. J. Exp. Med. **198:** 615–621.
17. KRIEHUBER, E. *et al.* 2001. Isolation and characterization of dermal lymphatic and blood endothelial cells reveal stable and functionally specialized cell lineages. J. Exp. Med. **194:** 797–808.
18. WICK, N. *et al.* 2007. Transcriptomal comparison of human dermal lymphatic endothelial cells ex vivo and in vitro. Physiol. Genomics **28:** 179–192.
19. WORBS, T. *et al.* 2007. CCR7 ligands stimulate the intranodal motility of T lymphocytes in vivo. J. Exp. Med. **204:** 489–495.
20. BROMLEY, S.K., S.Y. THOMAS & A.D. LUSTER. 2005. Chemokine receptor CCR7 guides T cell exit from peripheral lymph nodes and entry into afferent lymphatics. Nat. Immunol. **6:** 895–901.
21. DEBES, G.F. *et al.* 2005. Chemokine receptor CCR7 required for T lymphocyte exit from the peripheral tissues. Nat. Immunol. **6:** 889–894.

22. LEDGERWOOD, L.G. *et al.* 2008. The sphingosine 1-phosphate receptor 1 causes tissue retention by inhibiting the entry of peripheral tissue T lymphocytes into afferent lymphatics. Nat. Immunol. **9:** 42–53.

23. KABASHIMA, K. *et al.* 2007. CXCL12-CXCR4 engagement is required for migration of cutaneous dendritic cells. Am. J. Pathol. **171:** 1249–1257.

24. JOHNSON, L.A. *et al.* 2006. An inflammation-induced mechanism for leukocyte transmigration across lymphatic vessel endothelium. J. Exp. Med. **203:** 2763–2777.

25. MANCARDI, S. *et al.* 2003. Evidence of CXC, CC and C chemokine production by lymphatic endothelial cells. Immunology **108:** 523–530.

26. NIBBS, R.J. B. *et al.* 2001. The beta-chemokine receptor D6 is expressed by lymphatic endothelium and a subset of vascular tumors. Am. J. Pathol. **158:** 867–877.

27. MARTINEZ DE LA TORRE, Y. *et al.* 2005. Increased inflammation in mice deficient for the chemokine decoy receptor D6. Eur. J. Immunol. **35:** 1342–1346.

28. PALFRAMAN, R.T. *et al.* 2001. Inflammatory chemokine transport and presentation in HEV: a remote control mechanism for monocyte recruitment to lymph nodes in inflamed tissue. J. Exp. Med. **194:** 1361–1373.

29. GRETZ, J.E. *et al.* 2000. Lymph-borne chemokines and other low molecular weight molecules reach high endothelial venules via specialized conduits while a functional barrier limits access to the lymphocyte microenvironments in lymph node cortex. J. Exp. Med. **192:** 1425–1439.

30. SMITH, J.B., G.H. MCINTOSH & B. MORRIS. 1970. The traffic of cells through tissues: a study of peripheral lymph in sheep. J. Anat. **107:** 87–100.

31. ABADIE, V. *et al.* 2005. Neutrophils rapidly migrate via lymphatics after *Mycobacterium bovis* BCG intradermal vaccination and shuttle live bacilli to the draining lymph nodes. Blood **106:** 1843–1850.

32. MALETTO, B.A. *et al.* 2006. Presence of neutrophil-bearing antigen in lymphoid organs of immune mice. Blood **108:** 3094–3102.

33. ALON, R. *et al.* 1995. The integrin VLA-4 supports tethering and rolling in flow on VCAM-1. J. Cell Biol. **128:** 1243–1253.

34. VAN DINTHER-JANSSEN, A.C. *et al.* 1991. The VLA-4/VCAM-1 pathway is involved in lymphocyte adhesion to endothelium in rheumatoid synovium. J. Immunol. **147:** 4207–4210.

35. ROTHLEIN, R. *et al.* 1986. A human intercellular adhesion molecule (ICAM-1) distinct from LFA-1. J. Immunol. **137:** 1270–1274.

36. SPRINGER, T.A. 1990. Adhesion receptors of the immune system. Nature **346:** 425.

37. OPPENHEIMER-MARKS, N. *et al.* 1991. Differential utilization of ICAM-1 and VCAM-1 during the adhesion and transendothelial migration of human T lymphocytes. J. Immunol. **147:** 2913–2921.

38. MAKGOBA, M.W. *et al.* 1988. ICAM-1 a ligand for LFA-A-dependent adhesion of B, T and myeloid cells. Nature **6151:** 86–88.

39. KAVANAUGH, A.F. *et al.* 1991. Role of CD11/CD18 in adhesion and transendothelial migration of T cells. J. Immunol. **146:** 4149–4156.

40. GREENWOOD, J., Y. WANG & V.L. CALDER. 1995. Lymphocyte adhesion and transendothelial migration in the central nervous system: the role of LFA-1, ICAM-1, VLA-4 and VCAM-1. Immunology **86:** 408–415.

41. MCLAUGHLIN, F. *et al.* 1998. Tumour necrosis factor (TNF)-alpha and interleukin (IL)-1beta down-regulate intercellular adhesion molecule (ICAM)-2 expression on the endothelium. Cell Adhes. Commun. **6:** 381–400.

42. LEHMANN, J.C. U. *et al.* 2003. Overlapping and selective roles of endothelial intercellular adhesion molecule-1 (ICAM-1) and ICAM-2 in lymphocyte trafficking. J. Immunol. **171:** 2588–2593.

43. SLIGH, J.E. *et al.* 1993. Inflammatory and immune responses are impaired in mice deficient in intercellular adhesion molecule 1. Proc. Natl. Acad. Sci. USA **90:** 8529–8533.

44. XU, H. *et al.* 2001. The role of ICAM-1 molecule in the migration of Langerhans cells in the skin and regional lymph node. Eur. J. Immunol. **31:** 3085–3093.

45. SCHENKEL, A.R. *et al.* 2004. Locomotion of monocytes on endothelium is a critical step during extravasation. Nat. Immunol. **5:** 393–400.

46. JOHNSON-LEGER, C., M. AURRAND-LIONS & B.A. IMHOF. 2000. The parting of the endothelium: miracle, or simply a junctional affair? J. Cell Sci. **113:** 921–933.

47. SCHENKEL, A.R. *et al.* 2002. CD99 plays a major role in the migration of monocytes through endothelial junctions. Nat. Immunol. **3:** 143–150.

48. DEL MASCHIO, A. *et al.* 1999. Leukocyte recruitment in the cerebrospinal fluid of mice with experimental meningitis is inhibited by an antibody to junctional adhesion molecule (JAM). J. Exp. Med. **190:** 1351–1356.

49. SHAW, S.K. *et al.* 2001. Reduced expression of junctional adhesion molecule and platelet/endothelial cell adhesion molecule-1 (CD31) at human vascular endothelial junctions by cytokines tumour necrosis factor-alpha plus interferon-gamma does not reduce leukocyte transmigration under flow. Am. J. Pathol. **159:** 2281–2291.

50. MULLER, W.A. 2003. Leukocyte-endothelial-cell interactions in leukocyte transmigration and the inflammatory response. Trends Immunol. **24:** 326–333.

51. SHAW, S.K. *et al.* 2001. Real-time imaging of vascular endothelial-cadherin during leukocyte transmigration across endothelium. J. Immunol. **167:** 2323–2330.

52. MAMDOUH, Z. *et al.* 2003. Targeted recycling of PECAM from endothelial surface-connected compartments during diapedesis. Nature **421:** 748–753.

53. MARCHESI, V.T. & J.L. GOWANS. 1964. The migration of lymphocytes through the endothelium of venules in lymph nodes: an electron microscope study. Proc. R. Soc. Lon. Ser. B. **159:** 283–290.

54. FENG, D. *et al.* 1998. Neutrophils emigrate from venules by a transendothelial cell pathway in response to FMLP. J. Exp. Med. **187:** 903–915.

55. BARREIRO, O. *et al.* 2002. Dynamic interaction of VCAM-1 and ICAM-1 with moesin and ezrin in a novel endothelial docking structure for adherent leukocytes. J. Cell Biol. **157:** 1233–1245.

56. MILLAN, J. *et al.* 2006. Lymphocyte transcellular migration occurs through recruitment of endothelial ICAM-1 to

caveola- and F-actin-rich domains. Nat. Cell Biol. **8:** 113–123.

57. CARMAN, C.V. *et al.* 2003. Endothelial cells proactively form microvilli-like membrane projections upon intercellular adhesion molecule 1 engagement of leukocyte LFA-1. J. Immunol. **171:** 6135–6144.

58. CARMAN, C.V. & T.A. SPRINGER. 2004. A transmigratory cup in leukocyte diapedesis both through individual vascular endothelial cells and between them. J. Cell Biol. **167:** 377–388.

59. CARMAN, C.V. *et al.* 2007. Transcellular diapedesis is initiated by invasive podosomes. Immunity **26:** 784–797.

60. NIEMINEN, M. *et al.* 2006. Vimentin function in lymphocyte adhesion and transcellular migration. Nat. Cell Biol. **8:** 156–162.

61. WYBLE, C.W. *et al.* 1997. TNF-alpha and IL-1 upregulate membrane-bound and soluble E-selectin through a common pathway. J. Surg. Res. **73:** 107–112.

62. VON ANDRIAN, U.H. & C.R. MACKAY. 2000. T-cell function and migration. N. Engl. J. Med. **343:** 1020–1034.

63. BEVILACQUA, M.P. *et al.* 1989. Endothelial leukocyte adhesion molecule 1: an inducible receptor for neutrophils related to complement regulatory proteins and lectins. Science **243:** 1160–1165.

64. WELLER, P.F. *et al.* 1991. Human eosinophil adherence to vascular endothelium mediated by binding to vascular cell adhesion molecule 1 and endothelial leukocyte adhesion molecule 1. Proc. Natl. Acad. Sci. USA **88:** 7430–7433.

65. HIDALGO, A. *et al.* 2007. Complete identification of E-selectin ligands on neutrophils reveals distinct functions of PSCL-1, ESL-1 and CD44. Immunity **26:** 477–489.

66. KANNAGI, R. 2002. Regulatory roles of carbohydrate ligands for selectins in the homing of lymphocytes. Curr. Opin. Struct. Biol. **12:** 599–608.

67. SACKSTEIN, R. 2005. The lymphocyte homing receptors: gatekeepers of the multistep paradigm. Curr. Opin. Hematol. **12:** 444–450.

68. SIMON, S.I. *et al.* 2000. Neutrophil tethering on E-selectin activates 2 integrin binding to ICAM-1 through a mitogen-activated protein kinase signal transduction pathway. J. Immunol. **164:** 4348–4358.

69. CERA, M.R. *et al.* 2004. Increased DC trafficking to lymph nodes and contact hypersensitivity in junctional adhesion molecule-A-deficient mice. J. Clin. Invest. **114:** 729–738.

70. BALUK, P. *et al.* 2007. Functionally specialized junctions between endothelial cells of lymphatic vessels. J. Exp. Med. **204:** 2349–2362.

71. AURRAND-LIONS, M. *et al.* 2001. JAM-2, a novel immunoglobulin superfamily molecule, expressed by endothelial and lymphatic cells. J. Biol. Chem. **276:** 2733–2741.

72. JOHNSON-LEGER, C. *et al.* 2002. Junctional adhesion molecule-2 (JAM-2) promotes lymphocyte transendothelial migration. Blood **100:** 2479–2486.

73. IMHOF, B.A. & M. AURRAND-LIONS. 2004. Adhesion mechanisms regulating the migration of monocytes. Nat. Rev. Immunol. **4:** 423–444.

74. LAMAGNA, C. *et al.* 2005. Dual interaction of JAM-C with JAM-B and alpha(M)beta2 integrin: function in junc-

75. STAUNTON, D.E., M.L. DUSTIN & T.A. SPRINGER. 1989. Functional cloning of ICAM-2, a cell adhesion ligand for LFA-1 homologous to ICAM-1. Nature **339:** 61–64.

76. WETHMAR, K. *et al.* 2006. Migration of immature mouse DC across resting endothelium is mediated by ICAM-2 but independent of beta2-integrins and murine DC-SIGN homologues. Eur. J. Immunol. **36:** 2781–2794.

77. MULLER, W.A. *et al.* 1989. A human endothelial cell-restricted, externally disposed plasmalemmal protein enriched in intercellular junctions. J. Exp. Med. **170:** 399.

78. BUCKLEY, C.D. *et al.* 1996. Identification of alpha v beta 3 as a heterotypic ligand for CD31/PECAM-1. J. Cell Sci. **109:** 437–445.

79. JACKSON, D.E. 2003. The unfolding tale of PECAM-1. FEBS Lett. **540:** 7–14.

80. BOGEN, S.A. *et al.* 1992. Association of murine CD31 with transmigrating lymphocytes following antigenic stimulation. Am. J. Pathol. **141:** 843–854.

81. MULLER, W.A. *et al.* 1993. PECAM-1 is required for transendothelial migration of leukocytes. J. Exp. Med. **178:** 449–460.

82. SCHENKEL, A.R., T.W. CHEW & W.A. MULLER. 2004. Platelet endothelial cell adhesion molecule deficiency or blockade signficantly reduces leukocyte emigration in a majority of mouse strains. J. Immunol. **173:** 6403–6408.

83. LOU, O. *et al.* 2007. CD99 is a key mediator of the transendothelial migration of neutrophils. J. Immunol. **178:** 1136–1143.

84. WEGMANN, F. *et al.* 2006. ESAM supports neutrophil extravasation, activation of Rho, and VEGF-induced vascular permeability. J. Exp. Med. **203:** 1671–1677.

85. PREVO, R. *et al.* 2004. Rapid plasma membrane-endosomal trafficking of the lymph node sinus and high endothelial venule scavenger receptor/homing receptor stabilin-1 (FEEL-1/CLEVER-1). J. Biol. Chem. **279:** 52580–52592.

86. SALMI, M. *et al.* 2004. CLEVER-1 mediates lymphocyte transmigration through vascular and lymphatic endothelium. Blood **104:** 3849–3857.

87. STAHL, P.D. & R.A. EZEKOWITZ. 1998. The mannose receptor is a pattern recognition receptor involved in host defense. Curr. Opin. Immunol. **10:** 50–55.

88. IRJALA, H. *et al.* 2001. Mannose receptor is a novel ligand for L-selectin and mediates lymphocyte binding to lymphatic endothelium. J. Exp. Med. **194:** 1033–1041.

89. PETROVA, T.V. *et al.* 2002. Lymphatic endothelial reprogramming of vascular endothelial cells by the Prox-1 homeobox transcription factor. EMBO J. **21:** 4593–4599.

90. HIRAKAWA, S. *et al.* 2003. Identification of vascular lineage-specific genes by transcriptional profiling of isolated blood vascular and lymphatic endothelial cells. Am. J. Pathol. **162:** 575–586.

91. WETTERWALD, A. *et al.* 1996. Characterization and cloning of the E11 antigen, a marker expressed by rat osteoblasts and osteocytes. Bone **18:** 125–132.

92. KERJASCHKI, D. *et al.* 2004. Lymphatic neoangiogenesis in human kidney transplants is associated with immunolog-

ically active lymphocytic infiltrates. J. Am. Soc. Nephrol. **15:** 603–612.

93. MARTIN-VILLAR, E. *et al.* 2005. Characterization of human PA2.26 antigen (T1alpha-2, podoplanin), a small membrane mucin induced in oral squamous cell carcinomas. Int. J. Cancer **113:** 899–910.

94. WICKI, A. *et al.* 2006. Tumor invasion in the absence of epithelial-mesenchymal transition: podoplanin-mediated remodeling of the actin cytoskeleton. Cancer Cell **9:** 261–272.

95. RAMIREZ, M.I. *et al.* 2003. T1alpha, a lung type I cell differentiation gene, is required for normal lung cell proliferation and alveolus formation at birth. Dev. Biol. **256:** 61–72.

96. SCHACHT, V. *et al.* 2003. T1alpha/podoplanin deficiency disrupts normal lymphatic vasculature formation and causes lymphedema. EMBO J. **22:** 3546–3556.

97. BANERJI, S. *et al.* 1999. LYVE-1, a new homologue of the CD44 glycoprotein, is a lymph-specific receptor for hyaluronan. J. Cell Biol. **144:** 789–801.

98. PREVO, R. *et al.* 2001. Mouse LYVE-1 is an endocytic receptor for hyaluronan in lymphatic endothelium. J. Biol. Chem. **276:** 19420–19430.

99. JACKSON, D.G. 2003. The lymphatics revisited: new perspectives from the hyaluronan receptor LYVE-1. Trends Cardiovasc. Med. **13:** 1–7.

100. JACKSON, D.G. 2004. Biology of the lymphatic marker LYVE-1 and applications in research into lymphatic trafficking and lymphangiogenesis. Acta Pathol. Microbiol. Immunol. Scand. **112:** 526–538.

101. GALE, N.W. *et al.* 2006. Normal lymphatic development and function in mice deficient for the lymphatic hyaluronan receptor LYVE-1. Mol. Cell Biol. **27:** 595–604.

102. JOHNSON, L.A. *et al.* 2007. Inflammation-induced uptake and degradation of the lymphatic endothelial hyaluronan receptor LYVE-1. J. Biol. Chem. **282:** 33671–33680.

103. NIGHTINGALE, T.D., S. BANERJI & D.G. JACKSON. 2005. Functional regulation of the lymphatic endothelium hyaluronan receptor LYVE-1: the role of N-glycosylation. *In* Hyaluronan: Structure, Metabolism, Biological Activity, Therapeutic Applications, Vol. 2. E.A. Balazs & V.C. Hascall, Eds.: 615–618. Matrix Biology Institute. Edgewater, NJ.

104. HURTADO, J.C., Y.-J. KIM & B.S. KWON. 1997. Signals through 4-1BB are costimulatory to previously activated splenic T cells and inhibit activation-induced cell death. J. Immunol. **158:** 2600–2609.

105. BRADFIELD, P.F. *et al.* 2007. JAM-C regulates unidirectional monocyte transendothelial migration in inflammation. Blood **110:** 2545–2555.

Research Perspectives in Inherited Lymphatic Disease

An Update

ROBERT E. FERRELL AND DAVID N. FINEGOLD

Department of Human Genetics, Graduate School of Public Health, University of Pittsburgh, Pittsburgh, Pennsylvania, USA

Genetic studies of inherited lymphedema have provided the starting point for the molecular dissection of lymphatic development and disease. Here, we update the recent contribution of the study of inherited lymphedema and discuss the parallels between mouse models of lymphedema and inherited lymphedema. That the known mutations leading to lymphatic phenotypes explain fewer than half the cases of lymphedema means that the continued study of these disorders may reveal new pathways in lymphatic biology.

Key words: lymphedema; lymphangiogenesis; Milroy's disease; FOXC2

Introduction

Primary lymphedema, the accumulation of protein-rich fluid (lymph) in the interstitial spaces as a result of an anatomic or functional defect in lymphatic vessels or nodes, is a heterogeneous condition occurring as a non-syndromic mendelian disorder or as part of complex genetic syndromes. Since it has been recognized that mutations in the vascular endothelial growth factor receptor-3 cause lymphedema, and, more importantly, that the VEGF-C/VEGFR3 signaling system is central to both normal and pathologic lymphangiogenesis, further details have come to light regarding the regulation of normal and abnormal lymphatic development. Other recent reviews have summarized the molecular details of normal and pathologic lymphangiogenesis.[1–3] This review summarizes the advances that can be credited to the study of the mendelian phenotypes of inherited lymphedema.

Hereditary Lymphedema I (Milroy's Disease)

Milroy's disease, autosomal dominant lymphedema (OMIM 153100), with an early onset and without ma-

jor nonlymphatic manifestations, is predominantly due to mutations in the kinase domain of vascular endothelial growth factor receptor-3 (VEGFR3; FLT4). Kinase domain mutations in VEGFR3 cause both familial and sporadic cases of Milroy's disease.[4] FIGURE 1 summarizes the location and nature of VEGFR3 mutations documented in primary lymphedema. The demonstration that inactivating mutations in VEGFR3 and the subsequent elucidation in the VEGF-C/VEGFR3 pathway catalyzed the study of lymphangiogenesis in lymphatic disease and in normal lymphatic development, and raised the question of its importance in many human diseases.

The identification of the mutations has important therapeutic implications. Karkkainen *et al.*[5] demonstrated that the adeno-associated virus–mediated delivery of VEGF-C or VEGF-C165 ameliorated the cutaneous manifestation of lymphedema in the Chy mouse, a genetic model for Milroy's diesase. Saaristo *et al.*[6] demonstrated that VEGF-C156S, could induce lymphogenesis without the blood vascular side effects of VEGF-C in the Chy mouse model. Yoon *et al.*[7] showed that the delivery of naked plasmid DNA coding for human VEGF-C was effective in resolving lymphedema in both the rabbit ear and mouse tail model of secondary lymphedema. Beneficial effects of manipulating the VEGFC/VEGFR3 system in lymphedema or in other circumstances, like solid organ transplantation[8] and wound healing,[9] have been observed. However, these effects must be weighed against its effects in promoting tumor metastasis. Lymphangiogenesis has been implicated in the metastatic spread

Address for correspondence: Robert E. Ferrell, Ph.D., Department of Human Genetics, Graduate School of Public Health, University of Pittsburgh, Pittsburgh PA 15261. Voice: +1-412-624-3018; fax: +1-412-624-3020

rferrell@hgen.pitt.edu

Ann. N.Y. Acad. Sci. 1131: 134–139 (2008). © 2008 New York Academy of Sciences.
doi: 10.1196/annals.1413.012

VEGFR-3 Mutations Causing Hereditary Lymphedema

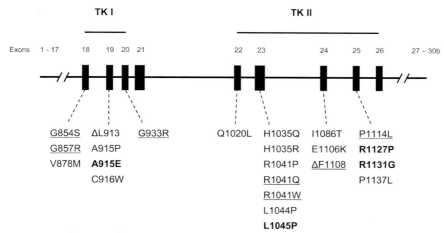

FIGURE 1. Schematic of the kinase domains, TKI and TKII, and human vascular endothelial growth factor receptor-3 and the mutations that have been observed in patients with Milroy's disease. Mutations that have been seen more than once are underlined, and unpublished mutations from our laboratory are boldface.

of a variety of tumors, including those of the breast, prostate, lung, and colorectum, and has been postulated to be an essential step in the spread of metastasis via the lymphatic system.[10,11] Clearly much remains to be learned about lymphangiogenesis and its manipulation in human genetic and nongenetic diseases.

Lymphedema–Distichiasis

Null mutations of FOXC2 (MFH1) have an extreme skeletal and cardiovascular phenotype[11–14] in mice. The recognition that haploinsufficiency of FOXC2 in humans causes the lymphedema–distichiasis phenotype (OMIM 153100) led to further careful study of the FOXC2± mice. Kriederman *et al.*[15] reported that these mice uniformly displayed distichiasis and exhibited both lymphatic vessel and lymph node hyperplasia. They noted retrograde lymph flow through incompetent interlymphatic values. Petrova *et al.*[16] noted that more than 80% of lymphatics were covered with pericytes/smooth muscle cells in biopsies from LD patients, these abnormal cells were found in only 10% of normal subjects. These investigators suggested that FOXC2 is essential for the establishment of a PC/SMC–free lymphatic network and did so by suppression of PDGF in lymphatic vessels. They suggested that the combination of defective lymphatic values, and impaired permeability of lymphatic capillaries due to the excess

recruitment of PC/SMCs through failure of inhibition of PDGF, was responsible for the characteristic impairment of lymphatic drainage seen in LD patients. Dagenais *et al.*[17] in a detailed analysis of the pattern of FOXC2 expression in the normal mouse, found that FOXC2 was expressed in lymphatic primordia, jugular lymph sacs, lymphatic collectors and capillaries, as well as in podocytes, developing eyelids, and other tissues associated with anomalies in LD patients.

Rosbotham *et al.*[18] noted the presence of varicose veins in both affected and unaffected members of a large LD kindred. Further examination of these unaffected members revealed evidence of subclinical lymphedema. Brice *et al.*[19] showed that varicose veins were present in half the subjects with LD in a large cohort of LD patients with FOXC2 mutations. This led Ng *et al.*[20] to study varicose veins in female twins recruited from the St. Thomas' Twin Registry. They reported strong evidence of linkage to the FOXC2 region of chromosome 16. This led Mellor *et al.*[21] to study venous reflux in the legs of 18 subjects known to have LD mutations in FOXC2. This study established a specific role for FOXC2 in primary venous value failure and emphasized the parallels in the development of both the venous and lymphatic system. A corroborating study would be to examine the potential lymphatic phenotype of patients having a family history with varicose veins not associated with distichiasis.

While most mutations causing LD are nonsense mutations, insertions, or deletions leading to truncation mutations eliminating the DNA binding domain, there have been several missense mutations identified, including the S125L mutation, which lacks transactivating activity *in vitro*, and an R121H mutation, which shows greatly reduced nuclear localization. Two others, T116R and S235I, do not occur in an obvious functional domain, but lead to a phenotype that is indistinguishable from the remaining LD mutations. Sholto-Douglas-Vernon *et al.*[22] reported two mutations leading to apparently LD phenotypes that did not occur in the coding sequence of FOXC2 and were not due to deletion. There remains the possibility of a promoter mutation or inactivation of FOXC2 due to a position effect, as was the original case reported by Fang *et al.*[23]

Cederberg *et al.*[24] established a role for FOXC2 in the regulation of the distribution and morphology of adipose tissue, and reported that overexpression of FOXC2 in white and brown adipose tissue of the mouse reduced the deposition of intra-abdominal fat and protected against insulin resistance. Ridderstrole *et al.*[25] examined the expression of FOXC2 mRNA in subjects with type-2 diabetes and found that there was a significant correlation between levels of expression in visceral fat and muscle and IOMA-IR. They also identified a polymorphism (C-512T) which partially explained this correlation. Several subsequent studies examined the C-512T polymorphism and confirmed that the T allele was significantly associated with obesity and with features of the metabolic syndrome. This, and the presence of abnormal lymphatics and adult onset obesity in PROX1± mice,[26] emphasizes the relationship between lymphatic function and obesity. Careful metabolic assessment of metabolism and lipid metabolism in carriers of FOXC2 mutations would be of interest.

Lymphedema and Ptosis and Yellow Nail Syndrome

The classification of lymphedema and ptosis (OMIM #15300) and yellow nail syndrome (OMIM #153300) as distinct autosomal dominant genetic syndromes was questioned by Finegold.[27] Recent papers support this speculation. In the report of Brice *et al.*[19] of 74 patients whose disease was ascertained on the basis of lymphedema–distichiasis and FOXC2 mutations, the occurrence of ptosis of varying levels of severity suggest that it may be one of the variable manifestations of FOXC2 mutation, rather than a genetically distinct condition.

Recently, Hogue *et al.*[28] reviewed yellow nail syndrome and presented 11 new cases. They noted the variety of diverse conditions in which yellow nail syndrome was reported, and remarked that the majority of these patients had a negative family history, including 10 of the 11 cases they presented. This led them to question whether yellow nail syndrome was a dominantly inherited condition or a sporadically acquired condition. Screening for FOXC2 mutations in a large group of families identified as having yellow nail syndrome would test the above speculation.

Noonan Syndrome

Mutation in three genes, PTPN11,[29] KRAS[30] and SOS1,[31,32] are reported to account for most cases of the Noonan syndrome (OMIM 136950). These three genes are involved in early steps of the RAS-ERK pathway.[32] Although lymphedema is reported as an occasional feature of Noonan syndrome, lymphedema is neither a prominent nor a specific feature of the Noonan phenotype mutations. Gene expression studies in lymphatic endothelial cells do not identify the RAS-ERK pathway as a prominent pathway affecting lymphogenesis or lymphatic function[33,34] and the pathway does not appear to be disrupted in any of the knockout mice with a lymphatic phenotype.[1]

Aagenaes' Syndrome (Lymphedema–Cholestasis) and Hennekam's Lymphangiectasis–Lymphedema

Despite the 2002 mapping[35] of Aagenaes' syndrome to a 6.6-cM region of chromosome 15q, there seems to be no progress in delineating the molecular basis of Aagenaes' syndrome (OMIM 214900). On the basis of the absence of linkage to this region, Fruhwirth *et al.*[36] presented evidence for a second locus in lymphedema–cholestasis syndrome, which they designated LCS2. However, the gene for this putative locus remains unknown.

Likewise, there seems to have been no progress in defining the genetic basis of Hennekam syndrome (OMIM 235510). These two recessive phenotypes may be amenable to homozygosity mapping techniques using high-density SNP genotyping arrays.

Hypotrichosis–Lymphedema–Telangiectasia Syndrome

Irrthum *et al.*[37] reported three families with hypotrichosis–lymphedema–telangiectasia (OMIM

TABLE 1. Mouse genes associated with lymphedema phenotypes from the Mouse Informatics Database[a]

Gene	Phenotype
Angiopoietin 2	Structural defects of the larger lymphatic vessels and abnormal patterning of the smaller lymphatics, as well as chylous ascites in some newborn pups
Ephrin	Severe or mild defects in remodeling of the lymphatic capillary plexus and lymphatic drainage system depending on the mutation
Elk	Congenital chylothorax with extremely dilated thoracic lymphatic vessels without lymphedema
Integrin alpha 9	Congenital chylothorax
Podoplainin	Multiple anomalies including congenital lymphedema of the skin especially in the legs and neck, abnormal lymphatic capillary networks, and intestinal lymphagiectasis
Prox-1	Multiple affected organ systems including significant edema secondary to failure of the development of the budding and sprouting of the lymphatic system

[a]http://www.informatics.jax.org/searches/allele_form.shtml

607823), an autosomal disorder characterized by chronic swelling of the extremities due to dysfunction of the lymphatics, based on homology to the *ragged* mutations in the mouse. Knockout of SOX18 in the mouse leads to a severe phenotype, including the absence of vibrissae and coat hairs, generalized edema, and cyanosis, and the pups seldom survive past weaning. The heterozygote has a mild coat-color defect, but no cardiovascular phenotype.[38] However, spontaneous mutations in SOX18, the cause of the *ragged* variants, exhibit edema, chylous ascites, and cyanosis at birth, as well as hair and skin defects.[39] Unfortunately, the lymphatic phenotype has not been examined in the same detail as the lymphatic phenotypes of FOXC2 and VEGFR3 have been studied.

Relevant Murine Models

In the three best molecularly characterized lymphedema syndromes, mutations in FLT4, FOXC2, and SOX18, homologous mouse phenotypes have been identified. The *CHY* mouse demonstrates swollen limbs and chylous ascites. The phenotype is the result of a chemically induced mutation in the mouse homologue of FLT4. The *ragged* phenotype has abnormal coat pigmentation as well as abnormal zigzag hairs. This phenotype is the result of a spontaneous mutation in the mouse Sox18. A small number of these mice also develop chylous ascites. Targeted mutation in the mouse FOXC2 is associated with hyperplastic lymphatics and abnormal lymphatic valves. These studies offer the possibility that other existing mouse phenotypes may provide additional insights into human inherited lymphedema.

TABLE 1 lists genes identified in the mouse informatics database associated with lymphedema or abnormal lymphatic morphology. Inherited forms of human lymphedema are not reported to be caused by these genes; however, these genes may still be attractive candidates for the mutation which is causal in inherited lymphedema.

While much has been learned about lymphatic development by studies of knockout mice, thus far the genes causal for inherited lymphedema have been identified through patient and family studies by more traditional genetic approaches. However, the current cadre of identified genes only accounts for a fraction of the cases of inherited lymphedema, offering the hope that the future study of the lymphedema phenotype will yield a wealth of new information on lymphatic development and function.

Conflicts of Interest

The authors declare no conflicts of interest.

References

1. CUENI, L.N. & M. DETMAR. 2006. New insights into the molecular control of the lymphatic vascular system and its role in disease. J. Invest. Dermatol. **126:** 2167–2177.
2. ADAMS, R.H. & K. ALITALO. 2007. Molecular regulation of angiogenesis and lymphangiogenesis. Nat. Rev. **8:** 464–478.
3. MAKINEN, T., C. NORRMEN & P.V. PETROVA. 2007. Molecular mechanisms of lymphatic vascular development. Mol. Life Sci. e-pub ahead of publication.
4. CARVER, C., G. BRICE, S. MANSOUR, *et al.* 2007. Three children with Milroy disease and *de novo* mutations in VEGFR3. Clin. Genet. **71:** 187–189.
5. KARKKAINEN, M.J., A. SAARISTO, L. JUSSILA, *et al.* 2001. A model for gene therapy of human hereditary lymphedema. Proc. Natl. Acad. Sci. USA **98:** 12677–12682.

6. SAARISTO, A., T. VEIKKOLA, T. TAMMELA, *et al.* 2002. Lymphangiogenic gene therapy with minimal blood vascular side effects. J. Exp. Med. **196:** 719–730.

7. YOON, Y., T. MURAYAMA, E. GRAVEREAUX, *et al.* 2003. VEGF-C gene therapy augments postnatal lymphangiogenesis and ameliorates secondary lymphedema. J. Clin. Invest. **111:** 717–725.

8. KERJASCHKI, D., N. HUTTARY, I. RAAB, *et al.* 2006. Lymphatic endothelial progenitor cells contribute to *de novo* lymphangiogenesis in human renal transplants. Nature Med. **12:** 230–234.

9. SAARISTO, A., T. TAMMELA, A. FARKKILA, *et al.* 2006. Vascular endothelial growth factor-C accelerates diabetic wound healing. Am. J. Pathol. **169:** 1080–1087.

10. HU, Y., I. RAJANTIE, K. PAJUSOLA, *et al.* 2005. Vascular endothelial growth factor receptor 3-mediated activation of lymphatic endothelium is crucial for tumor cell entry and spread via lymphatic vessels. Can. Res. **65:** 4739–4746.

11. ROBERTS, N., B. KLOOS, M. CASSELLA, *et al.* 2006. Inhibition of VEGFR-3 activation with the antagonistic antibody more potently suppresses lymph node and distant metastases than inactivation of VEGFR-2. Cancer Res. **66:** 2650–2657.

12. IIDA, K., H. KOSEKI, H. KAKINUMA, *et al.* 1997. Essentials roles of the winged helix transcription factor MFH-1 in aortic arch patterning and skeletogenesis. Development **124:** 4627–4638.

13. WINNIER, G.E., L. HARGETT & B.L.M. HOGAN. 1997. The winged helix transcription factor MFH1 is required for proliferation and patterning of paraxial mesoderm in the mouse embryo. Gene Dev. **11:** 926–940.

14. WINNIER, G.E., T. KUME, K. DENG, *et al.* 1999. Roles for the winged helix transcription factors MF1 and MFH1 in cardiovascular development revealed by nonallelic non-complementation of null alleles. Dev. Biol. **213:** 418–431.

15. KRIEDERMAN, B.M., T.L. MYLOYDE, M.H. WITTE, *et al.* 2003. FOXC2 haploinsufficient mice are a model for human autosomal dominant lymphedema-distichiasis syndrome. Hum. Mol. Genet. **12:** 1179–1185.

16. PETROVA, T.V., T. KARPANEN, C. NORRMEN, *et al.* 2004. Defective values and abnormal mural cell recruitment underlie lymphatic vascular failure in lymphedema-distichiasis. Nat. Med. **10:** 974–981.

17. DAGENAIS, S.L., R.L. HARTSOUGH, R.P. ERICKSON, *et al.* 2004. FOXC2 is expressed in developing lymphatic vessels and other tissues with lymphedema-distichiasis syndrome. Gene Expression Patterns **4:** 611–619.

18. ROSBOTHAM, J.L., G.W. BRICE, A.H. CHILD, *et al.* 2000. Distichiasis-lymphadema: clinical features, venous function and lymphoscintigraphy. Br. J. Dermatol. **142:** 148–152.

19. BRICE, G., S. MANSOUR R. BELL, *et al.* 2002. Analysis of the phenotypic abnormalities in lymphedema-distichiasis syndrome in 74 patients with FOXC2 mutations or linkage to 16p24. J. Med. Genet. **79:** 478–483.

20. NG, M.Y.M., T. ANDREW, T.D. SPECTOR, *et al.* 2004. Linkage to the FOXC2 region of chromosome 16 for varicose veins in otherwise healthy unselected sibling pairs. J. Med. Genet. **43:** 235–239.

21. MELLOR, R.H., G. BRICE, A.B.W. STANTON, *et al.* 2007. Mutations in FOXC2 are strongly associated with primary value failure in veins of the lower limb. Circulation **115:** 1912–1920.

22. SHOLTO-DOUGLAS-VERNON, C., R. BELL, G. BRICE, *et al.* 2005. Lymphedema-distichiasis and FOXC2: unreported mutations, *de novo* mutation estimate, families without coding mutations. Hum. Genet. **117:** 238–242.

23. FANG, J., S.L. DAGENAIS, R.P. ERICSON, *et al.* 2000. Mutations in FOXC2 (MFH1), a forkhead family transcription factor, are responsible for the hereditary lymphedema-distichiasis syndrome. Am. J. Hum. Genet. **67:** 1382–1388.

24. CEDERBERG, A., L.M. GRONNING & B. AHREN. 2001. FOXC2 is a winged helix gene that counteracts obesity, hypertriglyceridemia, and diet-induced insulin resistance. Cell **106:** 563–573.

25. RIDDERSTROLE, M., M. CARLSSON, M. KLANNERMARK, *et al.* 2002. FOXC2 mRNA expression and a 5′-untranslated region polymorphism of the gene are associated with insulin resistance. Diabetes **51:** 3554–3560.

26. HARVEY, N.L., R.S. SHRINIVASAN, M.E. DILLARD, *et al.* 2005. Lymphatic vascular defects promoted by Prox1 haploinsufficiency cause adult-onset obesity. Nat. Genet. **37:** 1072–1081.

27. FINEGOLD, D.N., M.A. KIMAK, E.C. LAWRENCE, *et al.* 2001. Truncating mutation in FOXC2 causes multiple lymphedema syndromes. Hum. Mol. Genet. **10:** 1185–1189.

28. HOQUE, S.R., S. MANSOUR, P.S. MORTIMER, *et al.* 2007. Yellow nail syndrome: not a genetic disorder? Eleven new cases and a review of the literature. Br. J. Dermatol. **156:** 1230–1234.

29. TARTAGLIA, M., E.L. MEHLER, R. GOLDBERG, *et al.* 2001. Mutations in PTPN11, encoding the protein tyrosine phosphatase SHP-2, cause Noonan syndrome. Nat. Genet. **29:** 465–468.

30. SCHUBBERT, S., M. ZENKER, S.L. ROWE, *et al.* 2006. Germline KRAS mutations cause Noonan syndrome. Nat. Genet. **38:** 331–398.

31. TARTAGLIA, M., L.A. PENNACCHIO, C. ZHAO, *et al.* 2007. Gain-of-function SOS1 mutations cause a distinctive form of Noonan syndrome. Nat. Genet. **39:** 75–79.

32. ROBERTS, A.E., T. ARAKI, K.D. SWANSON, *et al.* 2007. Germline gain-of-function mutations in SOS1 cause Noonan syndrome. Nat. Genet. **39:** 70–74.

33. PODGRABINSKA, S., P. BRAUN, P. VELASCO, *et al.*, 2003. Molecular characterization of lymphatic endothelial cells. Proc. Natl. Acad. Sci. USA **99:** 16069–16074.

34. HIRAKAWA, S., Y.W. HONG, N. HARVEY, *et al.* 2003. Identification of vascular lineage-specific genes by transcriptional profiling of isolated blood vascular and lymphatic endothelial cells. Am. J. Pathol. **162:** 575–586.

35. BULL, L.N., E. ROCHE, E.J. SONG, *et al.* 2002. Mapping of the locus for cholestasis- lymphedema syndrome (Aagenaes syndrome) to a 6.6-cM interval on chromosome 15q. Am. J. Hum. Genet. **67:** 994–999.

36. FRUHWIRTH, M., A.R. JANECKE, T. MULLER *et al.* 2003. Evidence for genetic heterogeneity in lymphedema-distichiasis syndrome. J. Pediatr. **142:** 441–447.

37. IRRTHUM, A., K. DEVRIENDT, D. CHITAYAT, *et al.*, 2003. Mutations in the transcription factor gene SOX18 underlie recessive and dominant forms of hypotrichosis-lymphedema-telangiectasia. Am. J. Hum. Genet. **72:** 1470–1478.

38. PENNISI, D., J. GARDNER, D. CHAMBERS, *et al.* 2000. Mutations in SOX18 underlie cardiovascular and hair follicle defects in ragged mice. Nat. Genet. **24:** 434–437.

39. JAMES, K., B. HOSBING, J. GARDNER, *et al.* 2003. SOX18 mutations in the ragged mouse alleles ragged like and opossum. Genesis **36:** 1–6.

Phenotypic Characterization
of Primary Lymphedema

FIONA CONNELL,[a] GLEN BRICE,[a] AND PETER MORTIMER[b]

[a] South West Thames Regional Genetics Unit, and

[b] Department of Cardiac and Vascular Sciences, St Georges University of London, London, SW17 0RE, United Kingdom

The phenotypic entities of primary lymphedema vary in age of onset, site of edema, associated features, inheritance patterns, and underlying genetic cause. Determining the representative phenotype for different types of genetically determined primary lymphedema has been successfully achieved with Milroy's disease and the lymphedema–distichiasis syndrome. Here we describe and illustrate their well-delineated phenotypes. Phenotype characterization facilitates the identification of causative genes, as has been demonstrated with *VEGFR3* and *FOXC2*, in Milroy's disease and lymphedema–distichiasis respectively. Other forms of primary lymphedema are discussed.

Key words: lymphedema–distichiasis; lymphoscintigraphy; phenotype; primary lymphedema; Milroy's disease

Introduction

Primary lymphedema is a chronic edema caused by a developmental abnormality of the lymphatic system.[1] The phenotypic entities of primary lymphedema vary in age of onset, site of edema, associated features, inheritance patterns, and underlying genetic cause. Determining the representative phenotype for different types of genetically determined primary lymphedema has been successfully achieved with Milroy's disease and lymphedema–distichiasis, and these conditions demonstrate how phenotype characterization facilitates the identification of causative genes.

The pathogenesis of most lymphatic disorders is not fully understood, but is an evolving research area. Methods of investigating the lymphatic system are also limited, but growing experience has enabled recognition of patterns pertaining to some different phenotypes.

This chapter describes the well-delineated phenotypes of Milroy's disease and lymphedema–distichiasis and addresses some of the pathogenic mechanisms that have been identified following accurate grouping of patients according to phenotype.

Address for correspondence: Glen Brice, South West Thames Regional Genetics Unit, St Georges University of London, Cranmer Terrace, London, UK SW17 0RE. Voice: +44-2087255192; fax: +44-2087253444.

gbrice@sgul.ac.uk

Clinical Phenotypes

Milroy's Disease

Milroy's disease was first recognized in 1892 by William Milroy, who described an inherited, painless, nonprogressive edema, of congenital onset, confined to the lower limbs.[2] In 1928 he published a follow-up paper demonstrating the benign nature of the condition.[3] Since that time the term "Milroy's disease" has sometimes been inaccurately used as an umbrella label for many types of primary lymphedema, when in fact the clinical phenotype has been well delineated.

Milroy's disease is an autosomal dominantly inherited condition. Genetic linkage of this disease was shown to be at a locus on chromosome 5q35.3. Mutations in vascular endothelial growth factor receptor-3, *VEGFR3* (also known as *FLT4*), which maps to this region, have been identified.[4–6] Subsequently, 19 mutations in *VEGFR3* have been reported.[4,6–9] Fourteen further mutations have been identified by our group (unpublished data). The mutations identified to date are all in the kinase domains of *VEGFR3* (coded by exons 17–26), but ascertainment bias may have contributed to this finding, as only Evans *et al.* analyzed the entire gene when studying 12 families with primary lymphedema.[8] VEGFR3 has a role in embryonic development of the cardiovascular system acting as a receptor for vascular endothelial growth factor C (VEGF-C) and VEGF-D.[10] Following Milroy's initial description, a family history was deemed necessary for the diagnosis of Milroy's disease, but *de novo* mutations

Ann. N.Y. Acad. Sci. 1131: 140–146 (2008). © 2008 New York Academy of Sciences.
doi: 10.1196/annals.1413.013

140

in the *VEGFR3* gene have now been reported in four patients with sporadic, congenital lymphedema.[11,12]

Not all patients with a Milroy's disease phenotype have *VEGFR3* mutations. There is uncertainty regarding the prevalence of kinase domain *VEGFR3* mutations in these patients, but in one study it has been found to be as high as 72% in patients with a typical Milroy's disease phenotype and a positive family history, and 64% if a positive family history is not a diagnostic criterion (S. Jeffery, personal communication). "Typical Milroy's disease phenotype" in this context is regarded as bilateral or unilateral lower limb lymphedema that was present at birth. The likelihood of detecting *VEGFR3* mutations in patients with other forms of primary lymphedema (e.g., patients with lymphedema affecting body parts other than the lower limbs or a syndromic or dysmorphic phenotype) is negligible. Brice *et al.* reported the clinical findings in 71 patients with clinical Milroy's disease and a *VEGFR3* mutation.[13] Exclusion of cases with no *VEGFR3* mutation meant that the clinical phenotype of Milroy's disease was refined. In this study 90% of patients with a *VEGFR3* mutation had lymphedema, demonstrating that there is incomplete penetrance in this condition. In another study, penetrance is reported to be 84%.[5] There is also wide inter- and intrafamilial variation in expression of the condition. Antenatal diagnosis of Milroy's disease has been reported as early as at 12 weeks' gestation, with the detection of pedal edema (and in one fetus, more extensive edema) on ultrasound scans; prenatal diagnosis can be carried out, but is rarely requested.[14–16] TABLE 1 summarizes this delineated phenotype and illustrates the typical Milroy's disease clinical signs.

Further characterization of the Milroy's phenotype through investigation with lymphoscintigraphy has revealed that lymphatic drainage routes and function are completely normal in the upper limbs despite the germline mutation. Lymphatic abnormalities are only demonstrable at sites of swelling. Here lymphoscintigraphy indicates failure of subcutaneous lymph absorption from the tissues by initial lymphatics (FIG. 1B). Contrary to the traditional view that aplasia of lymphatics is the mechanism by which Milroy's lymphedema arises, skin biopsy from the swollen feet of patients with *VEGFR3* mutations has revealed the presence of abundant skin lymphatics, but they are non-functional on fluorescence microlymphangiography (FML) (R. Mellor, personal communication). FML images the functional initial lymphatics in the skin after their uptake of FITC–dextran.[17] The mechanism by which Milroy's lymphedema develops therefore appears not to be due to an absence of initial lymphatics, but rather attributable to their functional failure. This is in direct contrast to knockout mice with similar *VEGFR3* mutations and lymphedema, where aplasia of skin lymphatics is observed.[18] This indicates the need for caution when inferring similarities between species' phenotypes, as manifestations may differ despite apparent identical genotypes.

The observation that patients with Milroy's lymphedema possess visible large superficial veins, even in infancy, prompted investigation by color Doppler venous duplex ultrasound. A high prevalence of superficial venous reflux was detected, suggesting, as with lymphedema–distichiasis (see below), that *VEGFR3* mutations are strongly associated with primary valve failure in leg veins (R. Mellor, personal communication). Venous reflux is, however, unlikely to contribute to the edema because the swelling manifests before the patient adopts the upright posture.

Lymphedema–Distichiasis

Lymphedema–distichiasis is a single gene disorder caused by *FOXC2* (forkhead transcription factor) mutations.[19] Distichiasis is an aberrant growth of eyelashes from the meibomian glands of the eyelid and comes from the Greek *distikhos*, meaning "two rows." Like Milroy's disease, it is an autosomal dominantly inherited condition. *FOXC2* mutations in lymphedema–distichiasis syndrome are reported as highly penetrant with variable expressivity.[20] In some rare familial cases of lymphedema–distichiasis no change can be found in *FOXC2*, suggesting heterogeneity or a mutation outside the *FOXC2* gene but having a "positional effect" on *FOXC2* expression.

The literature on *FOXC2* mutations has been confused by claims that sequence changes in this gene are responsible for other primary lymphedema disorders (Meige's disease, lymphedema–ptosis and lymphedema–yellow nail syndromes[21]). This assertion has been recently refuted.[22] Lymphedema with ptosis, described by Bloom[23] and previously thought to be a separate entity from lymphedema–distichiasis, may be allelic. The lymphedema–distichiasis syndrome was not recognized until 1964,[24] so would not have been recognized by Bloom. Families published as having "lymphedema with ptosis" may well have had distichiasis as this, even today, often goes undiagnosed because it can be difficult to detect and is asymptomatic in up to 25% of cases.[25] As many as 30% of those with a mutation in *FOXC2* have ptosis, and at least one family with reportedly isolated lymphedema with ptosis has been

TABLE 1. Clinical features of Milroy's disease

	Most commonly	Less frequently	Illustration
Age of onset of swelling	At birth	Becomes evident prior to age 1 year; very occasionally swelling develops in adulthood, but this is not typical	
Site of swelling Extent of swelling Type of swelling	Bilateral lower limb Below knee Woody, brawny texture	Unilateral lower limb Entire lower limb Softer, pitting edema	
Associated features	Prominent veins (especially long saphenous vein)		
	Deep toe creases		
	"Ski-jump" toenails		
	Papillomatosis (tends to be most severe over the second toe)		
	Cellulitis Hydrocoeles	Urethral abnormalities	

TABLE 2. Clinical features of lymphedema–distichiasis syndrome

Clinical finding	Description	Illustration
Distichiasis	Aberrant eyelashes arising from the meibomian glands, along posterior border of the lid margins.	
Lymphedema	Lower limb swelling (usually bilateral and predominantly asymmetrical) Onset often around late childhood/puberty (later in females). Severity can worsen over time. Can be complicated by cellulitus and/or papillomatosis.	
Associated features	Varicose veins Ptosis, photophobia Cleft palate Congenital heart disease (most commonly, tetralogy of Fallot) Spinal extradural cysts	

shown to have a *FOXC2* mutation.[21] In all cases where there appears to be lymphedema and ptosis, distichiasis should be specifically asked about and clinical examination by an experienced observer is recommended. In another study carried out to look for *FOXC2* mutations as a cause for primary lymphedema, of 23 probands reported to have isolated pubertal-onset primary lymphedema (Meige's disease), only one was found to have a mutation in *FOXC2*. Examination of the family members of the proband with the *FOXC2* mutation revealed that, although the proband did not have distichiasis, several relatives had evidence of distichiasis on examination, making a diagnosis of lymphedema–distichiasis in this family.[22]

Distichiasis is a congenital anomaly, with reported penetrance of 94–100% in lymphedema–distichiasis syndrome.[20,25] Distichiasis can be associated with corneal irritation, recurrent conjunctivitis, and photophobia, and these can be significant clues in the medical history. Lubrication, plucking, cryotherapy, electrolysis, and lid-splitting surgery are all treatment options for distichiasis.

In this condition, while distichiasis is present from birth, lymphedema usually develops only from puberty. Of interest. swelling tends to develop earlier in

life in males than females. Swelling is usually below knee and bilateral, but can extend into the thighs. Associated features in lymphedema–distichiasis include cleft palate and congenital heart disease, seen in 4% and 7% of lymphedema–distichiasis patients, respectively.[25] These are signs that can be sought on antenatal scans in at-risk individuals. Although prenatal diagnosis can be offered, in our experience, it is not often utilized.

Identification of *FOXC2* as the gene responsible for lymphedema–distichiasis syndrome enabled the associated clinical phenotype of the condition to be analyzed. Brice *et al.* defined the phenotype, having clinically characterized 74 affected subjects with *FOXC2* mutations or linkage to 16q24 (the *FOXC2* locus).[25] Erickson *et al.* also produced a report on the clinical heterogeneity of the disorder.[20] TABLE 2 summarizes and illustrates the clinical phenotype.

Further investigation of the lymphedema–distichiasis phenotype has revealed both lymph and venous reflux in lower limbs, suggesting primary valve failure. Lymphoscintigraphy, utilizing a subcutaneous tracer injection, reveals patent lymph drainage routes and uptake in ilioinguinal lymph nodes. Subsequently reflux occurs within the main collector lymphatics,

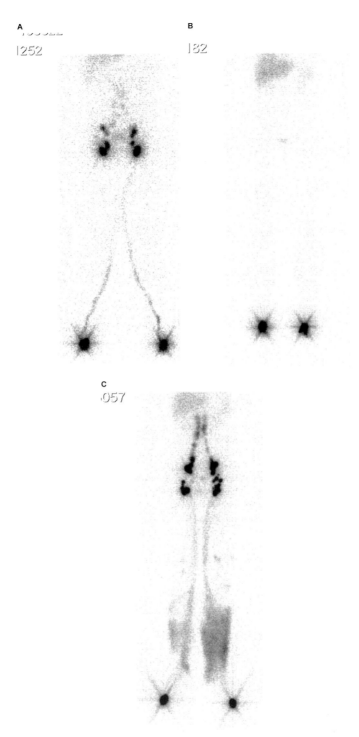

FIGURE 1. Lymphoscintigraphy. Normal **(A)**: showing uptake of trace bilaterally to groin lymph nodes; Milroy's disease **(B)**: no uptake of tracer from injection site. Majority of tracer remains in foot after 2h; lymphedema–distichiasis **(C)**: uptake of trace to both groin but with dermal backflow evident, particularly in left leg.

back to the initial lymphatics, thereby imaging the legs and feet in profile (FIG. 1). FML demonstrates that the initial lymphatic network in the skin is less functional when the leg is dependent (R. Mellor, personal communication). This was interpreted as indicating that higher intralymphatic pressures in the dependent position discouraged uptake of FITC–dextran. Skin biopsies, however, revealed that an abnormally large proportion of skin lymphatic vessels was covered with smooth muscle cells in those study participants with *FOXC2* mutations but not in family members without the mutation.[26] This complemented the findings in *FOXC2*$^{-/-}$, mice indicating that FOXC2 is essential for the morphogenesis of lymphatic valves and the establishment of a pericyte-free initial lymphatic network.[26]

Venous duplex ultrasound revealed superficial venous reflux in all patients with a *FOXC2* mutation, including three without lymphedema. Deep vein reflux was observed in 14 of 18 patients studied.[27] Thus, FOXC2 is required for normal venous function and more specifically for valve development and/or maintenance in humans. Unlike the case in Milroy's lymphedema, the finding of venous reflux in the lymphedema–distichiasis syndrome could be a major contributory factor in the onset and progression of swelling.

Other Primary Lymphedema Disorders

In addition to Milroy's disease and lymphedema–distichiasis, there are many other forms of primary lymphedema. Hypotrichosis–lymphedema–telangiectasia syndrome is the third primary lymphedema condition for which the genetic cause is known. Mutations in *SOX18* are causative. This extremely rare syndrome is characterized by the association of childhood-onset lymphedema in the legs, loss of hair, and telangiectasia, particularly in the palms. Inheritance is either autosomal dominant or autosomal recessive.[28]

Clinical classification of the different types of primary lymphedema is difficult without a recognized diagnostic pathway that encompasses more specific phenotypes. There is a historical tendency to divide primary lymphedema into three subgroups: lymphedema congenita, lymphedema praecox, and lymphedema tarda. However, classification purely on the age of onset of the swelling does not allow for refinement of phenotypes.

Meige's syndrome is typically a pubertal onset form of lymphedema, affecting the lower limbs, that tends to follow an autosomal dominant pattern of inheritance. There are no other defining clinical features of the condition and, thus with a nonspecific phenotype, obtaining a cohort of cases/families for targeted molecular investigation has not been fruitful in the search for a causative gene.

Our experience has shown us that, in addition to the many syndromes that are known to have lymphedema as a clinical feature (e.g., Hennekam's, Aagenaes's, microcephaly–chorioretinopathy–lymphedema, Mucke's, and the Noonan syndrome, and Turner's and Prader–Willi syndromes), there are forms of primary lymphedema that do not form part of a syndromic diagnosis, but demand recognition as separate entities.

Lymphedema can be generalized, affecting the whole body, or can affect arms and legs, face, conjunctiva, and genitalia in a segmental pattern, with any combination of body parts being affected. In the primary lymphedema disorders, facial, conjunctival, and genitalia lymphedema is usually seen in association with one or more limbs involved. There is also a significant cohort of patients with primary lymphedema of limbs, with/without other body parts affected, that have a systemic component to the lymphatic abnormality. These cases can present antenatally with signs of edema. Systemic involvement in primary lymphedema disorders includes intestinal lymphangiectasia, ascites, pleural effusions, pericardial effusions, and pulmonary lymphangiectasia. Diagnosis of such complications has important implications for patient management.

The benefit of characterizing the specific phenotype of these other types of primary lymphedema, will be that grouping patients will facilitate the identification of new pathogenic genes and enable the recognition of emerging new syndromes. The natural history of these conditions can also be monitored and the risk of transmission to future generations established. This has been demonstrated by the clinical phenotyping and molecular genetic research. done in partnership, into Milroy's disease and lymphedema–distichiasis.

Acknowledgments

We would like to acknowledge the help of the following members of the Lymphoedema Research Consortium: Dr. Sahar Mansour, Professor Steve Jeffrey, and Dr. Russell Mellor from St George's University of London, UK.

Conflicts of Interest

The authors declare no conflicts of interest.

References

1. MORTIMER, P.S. 1995. Managing lymphoedema. Clin. Exp. Dermatol. **13:** 499–505.

2. MILROY, W.F. 1892. An undescribed variety of hereditary edema. N. Y. Med. J. **56:** 503.

3. MILROY, W.F. 1928. Chronic hereditary edema: Milroy's disease. JAMA **91:** 1172–1175.

4. FERRELL, R.E., K.L. LEVINSON, J.H. ESMAN, *et al.* 1998. Hereditary lymphedema: evidence for linkage and genetic heterogeneity. Hum. Mol. Genet. **7:** 2073–2078.

5. EVANS, A.L., G. BRICE, V. SOTIROVA, *et al.* 1999. Mapping of primary congenital lymphedema to the 5q35.3 region. Am. J. Hum. Genet. **64:** 547–555.

6. KARKKAINEN, M.J., R.E. FERRELL, E.C. LAWRENCE, *et al.* 2000. Missense mutations interfere with VEGFR-3 signalling in primary lymphoedema. Nat. Genet. **25:** 153–159.

7. IRRTHUM, A., M.J. KARKKAINEN, K. DEVRIENDT, *et al.* 2000. Congenital hereditary lymphedema caused by a mutation that inactivates VEGFR3 tyrosine kinase. Am. J. Hum. Genet. **67:** 295–230.

8. EVANS, A.L., R. BELL, G. BRICE, *et al.* 2003. Identification of eight novel VEGFR-3 mutations in families with primary congenital lymphedema. J. Med. Genet. **40:** 697–703.

9. BUTLER, M.G., S.L. DAGENAIS, S.G. ROCKSON & T.W. GLOVER. 2007. A novel VEGFR3 mutation causes Milroy disease. Am. J. Med. Genet. A **143:** 1212–1217.

10. DUMONT, D.J., L. JUSSILA, J. TAIPALE, *et al.* 1998. Cardiovascular failure in mouse embryos deficient in VEGF receptor-3. Science **282:** 946–949.

11. GHALAMKARPOUR, A., S. MORLOT, A. RAAS-ROTHSCHILD, *et al.* 2006. Hereditary lymphedema type I associated with VEGFR3 mutation: the first de novo case and atypical presentations. Clin. Genet. **70:** 330–335.

12. CARVER, C., G. BRICE, S. MANSOUR, *et al.* 2007. Three children with Milroy disease and de novo mutations in VEGFR3. Clin. Genet. **71:** 187–189.

13. BRICE, G., A.H. CHILD, A. EVANS, *et al.* 2005. Milroy disease and the VEGFR-3 mutation phenotype. J. Med. Genet. **42:** 98–102.

14. FRANCESCHINI, P., D. LICATA, G. RAPELLO, *et al.* 2001. Prenatal diagnosis of Nonne-Milroy lymphedema. Ultrasound Obstet. Gynecol. **18:** 182–183.

15. MAKHOUL, I.R., P. SUJOV, N. GHANEM & M. BRONSHTEIN. 2002. Prenatal diagnosis of Milroy's primary congenital lymphedema. Prenat. Diagn. **22:** 823–826.

16. DANIEL-SPIEGEL, E., A. GHALAMKARPOUR, R. SPIEGEL, *et al.* 2005. Hydrops fetalis: an unusual prenatal presentation of hereditary congenital lymphedema. Prenat. Diagn. **25:** 1015–1018.

17. BOLLINGER, A., K. JÄGER, F. SGIER & J. SEGLIAS. 1981. Fluorescence microlymphography. Circulation **64:** 1195–1200.

18. KARKKAINEN, M.J., A. SAARISTO, L. JUSSILA, *et al.* 2001. A model for gene therapy of human hereditary lymphedema. Proc. Natl. Acad. Sci. **98:** 12677–12682.

19. FANG, J., S.L. DAGENAIS, R.P. ERICKSON, *et al.* 2000. Mutations in FOXC2 (MFH-1), a forkhead family transcription factor, are responsible for the hereditary lymphedema-distichiasis syndrome. Am. J. Hum. Genet. **67:** 1382–1388.

20. ERICKSON, R.P., S.L. DAGENAIS, M.S. CAULDER, *et al.* 2001. Clinical heterogeneity in lymphedema-distichiasis with FOXC2 truncating mutations. J. Med. Genet. **38:** 761–766.

21. FINEGOLD, D.N., M.A. KIMAK, E.C. LAWRENCE, *et al.* 2001. Truncating mutations in FOXC2 cause multiple lymphedema syndromes. Hum. Mol. Genet. **10:** 1185–1189.

22. REZAIE, T., R. GHOROGHCHIAN, R. BELL, *et al.* 2008. Primary non-syndromic lymphoedema (Meige disease) is not caused by mutations in FOXC2. Eur. J. Hum. Genet. **16**(3): 300–304.

23. BLOOM, D. 1941. Hereditary lymphedema (Nonne-Milroy-Meige): report of a family with hereditary lymphedema associated with ptosis of the eyelids in several generations. N. Y. State J. Med. **41:** 856–863.

24. FALLS, H.F. & E.D. KERTESZ, 1964. A new syndrome combining pterygium colli with developmental anomalies of the eyelids and lymphatics of the lower extremities. Trans. Am. Ophthal. Soc. **62:** 248–275.

25. BRICE, G., S. MANSOUR, R. BELL, *et al.* 2002. Analysis of the phenotypic abnormalities in lymphedema-distichiasis syndrome in 74 patients with FOXC2 mutations or linkage to 16q24. J. Med. Genet. **39:** 478–483.

26. PETROVA, T.V., T. KARPANEN, NORRMÉN C, *et al.* 2004. Defective valves and abnormal mural cell recruitment underlie lymphatic vascular failure in lymphedema distichiasis. Nat. Med. **10:** 974–981.

27. MELLOR, R.H., G. BRICE, A.W. STANTON, *et al.* 2007. Mutations in FOXC2 are strongly associated with primary valve failure in veins of the lower limb. Circulation **115:** 1912–1920.

28. IRRTHUM, A., K. DEVRIENDT, D. CHITAYAT, *et al.* 2003. Mutations in the transcription factor gene SOX18 underlie recessive and dominant forms of hypotrichosis-lymphedema-telangiectasia. Am. J. Hum. Genet. **72:** 1470–1478.

Estimating the Population Burden of Lymphedema

STANLEY G. ROCKSON AND KAHEALANI K. RIVERA

Stanford Center for Lymphatic and Venous Disorders, Division of Cardiovascular Medicine, Stanford University School of Medicine, Stanford, California, USA

Lymphedema is a complex, regional edematous state that ensues when lymph transport is insufficient to maintain tissue homeostasis. The disorder is remarkably prevalent, but the population implications of lymphatic dysfunction are not well-studied. Prevalence estimates for lymphedema are relatively high, yet its prevalence is likely underestimated. The ability to estimate the burden of disease poses profound implications for current and future lymphedema patients, but the challenge to correctly surmise the incidence and prevalence of lymphedema is complex and the relevant medical literature is scanty. In the absence of the highly desired, prospectively designed and rigorously performed relevant epidemiologic studies, it is instructive to look at the existing studies of lymphedema disease burden. In the current review, the extant literature is examined in the context of the disease setting in which tissue edema is encountered. Incidence or prevalence estimates are provided or inferred, and, where feasible, the size of the subject population is also identified. It is extremely attractive to contemplate that future approaches will entail formal, prospectively designed studies to objectively quantitate incidence and prevalence statistics for individual categories, as well as for the global lymphedema population.

Key words: cancer-related lymphedema; lymphedema epidemiology; trauma-related lymphedema; tissue edema

Lymphedema, the regional, complex edematous state that ensues when lymph transport is insufficient to maintain tissue homeostasis, is remarkably prevalent, but the population implications of lymphatic dysfunction are not well studied.[1–3] Prevalence estimates for lymphedema are relatively high, yet its prevalence is likely underestimated.[4] Attempts to identify the population impact of lymphedema are hampered by the fact that this chronic, debilitating disease is frequently underrecognized or misdiagnosed: treatment delays are common and many patients never receive treatment.[4] Recent prevalence estimates suggest that chronic edema is present between 1.33 per thousand[5] to 1.44 per thousand.[6] Nevertheless, underestimation of disease prevalence emanates, at least in part, from the problem of ascertainment of disease by health care professionals: not all patients are likely to receive treatment for the condition,[6] a factor that is particularly relevant to lymphedema. Another confounding attribute is the variable manner in which lymphedema

is clinically detected and defined.[3,7] In the absence of a repeatable, valid, and accepted definition for the presence of lymphedema,[3] the reported incidence and prevalence estimates for lymphedema must be scrutinized with regard to the methods in which tissue edema is sought, identified, and quantitated.

The ability to estimate the population burden of disease poses profound implications for current and future lymphedema patients, having an impact upon risk stratification of an elusive disease, and profound implications for insurance and reimbursement issues. Furthermore, the motivation of pharmaceutical and biotechnology sectors to undertake the development of new treatment strategies is heavily linked to the accurate perception of the disease burden.[8]

The challenge to correctly surmise the incidence and prevalence of lymphedema is complex and the relevant medical literature is scanty.[3] Nevertheless, in the absence of the highly desired, prospectively designed and rigorously performed relevant epidemiologic studies, it is instructive to examine, in some detail, the existing reports that enumerate the lymphedema disease burden. In the current review, the extant literature will be examined in the context of the disease setting in which tissue edema is encountered. Incidence or prevalence estimates are provided or inferred, and,

Address for correspondence: Stanley G. Rockson, M.D., Stanford Center for Lymphatic and Venous Disorders, Division of Cardiovascular Medicine, Stanford University School of Medicine, 300 Pasteur Drive, Stanford, CA 94305. Voice: +1-650-725-7571; fax: +1-650-725-1599.
srockson@cvmed.stanford.edu

Ann. N.Y. Acad. Sci. 1131: 147–154 (2008). © 2008 New York Academy of Sciences.
doi: 10.1196/annals.1413.014

TABLE 1. Primary lymphedema

Diagnosis	Incidence/Prevalence	References
Primary lymphedema	1.15/100,000 persons < age 20 y	12
Genital lymphedema	49% of male genital edema is accompanied by primary lower extremity lymphedema ($n = 33$ males)	22
	39% had hydrocele	
Lymphedema–distichiasis syndrome	94.2% penetrance of lymphedema ($n = 74$); 49% varicose veins of early onset	13
Turner's syndrome	30% prevalence of lymphedema ($n = 23$)	15–19
	Congenital lymphedema in 17% ($n = 23$)	
	15.4% had cystic hygroma	
	Fetal diagnosis of cystic hygroma in 75%	
Aagenaes' syndrome	100% of affected individuals	14
Trisomy 18	15.4% had cystic hygroma	17

where feasible, the size of the subject population is also identified (TABLES 1–3).

The simplest schema for the classification of lymphedema patients relies upon a differentiation between causes that are designated as *primary* or *secondary* (also known as *acquired*).[1]

Primary Lymphedema

Primary lymphedema may be present at birth or develop at predictable points in the patients' natural history. Therefore, the primary lymphedemas are often further classified according to the age of the patient when the edema is first detected. Congenital lymphedema is apparent at birth or becomes recognized within the first 2 years of life. *Lymphedema praecox* is most commonly detected at the time of puberty, but may appear as late as the third decade of life. *Lymphedema tarda* is typically first detected after age 35.

Reported estimates of the incidence of primary lymphedema suggest that this condition is neither common nor rare.[9] While current published figures for the incidence and prevalence are scanty, small observational analyses have suggested that, for example, 8% of newly diagnosed lymphedema clinic patients had primary forms of the disease.[10] In published observations of chronic outpatient lymphedema care, among those with non-cancer-related disease, 28% were diagnosed with primary lymphedema.[11]

Prevalence estimates for all forms of primary lymphedema (TABLE 1) suggest that the disease, without regard to pathogenesis, may affect 1.15/100,000 individuals younger than 20 years of age.[12]

Among the disease entities within this category, there are several identified heritable diseases.[2] When cases of congenital lymphedema cluster in families, an autosomal dominant pattern of transmission is most frequently described, such as in the lymphedema–distichiasis syndrome,[13] whereas Aage-naes' syndrome[14] represents an autosomal recessive expression of disease. In addition to specific gene mutations, Turner's syndrome[15–19] and trisomy 18[17] should be considered in this context. While genetically predetermined forms of the primary lymphedema are relatively frequently encountered, sporadic instances of primary disease are more common.[20] Specific categories for which prevalence estimates are available include fetal cystic hygroma[21] and genital lymphedema.[22]

A disease category that, clinically, can readily be confused with primary lymphedema is the entity known as *lipedema*. This condition, both confused with and mistaken for lymphedema, lacks confident prevalence estimates.[23]

Secondary Lymphedema

The obliteration of previously normal lymphatic channels is the hallmark of the acquired form of lymphedema. The most common cause of acquired lymphedema in developed countries, including the United States, is iatrogenic, reflecting predominantly the large patient group in whom lymphatic trauma is a direct consequence of surgical and radiotherapeutic interventions for cancer.[24] Lymphedema can also be acquired from other forms of lymphatic vascular trauma. These include burns and large or circumferential wounds to the extremity, but relative prevalence estimates are difficult to ascertain.

Cancer-related Lymphedema

In the context of cancer therapeutics, lymphedema is most commonly associated with surgical excision of lymph nodes or their irradiation (TABLE 2). In survivors of late cervical, axillary, or inguinal lymphadenectomy, the 14.9% observed incidence of late lymphedema was the most commonly encountered complication, particularly after groin dissection,[25] where an incidence of

TABLE 2. Cancer-related lymphedema

Diagnosis	Incidence/Prevalence	References
Breast cancer	Node-negative disease ($n = 1031$): 5% after sentinel node biopsy	56
	13% after standard axillary dissection	
	22% after surgery and radiotherapy	54
	Sentinel node biopsy: 6.9% at 6 months ($n = 4975$)	57
Malignant melanoma	26% lymphedema after groin lymph node dissection ($n = 204$)	60
	10% after prophylactic lymph node dissection ($n = 44$); 23% after therapeutic lymph node dissection ($n = 64$)	61
	Lymphedema 28.3% ($n = 367$)	63
Cervical cancer	Prevalence in Stage I-IIA 31% ($n = 228$)	105
	Early-stage disease: 21% in first year ($n = 167$)	106
	Stage IB-IIA: radical hysterectomy alone, 5%; adjuvant external irradiation, 22% ($n = 320$)	28
	Stage I-IIA: radical hysterectomy, 10%; surgery and preoperative radiation, 11% ($n = 233$)	68
	Stage IB-IIA, radical hysterectomy and postoperative radiation: 42% ($n = 179$)	107
	T1b-2b cervical cancer, surgery, pelvic lymphadenectomy and postoperative radiation: 42% at 5 y and 49% at 10 y ($n = 128$)	69 69
Endometrial cancer	Surgery $\pm$ radiotherapy 4.6% ($n = 517$)	72
	Early disease 0.7% ($n = 168$)	70
	Early disease 1.8% ($n = 396$)	71
	Surgical therapy 3.4% ($n = 122$)	108
	Uterine corpus cancer 2.4% ($n = 1289$)	74
	Uterine cancer 17.7% ($n = 141$)	73
Vulvar cancer	Squamous cell carcinoma of vulva 26% ($n = 61$)	80
	Irradiation or lymphadenectomy 16% ($n = 48$)	76
	Resection and inguinofemoral lymphadenectomy 28% ($n = 187$)	79
	Radical vulvectomy 48%; modified vulvectomy 12% ($n = 149$)	78
	Stage 1–IV 13% ($n = 60$)	77
	Squamous cell carcinoma 47% ($n = 83$)	75
Prostate cancer	Stage A2–C 3.1% ($n = 289$)	84
	Stage A2–C 7.7% ($n = 65$)	82
	Radiotherapy and staging lymphadenectomy 13% ($n = 16$)	83
	Prostatectomy 2%; surgery and radiotherapy 9% ($n = 442$)	81
Penile cancer	Penile cancer 23% ($n = 53$)	87
	Surgery with groin dissection 33% chronic edema ($n = 67$)	85
	Inguinal (25%) and ilioinguinal (29%) lymphadenectomy ($n = 234$)	86
Soft tissue sarcoma	Survivors of therapeutic intervention: 30% with lymphedema ($n = 54$)	109
	Limb-sparing surgery and radiotherapy 19% ($n = 145$)	90
	Soft tissue sarcoma 3% ($n = 156$)	88
	Pre- (16%) versus postoperative (23%) radiotherapy ($n = 129$)	89

TABLE 3. Trauma-related and iatrogenic lymphedema

Diagnosis	Incidence/Prevalence	References
Peripheral arterial disease	30% of Stage II patients	98
	80% of Stage III and IV patients	
Varicose vein surgery	0.5% of patients (retrospective questionnaire reflecting >184,000 surgeries)	99
Saphenous vein harvesting for aortocoronary bypass	10% incidence of lymphedema ($n = 50$)	100
Burns	1% prevalence in a burn unit (retrospective file review)	101
Circumcision and buried penis repair	1.3% incidence of lymphedema of the penile shaft ($n = 83$)	102
Intrathecal infusion for analgesia of non-cancer-related pain	Development of lymphedema in 22% after pump insertion ($n = 23$)	103
Sirolimus administration after organ transplantation	Lymphedema in 6% (file review of 18 patients)	104

up to 40% has been observed. These surgical complications of tumor nodal metastasis staging and nodal therapeutics do occur, most commonly, in association with breast cancer, malignant melanoma, gynecologic malignancy (cervical, endometrial, and vulvar), urologic malignancy (penile and prostate), and soft tissue sarcomas. The reported frequency of leg edema after pelvic or genital cancer surgeries, particularly when there has been inguinal and pelvic lymph node dissection or irradiation, varies between 1% and 47%.[26,27] Pelvic irradiation increases the frequency of leg lymphedema after cancer surgery.[28,29]

Breast Cancer–related Lymphedema

Breast cancer–associated lymphedema is the most extensively studied cause of acquired lymphatic vascular insufficiency.[30] Clearly, this reflects the fact that the problem of breast cancer–associated lymphedema is the one most commonly encountered. Axillary lymph node dissection and adjuvant radiation therapy are both predisposing factors, particularly when the axilla is included in the radiation field.[31] Ever since Halsted first described this seemingly unavoidable complication of breast cancer intervention,[32] the medical literature has been replete with descriptions of this phenomenon, persisting into the modern surgical and radiotherapeutic era[5,33–50] (TABLE 2). The accrual of new cases is linear-to-exponential in the first 3 years after interventions; thereafter, new case appearance diminishes in number, but persists throughout the natural history of the survival period. Incidence estimates vary broadly, but the most recent observations suggest that lymphedema of the arm is likely to occur in approximately 20% of breast cancer patients after axillary clearance.[5,44] Factors that influence the likelihood of lymphedema development include the extent of axillary dissection[42,44,47,48,51,52] and the use of radiotherapy,[37,38,41–44] particularly when the two treatment modalities are used adjunctively.[33,34,39,47,49,53,54] Systemic factors[1,31,37,45,48] and other surgical variables[33,37–39,44,45,51] may play a role in the incidence of breast cancer–related lymphedema.

Even with recent improvements in surgical and radiotherapeutic techniques, lymphedematous complications cannot be obviated and are, in fact, not uncommon.[40,43,50,55] While incidence of lymphedema is diminished with the sentinel node technique, the risk is certainly not eliminated.[56,57]

Lymphedema and Malignant Melanoma

In malignant melanoma, lymph node dissection for tumor nodal metastasis staging engenders substantial risk for the development of lymphedema (TABLE 2).

Early reports have suggested an incidence approaching 80%,[58] but more recent reports suggest a more modest, yet quite substantial risk that ranges from 6% to 29%.[59–66] As is the case for breast cancer staging and therapeutics, sentinel node biopsy substantially reduces the likelihood of developing lymphedema.[67] Furthermore, prophylactic lymph node dissection appears to be associated with a lesser incidence of lymphedema than therapeutic lymph node excision.[61]

Lymphedema and Gynecologic Malignancy

The treatment of cervical, endometrial, and vulvar malignancies is associated with a significant incidence of acquired lymphedema (TABLE 2). In cervical cancer treated with radical hysterectomy alone, the incidence has been observed to be as low as 5–10%[28,68]; however, adjunctive pelvic irradiation will augment the risk to as high as 49% at 10 years after treatment.[69] The reported frequency of lymphedema after uterine cancer therapy is somewhat more modest,[70–74] while the treatment of vulvar malignancies occasions a comparable degree of lymphedema.[75–80]

Lymphedema and Urologic Malignancy

Groin dissection, nodal staging, and radiotherapy play a substantial role in the treatment of urologic malignancy as well. In prostate cancer, with some dependence upon the grade of neoplastic disease, the incidence of lymphedema has been reported as 3–8%[81–84]; the adjunctive use of radiotherapy has a three- to fourfold augmenting effect upon the likelihood of developing lymphedema.[81,83] In the treatment of penile cancer, where inguinal and ilio-inguinal lymphadenectomy are common, lymphedema has been observed in 23–33% of the small series of patients studied.[85–87]

Lymphedema and Soft Tissue Sarcoma

Therapy of soft tissue sarcoma often entails wide local excision and radiotherapy, thereby invoking the classic risk factors for lymphedema development (TABLE 2). Soft-tissue sarcoma has a reported association of 3%[88] to 30% with acquired lymphedema development.[109] As in the other settings of antineoplastic therapy, the use of irradiation, either pre-[89] or postoperatively,[89,90] seems to increase substantially the risk of development of lymphedema.

Infection

Recurrent episodes of bacterial lymphangitis lead to thrombosis and fibrosis of the lymphatic channels and are one of the most common causes of lymphedema.[91] The etiologic bacteria are almost always streptococci that are prone to enter through breaks in the skin or fissures induced by trichophytosis. Recurrent bacterial lymphangitis is also a frequent complicating factor of lymphedema from any cause. Incidence and prevalence estimates for this cause of lymphedema are generally lacking.

Lymphatic filariasis, a nematode infection endemic to 83 countries within the endemic regions of Asia, Africa, and the Americas, is the most common cause of secondary lymphedema in the world. Common tropical filariae include *Wuchereria bancrofti* and *Brugia malayi* or *timori*.[92,93] Other *Brugia* species are found in North America and occasionally cause lymphatic obstruction.[94] In endemic areas of the world, up to 54% of the population may have microfilariae detectable in the blood.[95] The microfilariae are transmitted by an obligate mosquito vector and induce recurrent lymphangitis and, eventually, fibrosis of lymph nodes.

Lymphatic filariasis is estimated to infect more than 129 million people in tropical and subtropical areas throughout the world.[96,97] In most of the infected individuals, the condition is subclinical, but the acute manifestations can include filarial adenolymphangitis, acute dermatolymphangioadenitis, and tropical eosinophilia.[96] It has been estimated that 14 million people suffer from lymphedema and elephantiasis of the leg caused by lymphatic filariasis. Filariasis is recognized as one of the world's most disabling diseases.[97]

Traumatic and Iatrogenic Causes

There are a number of non-cancer-related settings in which the acquisition of lymphedema is described and predictable (TABLE 3). Many of these circumstances relate to disorders of the nonlymphatic vasculature. Although the observations are limited in scope and number, lymphedema has been described in patients with occlusive peripheral arterial disease (up to 80% of patients)[98] and after varicose vein surgery (0.5%)[99] and saphenous vein harvesting for aortocoronary bypass surgery 10%.[100]

Lymphedema has been described in burn patients,[101] as well as in those who undergo non-cancer-related penile surgery.[102] One very interesting iatrogenic context for acquired lymphedema is in drug administration: both intrathecal pump insertion[103] and sirolimus administration[104] have been associated with the development of lymphedema in subsets of these patients.

It is notable, in contrast to these iatrogenic settings for acquired lymphedema, that there were no instances of lymphedema after prophylactic mastectomy in a series of 1356 patients.

Conclusion

This review of the epidemiology of lymphedema is intended to focus upon the clinical settings in which the incidence and prevalence of lymphedema have been perceived to be substantial. The settings in which this disorder does occur suggests that the population-based problem of lymphedema is at least substantial, and, quite possibly, profound. The prevalence of lymphatic filariasis and neoplastic diseases alone implies a large global burden of lymphatic disease. The data presented in this review represent, regrettably, only the aggregate of many small observations, often retrospective, and almost never rigorously undertaken. It is extremely heartening to contemplate future approaches that will entail formal, prospectively designed studies to objectively quantitate incidence and prevalence statistics for individual categories, as well as for the global lymphedema population. These statistics will facilitate future stratification of risk and will encourage more rapid development of diagnostic and therapeutic strategies to reduce the burden of disease.

Conflicts of Interest

The authors declare no conflicts of interest.

References

1. SZUBA, A. & S. ROCKSON. 1997. Lymphedema: anatomy, physiology and pathogenesis. Vasc. Med. **2:** 321–326.
2. ROCKSON, S.G. 2001. Lymphedema. Am. J. Med. **110:** 288–295.
3. WILLIAMS, A.F., P.J. FRANKS & C.J. MOFFATT. 2005. Lymphoedema: estimating the size of the problem. Palliat. Med. **19:** 300–313.
4. SZUBA, A. *et al.* 2003. The third circulation: radionuclide lymphoscintigraphy in the evaluation of lymphedema. J. Nucl. Med. **44:** 43–57.
5. MOFFATT, C.J. *et al.* 2003. Lymphoedema: an underestimated health problem. Q. J. Med. **96:** 731–738.
6. PETLUND, C. 1990. Prevalence and incidence of chronic lymphoedema in a western European country. *In* Progress in Lymphology, Vol. XII. M. Nishi, S. Uchina & S. Yabuki, Eds.: 391–394. Elsevier Science Publishers B.V. Amsterdam, the Netherlands.
7. ERICKSON, V. *et al.* 2001. Arm edema in breast cancer patients. J. Natl. Cancer Inst. **93:** 96–111.
8. ROCKSON, S.G. 2005. Therapeutics for lymphatic disease: the role of the pharmaceutical and biotechnology sector. Lymphat. Res. Biol. **3:** 103–104.
9. ROCKSON, S. 2000. Primary lymphedema. *In* Current Therapy in Vascular Surgery. C. Ernst & J. Stanley, Eds.: Mosby. Philadelphia, PA.

10. SITZIA, J. *et al.* 1998. Characteristics of new referrals to twenty-seven lymphoedema treatment units. Eur. J. Cancer Care (Engl.) **7:** 255–262.

11. WILLIAMS, A.E., S. BERGL & R.G. TWYCROSS. 1996. A 5-year review of a lymphoedema service. Eur. J. Cancer Care (Engl.) **5:** 56–59.

12. SMELTZER, D.M., G.B. STICKLER & A. SCHIRGER. 1985. Primary lymphedema in children and adolescents: a follow-up study and review. Pediatrics **76:** 206–218.

13. BRICE, G. *et al.* 2002. Analysis of the phenotypic abnormalities in lymphoedema-distichiasis syndrome in 74 patients with FOXC2 mutations or linkage to 16q24. J. Med. Genet. **39:** 478–483.

14. AAGENAES, O. & S. MEDBO. 1993. [Hereditary intrahepatic cholestasis with lymphedema–Aagenaes syndrome]. Tidsskr. Nor. Laegeforen. 113: 3673–3677.

15. FINDLAY, C.A., M.D. DONALDSON & G. WATT. 2001. Foot problems in Turner's syndrome. J. Pediatr. **138:** 775–777.

16. RUIBAL FRANCISCO, J.L. *et al.* 1997. [Turner's syndrome: relationship between the karyotypes and malformations and associated diseases in 23 patients]. An. Esp. Pediatr. **47:** 167–171.

17. TANRIVERDI, H.A. *et al.* 2005. Outcome of cystic hygroma in fetuses with normal karyotypes depends on associated findings. Eur. J. Obstet. Gynecol. Reprod. Biol. **118:** 40–46.

18. CHEN, C.P. *et al.* 1996. Cytogenetic evaluation of cystic hygroma associated with hydrops fetalis, oligohydramnios or intrauterine fetal death: the roles of amniocentesis, postmortem chorionic villus sampling and cystic hygroma paracentesis. Acta Obstet. Gynecol. Scand. **75:** 454–458.

19. LOSCALZO, M.L. *et al.* 2005. Association between fetal lymphedema and congenital cardiovascular defects in Turner syndrome. Pediatrics **115:** 732–735.

20. WOLFE, J.H.N. & J.B. KINMONTH. 1981. The prognosis of primary lymphedema of the lower limbs. Arch. Surg. **116:** 1157–1160.

21. DOUVIER, S. *et al.* 1991. [Retrocervical cystic hygroma: diagnosis, prognosis and management: a series of 13 cases]. J. Gynecol. Obstet. Biol. Reprod. (Paris) **20:** 183–190.

22. VIGNES, S. & P. TREVIDIC. 2005. [Lymphedema of male external genitalia: a retrospective study of 33 cases]. Ann. Dermatol. Venereol. **132:** 21–25.

23. RUDKIN, G.H. & T.A. MILLER. 1994. Lipedema: a clinical entity distinct from lymphedema. Plast. Reconstr. Surg. **94:** 841–847; discussion 848–849.

24. SZUBA, A. & S. ROCKSON. 1998. Lymphedema: a review of diagnostic techniques and therapeutic options. Vasc. Med. **3:** 145–156.

25. SHAW, J.H. & E.M. RUMBALL. 1990. Complications and local recurrence following lymphadenectomy. Br. J. Surg. **77:** 760–764.

26. WERNGREN-ELGSTROM, M. & L.D. 1994. Lymphoedema of the lower extremities after surgery and radiotherapy for cancer of the cervix. Scand. J. Plast. Reconstr. Surg. Hand. Surg. **28:** 289–293.

27. FIORICA, J.V. *et al.* 1990. Morbidity and survival patterns in patients after radical hysterectomy and postoperative adjuvant pelvic radiotherapy. Gynecol. Oncol. **36:** 343–347.

28. SOISSON, A.P. *et al.* 1990. Adjuvant radiotherapy following radical hysterectomy for patients with stage IB and IIA cervical cancer. Gynecol. Oncol. **37:** 390–395.

29. STARRITT, E. *et al.* 2004. Lymphedema after complete axillary node dissection for melanoma: assessment using a new, objective definition. Ann. Surg. **240:** 866–874.

30. ROCKSON, S.G. 2006. Addressing the unmet needs in lymphedema risk management. Lymphat. Res. Biol. **4:** 42–46.

31. ROCKSON, S.G. 1998. Precipitating factors in lymphedema: myths and realities. Cancer **83:** 2814–2816.

32. HALSTED, W.S. 1921. The swelling of the arm after operation for cancer of the breast, elephantiasis chirurgica, its causes and prevention. Bull. Johns Hopkins Hosp. **32:** 309–313.

33. KISSIN, M. *et al.* 1986. Risk of lymphedema following the treatment of breast cancer. Br. J. Surg. **73:** 580–584.

34. AITKEN, R.J. *et al.* 1989. Arm morbidity within a trial of mastectomy and either nodal sample with selective radiotherapy or axillary clearance. Br. J. Surg. **76:** 568–571.

35. BADR EL DIN, A. *et al.* 1989. Local postoperative morbidity following pre-operative irradiation in locally advanced breast cancer. Eur. J. Surg. Oncol. **15:** 486–489.

36. HOE, A.L. *et al.* 1992. Incidence of arm swelling following axillary clearance for breast cancer. Br. J. Surg. **79:** 261–262.

37. SEGERSTROM, K. *et al.* 1992. Factors that influence the incidence of brachial oedema after treatment of breast cancer. Scand. J. Plast. Reconstr. Surg. Hand. Surg. **26:** 223–227.

38. MORTIMER, P. *et al.* 1996. The prevalence of arm oedema following treatment for breast cancer. Q. J. Med. **89:** 377–380.

39. SCHUNEMANN, H. & N. WILLICH. 1997. [Lymphedema after breast carcinoma: a study of 5868 cases]. Dtsch. Med. Wochenschr. **122:** 536–541.

40. PETREK, J.A. & M.C. HEELAN. 1998. Incidence of breast carcinoma-related lymphedema. Cancer **83:** 2776–2781.

41. BERLIN, E. *et al.* 1999. Postmastectomy lymphoedema: treatment and a five-year follow-up study. Int. Angiol. **18:** 294–298.

42. HOJRIS, I. *et al.* 2000. Late treatment-related morbidity in breast cancer patients randomized to postmastectomy radiotherapy and systemic treatment versus systemic treatment alone. Acta Oncol. **39:** 355–372.

43. TENGRUP, I. *et al.* 2000. Arm morbidity after breast-conserving therapy for breast cancer. Acta Oncol. **39:** 393–397.

44. HERD-SMITH, A. *et al.* 2001. Prognostic factors for lymphedema after primary treatment of breast carcinoma. Cancer **92:** 1783–1787.

45. PETREK, J.A. *et al.* 2001. Lymphedema in a cohort of breast carcinoma survivors 20 years after diagnosis. Cancer **92:** 1368–1377.

46. SENER, S.F. *et al.* 2001. Lymphedema after sentinel lymphadenectomy for breast carcinoma. Cancer **92:** 748–752.

47. KWAN, W. *et al.* 2002. Chronic arm morbidity after curative breast cancer treatment: prevalence and impact on quality of life. J. Clin. Oncol. **20:** 4242–4248.

48. MERIC, F. *et al.* 2002. Long-term complications associated with breast-conservation surgery and radiotherapy. Ann. Surg. Oncol. **9:** 543–549.

49. NAGEL, P.H. *et al.* 2003. Arm morbidity after complete axillary lymph node dissection for breast cancer. Acta. Chir. Belg. **103:** 212–216.

50. QUERCI DELLA ROVERE, G. *et al.* 2003. An audit of the incidence of arm lymphoedema after prophylactic level I/II axillary dissection without division of the pectoralis minor muscle. Ann. R. Coll. Surg. Engl. **85:** 158–161.

51. PEZNER, R.D. *et al.* 1986. Arm lymphedema in patients treated conservatively for breast cancer: relationship to patient age and axillary node dissection technique. Int. J. Radiat. Oncol. Biol. Phys. **12:** 2079–2083.

52. RAMPAUL, R.S. *et al.* 2003. Incidence of clinically significant lymphoedema as a complication following surgery for primary operable breast cancer. Eur. J. Cancer **39:** 2165–2167.

53. COEN, J.J. *et al.* 2003. Risk of lymphedema after regional nodal irradiation with breast conservation therapy. Int. J. Radiat. Oncol. Biol. Phys. **55:** 1209–1215.

54. TOLEDANO, A. *et al.* 2006. Concurrent administration of adjuvant chemotherapy and radiotherapy after breast-conserving surgery enhances late toxicities: long-term results of the ARCOSEIN multicenter randomized study. Int. J. Radiat. Oncol. Biol. Phys. **65:** 324–332.

55. KORNBLITH, A.B. *et al.* 2003. Long-term adjustment of survivors of early-stage breast carcinoma, 20 years after adjuvant chemotherapy. Cancer **98:** 679–689.

56. MANSEL, R.E. *et al.* 2006. Randomized multicenter trial of sentinel node biopsy versus standard axillary treatment in operable breast cancer: the ALMANAC Trial. J. Natl. Cancer Inst. **98:** 599–609.

57. WILKE, L.G. *et al.* 2006. Surgical complications associated with sentinel lymph node biopsy: results from a prospective international cooperative group trial. Ann. Surg. Oncol. **13:** 491–500.

58. PAPACHRISTOU, D. & J.G. FORTNER. 1977. Comparison of lymphedema following incontinuity and discontinuity groin dissection. Ann. Surg. **185:** 13–16.

59. KARAKOUSIS, C.P., M.A. HEISER & R.H. MOORE. 1983. Lymphedema after groin dissection. Am. J. Surg. **145:** 205–208.

60. URIST, M.M. *et al.* 1983. Patient risk factors and surgical morbidity after regional lymphadenectomy in 204 melanoma patients. Cancer **51:** 2152–2156.

61. INGVAR, C., C. ERICHSEN & P.E. JONSSON. 1984. Morbidity following prophylactic and therapeutic lymph node dissection for melanoma–a comparison. Tumori **70:** 529–533.

62. KARAKOUSIS, C.P. & D.L. DRISCOLL. 1994. Groin dissection in malignant melanoma. Br. J. Surg. **81:** 1771–1774.

63. VROUENRAETS, B.C. *et al.* 1995. Long-term morbidity after regional isolated perfusion with melphalan for melanoma of the limbs: the influence of acute regional toxic reactions. Arch. Surg. **130:** 43–47.

64. HUGHES, T.M. & J.M. THOMAS. 1999. Combined inguinal and pelvic lymph node dissection for stage III melanoma. Br. J. Surg. **86:** 1493–1498.

65. STROBBE, L.J. *et al.* 1999. Positive iliac and obturator nodes in melanoma: survival and prognostic factors. Ann. Surg. Oncol. **6:** 255–262.

66. SERPELL, J.W., P.W. CARNE & M. BAILEY. 2003. Radical lymph node dissection for melanoma. Aust. N.Z. J. Surg. **73:** 294–299.

67. WRONE, D.A. *et al.* 2000. Lymphedema after sentinel lymph node biopsy for cutaneous melanoma: a report of 5 cases. Arch. Dermatol. **136:** 511–514.

68. SNIJDERS-KEILHOLZ, A. *et al.* 1999. Adjuvant radiotherapy following radical hysterectomy for patients with early-stage cervical carcinoma (1984–1996). Radiother. Oncol. **51:** 161–167.

69. CHATANI, M. *et al.* 1998. Adjuvant radiotherapy after radical hysterectomy of the cervical cancer: prognostic factors and complications. Strahlenther. Onkol. **174:** 504–509.

70. ORR, J.W., JR. *et al.* 1991. Surgical staging of uterine cancer: an analysis of perioperative morbidity. Gynecol. Oncol. **42:** 209–216.

71. ORR, J.W., JR., J.L. HOLIMON & P.F. ORR. 1997. Stage I corpus cancer: is teletherapy necessary? Am. J. Obstet. Gynecol. **176:** 777–788; discussion 788–789.

72. NUNNS, D. *et al.* 2000. The morbidity of surgery and adjuvant radiotherapy in the management of endometrial carcinoma. Int. J. Gynecol. Cancer **10:** 233–238.

73. RYAN, M. *et al.* 2003. Aetiology and prevalence of lower limb lymphoedema following treatment for gynaecological cancer. Aust. N.Z. J. Obstet. Gynaecol. **43:** 148–151.

74. ABU-RUSTUM, N.R. *et al.* 2006. The incidence of symptomatic lower-extremity lymphedema following treatment of uterine corpus malignancies: a 12-year experience at Memorial Sloan-Kettering Cancer Center. Gynecol. Oncol. **103**(2): 714–718.

75. HABERTHUR, F., A.C. ALMENDRAL & B. RITTER. 1993. Therapy of vulvar carcinoma. Eur. J. Gynaecol. Oncol. **14:** 218–227.

76. PETEREIT, D.G. *et al.* 1993. Inguinofemoral radiation of N0,N1 vulvar cancer may be equivalent to lymphadenectomy if proper radiation technique is used. Int. J. Radiat. Oncol. Biol. Phys. **27:** 963–967.

77. BELL, J.G., J.S. LEA & G.C. REID. 2000. Complete groin lymphadenectomy with preservation of the fascia lata in the treatment of vulvar carcinoma. Gynecol. Oncol. **77:** 314–318.

78. LEMINEN, A., M. FORSS & J. PAAVONEN. 2000. Wound complications in patients with carcinoma of the vulva: comparison between radical and modified vulvectomies. Eur. J. Obstet. Gynecol. Reprod. Biol. **93:** 193–197.

79. GAARENSTROOM, K.N. *et al.* 2003. Postoperative complications after vulvectomy and inguinofemoral lymphadenectomy using separate groin incisions. Int. J. Gynecol. Cancer **13:** 522–527.

80. JUDSON, P.L. *et al.* 2004. A prospective, randomized study analyzing sartorius transposition following inguinal-femoral lymphadenectomy. Gynecol. Oncol. **95:** 226–230.

81. ANSCHER, M.S. & L.R. PROSNITZ. 1987. Postoperative radiotherapy for patients with carcinoma of the prostate undergoing radical prostatectomy with positive surgical margins, seminal vesicle involvement and/or penetration through the capsule. J. Urol. **138:** 1407–1412.

82. BOILEAU, M.A. *et al.* 1988. Interstitial gold and external beam irradiation for prostate cancer. J. Urol. **139:** 985–988.

83. AMDUR, R.J. *et al.* 1990. Adenocarcinoma of the prostate treated with external-beam radiation therapy: 5-year minimum follow-up. Radiother. Oncol. **18:** 235–246.

84. GRESKOVICH, F.J. *et al.* 1991. Complications following external beam radiation therapy for prostate cancer: an analysis of patients treated with and without staging pelvic lymphadenectomy. J. Urol. **146:** 798–802.

85. PILEPICH, M.V. *et al.* 1984. Treatment-related morbidity in phase III RTOG studies of extended-field irradiation for carcinoma of the prostate. Int. J. Radiat. Oncol. Biol. Phys. **10:** 1861–1867.

86. RAVI, R. 1993. Prophylactic lymphadenectomy vs observation vs inguinal biopsy in node-negative patients with invasive carcinoma of the penis. Jpn. J. Clin. Oncol. **23:** 53–58.

87. BEVAN-THOMAS, R., J.W. SLATON & C.A. PETTAWAY. 2002. Contemporary morbidity from lymphadenectomy for penile squamous cell carcinoma: the M.D. Anderson Cancer Center Experience. J. Urol. **167:** 1638–1642.

88. KEUS, R.B. *et al.* 1994. Limb-sparing therapy of extremity soft tissue sarcomas: treatment outcome and long-term functional results. Eur. J. Cancer. **30A:** 1459–1463.

89. DAVIS, A.M. *et al.* 2005. Late radiation morbidity following randomization to preoperative versus postoperative radiotherapy in extremity soft tissue sarcoma. Radiother. Oncol. **75:** 48–53.

90. STINSON, S.F. *et al.* 1991. Acute and long-term effects on limb function of combined modality limb sparing therapy for extremity soft tissue sarcoma. Int. J. Radiat. Oncol. Biol. Phys. **21:** 1493–1499.

91. SCHIRGER, A. 1983. Lymphedema. Cardiovasc. Clin. **13:** 293–305.

92. TAN, T.J., E. KOSIN & T.H. TAN. 1985. Lymphographic abnormalities in patients with *Brugia malayi* filariasis and "idiopathic tropical eosinophilia". Lymphology **18:** 169–172.

93. OTTESEN, E.A. 1985. Efficacy of diethylcarbamazine in eradicating infection with lymphatic-dwelling filariae in humans. Rev. Infect. Dis. **7:** 341–356.

94. BAIRD, J.K. *et al.* 1986. North American brugian filariasis: report of nine infections of humans. Am. J. Trop. Med. Hyg. **35:** 1205–1209.

95. SUPALI, T. *et al.* 2002. High prevalence of *Brugia timori* infection in the highland of Alor Island, Indonesia. Am. J. Trop. Med. Hyg. **66:** 560–565.

96. KEISER, P.B. & T.B. NUTMAN. 2002. Update on lymphatic filarial infections. Curr. Infect. Dis. Rep. **4:** 65–69.

97. TAYLOR, M.J. 2002. *Wolbachia* endosymbiotic bacteria of filarial nematodes: a new insight into disease pathogenesis and control. Arch. Med .Res. **33:** 422–424.

98. BALZER, K. & I. SCHONEBECK. 1993. [Edema after vascular surgery interventions and its therapy]. Z. Lymphol. **17:** 41–47.

99. OUVRY, P.A., H. GUENNEGUEZ & P.A. OUVRY. 1993. [Lymphatic complications from variceal surgery]. Phlebologie **46:** 563–568.

100. CARRIZO, G.J., J.J. LIVESAY & L. LUY. 1999. Endoscopic harvesting of the greater saphenous vein for aortocoronary bypass grafting. Tex. Heart. Inst. J. **26:** 120–123.

101. HETTRICK, H. *et al.* 2004. Incidence and prevalence of lymphedema in patients following burn injury: a five-year retrospective and three-month prospective study. Lymphat. Res. Biol. **2:** 11–24.

102. FRENKL, T.L., S. AGARWAL & A.A. CALDAMONE. 2004. Results of a simplified technique for buried penis repair. J. Urol. **171:** 826–828.

103. ALDRETE, J.A. & J.M. COUTO DA SILVA. 2000. Leg edema from intrathecal opiate infusions. Eur. J. Pain **4:** 361–365.

104. IBANEZ, J.P. *et al.* 2005. Sirolimus in pediatric renal transplantation. Transplant. Proc. **37:** 682–684.

105. HONG, J.H. *et al.* 2002. Postoperative low-pelvic irradiation for stage I-IIA cervical cancer patients with risk factors other than pelvic lymph node metastasis. Int. J. Radiat. Oncol. Biol. Phys. **53:** 1284–1290.

106. GERDIN, E., S. CNATTINGIUS & P. JOHNSON. 1995. Complications after radiotherapy and radical hysterectomy in early-stage cervical carcinoma. Acta. Obstet. Gynecol. Scand. **74:** 554–561.

107. YEH, S.A. *et al.* 1999. Postoperative radiotherapy in early stage carcinoma of the uterine cervix: treatment results and prognostic factors. Gynecol. Oncol. **72:** 10–15.

108. TOZZI, R. *et al.* 2005. Analysis of morbidity in patients with endometrial cancer: is there a commitment to offer laparoscopy? Gynecol. Oncol. **97:** 4–9.

109. ROBINSON, M.H. *et al.* 1991. Limb function following conservation treatment of adult soft tissue sarcoma. Eur. J. Cancer **27:** 1567–1574.

The Clinical Spectrum of Lymphatic Disease

KAVITA RADHAKRISHNAN AND STANLEY G. ROCKSON

Stanford Center for Lymphatic and Venous Disorders, Division of Cardiovascular Medicine, Stanford University School of Medicine, Stanford, California, USA

Lymphatic disease is quite prevalent, and often not well clinically characterized. Beyond lymphedema, there is a broad array of human disease that directly or indirectly alters lymphatic structure and function. The symptomatic and objective presentation of these patients can be quite diverse. In this review, we have attempted to provide a systematic overview of the subjective and objective spectrum of lymphatic disease, with consideration of all of the categories of disease that primarily or secondarily impair the functional integrity of the lymphatic system. Lymphedema is discussed, along with chromosomal disorders, lymphangioma, infectious diseases, lymphangioleiomyomatosis, lipedema, heritable genetic disorders, complex vascular malformations, protein-losing enteropathy, and intestinal lymphangiectasia.

Key words: chromosomal disorders; lymphangioma; lymphangioleiomyomatosis; protein-losing enteropathy (PLE)

Introduction

Given the central role of the lymphatic system in circulatory, metabolic, and immune-related homeostasis, it is not surprising that genetic, developmental, and acquired disorders of this system manifest through a broad array of predictable sequelae. The lymphatic-specific manifestations range from blunted immune responses[1] and impaired metabolic status[2] to the appearance of the debilitating and disfiguring form of regional swelling generally termed *lymphedema*.[3]

Lymphatic disease is quite prevalent, and often not well clinically characterized.[3,4] Acquired disease of the lymphatics most often takes the form of lymphatic circulatory disruption, typically resulting from trauma, infection, neoplasia, or from iatrogenic causes.[3,5] When regional lymphatic flow is insufficient to maintain tissue homeostasis, interstitial fluid accumulates and swelling ensues.

Beyond lymphedema, there is a broad array of human disease that directly or indirectly alters lymphatic structure and function. Not surprisingly, the symptomatic (TABLE 1) and objective (TABLE 2) presentation of these patients can be quite diverse. Diagnosis and differential diagnosis poses distinct challenges. In this review, we have attempted to provide a systematic overview of the categories of disease that primarily or secondarily impair the functional integrity of the lymphatic system.

Lymphedema

Heritable congenital lymphedema of the lower extremities was first described by Nonne in 1891.[6] In 1892, Milroy[7] described the familial distribution of congenital lymphedema, noting the involvement of 26 persons in a single family, spanning six generations.[8] **Nonne–Milroy's lymphedema** is characterized by unilateral or bilateral swelling of the legs, arms, and/or face with gradual and irreversible fibrotic changes. Additional, distinct variants of heritable lymphedema have subsequently been described. In 1898, Meige reported cases of lymphedema in which the age of onset was after puberty, and which often appeared alongside acute cellulitis.[8] In 1964 another variety of pubertal-onset lymphedema[9] was reported, in which the affected individuals had distichiasis (i.e., an auxiliary set of eyelashes).[10] Patients with the Nonne–Milroy's syndrome, **Meige's syndrome**, and the **lymphedema–distichiasis syndrome** typically present with pitting edema that is commonly limited to the legs. Although autosomal dominant transmission characterizes heritable lymphedema, the molecular mechanism and the pathogenesis differ among the various entities identified.

Address for correspondence: Stanley G. Rockson, M.D., Division of Cardiovascular Medicine, Falk Cardiovascular Research Center, Stanford University School of Medicine, Stanford, CA 94306. Voice: +1-650-725-7571; fax: +1-650-725-1599.

srockson@cvmed.stanford.edu

Ann. N.Y. Acad. Sci. 1131: 155–184 (2008). © 2008 New York Academy of Sciences.
doi: 10.1196/annals.1413.015

TABLE 1. Symptomatic correlates of lymphatic disease

Symptom	Lymphatic disease
Edema	Gorham's disease
	Hennekam's syndrome
	Intestinal lymphangiectasia
	Klinefelter's syndrome
	Klippel–Trenaunay syndrome
	Lymphedema–distichiasis
	Lymphedema–hypoparathyroidism
	Meige's disease (lymphedema tarda)
	Milroy's disease
	Neurofibromatosis
	Protein-losing enteropathy
	Stewart–Treves syndrome
	Triploidy syndrome
	Turner's syndrome
	Yellow nail syndrome
Decreased appetite	Noonan syndrome
Recurrent vomiting	Noonan syndrome
	Intestinal lymphangiectasia
Difficulty swallowing	Noonan syndrome
Joint pain	Blue rubber bleb nevus syndrome
	Noonan syndrome
Muscle pain	Noonan syndrome
Skin lesions	Blue rubber bleb nevus syndrome
	Maffucci syndrome
	Neurofibromatosis
Back pain	Neurofibromatosis
Diarrhea	Intestinal lymphangiectasia
	Protein-losing enteropathy
Weight loss	Protein-losing enteropathy
Abdominal pain	Protein-losing enteropathy
Infertility	Turner's syndrome
Jaundice	Aagenaes' syndrome
Severe itching	Aagenaes' syndrome
Dry eyes	Turner's syndrome
Absent menstruation	Turner's syndrome
Discoloration of urine	Filariasis
Wheezing	Lymphangioleiomyomatosis
	Lymphangiomatosis
	Pulmonary lymphangiectasia
Dyspnea	Intestinal lymphangiectasia
	Lymphangioleiomyomatosis
Nausea	Lymphangioleiomyomatosis
	Pulmonary lymphangiectasia
Hemoptysis	Lymphangioleiomyomatosis
Increased appetite	Prader–Willi syndrome
Fever	Filariasis
Bloating	Lymphangioleiomyomatosis
Abdominal distension	Lymphangioleiomyomatosis
Sputum production	Lymphangioleiomyomatosis
Chest pain	Lymphangioleiomyomatosis
Noisy respiration	Lymphangioleiomyomatosis
One or more bone fractures	Cystic angiomatosis

TABLE 2. Objective correlates of lymphatic disease

Objective correlate	Lymphatic disease
Facial abnormalities	Hennekam's syndrome
	Noonan syndrome
	Triploidy syndrome
Webbing of the neck	Noonan syndrome
Heart murmur	Noonan syndrome
Mental retardation	Hennekam's syndrome
	Noonan syndrome
Distichiasis	Lymphedema–distichiasis
Ascites	Protein-losing enteropathy
"Shield chest"	Turner's syndrome
Short stature	Edwards' syndrome
	Intestinal lymphangiectasia
	Maffucci's syndrome
	Turner's syndrome
Absent/incomplete puberty	Klinefelter's syndrome
	Turner's syndrome
Delayed puberty	Hennekam's syndrome
Gynecomastia	Klinefelter's syndrome
Holoprosencephaly	Edwards' syndrome
	Patau syndrome
Hypotelorism	Patau syndrome
	Triploidy syndrome
Microphthalmia	Patau syndrome
	Triploidy syndrome
Anophthalmia	Patau syndrome
Rocker-bottom feet	Patau syndrome
Cutis aplasia	Patau syndrome
Omphalocele	Patau syndrome
Apneic episodes	Edwards' syndrome
Microcephaly	Edwards' syndrome
Muscular hypotonia	Triploidy syndrome
Low-set ears	Triploidy syndrome
Yellow nails	Yellow nail syndrome
Enlarged liver	Aagenaes' syndrome
Hypotonia	Prader–Willi syndrome
Hypomentia	Prader–Willi syndrome
Hypogonadism	Prader–Willi syndrome
Obesity	Prader–Willi syndrome
Râles	Lymphangioleiomyomatosis
	Pulmonary lymphangiectasia
Enlarged lymph nodes	Lymphangioleiomyomatosis
Unequal arm/leg	Maffucci's syndrome
Bone deformities	Maffucci's syndrome
Hematemesis	Blue rubber bleb nevus syndrome
Melena	Blue rubber bleb nevus syndrome
Rectal bleeding	Blue rubber bleb nevus syndrome
Vesicles	Lymphangioma circumscriptum
Hydrops fetalis	Cystic hygroma
Hernias	Patau syndrome

In numerous families afflicted with Nonne–Milroy's lymphedema, the disorder has been linked to a mutation in flt4, the gene that encodes the vascular endothelial growth factor receptor 3 (VEGFR-3).[11] Insufficient tyrosine kinase signaling by VEGFR-3 leads to the faulty development of aplastic or hypoplastic peripheral lymphatic channels.[11]

Mutations leading to haploinsufficiency of the nuclear transcription factor, FOXC2, have been implicated in the pathogenesis of lymphedema–distichiasis.[12] More recently, the molecular defect has been linked to abnormal morphogenesis of the lymphatic valve apparatus,[13] which helps to explain the clinical pattern of lymph reflux that is observed in these patients. Valve defects and abnormal pericyte/lymphatic endothelial cell interactions characterize the FOXC2 defect of lymphedema–distichiasis.

Chromosomal Disorders

Chromosomal disorders can lead to the birth of viable individuals with multiple organic defects (TABLE 3). These disorders are uncommon; hence the chromosomal basis can be readily overlooked or misdiagnosed. Confirmatory identification can be achieved only through detailed cytogenetic studies. Many of these disorders severely distort lymphatic function. **Turner's syndrome** and **Klinefelter's syndrome** are linked to the sex chromosomes, while **Edwards' syndrome** and **Patau syndrome** are linked to autosomal chromosomes. **Triploidy syndrome** denotes the presence of an extra copy of all the chromosomes.

Turner's syndrome, first described in 1938,[14] is the most common sex-linked chromosomal aberration in females.[15] The patients described by Turner were later identified as having either a partial or complete absence of one X-chromosome. More recently, the classification for Turner's syndrome has been broadened to include patients who display mosaicism, or possess chromosomal complements, such as 45,X/46,XX.[16] Patients with Turner's syndrome are short in stature and typically present with amenorrhea.[17] This syndrome is typically associated with gonadal dysgenesis, leading to drastically reduced levels of the female sex hormones.[16] Neonates with Turner's syndrome often display congenital lymphedema of the hands and feet.[18] As a consequence of the sex chromosome loss, these patients are at high risk for a variety of disorders, including diabetes, hypothyroidism, osteoporosis, and congenital heart disease.[16] Exogenous administration of estrogens and related derivatives has been proposed as potential therapy. Administration of growth hormone increases final height in females with Turner's syndrome but, paradoxically, growth hormone in conjunction with low-dose estrogen therapy decreases the final height in these patients.[19]

Klinefelter's syndrome, first described in 1942,[20] is a disease of males, with an observed frequency of approximately 1:600.[21] It is caused by the presence of more than one X and/or Y chromosome.[22] The syndrome is characterized by hypogonadism that presents with extremely variable phenotypes. The only common symptom is infertility.[22] The genotypes are variable, including 47,XXY, 48,XXYY, 48,XXXY, 49,XXXYY, and 49,XXXXY. Clinical characteristics include small testes, azoospermia, tall stature, and gynecomastia. These findings can be reconciled by the hormonal findings, namely, low testosterone levels, and absent feedback inhibition from testosterone, leading to an abundance of circulating FSH, LH, and estradiol levels.[22] In situations where cognitive facilities are affected, androgen therapy has shown promising results in promoting language development and learning abilities, and a more normal adolescent development.[22]

Triploidy syndrome is estimated to occur in 15–20% of spontaneous abortions due to chromosomal abnormalities. Triploidy can result from an extra set of maternal or paternal chromosomes, as well as from double fertilization.[23] The few live births with this syndrome have a short life span, averaging 20 hours,[24] although survival up to 10.5 months has been reported.[25] In most cases an enlarged placenta results, along with an edematous fetus, dysplastic cranial bones, ocular defects, and abnormalities in limbs, male genitalia, and various internal organs.[26]

Edwards' syndrome, also known as trisomy 18, was first described in 1960.[27] After trisomy 21, it is the second most common autosomal trisomy, with a prevalence of 1 in 6000–8000 live births. Patients' anomalies include cranial abnormalities, webbing of the neck, malformed ears, and mental retardation.[27] Only 5–10% of children survive for more than a year because feeding difficulties, cardiac and renal malformations, and other defects. The clinical presentations of **Edwards'** and **Patau syndrome** may appear similar to physicians who do not frequently encounter these syndromes. Patau syndrome, or **trisomy 13**, is rare and the most lethal of the viable trisomies.[28] The median survival for newborns is less than 3 days. Though clinical features of patients vary, mental deficiency is a consistent marker of the disease.[28] Holoprosencephaly is also associated with Patau syndrome, which disrupts mid face development.[28] Cardiac abnormalities can be present[29] and cardiopulmonary arrest is a major cause of death.

TABLE 3. Chromosomal disorders

Disease condition	Symptoms	Signs	Laboratory findings	Genetic defect inheritance	Pathology	Associated conditions	Diagnostic methods	Differential diagnosis
Turner's syndrome	Short stature; swollen hands/feet at birth	Ovarian failure; hypoplastic or hyperconvex nails; underdeveloped breasts and genitalia; webbed neck; short stature, low hairline in back; simian crease (a single crease in the palm); and abnormal bone development of the chest	Elevated liver enzymes	X-linked dominant inheritance	Absence of one set of genes from the short arm of one X chromosome	Hydronephrosis; pyelonephritis; idiopathic hypertension; diabetes; osteoporosis; congenital lymphedema; webbed neck; nail dysplasia; high palate; short fourth metacarpal; hearing loss; hypothyroidism; liver function abnormalities	Karyotype (only one X)	Noonan
	Webbed neck; drooping eyelids; "shield-shaped" broad, flat chest; absent or incomplete development at puberty, including sparse pubic hair and small breasts; infertility						Ultrasound (kidney)	
	Dry eyes						Echo/MRI; audiology	
	Absent menstruation; absent normal moisture in vagina; painful intercourse							

TABLE 3. Continued

Disease condition	Symptoms	Signs	Laboratory findings	Genetic defect inheritance	Pathology	Associated conditions	Diagnostic methods	Differential diagnosis
Klinefelter's syndrome (addition of more than 1 extra X and/or Y chromosome)	Infertility; gynecomastia (development of breasts in males)	Lack of secondary sexual characteristics; lack of facial/body/sexual hair; high-pitched voice; female type of fat distribution; testicular dysgenesis	Cytogenetic studies Hormone testing: high plasma FSH, LH, estradiol levels, low plasma testosterone Increase in testosterone in response to hCG Increased urinary gonadotropins (abnormal Leydig cell function) Serum osteocalcin levels decreased; hydroxyl-proline/creatinine ratio increased (increased resorption, decreased deposition)		Small, firm testes with seminiferous tubular hyalinization; sclerosis; degenerated Leydig cells; histology of gynecomastic breasts; hyperplasia of interductal tissue	Mitral valve prolapse Varicose veins	ECG to detect mitral valve prolapse Radiographs to detect lower bone mineral density, radioulnar synotosis, taurodontism	

Continued

TABLE 3. Continued

Disease condition	Symptoms	Signs	Laboratory findings	Genetic defect inheritance	Pathology	Associated conditions	Diagnostic methods	Differential diagnosis
Patau syndrome	Scalp defects	Holoprosencephaly (brain does not divide completely into halves)	Cytogenetics			Cardiac defects		
Trisomy chromosome 13	Cleft lip/palate	Hypotelorism	Prenatally FISH on interphase cells			Patent ductus arteriosus		
	Facial defects (absent or malformed nose)	Microphthalmia				Ventricular septal defect		
	Hernias	Anophthalmia Rocker-bottom feet Microphthalmia cutis aplasia Omphalocele				Atrial septal defect Dextrocardia Capillary hemangiomata Polycystic kidneys/other renal malformations		
Edwards' syndrome (trisomy 18)	Stop breathing; poor feeding	Apneic episodes, marked failure to thrive; severe growth retardation, mental retardation; malformations (e.g., microcephaly, cerebellar hypoplasia, hypoplasia/aplasia of corpus callosum, holoprosencephaly)	Conventional cytogenetic studies			Cardiac ventricular septal defects with poly-valvular heart disease	ECG for cardiac defects	Arthrogryposis

TABLE 3. Continued

Disease condition	Symptoms	Signs	Laboratory findings	Genetic defect inheritance	Pathology	Associated conditions	Diagnostic methods	Differential diagnosis
						Atrial septal defects; patent ductus arteriosus; overriding aorta; hypoplastic left heart syndrome; tetralogy of Fallot; transposition of great arteries Pulmonary hypoplasia; abnormal lobation of lung Thyroid hypoplasia		
Triploidy syndrome		General dysmaturity; muscular hypotonia; large posterior fontanel; hypertelorism; microphthalmia; colobomata; cutaneous syndactaly Abnormalities of the skull, face, limbs, genitalia (male karyotype), various internal organs Fetal hypoplasia, microstomia, low-set ears			Triploid cell lines may have disappeared from peripheral blood so evidence of triploidy can only be found in the cultured skin fibroblasts			

Lymphangioma

Lymphangioma (TABLE 4) is a congenital lymphatic malformation that arises during embryologic development. Lymphangiomata may arise from segments of lymphatic vascular tissue that either fail appropriately anastomose, or may represent portions of lymph sacs that become grouped together during development.[30] Lymphangiomas are normally detected within the first 2 years of life.[30] The presence of multiple or widespread lymphatic vascular malformations of this type can be termed **lymphangiomatosis**. MRI is the most useful diagnostic approach, since it permits analysis of the lymphatic system within various tissue layers.[30] Surgical excision is the commonly employed therapeutic approach. The lesions are classified by size and depth of formation, with the smaller, superficial form designated as **lymphangioma circumscriptum,** while the deeper lesions are divided into **cavernous lymphangiomas** and **cystic hygromas**.

Lymphangioma circumscriptum is characterized by the presence of numerous superficial vesicles that are approximately 1–2 mm in diameter and often filled with clear fluid.[31] Subcutaneous lymphatic sacs are connected via dilated lymphatic channels to these thin-walled vesicles, without connections to the normal lymphatic system.[32] The defining presentation of lymphangioma circumscriptum is the oozing of colorless fluid.[31]

Cavernous lymphangiomas are large, loosely defined masses of soft tissue with lymphatic dilation in the dermis, subcutaneous tissue, and intermuscular septa.[30] The overlying skin generally remains uninvolved, although skin changes, such as hyperplasia and hyperpigmentation, may occur.

Cystic hygromas, which are fluid-filled lesions, are caused by the failure of jugular lymphatic sacs to connect to and to drain into the jugular veins, thereby leading to the stagnation of lymphatic fluid. These sacs then enlarge, and the fluid they contain fills lymphatic vessels and connective tissues, forming the definitive lesion. They are similar to cavernous lymphangiomas, although the hygromas are often encased within a fibrous capsule.[30] They commonly occur near lymphatico–venous junctions, and force fluid to accumulate in dilated lymphatics, leading to progressive lymphedema.[33] Cystic hygromas are commonly associated with other conditions, including Turner's syndrome, Klinefelter's syndrome, and various trisomies.[30] Surgical excision is only employed for superficial lesions. Cystic hygromas that are more deeply rooted require nonsurgical treatments, including intralesional injection of sclerosing agents, such as bleomycin[34] and OK-432,[35] both of which can induce regression of the hygromas.

Infectious Diseases

Lymphatic dysfunction can arise as a consequence of invading pathogens. Lymphatic filariasis and lymphangitis (TABLE 5) are two such conditions, initiated by organisms that infiltrate and infect the lymphatic system, inhibiting lymph flow and impairing normal immune function.

Globally, more than 129 million patients are afflicted by filariasis. This condition, frequently disfiguring and disabling, is characterized by markedly impaired lymphatic function and lymphangiectasia. The prevalence of lymphatic filariasis lags behind only malaria and tuberculosis in the magnitude of its impact on the global burden of disease.[36] Patients are infected by filariae, or parasitic worms, which take up residence in the lymphatic structures. The offspring of filariae circulate in blood.[36] The resulting compromised lymphatics mediate adenolymphangitis, which results in fibrosis and stenosis of the lymph nodes and limits the formation of new lymph channels. While the common clinical presentations often include hydrocele, lymphedema, and tropical pulmonary eosinophilia, infected patients can remain asymptomatic.[37] Diagnosis hinges upon detection of microfilariae in the blood and localization of obstructing lesions in the lymphatics using a combination of ultrasound and Doppler.[36] Traditionally, treatment has centered around antiparasitic medications, including diethylcarbamazine, albendazole, and ivermectin.[38,39] Newer approaches include antifilarial chemotherapy, along with antibiotics such as tetracyclines.[40]

Lymphangitis is caused by the inflammation of lymphatic channels through tissue infection. Pathogenic organisms include bacteria, fungi, viruses, and protozoa.[41] Patients present with fever, chills, muscular pain, and headache.[42] The distinguishing characteristic of this condition is the presence of erythematous, irregular cutaneous streaks in the affected part of the body. The lymph nodes can be enlarged and tender.[41] Patients generally have a history of trauma, often minor, or skin infection.[43] The treatment approach mandates the utilization of appropriate antimicrobials.[41]

Lymphangioleiomyomatosis

Lymphangioleiomyomatosis (LAM) is characterized by the spread of abnormal smooth muscle cells (LAM

TABLE 4. Lymphangioma

Disease condition	Symptoms	Signs	Laboratory findings	Genetic defect inheritance	Pathology	Diagnostic methods	Differential diagnosis
Lymphangioma (uncommon, hamartomatous, congenital malformations of lymphatic system that involve skin and subcutaneous tissues; superficial vesicles)			Factor VIII-related antigen is present in hemangiomas but negative or weakly positive in lymphangiomas		Vesicles represent dilated lymph channels that cause the dermis to expand. The lumen is filled with lymphatic fluid and often contains red blood cells, lymphocytes, macrophages, neutrophils; lined by flat endothelial cells	MRI biopsy	Dabska's low-grade angiosarcoma; dermatitis herpetiformis; herpes simplex; herpes zoster; lipomas; lymphangiectasia; malignant melanoma; metastatic carcinoma of the skin; neurofibromatosis; Stewart–Treves syndrome
Lymphangioma circumscriptum (more deep-seated)	Verrucous changes; clear	Persistent, multiple clusters of translucent vesicles that contain clear lymph fluid					
Cavernous lymphangioma	Solitary rubbery nodule with no skin changes	Superficial saccular dilations from underlying lymphatic vessels that occupy papilla and push upward against the overlying epidermis			Large, irregular channels in reticular dermis, lined by single layer of endothelial cells		

TABLE 5. Infectious diseases

Disease condition	Symptoms	Signs	Laboratory findings	Pathology	Associated conditions	Diagnostic methods	Differential diagnosis
Lymphatic filariasis	Fever; inguinal or axillary lymphadenopathy; testicular and/or inguinal pain; skin exfoliation; and limb or genital edema; cloudy, milk-like urine	Episodic attacks of fever associated with inflammation of the inguinal lymph nodes, testis, spermatic cord, lymphedema, or a combination of these; abscess formation at nodes; cellular invasion with plasma cells/eosinophils/macrophages with hyperplasia of lymphatic endothelium; lymphatic damage and chronic leakage of protein-rich lymph in the tissues; thickening of skin and chronic infections contribute to the appearance of elephantiasis	Detection of microfilariae in blood	Fibrosis of affected lymph nodes; stenosis of lymphatics with limited collateral channel formation; cutaneous changes: hyperkeratosis, acanthosis, lymph and fatty tissue, loss of elastin fibers, and fibrosis	Bacterial/fungal lymphadenitis; relapsing cellulitis	Chest radiograph	Angioedema; asthma; Hodgkin's disease; hydrocele; leprosy; lymphedema; lymphoma (non-Hodgkin's); Milroy's disease, scrotal/testicular trauma
Lymphangitis	Erythematous cutaneous streaks; fever, chills, malaise; headache; anorexia; myalgia; recent skin trauma	Erythematous and irregular linear streaks extend from primary infection site toward draining regional nodes; tenderness and heat; blistering of skin; lymph nodes swollen and tender; children may be febrile/tachycardic	CBC (marked leukocytosis) and blood culture (leading-edge culture or aspiration of pus)			Ultrasound (lymphatic obstruction of inguinal and scrotal lymphatics); Lymph node or skin nodule biopsy	Contact dermatitis

cells) through both the pulmonary interstitium and the axial lymphatics, leading to the cystic destruction of the lung along with lymphatic wall thickening[44] (TABLE 6). LAM is also characterized by the presence of pulmonary cysts and angiomyolipomas, tumors comprised of LAM cells, adipose tissue, and underdeveloped blood vessels.[45] LAM is an extremely rare disease, found in fewer than 1 in a million individuals. It affects mainly women of childbearing age.[46] The chief symptoms and clinical presentations associated with LAM are pulmonary, including pneumothorax, progressive dyspnea, chylous pleural effusions, cough, hemoptysis, and chyloptysis.[44] Non-pulmonary findings include lymphangioleiomyomas, the large cystic masses commonly found in the abdominal and retroperitoneal regions, and chylous ascites. The pulmonary cysts can be detected through high-resolution chest tomography. The key diagnostic tool is tissue biopsy of either the lungs or lymphatics with immunohistochemical staining for the antigen HMB45.[47] The target of this antigen is the melanoma-related glycoprotein 100, which is specific for LAM cells. Treatment is often focused on preventing or minimizing pneumothorax through pleurodesis and pleurectomy.[48] Additionally, embolization of angiomyolipomas is performed if necessary. Because LAM presents in young females and is found to be heightened in the presence of increased estrogen, it was believed that progesterone would have therapeutic effects. However, it has been shown that progesterone can only slow the progression of the disease, while also producing numerous negative side effects.[49,50] Current research is focused on determining whether rapamycin may be helpful in treating angiomyolipomas.[44]

Lipedema

Lipedema (TABLE 7) was first described by Allen and Hines[51] in 1940 as a bilateral, gradual accumulation of fatty deposition in the lower extremities and buttocks.[52] The body habitus superficially resembles that of bilateral lower extremity lymphedema, although the involvement of the two limbs is substantially more symmetrical than in lymphedema, and there is almost always sparing of the feet. The condition is found almost exclusively in females. Affected individuals often describe a family history of large legs.[53] Lipedema is further characterized by the presence of normal cutaneous architecture, lacking the fibrotic changes often seen in lymphedema.[54] Malleolar fat pads in lipedema patients are prominent, while they are normal in lymphedema patients. Furthermore, the edema is characteristically non-pitting.[53] Patients often complain of

severe pain in, or aching of, the lower extremities, often below the knees.[53] Histologic sampling reveals edematous adipose cells that are sometimes hyperplastic.[55] The chief diagnostic finding that distinguishes lipedema from lymphedema is the presence of normal dynamic lymphatic function by lymphoscintigraphy.[53] However, the microlymphatic function can become distorted in lipedema, and a component of secondary lymphedema often supervenes. The usual elements of complex decongestive physiotherapy, which have an ameliorating effect upon lymphedema, add little value for patients with lipedema.[56] Suction lipectomy is a promising treatment that has been demonstrated to significantly reduce the size of extremities.[56]

Heritable Disorders

There is an array of syndromic heritable disorders that are associated with dysfunction of the lymphatic system (TABLE 8). Often, these syndromes are also associated with abnormal facial and mental development. Because these disorders are rare, insights into the expression of disease are often limited or incomplete. A useful organizational schema is to classify the disorders by their autosomal recessive (**Hennekam's syndrome**, the **Prader–Willi syndrome**, and **Aagenaes' syndrome**) or autosomal dominant (**Noonan syndrome**, **Adams–Oliver syndrome**, and **neurofibromatosis**) modes of genetic transmission.

Autosomal Recessive

Hennekam's syndrome was first described in 1989 following the study of an inbred family from a small fishing town in the Netherlands. The condition is characterized by lymphangiectasia, severe lymphedema, facial abnormalities, and mental retardation.[57] Since the original description of the syndrome, only 24 patients have been diagnosed with this condition; nevertheless, since these descriptions span 11 widely separated countries, the genetic anomaly is believed to be diffuse.[58] The facial disfiguration representative of this syndrome includes a flat face and nasal bridge, small mouth, ear defects, and widely spaced eyes.[57] Gabrielli *et al.* have proposed that these facial anomalies may be the consequence of jugular lymphatic obstruction.[59] Such lymphatic obstruction would result in facial lymphedema with resultant cutaneous overdistention, and facial distortion. Others have theorized that the associated intestinal lymphangiectasia might promote protein loss, thereby causing peripheral edema, ascites, and the loss of lymphocytes and vitamins,[58] all of which are seen in patients with

TABLE 6. Lymphangioleiomyomatosis and lymphangiomatosis

Disease condition	Symptoms	Signs	Genetic defect inheritance	Pathology	Associated conditions	Diagnostic methods	Differential diagnosis
Lymphangio-leiomyomatosis	Dyspnea; hemoptysis; chyloptysis; nausea, bloating, abdominal distension; cough; sputum production; wheezing; chest pain gurgling in the chest	Râles; pneumothorax; chylothorax; chylous pleural effusions; lymphadenopathy	Sporadic		Renal hamartomas; lymphedema; chylopericardium; cystic soft tissue masses; uterine fibroids; pulmonary hemorrhage; hemosiderosis	Biopsy: Open Lung Transbronchial + HMB45 staining (LAM cells) Tissue biopsy of involved lymphatics Other: High-resolution computerized tomography (HRCT) of thorax/abdomen Pulmonary function test (airflow obstructions, impaired gas transfer) Chest radiograph	Asthma; spontaneous pneumothorax; emphysema; tuberous sclerosis; interstitial pulmonary fibrosis; pulmonary lymphangiectasia; bronchiolitis; leiomyosarcoma
Lymphangioma-tosis	Presents in late childhood; can occur in any tissue in which lymphatics are normally found; predilection for thoracic and neck involvement; wheezes (misdiagnosed as asthma)			Multiple lymphangiomas (well-differentiated lymphatic tissue that present as multicystic or sponge-like accumulations; benign proliferations of the lymphatic channels with abnormal connections to the lymphatic system); anastomosing endothelial lined spaces along pulmonary lymphatic routes accompanied by asymmetrically spaced bundles of spindle cells	Pericardial or pleural effusions; chylous effusions; chyloptysis; hemoptysis; chylopericardium; chylous ascites; protein-wasting enteropathy; peripheral lymphedema; hemihypertrophy; lymphopenia; disseminated intravascular coagulopathy; coexistence of lytic bone lesions and chylothorax	Bone biopsy lymphangiography chest radiograph CT imaging MRI	Asthma

TABLE 7. Lipedema

Disease condition	Symptoms	Signs	Genetic defect inheritance	Pathology	Associated conditions	Diagnostic methods	Differential diagnosis
Lipedema	Insidious onset in adolescence with progression; lower extremity edema with foot sparing; peau d'orange; easy bruising; pain; varicose veins; weight gain	Non-pitting edema; absent Stemmer's sign	Sporadic	Fibro-sclerosis; damage to the deep venous system	Cellulitis; degenerative arthrosis	Lymphoscintigraphy normal	Lymphedema; obesity; elephantiasis; myxedema

this syndrome. Interestingly, differences in presentation and symptoms are vast, suggesting differential expressivity of a single gene.[58]

The **Prader–Willi syndrome** shares with Hennekam's syndrome the attributes of facial anomaly and mental retardation. Patients generally also manifest neonatal hypotonia, hypogonadism, hyperphagia, and small hands and feet.[60] A short, young woman with many of these characteristics was first described in 1864, but Andrea Prader officially described the syndrome in 1956.[61] This syndrome is an example of genomic imprinting, where the clinical expression of the disease is dependent on the parent from whom the abnormality was inherited. The syndrome has been linked to chromosome 15, and is paternally inherited.[62] If the abnormal chromosome comes from the mother, then the phenotypic expression differs, a condition described as **Angelman's syndrome**.

Aagenaes' syndrome is another autosomal recessive lymphatic condition. It consists of cholestatic liver disease in conjunction with generalized lymphedema.[63] The edema is most commonly found in the lower extremities, as well as in the hands and scrotum.[63] This syndrome was first studied by Aagenaes in 1968, and most patients who bear the diagnosis are from a single region in southwestern Norway.[64] Most patients follow the general pattern of lymphedema progression and simultaneous cholestatic regression with age.[64] The pathology of the two components of the disease is distinct, with the liver disease attributed to giant-cell transformation, and the lymphedema to lymphatic vessel hypoplasia.[64] Symptoms include itching and jaundice, and there is often a growth delay during childhood.[64] Nutritional and vitamin supplements have been shown to have therapeutic value for patients.[64]

Autosomal Dominant

In 1883 a medical student, Kobylinski, described a young patient with a webbed neck. This is the first documentation of the **Noonan syndrome**, a congenital syndrome consisting of a webbed neck, mental retardation, short stature,[65] and cardiac defects, particularly pulmonary valve stenosis.[66] Additionally, many patients have lymphatic vessel dysplasia, neonatal lymphedema, intestinal or pulmonary lymphangiectasia, and cystic hygromas.[65] Analysis of patients with this condition reveals that the genetic abnormality can either occur sporadically or as an autosomally dominant form of genetic transmission. Collins and Turner found that while the sporadic mutation was distributed evenly among socioeconomic groups, the familial mutation was predominantly found among

TABLE 8. Heritable disorders

Disease condition	Symptoms	Signs	Laboratory findings	Genetic defect inheritance	Pathology	Associated conditions	Diagnostic methods	Differential diagnosis
Adams–Oliver syndrome	Defects of the scalp and cranium associated with distal limb anomalies and occasional mental retardation			Most autosomal dominant; some sporadic autosomal recessive		**Cardiovascular system:** Tetralogy of Fallot and pulmonary atresia **Head and neck:** Acrania of the flat bones with normal bones of the cranial base; hypoplastic facies and microcephaly **Hand and foot:** Hypoplastic nails; simple syndactyly; bony syndactyly; transverse reduction defects; ectrodactyly; polydactyly; brachydactyly **Spine:** Occasional spina bifida **Skin:** Aplasia cutis congenita of the scalp **Nervous system:** Hydrocephalus, epilepsy	Radiographs; echocardiography	

TABLE 8. Continued

Disease condition	Symptoms	Signs	Laboratory findings	Genetic defect inheritance	Pathology	Associated conditions	Diagnostic methods	Differential diagnosis
Noonan syndrome	Decreased appetite; frequent or forceful vomiting; dysphagia; severe joint or muscle pain	Facial abnormalities, webbing of the neck; chest deformities; heart murmur; mental retardation	Thrombocytopenia; prolonged activated partial thromboplastin time	Autosomal dominant or sporadic		Pulmonary valvular stenosis; cryptorchidism lymphedema; osteoporosis; vasculitis progressive high-frequency sensorineural hearing loss; pleural effusions; hydrops fetalis	Bleeding diatheses (factor XI deficiency); karyotype to distinguish from Turner's syndrome (XO); electro- and echocardiography; audiologic evaluation	Turner syndrome; Trisomy 21; Escobar syndrome
Neuro-fibromatosis	Coffee-colored macular lesions; freckling in non-sun-exposed areas; back pain	Neurofibromas; optic glioma; hamartomas on iris; distinctive bony lesions		Autosomal dominant (no family history in 50%); high mutation rate	Vasculopathy (arterial stenoses due to intimal cellular proliferation); fibromuscular hyperplasia of arteries leads to renal artery stenosis; cerebral infarction; aneurysm (rare)	Impaired respiration or degluitition; interstitial pulmonary fibrosis; renal artery stenosis; cerebral infarction; short stature; scoliosis; hypertension developmental delay	MRI CT imaging	
Aagenaes' syndrome (cholestasis with malabsorption)	Predominantly in Norwegian patients; jaundice; severe pruritis	Hepatomegaly		Possibly autosomal recessive	Generalized lymphatic anomaly (lymph-edema due to lymph vessel hypoplasia); giant-cell hepatitis with fibrosis of portal tract	Hepatic cirrhosis; growth retardation; rickets; peripheral neuropathy; recurrent cellulitis; hyperlipidemia; bleeding (vitamin K deficiency)	Liver biopsy (giant-cell transformations)	

Continued

TABLE 8. Continued

Disease condition	Symptoms	Signs	Laboratory findings	Genetic defect inheritance	Pathology	Associated conditions	Diagnostic methods	Differential diagnosis
Hennekam's syndrome	Edema; facial anomalies moderate developmental problems	Lymphedema; lymphangiecta-sia; facial anomalies; delayed onset of puberty; moderate mental retardation	Hypoproteinemia Hypogamma-globulinemia Hypoalbuminemia Hypocalcemia Leukopenia with lymphopenia	In one report of 10 familial cases, equal sex ratio, increased parental consanguinity, no vertical transmission; consistent with autosomal recessive		Papillae nervi optici Slight tortuosity of the veins Yellow macula Conduction deafness Narrow upper thorax Congenital heart defect Umbilical hernias Contractures Syndactyly Brain cyst Pyloric stenosis Hearing loss Blood vessel anomalies Atonic seizures	Audiometry	

TABLE 8. Continued

Disease condition	Symptoms	Signs	Laboratory findings	Genetic defect/inheritance	Pathology	Associated conditions	Diagnostic methods	Differential diagnosis
Prader–Willi syndrome	Neonatal hypotonia; fetus small for gestational age; undescended testes; delayed motor development; slow mental development very small hands and feet in comparison to body rapid weight gain; insatiable appetite, food craving; almond-shaped eyes; narrow bifrontal skull; morbid obesity; skeletal (limb) abnormalities, striae	Hypotonia; hypomentia; hypogonadism; obesity	Genetic testing, including chromosomal analysis for methylation patterns in PWS region; Southern blot hybridization/PCR; analysis for underlying uniparental disomy; fluorescent *in situ* hybridization (FISH) can confirm prenatal diagnosis; evaluation for hypogonadism: measurements of insulin-like growth factor-1 (IGF-1) and IGFBP-3; assessment of thyroid/adrenal and pituitary status	Genomic imprinting; differential gene expression based upon parent of origin (loss of paternal gene or maternal disomy)		Cardiopulmonary compromise; diabetes; orthopedic problems	Cranial MRI (to evaluate for hypopituitarism); serial dual-energy X-ray absorptiometry (DEXA) scanning for detection/monitoring of osteoporosis; scoliosis studies; chest X-ray; abdominal ultrasonography; CT and gastrointestinal imaging	Infantile hypotonia; neonatal sepsis; developmental delay/mental retardation, obesity, and hypogonadism; Albright hereditary osteodystrophy; Bardet-Biedl syndrome; Cohen syndrome; Borjeson-Forssmann-Lehmann syndrome; Some patients with Fragile X syndrome; Possible 6q or 1p deletions

lower groups.[67] In order to ascertain the pathophysiology of the condition, patients were lymphangiographically investigated. The documented abnormalities included lymphatic vessel aplasia and hypoplasia and diminished lymphatic flow.[65] It has been hypothesized that the webbed neck phenotype results from the regression of cystic hygromas when lymphatic obstruction is alleviated.[65] Additionally, embryonic lymphedema may prevent proper migration of tissues during development, leading to the anomalies, such as the cryptorchidism, hypertelorism, and low-set ears that are common in these patients.[65] Prenatal diagnosis is often accomplished ultrasonographically, with detection of cystic hygromas and edema.[65]

Adams–Oliver syndrome comprises congenital cutis aplasia, or absence of all skin layers, which generally manifests as scalp and skull defects, along with distal limb abnormalities.[68] Although patients often have severe defects of the skull, central nervous system defects have not been reported. Intelligence and intellectual development is normal.[69] While the most common mode of inheritance is autosomal dominant,[70] sporadic cases and autosomal recessive inheritance have also been documented.[70] Some affected individuals also have associated cardiac defects. Pousti *et al.* have postulated that the genetic defect decreases the stability of embryonic blood vessels, thereby disrupting vascular development, particularly in the cranial vertex and limbs.[69]

Neurofibromatosis is a single-gene disorder of the nervous system. The responsible gene has been mapped to chromosome 17.[71,72] The protein involved is neurofibromin, a tumor suppressor, and an inhibitor of cellular growth and differentiation in neurons, glial cells, and Schwann cells.[73] This loss of inhibition results in uncontrolled cell growth of central and peripheral nervous system cells, and can lead to the formation of the tumors (neurofibromas) when both alleles of the gene are lost, as well as non-tumor manifestations if a single allele is mutated.[74] Clinical characteristics include hyperpigmented macules, neurofibromata, and benign intracranial calcifications.[74] Neurofibromatosis is inherited in an autosomal dominant manner, but clinical expression varies greatly.[74] Useful diagnostic modalities include MRI and CT imaging. Currently, clinical trials are under way to ascertain the effect of pirfenidone, an antifibrotic compound, on the neurofibromas, while other investigators are attempting to develop new methods of measuring the growth of neurofibromas through volumetric MRI.[75]

Complex Vascular Malformations

Various disorders represent the result of abnormal development of, or insult to, the blood vascular and lymphatic vascular systems (TABLE 9). These diseases often have a superficial component, and present as irregularities of the skin, in the form of nodules or lesions. Since the pathology of these conditions is complicated, therapies are focused upon alleviating the dermal afflictions.

Cystic angiomatosis is a congenital condition[76] of unknown etiology, defined by the presence of numerous cystic skeletal lesions.[77] The lesions are generally round or oval, and they vary widely in size.[77] Although the clinical course is varied, the lesions most frequently present during the first few decades of life.[78] The cystic lesions may be due to dilated blood vessels or lymphatic channels, or both.[76,77] The cysts are encircled by a single, flat layer of endothelial cells.[76] Patients present with soft tissue masses, and sometimes have pain and swelling due to pathologic fracture.[77,79,80] Cystic angiomas are easily detectable on radiographs since they represent areas of destroyed bone that are sharply defined by a sclerotic rim.[77] Biopsies, often in affected areas of the rib, are performed to confirm the diagnosis.[76] Chemotherapy and radiotherapy have been attempted, but generally have been ineffectual.[76]

Maffucci's syndrome was first described in 1881; since then, fewer than 200 cases have been reported.[81] The syndrome is characterized by the presence of hard subcutaneous enchondromas and hemangiomas[82] due to mesodermal dysplasia.[83] Most of these tumors are benign, with a 15–20% incidence of malignant trasnsformation.[82] Dyschondroplasia, improper formation of bone in cartilage, is seen.[84] Maffucci's syndrome often impairs lymphatic system function, leading to edema and secondary infection.[85] Most individuals afflicted with this syndrome are phenotypically normal at birth; lesions appear during childhood and may progressively worsen.[83]

Diffuse **hemangiomatosis** is defined by the presence of non-malignant, visceral hemangiomas that affect at least three organ systems.[86] The lesions are attributed to a congenital defect. The vascular hamartomas are postulated to arise as a consequence of deficiencies in pericytes in the vascular wall.[86] This condition is extremely rare, with fewer than 70 reported cases.[87] All described patients have lesions of the skin, liver, brain, lungs, and gastrointestinal tract.[87] Therapy with corticosteroids and interferon-α has been attempted.[87] The mechanism of the favorable response to these medications is not known.[87]

TABLE 9. Complex vascular malformations

Disease condition	Symptoms	Signs	Laboratory findings	Genetic defect inheritance	Pathology	Associated conditions	Diagnostic methods	Differential diagnosis
Proteus syndrome	Partial gigantism; long face; wide nasal bridge; mouth open at rest; upper body wasting; learning disabilities; occasional seizures	Cutaneous and subcutaneous lesions, including vascular malformations, lipomas; hyperpigmentation; and several types of nevi		Somatic mosaicism for a dominant, unidentified lethal mosaicism	Connective tissue nevi resemble tightly compacted, collagen-rich connective tissue; epidermal nevi generally exhibit a combination of hyperkeratosis, parakeratosis, acanthosis, and papillomatosis		Radiographs; CT/MRI	Neurofibromatosis
Maffucci's syndrome	Soft, blue-colored growths in distal aspects of extremities; short stature; unequal arm/leg length	Enchondromas with multiple angiomas; bony deformities; dark, irregularly shaped hemangiomas		Sporadic, manifests early in life (~4–5 years); 25% of cases are congenital	Thrombi often form within vessels and develop into phleboliths: these appear as calcified micro-vessels; chondrosarcomas diagnosed by poorly differentiated pleiomorphic chondrocytes	Enchondromas develop from mesodermal dysplasia; unequal leg length; pathologic fractures, malunion of fractures; chondrosarcoma, hemangiosarcoma, lymphangiosarcoma	CT/MRI bone biopsy if enchondromas evolve	Klippel–Trenaunay–Weber syndrome Kaposi's sarcoma; Klippel–Trenaunay–Weber syndrome; Proteus syndrome

Continued

TABLE 9. *Continued*

Disease condition	Symptoms	Signs	Laboratory findings	Genetic defect inheritance	Pathology	Associated conditions	Diagnostic methods	Differential diagnosis
Blue rubber bleb nevus syndrome	Multiple skin lesions: protuberant, dark blue, compressible blebs	Lesions are asymptomatic, but may be painful or tender; hyperhidrosis of skin overlying the lesion; fatigue (from blood loss); hematemesis, melena, or frank rectal bleeding; joint pain; blindness (cerebral or cerebellar cavernomas that hemorrhage into occipital lobes)	Fecal occult blood; CBC; screen for iron deficiency anemia; urinalysis (hematuria may be caused by bladder lesions)	Sporadic, autosomal dominant inheritance also reported	Vascular tissue with tortuous, blood-filled ectatic vessels, lined by single layer of endothelium, with surrounding thin connective tissue; dystrophic calcification may be present		Imaging: radiographs are useful in suspected bone or joint involvement MRI; endoscopy for gastrointestinal lesions	Kaposi's sarcoma; Klippel–Trenaunay–Weber syndrome; Maffucci's syndrome; venous lakes

TABLE 9. *Continued*

Disease condition	Symptoms	Signs	Laboratory findings	Genetic defect inheritance	Pathology	Associated conditions	Diagnostic methods	Differential Diagnosis
Diffuse hemangiomatosis	Neonatal premonitory findings: small red macule; telangiectasia, or blue macule at the hemangioma site	Neonatal visceral, non-malignant hemangiomas; vascular hamartomas		Congenital defect	Lesions have dilated thin-walled channels lined by a single layer of flattened endothelial cells, with few focal areas of endothelial proliferation; no other cellular hyperplasia or pleomorphism; well-formed vascular channels; abnormal capillaries coursing in their normal situation through muscle suggests that hemangiomas are hamartomas	Thrombocytopenia and hemangioma; pneumothorax; sclerosis; gastrointestinal hemorrhage; anemia; central nervous system involvement; hydrocephalus; hemorrhage; heart failure; scarring; ulceration	Biopsy	
Cystic angiomatosis	Soft tissue masses, localized pain and swelling related to pathologic fracture	Dyspnea with or without cyanosis, ascites, splenomegaly; hepatomegaly; anemia; soft tissue masses		Vascular malformation of congenital origin	Dilated, cavernous thin walled vascular channels lined by flat endothelial cells (similar to LAM)	Osler–Weber–Rendu syndrome	Radiologic: round or ovoid geographic osteolytic lesions with sclerotic borders, little residual central trabeculation, no periosteal reaction or significant matrix formation	Multifocal Langerhans cell histiocytosis; hyperparathyroidism; metastatic carcinoma; lymphoma; mastocytosis; sarcoidosis; phakomatoses; Maffucci's syndrome

Continued

TABLE 9. *Continued*

Disease condition	Symptoms	Signs	Laboratory findings	Genetic defect inheritance	Pathology	Associated conditions	Diagnostic methods	Differential diagnosis
Klippel–Trenaunay syndrome		Capillary hemangioma/port-wine stain: distinct, linear border; nevus flammeus; large, lateral, superficial vein beginning at foot/lower leg to entry point in the thigh/gluteal area; bony/soft tissue hypertrophy: limb hypertrophy/ length discrepancies				Lymphedema; spina bifida; hypospadias; polydactyly; syndactyly; oligodactyly; hyperhidrosis; hypertrichosis; paresthesia; decalcification of involved bones; chronic venous insufficiency; dermatitis; poor wound healing; ulceration; thrombosis; emboli	Doppler ultrasound: differentiation of vascular tumors from vascular malformations CT/MRI effective for visualizing extent of lesions and infiltration of deeper tissues Plain film radiography: measurement of long bones Color Doppler ultrasound/duplex scanning: visualize patency of deep venous system Angiography	Maffuci syndrome; Proteus syndrome

TABLE 9. Continued

Disease condition	Symptoms	Signs	Laboratory findings	Genetic defect inheritance	Pathology	Associated conditions	Diagnostic methods	Differential diagnosis
Cystic hygromas	Single or multiple fluid-filled lesions that occur at sites of lymphatic–venous connection; primarily in neck and axilla	Lymphedema Hydrops fetalis	Hypoproteinemia	Congenital; autosomal recessive	Dilated, disorganized lymph channels due to failure of lymph sacs to establish venous drainage	Turner's syndrome; autosomal trisomies (trisomy 21); Klinefelter's syndrome; Noonan syndrome; Proteus syndrome; hereditary lymphedema; effusions: pleura, pericardium, abdomen	Chest X-ray: evaluate mediastinum/chest MRI: evaluation of the cystic component of hygroma CT to detect hilar involvement	Hemangiomas; lymphangiomas; vascular malformations; neck tumors; encephalocele; meningocele; branchial cyst; lipoma; laryngocele; thyroglossal duct cyst; dermoid cyst; teratoma (rare)

Continued

TABLE 9. *Continued*

Disease condition	Symptoms	Signs	Laboratory findings	Genetic defect inheritance	Pathology	Associated conditions	Diagnostic methods	Differential diagnosis
Gorham's disease	Dull aching pain or insidious onset (limitation of motion, progressive weakness); swelling	Massive bone loss		No familial predisposition	Non-malignant proliferation of thin-walled vessels; proliferative vessels may be capillary/sinusoidal or cavernous; wide capillary-like vessels	Fracture; hemangiomas; angiomatosis of blood vessels/sometimes lymphatic vessels; chylous pericardial and pleural effusions; vertebral disease	Radiographs: regional osseous destruction Radioisotope bone scan: increased vascularity initially, with eventual areas of decreased uptake where osseous tissue is diminished CT/MRI blood tests and radiographs to exclude infection, cancer, inflammatory, and endocrine disorders	Massive osteolysis; acro-osteolysis of Hajdu and Cheney syndrome; idiopathic multicentric osteolysis; multicentric osteolysis with nephropathy; hereditary multicentric osteolysis; neurogenic osteolysis; acro-osteolysis of Joseph; acro-osteolysis of Shinz; Farber's disease; Winchester's syndrome; osteolysis with detritic synovitis

Gorham's disease is the uncontrolled growth of non-malignant vascular channels that lead to lysis of the affected bone.[88,89] The condition is associated with angiomatosis of blood and lymphatic vessels.[89] The shoulder[90] and pelvis[91] are most frequently affected in this disease. Chylous pericardial and pleural effusions are associated with this condition, and chylothorax can sometimes result from dilation of lymphatic vessels with reflux into pleural cavity.[89] Treatment involves surgery, with resection or bone reconstruction, and radiation.[89]

Proteus syndrome is a congenital overgrowth of numerous body tissues and cell lines.[92] Named for the character in Greek mythology who had the ability to change his shape at will, this condition is polymorphic in nature.[93] It is characterized by subcutaneous tumors, hyperostosis, hyperplastic connective tissue in the soles and palms, pigmented nevi, and partial gigantism of hands or feet.[92,93] Cell components appear normal, although there are signs of hyperplasia or disorganization of cells.[92] The condition is rare, with only a few hundred estimated afflicted persons. It is sporadic and mosaic, in that individuals have certain cells with mutations and others that are normal.[92] There is currently no molecular marker for this condition, which is often mistaken for **Klippel–Trenaunay syndrome** (KTS), **Maffuci's syndrome**, or **neurofibromatosis**, among others. Since orthopedic complications often arise, particularly scoliosis, treatment is generally surgical or takes the form of physical therapy.[92] Future directions will involve characterizing the molecular defect responsible for the condition, thereby paving the way for accurate diagnosis and pharmacologic therapy.[92]

Klippel–Trenaunay syndrome represents a combination of vascular malformations, including capillary anomalies (port wine stain), varicose veins, and the hypertrophy of bone and soft tissue.[94] While Klippel–Trenaunay syndrome generally manifests in a single extremity, it can also affect multiple limbs or the entire body.[94] Histologically, the condition manifests as dilated telangiectatic vessels in the upper dermis which do not spontaneously regress.[95]

Recent studies to pinpoint the genetic abnormality leading to this condition have been insightful. Some Klippel–Trenaunay syndrome patients have a mutation in the VGFQ gene, which, analogously to vascular endothelial growth factor is an angiogenic factor. These mutations are either chromosomal translocations or point mutations, and both tend to enhance the effect of the protein.[96] The condition is phenotypically diverse, and therefore it is hypothesized that it is genetically heterogeneous as well, and that other genes may also be implicated.[96] Complications arising from Klippel–Trenaunay syndrome include pain and lymphedema. Doppler ultrasounds are employed to distinguish Klippel–Trenaunay syndrome from hemangiomas, while CT and MRI is used to determine the depth of tissue involvement.[97–99] Lymphoscintigraphy is employed when the lymphatics are thought to be involved,[100] particularly when patients present with lymphedema. Treatment is aimed at providing symptomatic relief in the form of elevation and compression stockings for edema.[101,102]

Blue rubber bleb nevus consists of vascular nevi of the skin and hemangiomas of the gastrointestinal tract that lead to hemorrhage and anemia.[103] The name arises from the fact that the nevi are blue and rubbery, and also soft and easy to compress.[104] The venous malformations are either congenital or present in the first years of life, and progress in both size and number over time.[103,105] The condition is sporadic, though certain instances of autosomal dominant inheritance have been reported.[103] Deformities of surrounding bone may occur as the result of increased pressure from hemangiomas.[106] Treatment includes transfusions and iron replacements[103] for gastrointestinal blood loss, and endoscopy is employed for less-invasive treatments, such as sclerotherapy, for the lesions.[105] Pharmacologic treatments including corticosteroids and interferon-α have been found ineffective: lesions regress, but then return once treatment is stopped.[105]

Protein-Losing Enteropathy and Intestinal Lymphangiectasia

Loss of lymphatic fluid and plasma protein within the lumen of the gastrointestinal tract can lead to edema and hypoproteinemia (TABLE 10). These phenomena are encountered in a variety of afflictions, including protein-losing enteropathy (PLE) and intestinal lymphangiectasia. The mechanisms that predispose to this form of protein loss are not yet fully understood; however, patients with PLE typically have local lymphatic obstruction and stasis,[107] while those with lymphangiectas have dilated lymphatic vessels in the intestinal villi.[108]

Protein-Losing Enteropathy

Patients with PLE have excessive protein loss into the gastrointestinal lumen leading to hypoproteinemia.[107] PLE is associated with numerous disorders, including inflammatory bowel disease, infection, celiac disease, intestinal lymphangiectasia, thoracic duct obstruction, and cardiac disease.[109] Generally, obstruction of lymphatic vasculature yields increased hydrostatic pressure throughout the lymphatic system of the gastrointestinal

TABLE 10. Protein-losing enteropathy and intestinal lymphangiectasia

Disease condition	Symptoms	Signs	Genetic defect inheritance	Pathology	Associated conditions	Diagnostic methods	Differential diagnosis
Protein-losing enteropathy		Edema, diarrhea	Sporadic		Inflammatory bowel disease; infection; celiac disease; intestinal lymphangiectasia; thoracic duct obstruction; and cardiac disease	Documentation of hypoalbuminemia without proteinuria; reduced plasma gamma globulins, cholesterol, alpha-1 antitrypsin; lymphopenia; nuclear studies	Intestinal lymphangiectasia
Intestinal lymphangiectasia		Severe edema, ascites; pleural effusion; hypoproteinemia		Thickening of small bowel wall; dilatation of intestinal micro-lymphatics	Inflammatory and neoplastic disease	CT imaging	

tract, resulting in lymph stasis. Protein-rich lymphatic fluid is consequently lost within the lumen of the gastrointestinal tract through the lacteals in the intestinal microvilli. Protein loss in patients with PLE is non-selective, in contradistinction to glomerular diseases, where loss is size-dependent,[110] and includes plasma proteins, albumin, globulins, and transferrin.[109] If loss of albumin exceeds its rate of synthesis, edema develops. Other clinical manifestations include ascites and pleural and pericardial effusions.

Diagnosis often relies upon identification of the characteristic laboratory abnormalities, which include: hypoalbuminemia without proteinuria; reduced plasma concentrations of gamma globulins, cholesterol, and alpha-1 antitrypsin; lymphopenia; and malabsorption of fat and fat-soluble vitamins.[109] While intravenously administered radioactive macromolecules, such as Cr-51, In-111, and I-125, are used to tag and quantify protein loss,[109] abdominal scintigraphy can additionally demonstrate sites of protein loss.[107] The most commonly employed diagnostic tool is the measurement of endogenous proteins. As an example, both fecal concentrations and clearance of alpha-1 antitrypsin are much higher in PLE patients than in unaffected individuals.

Recommended medical care depends on the underlying cause. For patients with lymphatic obstruction, severely reducing dietary fat intact, along with supplementation of medium chain triglycerides, can reduce the hydrostatic pressure within the lymphatic system and thereby decrease protein loss. Intravenous albumin replacement,[111] small bowel resection,[112] or high-dose steroid therapy may also prove beneficial.[113,114] For patients with congenital heart disease, recent studies suggest that heparin may reduce leakage of protein into the intestinal lumen.[111]

Intestinal Lymphangiectasia

Intestinal lymphangiectasia is a rare condition characterized by severe edema, thickening of small-bowel wall, PLE, ascites, and pleural effusion.[115] If lymphatic fluid and proteins are lost into the gastrointestinal tract, patients may present with generalized edema due to hypoproteinemia.[108] The condition may be primary, resulting from a congenital lymphatic vascular disorder, or secondary, as a consequence of inflammatory or neoplastic involvement of the lymphatic system.[116] The pathogenesis remains unclear. Yang and Jung propose that intestinal lymphangiectasia may develop when lymphatic obstruction involves a segment of the bowel.[117] Holt proposes that dilated intestinal lymphatics may rupture, producing a leak of lymph into the intestinal lumen.[118] Typically dilatation of lymphatic channels in the intestinal villi leads to a malabsorption of fat because long-chain fatty acids can no longer be adequately processed by these abnormal lymphatic vessels.[108]

CT imaging is often employed in the diagnosis of intestinal lymphangiectasia. The images reveal diffuse, nodular thickening of the bowel wall with ascites[119] and hypo-dense streaks in the small bowel, reflecting the markedly dilated lymphatics.[120] Replacement of dietary long chain fatty acids with a medium chain triglyceride formula reduces intestinal protein losses.[118]

Conflicts of Interest

The authors declare no conflicts of interest.

References

1. RISTEVSKI, B., H. BECKER, M. CYBULSKY, *et al.* 2006. Lymph, lymphocytes, and lymphatics. Immunol. Res. **35:** 55–64.

2. HARVEY, N.L., R.S. SRINIVASAN, M.E. DILLARD, *et al.* 2005. Lymphatic vascular defects promoted by Prox1 haploinsufficiency cause adult-onset obesity. Nat. Genet. **10:** 1072–1081.

3. ROCKSON, S.G. 2001. Lymphedema. Am. J. Med. **110:** 288–295.

4. SZUBA, A. & S.G. ROCKSON. 1997. Lymphedema: anatomy, physiology and pathogenesis. Vasc. Med. **2:** 321–326.

5. ROCKSON, S.G. 2006. Lymphedema. Curr. Treat. Options Cardiovasc. Med. **8:** 129–136.

6. NONNE, M. 1891. Vier Fälle von Elephantiasis congenita hereditaria. Virchows Archiv. **125:** 189–196.

7. MILROY, W. 1892. An undescribed variety of hereditary oedema. N.Y. Med. J. **56:** 505–508.

8. BRICE, G., A.H. CHILD, A. EVANS, *et al.* 2005. Milroy disease and the VEGFR-3 mutation phenotype. J. Med. Genet. **42:** 98–102.

9. ROSBOTHAM, J.L., G.W. BRICE, A.H. CHILD, *et al.* 2000. Distichiasis-lymphoedema: clinical features, venous function and lymphoscintigraphy. Br. J. Dermatol. **142:** 148–152.

10. FALLS, H.F. & E.D. KERTESZ. 1964. A new syndrome combining pterygium colli with developmental anomalies of the eyelids and lymphatics of the lower lxtremities. Trans. Am. Ophthalmol. Soc. **62:** 248–275.

11. IRRTHUM, A., M.J. KARKKAINEN, K. DEVRIENDT, *et al.* 2000. Congenital hereditary lymphedema caused by a mutation that inactivates VEGFR3 tyrosine kinase. Am. J. Hum. Genet. **67:** 295–301.

12. KRIEDERMAN, B.M., T.L. MYLOYDE, M.H. WITTE, *et al.* 2003. FOXC2 haploinsufficient mice are a model for human autosomal dominant lymphedema-distichiasis syndrome. Hum. Mol. Genet. **12:** 1179–1185.

13. PETROVA, T.V., T. KARPANEN, C. NORRMEN, *et al.* 2004. Defective valves and abnormal mural cell recruitment underlie lymphatic vascular failure in lymphedema distichiasis. Nat. Med. **10:** 974–981.

14. TURNER, H.H. 1938. Classic pages in obstetrics and gynecology: a syndrome of infantilism, congenital webbed neck, and cubitus valgus. Endocrinology **23:** 566–574.

15. SAENGER, P. 1996. Turner's syndrome. N. Engl. J. Med. **335:** 1749–1754.

16. GRAVHOLT, C.H., J. FEDDER, R.W. NAERAA, *et al.* 2000. Occurrence of gonadoblastoma in females with Turner syndrome and Y chromosome material: a population study. J. Clin. Endocrinol. Metab. **85:** 3199–3202.

17. SYBERT, V.P. & E. MCCAULEY. 2004. Turner's syndrome. N. Engl. J. Med. **351:** 1227–1238.

18. SAENGER, P. 1993. Clinical review 48: the current status of diagnosis and therapeutic intervention in Turner's syndrome. J. Clin. Endocrinol. Metab. **77:** 297–301.

19. QUIGLEY, C.A., B.J. CROWE, D.G. ANGLIN, *et al.* 2002. Growth hormone and low dose estrogen in Turner syndrome: results of a United States multi-center trial to near-final height. J. Clin. Endocrinol. Metab. **87:** 2033–2041.

20. KLINEFELTER, H.F. JR., E.C. REIFENSTEIN JR. & F. ALBRIGHT. 1942. Syndrome characterized by gynecomastia, aspermatogenesis without a-Leydigism and increased excretion of follicle-stimulating hormone. J. Clin. Endocrinol. Metab. **2:** 615–624.

21. KAMISCHKE, A., A. BAUMGARDT, J. HORST, *et al.* 2003. Clinical and diagnostic features of patients with suspected Klinefelter syndrome. J. Androl. **24:** 41–48.

22. MANDOKI, M.W., G.S. SUMNER, R.P. HOFFMAN, *et al.* 1991. A review of Klinefelter's syndrome in children and adolescents. J. Am. Acad. Child. Adolesc. Psychiatry **30:** 167–172.

23. WERTELECKI, W., J.M. GRAHAM, JR. & F.R. SERGOVICH. 1976. The clinical syndrome of triploidy. Obstet. Gynecol. **47:** 69–76.

24. SCHWAIBOLD, H., I. DULISCH, C.H. WITTEKIND, *et al.* 1990. Triploidy syndrome in a liveborn female. Teratology **42:** 309–315.

25. SHERARD, J., C. BEAN, B. BOVE, *et al.* 1986. Long survival in a 69,XXY triploid male. Am. J. Med. Genet. **25:** 307–312.

26. DOSHI, N., U. SURTI & A.E. SZULMAN. 1983. Morphologic anomalies in triploid liveborn fetuses. Hum. Pathol. **14:** 716–723.

27. EDWARDS, J.H., D.G. HARNDEN, A.H. CAMERON, *et al.* 1960. A new trisomic syndrome. Lancet **1:** 787–790.

28. MALIK, R., V.K. PANDYA, S. MALIK, *et al.* 2006. Holoprosencephaly: a feature of Patau syndrome. Indian J. Radiol. Imaging. **16:** 87–89.

29. PATAU, K., D.W. SMITH, E. THERMAN, *et al.* 1960. Multiple congenital anomaly caused by an extra autosome. Lancet **1:** 790–793.

30. FAUL, J.L., G.J. BERRY, T.V. COLBY, *et al.* 2000. Thoracic lymphangiomas, lymphangiectasis, lymphangiomatosis, and lymphatic dysplasia syndrome. Am. J. Respir. Crit. Care Med. **161:** 1037–1046.

31. MORDEHAI, J., E. KURZBART, D. SHINHAR, *et al.* 1998. Lymphangioma circumscriptum. Pediatr. Surg. Int. **13:** 208–210.

32. WHIMSTER, I.W. 1976. The pathology of lymphangioma circumscriptum. Br. J. Dermatol. **94:** 473–486.

33. CHERVENAK, F.A., G. ISAACSON, K.J. BLAKEMORE, *et al.* 1983. Fetal cystic hygroma: cause and natural history. N. Engl. J. Med. **309:** 822–825.

34. OKADA, A., A. KUBOTA, M. FUKUZAWA, *et al.* 1992. Injection of bleomycin as a primary therapy of cystic lymphangioma. J. Pediatr. Surg. **27:** 440–443.

35. MIKHAIL, M., R. KENNEDY, B. CRAMER, *et al.* 1995. Sclerosing of recurrent lymphangioma using OK-432. J. Pediatr. Surg. **30:** 1159–1160.

36. NUTMAN, T.B. 2001. Lymphatic filariasis: new insights and prospects for control. Curr. Opin. Infect. Dis. **14:** 539–546.

37. KUMARASWAMI, V. 2000. The clinical manifestations of lymphatic Filariasis. *In* Lymphatic Filariasis. Tropical Medicine: Science and Practice, Vol. 1. T.B. Nutman, Ed.: 103–126. Imperial College Press. London.

38. SHENOY, R.K., A. JOHN, B.S. BABU, *et al.* 2000. Two-year follow-up of the microfilaraemia of asymptomatic brugian filariasis, after treatment with two, annual, single doses of ivermectin, diethylcarbamazine and albendazole, in various combinations. Ann. Trop. Med. Parasitol. **94:** 607–614.

39. DUNYO, S.K., F.K. NKRUMAH & P.E. SIMONSEN. 2000. A randomized double-blind placebo-controlled field trial of ivermectin and albendazole alone and in combination for the treatment of lymphatic filariasis in Ghana. Trans. R. Soc. Trop. Med. Hyg. **94:** 205–211.

40. HOERAUF, A., L. VOLKMANN, K. NISSEN-PAEHLE, *et al.* 2000. Targeting of *Wolbachia* endobacteria in *Litomosoides sigmodontis*: comparison of tetracyclines with chloramphenicol, macrolides and ciprofloxacin. Trop. Med. Int. Health **5:** 275–279.

41. RYNCARZ, R., E.C. HEASLEY & T.J. BABINCHAK. 1999. The clinical spectrum of nodular lymphangitis. Hosp. Phys. **35:**: 63–66.

42. CRAVEN, R.B. & A.M. BARNES. 1991. Plague and tularemia. Infect. Dis. Clin. North. Am. **5:** 165–175.

43. KOSTMAN, J.R. & M.J. DINUBILE. 1993. Nodular lymphangitis: a distinctive but often unrecognized syndrome. Ann. Intern. Med. **118:** 883–888.

44. JOHNSON, S. 1999. Rare diseases. 1. Lymphangioleiomyomatosis: clinical features, management and basic mechanisms. Thorax **54:** 254–264.

45. TRAVIS, W.D., J. USUKI, K. HORIBA, *et al.* 1999. Histopathologic studies on lymphangioleiomyomatosis. *In* LAM and Other Diseases Characterized by Smooth Muscle Proliferation. J. Moss, Ed.: 171–217. Marcel Dekker. New York.

46. KALASSIAN, K.G., R. DOYLE, P. KAO, *et al.* 1997. Lymphangioleiomyomatosis: new insights. Am. J. Respir. Crit. Care. Med. **155:** 1183–1186.

47. JOHNSON, S.R., C.A. CLELLAND, J. RONAN, *et al.* 2002. The TSC-2 product tuberin is expressed in lymphangioleiomyomatosis and angiomyolipoma. Histopathology **40:** 458–463.

48. FERRANS, V.J., Z.X. YU, W. NELSON, *et al.* 2000. Lymphangioleiomyomatosis (LAM): a review of clinical and morphological features. J. Nippon. Med. Sch. **67:** 311–329.

49. CHU, S.C., K. HORIBA, J. USUKI, *et al.* 1999. Comprehensive evaluation of 35 patients with lymphangioleiomyomatosis. Chest. **115:** 1041–1052.

50. URBAN, T., R. LAZOR, J. LACRONIQUE, *et al.* 1999. Pulmonary lymphangioleiomyomatosis: a study of 69 patients. Medicine **78:** 321–337.

51. ALLEN, E.V. & E.A. HINES. 1940. Lipedema of the legs: a syndrome characterized by fat legs and orthostatic edema. Mayo. Clin. Proc. **15:** 184–187.

52. WOLD, L.E., E.A. HINES, JR. & E.V. ALLEN. 1951. Lipedema of the legs; a syndrome characterized by fat legs and edema. Ann. Intern. Med. **34:** 1243–1250.

53. WARREN, A.G., B.A. JANZ, L.J. BORUD, *et al.* 2007. Evaluation and management of the fat leg syndrome. Plast. Reconstr. Surg. **119:** 9e–15e.

54. FONKALSRUD, E.W. 1977. A syndrome of congenital lymphedema of the upper extremity and associated systemic lymphatic malformations. Surg. Gynecol. Obstet. **145:** 228–234.

55. BILANCINI, S., M. LUCCHI, S. TUCCI, *et al.* 1995. Functional lymphatic alterations in patients suffering from lipedema. Angiology **46:** 333–339.

56. RUDKIN, G.H. & T.A. MILLER. 1994. Lipedema: a clinical entity distinct from lymphedema. Plast. Reconstr. Surg. **94:** 841–849.

57. HENNEKAM, R.C., R.A. GEERDINK, B.C. HAMEL, *et al.* 1989. Autosomal recessive intestinal lymphangiectasia and lymphedema, with facial anomalies and mental retardation. Am. J. Med. Genet. **34:** 593–600.

58. VAN BALKOM, I.D., M. ALDERS, J. ALLANSON, *et al.* 2002. Lymphedema-lymphangiectasia-mental retardation (Hennekam) syndrome: a review. Am. J. Med. Genet. **112:** 412–421.

59. GABRIELLI, O., C. CATASSI, A. CARLUCCI, *et al.* 1991. Intestinal lymphangiectasia, lymphedema, mental retardation, and typical face: confirmation of the Hennekam syndrome. Am. J. Med. Genet. **40:** 244–247.

60. MCENTAGART, M.E., T. WEBB, C. HARDY, *et al.* 2000. Familial Prader-Willi syndrome: case report and a literature review. Clin. Genet. **58:** 216–223.

61. PRADER, A., A. LABHART & H. WILLI. 1956. Ein syndrome von adipositas, Kleinwuchs, Kryptorchismus and Oligophreniech myatonieartigen Zustand in neugoborenanalter. Schweiz. Med. Wochenschr. **86:** 1260–1261.

62. KHAN, N.L. & N.W. WOOD. 1999. Prader-Willi and Angelman syndromes: update on genetic mechanisms and diagnostic complexities. Curr. Opin. Neurol. **12:** 149–154.

63. BULL, L.N., E. ROCHE, E.J. SONG, *et al.* 2000. Mapping of the locus for cholestasis-lymphedema syndrome (Aagenaes syndrome) to a 6.6-cM interval on chromosome 15q. Am. J. Hum. Genet. **67:** 994–999.

64. DRIVDAL, M., T. TRYDAL, T.A. HAGVE, *et al.* 2006. Prognosis, with evaluation of general biochemistry, of liver disease in lymphoedema cholestasis syndrome 1

(LCS1/Aagenaes syndrome). Scand. J. Gastroenterol. **41:** 465–471.

65. WITT, D.R., H.E. HOYME, J. ZONANA, *et al.* 1987. Lymphedema in Noonan syndrome: clues to pathogenesis and prenatal diagnosis and review of the literature. Am. J. Med. Genet. **27:** 841–856.

66. MENDEZ, H.M. & J.M. OPITZ. 1985. Noonan syndrome: a review. Am. J. Med. Genet. **21:** 493–506.

67. COLLINS, E. & G. TURNER. 1973. The Noonan syndrome: a review of the clinical and genetic features of 27 cases. J. Pediatr. **83:** 941–950.

68. ADAMS, F.H. & C.P. OLIVER. 1945. Hereditary deformities in man due to arrest development. J. Hered. **36:** 3–8.

69. POUSTI, T.J. & R.A. BARTLETT. 1997. Adams-Oliver syndrome: genetics and associated anomalies of cutis aplasia. Plast. Reconstr. Surg. **100:** 1491–1496.

70. KOIFFMANN, C.P., A. WAJNTAL, B.J. HUYKE, *et al.* 1988. Congenital scalp skull defects with distal limb anomalies (Adams-Oliver syndrome–McKusick 10030): further suggestion of autosomal recessive inheritance. Am. J. Med. Genet. **29:** 263–268.

71. SEIZINGER, B.R., G.A. ROULEAU, L.J. OZELIUS, *et al.* 1987. Linkage analysis in von Recklinghausen neurofibromatosis (NF1) with DNA markers for chromosome 17. Genomics **1:** 346–348.

72. FOUNTAIN, J.W., M.R. WALLACE, A.M. BRERETON, *et al.* 1989. Physical mapping of the von Recklinghausen neurofibromatosis region on chromosome 17. Am. J. Hum. Genet. **44:** 58–67.

73. DASTON, M.M., H. SCRABLE, M. NORDLUND, *et al.* 1992. The protein product of the neurofibromatosis type 1 gene is expressed at highest abundance in neurons, Schwann cells, and oligodendrocytes. Neuron **8:** 415–428.

74. BRILL, C.B. 1989. Neurofibromatosis. clinical overview. Clin. Orthop. Relat. Res. **Aug**; (245): 10–15.

75. POUSSAINT, T.Y., D. JARAMILLO, Y. CHANG, *et al.* 2003. Interobserver reproducibility of volumetric MR imaging measurements of plexiform neurofibromas. AJR Am. J. Roentgenol. **180:** 419–423.

76. SCHAJOWICZ, F., C.L. AIELLO, M.V. FRANCONE, *et al.* 1978. Cystic angiomatosis (hamartous haemolymphagiomatosis) of bone: a clinicopathological study of three cases. J. Bone Joint Surg. Br. **60:** 100–106.

77. BOYLE, W.J. 1972. Cystic angiomatosis of bone: a report of three cases and review of the literature. J. Bone Joint Surg. Br. **54:** 626–636.

78. LEVEY, D.S., L.M. MACCORMACK, D.J. SARTORIS, *et al.* 1996. Cystic angiomatosis: case report and review of the literature. Skeletal Radiol. **25:** 287–293.

79. SECKLER, S.G., H. RUBIN & J.G. RABINOWITZ. 1964. Systemic cystic angiomatosis. Am. J. Med. **37:** 976–986.

80. BRUNZELL, J.D., S.W. SHANKLE & J.E. BETHUNE. 1968. Congenital generalized lipodystrophy accompanied by cystic angiomatosis. Ann. Intern. Med. **69:** 501–516.

81. MAFFUCCI, A. 1881. Di un caso di encondroma ed angioma multiplo. Contribuzione alla genesi embrionale dei tumori. Movimento Medico-Chirurgico **3:** 399–412.

82. KERR, H.D., J.C. KEEP & S. CHIU. 1991. Lymphangiosarcoma associated with lymphedema in a man with Maffucci's syndrome. South Med. J. **84:** 1039–1041.

83. JERMANN, M., K. EID, T. PFAMMATTER, *et al.* 2001. Maffucci's syndrome. Circulation **104:** 1693.

84. STRANG, C. & I. RANNIE. 1950. Dyschondroplasia with haemangiomata (Maffucci's syndrome): report of a case complicated by intracranial chondrosarcoma. J. Bone Joint Surg. Br. **32-B:** 376–383.

85. KINMOUTH, J.B. 1982. The Lymphatics: Surgery, Lymphography and Diseases of the Chyle and Lymph Systems. Edward Arnold Publishers Ltd. London.

86. HOLDEN, K.R. & F. ALEXANDER. 1970. Diffuse neonatal hemangiomatosis. Pediatrics **46:** 411–421.

87. LOPRIORE, E. & D.G. MARKHORST. 1999. Diffuse neonatal haemangiomatosis: new views on diagnostic criteria and prognosis. Acta. Paediatr. **88:** 93–97.

88. LEE, W.S., S.H. KIM, I. KIM, *et al.* 2002. Chylothorax in Gorham's disease. J. Korean Med. Sci. **17:** 826–829.

89. PATEL, D.V. 2005. Gorham's disease or massive osteolysis. Clin. Med. Res. **3:** 65–74.

90. REMIA, L.F., J. RICHOLT, K.M. BUCKLEY, *et al.* 1998. Pain and weakness of the shoulder in a 16-year-old boy. Clin. Orthop. Relat. Res. **Feb**; (347): 268–271, 287–290.

91. HEJGAARD, N. & P.R. OLSEN. 1987. Massive Gorham osteolysis of the right hemipelvis complicated by chylothorax: report of a case in a 9-year-old boy successfully treated by pleurodesis. J. Pediatr. Orthop. **7:** 96–99.

92. BIESECKER, L. 2006. The challenges of Proteus syndrome: diagnosis and management. Eur. J. Hum. Genet. **14:** 1151–1157.

93. CLARK, R.D., D. DONNAI, J. ROGERS, *et al.* 1987. Proteus syndrome: an expanded phenotype. Am. J. Med. Genet. **27:** 99–117.

94. KIHICZAK, G.G., J.G. MEINE, R.A. SCHWARTZ, *et al.* 2006. Klippel-Trenaunay syndrome: a multisystem disorder possibly resulting from a pathogenic gene for vascular and tissue overgrowth. Int. J. Dermatol. **45:** 883–890.

95. MULLIKEN, J.B. & J. GLOWACKI. 1982. Hemangiomas and vascular malformations in infants and children: a classification based on endothelial characteristics. Plast. Reconstr. Surg. **69:** 412–422.

96. TIAN, X.L., R. KADABA, S.A. YOU, *et al.* 2004. Identification of an angiogenic factor that when mutated causes susceptibility to Klippel-Trenaunay syndrome. Nature **427:** 640–645.

97. DUBOIS, J., L. GAREL, A. GRIGNON, *et al.* 1998. Imaging of hemangiomas and vascular malformations in children. Acad. Radiol. **5:** 390–400.

98. DUBOIS, J. & L. GAREL. 1999. Imaging and therapeutic approach of hemangiomas and vascular malformations in the pediatric age group. Pediatr. Radiol. **29:** 879–893.

99. KERN, S., C. NIEMEYER, K. DARGE, *et al.* 2000. Differentiation of vascular birthmarks by MR imaging: an investigation of hemangiomas, venous and lymphatic malformations. Acta. Radiol. **41:** 453–457.

100. BERRY, S.A., C. PETERSON, W. MIZE, *et al.* 1998. Klippel-Trenaunay syndrome. Am. J. Med. Genet. **79:** 319–326.

101. RING, D.S. & S.B. MALLORY. 1992. What syndrome is this? Klippel-Trenaunay syndrome. Pediatr. Dermatol. **9:** 80–82.

102. ENJOLRAS, O. & J.B. MULLIKEN. 1993. The current management of vascular birthmarks. Pediatr. Dermatol. **10:** 311–313.

103. RODRIGUES, D., M.L.M. BOURROUS, A.P.S. FERRER, *et al.* 2000. Blue rubber bleb nevus syndrome. Rev. Hosp. Clin. Fac. Med. Sao Paulo **55:** 29–34.

104. MOODLEY, M. & P. RAMDIAL. 1993. Blue rubber bleb nevus syndrome: case report and review of the literature. Pediatrics **92:** 160–162.

105. ERTEM, D., Y. ACAR, E. KOTILOGLU, *et al.* 2001. Blue rubber bleb nevus syndrome Pediatrics **107:** 418–420.

106. McCARTHY, J.C., M.J. GOLDBERG & S. ZIMBLER. 1982. Orthopaedic dysfunction in the blue rubber-bleb nevus syndrome. J. Bone Joint Surg. Am. **64:** 280–283.

107. CHIU, N.T., B. LEE, S. HWANG, *et al.* 2001. Protein-losing enteropathy: diagnosis with (99m)Tc-labeled human serum albumin scintigraphy. Radiology **219:** 86–90.

108. HILLIARD, R.I., J.B. McKENDRY & M.J. PHILLIPS. 1990. Congenital abnormalities of the lymphatic system: a new clinical classification. Pediatrics **86:** 988–994.

109. BHATNAGAR, A., R. KASHYAP, U.P. CHAUHAN, *et al.* 1995. Diagnosing protein losing enteropathy: a new approach using Tc-99m human immunoglobulin. Clin. Nucl. Med. **20:** 969–972.

110. STROBER, W., R.D. WOCHNER, P.P. CARBONE, *et al.* 1967. Intestinal lymphangiectasia: a protein-losing enteropathy with hypogammaglobulinemia, lymphocytopenia and impaired homograft rejection. J. Clin. Invest. **46:** 1643–1656.

111. DONNELLY, J.P., A. ROSENTHAL, V.P. CASTLE, *et al.* 1997. Reversal of protein-losing enteropathy with heparin therapy in three patients with univentricular hearts and Fontan palliation. J. Pediatr. **130:** 474–478.

112. WARSHAW, A.L., T.A. WALDMANN & L. LASTER. 1973. Protein-losing enteropathy and malabsorption in regional enteritis: cure by limited ileal resection. Ann. Surg. **178:** 578–580.

113. RYCHIK, J., D.A. PICCOLI & G. BARBER. 1991. Usefulness of corticosteroid therapy for protein-losing enteropathy after the Fontan procedure. Am. J. Cardiol. **68:** 819–821.

114. ROTHMAN, A. & J. SNYDER. 1991. Protein-losing enteropathy following the Fontan operation: resolution with prednisone therapy. Am. Heart J. **121:** 618–619.

115. VARDY, P.A., E. LEBENTHAL & H. SHWACHMAN. 1975. Intestinal lymphagiectasia: a reappraisal. Pediatrics **55:** 842–851.

116. FOX, U. & G. LUCANI. 1993. Disorders of the intestinal mesenteric lymphatic system. Lymphology **26:** 61–66.

117. YANG, D.M. & D.H. JUNG. 2003. Localized intestinal lymphangiectasia: CT findings. AJR Am. J. Roentgenol. **180:** 213–214.

118. HOLT, P.R. 1964. Dietary treatment of protein loss in intestinal lymphangiectasia: the effect of eliminating dietary long chain triglycerides on albumin metabolism in this condition. Pediatrics **34:** 629–635.

119. FAKHRI, A., E.K. FISHMAN, B. JONES, *et al.* 1985. Primary intestinal lymphangiectasia: clinical and CT findings. J. Comput. Assist. Tomogr. **9:** 767–770.

120. PURI, A.S., R. AGGARWAL, R.K. GUPTA, *et al.* 1992. Intestinal lymphangiectasia: evaluation by CT and scintigraphy. Gastrointest. Radiol. **17:** 119–121.

Congenital Lymphatic Malformations

FRANCINE BLEI

Departments of Pediatrics and Surgery (Plastic), Pediatric Hematology/Oncology, and Vascular Anomalies Program, New York University School of Medicine, and the Stephen D. Hassenfeld Center for Children with Cancer and Blood Disorders, New York University Medical Center, New York, New York 10016, USA

"Vascular anomalies" represents a spectrum of vascular lesions, of unclear etiology and often with unpredictable behavior. Patients with vascular anomalies represent a unique population, in that they have focal aberrations of vascular development (in vascular malformations) or vascular proliferation (in hemangiomas). The etiology of these disorders is unclear, and likely represents a multifactorial process. Vascular anomalies are an attractive model for the study of human disorders of vasculogenesis (development of the vasculature) and angiogenesis (new vessel growth from existing vessels).

Key words: angiogenesis; hemangiomas; lymphatic malformations; vascular malformations; vasculogenesis

The functional separation of vascular anomalies into proliferative lesions versus static malformations represents an important advance (TABLE 1 and FIG. 1), as their prognosis and management varies. Early detection, proper evaluation, and appropriate diagnosis are essential, as these entities are medically very different. TABLE 1 places lymphatic malformations and lymphangiomatosis, which are the focus of this review, in the context of vascular anomalies. The lymphedemas, which some consider malformations of the dermal lymphatics, are discussed in a separate chapter in this volume.

because of intralesional hemorrhage and/or combined lymphatic–venous or capillary malformations). Lymphatic malformations can occur in any location, and are present at birth but not always evident. They can be isolated or associated with other vascular malformations and/or syndromes (TABLE 2). The superficial component of lymphatic malformations may resemble warts, with thick, fluid-filled vesicular blebs, acanthosis, and hyperkeratosis. These blebs may ooze lymphatic or serosanguinous fluid. Extensive lymphatic malformations of the limbs may be associated with lymphedema, and show irregular contour and quality of the skin.

Lymphatic Malformations

Lymphatic malformations are classified as truncal versus extratruncal, microcystic versus macrocystic, or combined (e.g., with capillary and/or venous malformations). They can be focal or diffuse superficial ("lymphangioma circumscriptum"), subcutaneous, or deep, and they can intertwine within muscles or organs. Nomenclature of lymphatic malformations is confusing, as they are often (incorrectly) referred to cystic hygromas and lymphangiomas, hemangiomas (as some lymphatic malformations appear red or blue

Lymphatic Development

Florence Sabin, in 1902, proposed the centrifugal theory of lymphatic development, whereby primitive primary lymph sacs bud from venous endothelia during early development, with endothelial sprouting of these sacs to form the peripheral lymphatic system.[1] In contrast, the centripetal theory proposed by Huntington and McClure supported an independent mesenchymal origin of lymphatics which connected to the sacs.[2] More recent research on the molecular basis of embryonic lymphangiogenesis supports both theories—superficial lymphatics developing from mesenchymal lymphoblasts and deep lymphatics from veins (FIG. 2). Mice with VEGF-3 mutations have hypoplastic superficial lymphatics with a lymphedema phenotype,[3,4] and transgenic mice overexpressing VEGF-C (the ligand of VEGF-3) resulted in lymphatic

Address for correspondence: Francine Blei, Associate Medical Director, Stephen D. Hassenfeld Center for Children with Cancer and Blood Disorders of NYU Medical Center, 160 East 32nd Street, 2nd floor, L3 Medical, New York, NY 10016. Voice: +1-212-263-9912; fax: +1-212-263-8410.

francine.blei@nyumc.org

Ann. N.Y. Acad. Sci. 1131: 185–194 (2008). © 2008 New York Academy of Sciences.
doi: 10.1196/annals.1413.016

TABLE 1. Functional classification of vascular anomalies[a]

Proliferative vascular lesions	Static[b] vascular lesions (vascular malformations)
Hemangiomas	**Simple or combined**
Kaposiform hemangioendothelioma	Arterial
Tufted angioma	Venous
Kaposi's sarcoma	Capillary
Angiosarcoma	Lymphatic
	Macrocystic versus microcystic
	Lymphangiomatosis—focal versus diffuse
	Lymphangiectasia

[a]Modified from classification of the International Society of Vascular Anomalies.

[b]Vascular malformations generally do not proliferate, but they do respond to environmental triggers (e.g., infection, trauma, hormonal changes), which can aggravate the lesions.

vasculature hyperplasia.[5] The homeobox gene Prox-1 was identified as the first lymphatic-specific marker central to lymphatic budding, sprouting, and migration.[6] Oliver and Harvey propose a stepwise model of lymphangiogenesis whereby lymphatic vasculature development is initiated by Prox1 expression in a subset of vascular endothelial cells, which subsequently assume a lymphatic phenotype.[7] Other studies have demonstrated the role of neuropilin 2, angiopoietin-Tie and ephrin-Eph signaling systems, and FOXC2 in lymphangiogenesis.[3,8–13]

The pathogenesis of lymphatic vascular malformations is not entirely apparent, however: they are likely the result of developmental defects during embryonic lymphangiogenesis.[14] Recent basic research has greatly enhanced our understanding of lymphatic development and biology. Several excellent reviews on this topic summarize this work in the context of lymphatic diseases and the molecular regulation of lymphangiogenesis.[12,15–19] Identification of specific lymphatic markers, such as LYVE-1, D2–40, podoplanin, VEGF-C, and VEGF-D, has advanced the characterization of lymphatic tissue.[6,19–26] An animal model of lymphatic malformations, induced via intraperitoneal injection of Freund's incomplete adjuvant, may enable further studies on endovascular therapies for these lesions.[24] In contrast, intraperitoneal injection of Freund's incomplete adjuvant resulted in a vascularized oil granuloma rather than a lymphatic malformation.[25]

Histologically, lymphatic malformations are thin-walled fluid-filled cysts with surrounding connective tissue–containing stroma. Lymphatic endothelium is confirmed by immunochemical positivity with D2–40 and other lymphatic-specific markers.

Clinical Course

There is great variability in clinical course, associated complications and treatments of lymphatic malformations (TABLE 2 and FIG. 3).[26–28] Congenital lymphatic malformations of the head and neck can involve the tongue, orbit, face, and/or neck, and complications may be due to pressure on adjacent structures and soft tissue/skeletal overgrowth (e.g., macroglossia, prognathism, malocclusion) affecting speech, dentition, oral hygiene, and cosmesis. Trunk and extremity lymphatic malformations may impart functional impairment, limb length/girth discrepancies, and lymphedema. Large lymphatic malformations may be associated with hypo-proteinemia, necessitating albumin and immunoglobulin replacement, and lymphopenia with low levels of lymphocyte subsets, possibly on account of sequestration of circulating lymphocytes within the malformation.[29] Infection (local cellulites or systemic sepsis) and/or intralesional bleeding are frequent complications. Chylous pleural effusions and/or ascites may be present. An anatomic staging system by de Serres *et al.* may correlate with prognosis.[30,31] Management options range from close monitoring, supportive care, paracentesis, empirical antibiotic treatment for cellulites/sepsis, endovascular sclerotherapy, surgical excision, and/or Nd:YAG laser therapy.[32,33]

Prenatally Diagnosed Lymphatic Malformations

Lymphedema and larger lymphatic malformations can be diagnosed via prenatal ultrasound as early as the first trimester. Fetal cystic hygroma with multiloculated septated cervicofacial cystic masses may be associated with lymphedema and peritoneal effusions and may be detected by fetal ultrasound.[34,35] Descamps and colleagues reviewed 25 cases and provided a literature review of more than 970 case reports of "cystic hygroma" diagnosed via prenatal ultrasound performed between 10 and 23 weeks' gestation. Spontaneous resolution of the malformation occurred in several cases. Fetal karyotype analysis demonstrated 62% of cystic hygroma-type lymphatic malformations associated with chromosomal abnormalities, Turner's syndrome being the most common (33%). Factors associated with a poor outcome were structural malformations and abnormal karyotype.[36] Similarly, Tanriverdi *et al.* and Ganapathy *et al.* reported prenatally detected cystic hygromas

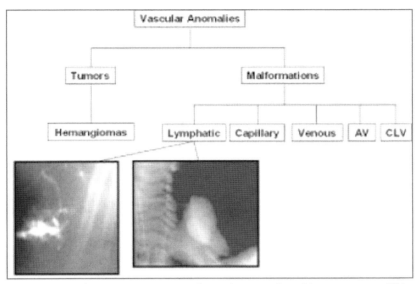

FIGURE 1. Schematic representation of vascular anomalies. AV, arteriovenous; CLV, capillary-lymphatic-venous. (From Smith.[33] Reproduced by permission.)

TABLE 2. Complications and management of lymphatic malformations

Lymphatic malformations—complications
Orthopedic
  Limb length/girth/hand/foot discrepancy
  Bony overgrowth
Dental
  Malocclusion
Hematologic
  Lymphopenia
  Bleeding
Infections
  Cellulitis/sepsis
Obstruction
  Airway
  Vision
Fluid leakage
  Leaking blebs
  Chylous ascites/pleural effusion
Lymphatic malformations—management
Supportive
  Hygiene—oral care, decrease bacterial flora
  Complete decompressive lymphatic massage therapy
  Compression therapy—custom-made gradient support stocking
  Pneumatic therapy
  Shoeing—custom-made, discrepant sizing
Surgical
  Tracheotomy
  Pleurocentesis, paracentesis
Endovascular
  Sclerotherapy
Medical
  Medical therapy—interferon, octeotride, other

with spontaneous regression in the minority of cases, and a high mortality in those fetuses having associated abnormalities. The data also suggest these pregnancies be closely monitored.[37,38] Howarth and colleagues studied outcomes from 99 pregnancies with prenatally diagnosed cystic hygroma, reporting a prevalence of 1 in 1775 live births. Of the pregnancies karyotyped, 61% demonstrated abnormalities, most often Turner's syndrome (33%). Many structural malformations were identified. Of note, while the rate of normal outcome from pregnancies complicated by prenatally diagnosed cystic hygroma was low (<10% in this study), postnatally detected cystic hygromas (not present in early gestation) appear to have a more promising outlook (TABLE 3).[39]

In utero sclerotherapy treatment of large prenatally detected macrocystic lymphatic malformations have been reported.[40–42] Congenital pulmonary lymphangiectasia can present with hydrops fetalis, prenatal pleural effusion, chylothorax, and perinatal demise.[43] *In utero* thoracocentesis has also been reported.[43] Some authors suggest that congenital pulmonary lymphangiectasia or Hennekams's syndrome (lymphedema, lymphangiectasia, and developmental delay) be considered in cases of fetal bilateral pleural effusions and nonimmune hydrops fetalis.[44–47]

Fetal airway obstruction can be anticipated when large cervicofacial or thoracic malformations are detected *in utero*. When necessary, delivery with *ex utero* intrapartum treatment (EXIT) procedures can be organized, with partial delivery of the fetus via cesarean

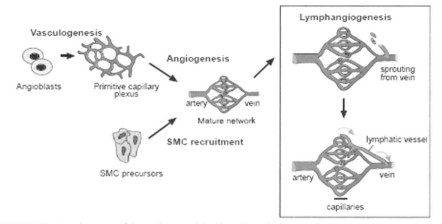

FIGURE 2. Development of the embryonic blood and lymphatic vessels. (From Makinen and Alitalo.[15] Reproduced by permission.)

section, and subsequent management of the neonatal airway while oxygenation is provided via the placenta.[48–50]

Syndromic Lymphatic Malformations

TABLE 4 summarizes syndromes associated with lymphatic dysplasia. Klippel–Trenaunay and Proteus syndromes may be associated with vascular anomalies, including lymphatic malformations. As postulated by Happle, these disorders represent "lethal gene surviving by mosaicism disorders of somatic mosaicism, as germline mutations would likely be incompatible with life."[51–54]

The Klippel–Trenaunay syndrome is characterized by capillary malformation, ipsilateral soft tissue (+/−bony) overgrowth, and underlying vascular malformation (venous +/− lymphatic). Maari and Frieden, in a retrospective review of patients with Klippel–Trenaunay syndrome found a correlation between a cutaneous "geographic" vascular stain and associated lymphatic malformation and complications.[55] A candidate genes for the Klippel–Trenaunay syndrome is5q3.[56–60]

Asymmetrical hypertrophy, macrodactyly, epidermal nevi, vascular malformations (capillary, venous, lymphatic), and exostoses are features of the Proteus syndrome. Chromosome 16 and PTEN (10q23) have been identified as putative candidate genes associated with this disorder.[61–65]

Gorham's disease (also know as Gorham–Stout disease, disappearing/vanishing/bone disease) is characterized by a proliferation of vessels (often lymphatic) resulting in progressive destruction and resorption of osseous matrix. There may be associated chylous pleu-

ral and/or pericardial effusions or neurologic sequelae (in association with spinal lesions).[66–69] Alkaline phosphatase, acid phosphatase, and/or an increase in the sensitivity to circulating factors including PDGF and IL-6 promoting osteoclast formation and bone resorption may play a role in the pathogenesis of this disease.[70–73]

Lymphangiomatosis

Primary intestinal lymphangiectasia is characterized by dilated and tortuous intestinal lymphatics due to damaged lymphatic drainage resulting in chylous leakage with protein-losing enteropathy, hypoalbuminemia, and edema. Involvement of extraintestinal lymphatics leads to chylous ascites, pleural effusions, and pericardial effusions. *Disseminated lymphangiomatosis*, seen mostly in children and young adults, results in pleural effusions, chylous ascites, and intestinal lymphangiomatosis. Surgical removal of thoracic lesions may be achieved, but the prognosis is guarded overall. Lymphopenia is also common and indicates failure of lymphocytes in the lymphatics to return to the general circulation, presumably on account of the structural changes of the lymphatics. Endoscopy, including capsule endoscopy, is key to diagnosis.[74–78] Dietary replacement of fat with medium chain triglycerides (MCTs) is recommended as MCTs are absorbed by the intestinal mucosa and metabolized and excreted directly into the portal venous system rather than intestinal lacteals. Pleurocentesis, paracentesis, sclerotherapy, systemic medical therapy, and radiotherapy are in the armamentarium, with variable effectiveness and sequelae.[79–81] Octeotride, a serotonin agonist, is reported as giving variable results.[82–85] Octeotride binds to somatostatin receptors in the intestinal lymphatics and decreases bile secretion into and chyle out of the

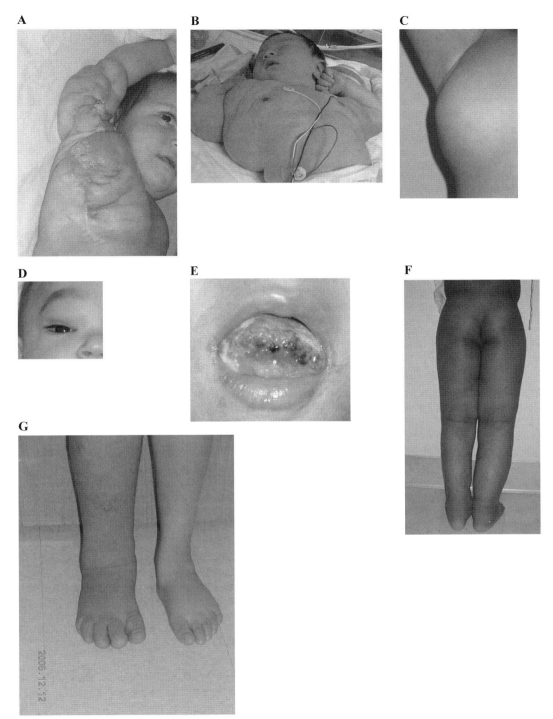

FIGURE 3. Lymphatic malformations. **(A–C)** truncal lymphatic malformations; **(D)** lymphatic malformation of orbit; **(E)** lymphatic malformation of tongue; **(F, G)** lymphedema in patient with intestinal lymphangiectasia **(F)** and lymphatic malformation **(G)**.

TABLE 3. Differences between prenatally and postnatally detected cystic hygromata[a]

Prenatal Diagnosis	Postnatal diagnosis
Diagnosed early in pregnancy	Diagnosed late in pregnancy, if at all
High rate of aneuploidy (>50%)	Low rate of aneuploidy (~1%)
Spontaneous resolution is the norm	Persists unless treated
Nuchal	Variety of sites
Marker for aneuploidy, syndromicity, congenital malformations	Generally isolated finding

[a]Adapted from Howarth *et al.*[39]

TABLE 4. Syndromes associated with lymphatic malformations

Syndrome	Findings	References
Klippel–Trenaunay syndrome	Vascular stain	55, 109
	Soft tissue and/or bony hypertrophy	
	Venous varicosities	
Proteus syndrome	Asymmetric overgrowth of body parts	63
	Cerebriform connective tissue nevi	
	Epidermal nevi	
	Vascular malformations (capillary, venous, lymphatic)	
	Dysregulated adipose tissue	
Osteolysis—Gorham-Stout syndrome	Lymphangiomatosis	69
	Osteolysis	
Hennekam's syndrome	Lymphedema	110
	Lymphangiectasia developmental delay	
Noonan syndrome	Lymphatic dysplasias	111, 112
Turner's syndrome	Lymphatic vessel hypoplasia	113
Trisomy 21	Generalized lymphatic dysplasia	114

intestine. Tested *in vitro*, Octreotide did not affect lymphatic function when applied to the mesenteric lymphatic vessel preparation.[86]

Therapies

Surgical Resection

When possible, lesions are removed surgically. In one series, the recurrence rate was 22% after primary resection, with prognosis correlated to a single anatomic site—the neck. Neonatal age and associated symptoms of infection, dyspnea, dysphagia, or hemorrhage were associated with a poorer prognosis.[87] Others have shown that persons with cervicofacial lymphatic malformations, mediastinal involvement, and multifocal lesions have a higher rate of postoperative problems, including infection, airway compromise requiring tracheostomy, poor dentition, abnormal speech, and intermittent bleeding.[31,88,89] Vascular anomalies specific to the thoracic duct, which are reviewed by Fishman *et al.*, are rare and there is an emphasis on the importance of precise anatomic demarcation of the malformation and discussion of surgical options.[90]

Sclerotherapy

Macrocystic lymphatic malformations may be amenable to endovascular sclerotherapy with bleomycin, doxycycline, or picibanil.[91–94] Picibanil (OK-432), produced from a strain of *Streptococcus pyogenes*, was first described by Ogita *et al.*[95] as a sclerosing agent for lymphatic malformations. Several isolated case reports and series have subsequently been reported.[96–98] A randomized prospective multicenter study of 30 patients with predominantly cervicofacial lymphatic malformations had successful outcomes in the majority of macrocystic lesions.[99] Similar findings

have been reported by others.[100–102] A retrospective review of percutaneous sclerotherapy (with ethanol, sodium tetradecyl sulfate 3% (STS), OK-432, and/or doxycycline) showed the best response in macrocystic lymphatic malformations.[103] Some authors report improvement with interferon alfa therapy.[79,104,105]

Optimally, patients should be followed by a multidisciplinary team of physicians and allied health professionals familiar with lymphatic malformations. This team includes medical specialists to monitor infectious, hematologic, gastrointestinal, and immunologic issues, as well as general medical care. A physiatrist is often involved to recommend appropriate orthotics and physical and/or occupational therapy regimens. Surgical specialists in otolaryngology, head and neck surgery, soft tissue and reconstructive surgery, and orthopedic surgery are often required. Dermatologists can provide laser therapy for lymphatic blebs as well as handle cutaneous complications. Dental needs must also be addressed. Radiologists include a general radiologist, nuclear medicine specialists (for lymphscintography) and interventional radiologists for sclerotherapy and/or venography. Adjunctive staff includes geneticists/genetic counselors, physical therapists, psychologists, social workers, and nutritionists.

Evaluation of a patient with a lymphatic malformation requires a thorough intake, including family, prenatal, and perinatal history. Physical examination should be complete, assessing for dysmorphic features, asymmetry of limb/hand/foot girth and length, transillumination, and assessment for blebs. Radiologic evaluation can start with ultrasound and then proceed to MRI and, if considered necessary, lymphoscintogram. A baseline laboratory evaluation is recommended for assessment of infection and/or lymphopenia, hypoproteinemia, or other anticipated problems.

Complications of lymphatic malformations depend on the anatomic location of the lesion. Head and neck lesions may be associated with airway compromise, necessitating tracheotomy, sleep apnea, difficulty with phonation, halitosis, intraoral bleeding, facial deformity, dental and orthodontic problems, bony overgrowth, ophthalmologic disturbances (e.g., proptosis, astigmatism, ptosis with deprivational amblyopia), and cosmetic deformities. One series identified a strong association between orbital lymphatic malformations and intracranial vascular (70%) and/or structural abnormalities (>10%).[106]

Infection can be local (focal cellulites) or systemic. Although controversial, some patients benefit from low-dose prophylactic antibiotics. Interferon alfa has been reported as an effective therapy for some patients with lymphatic malformations or lymphangiomatosis, resulting in radiologic and clinical improvement.[104,105,107] Patients with more aggressive disorders of lymphatic development (such as Gorham's lymphangiomatosis) have increased morbidity and a less predictable clinical course. Therapy is often multimodal and empirical, requiring surgery, drainage procedures, biphosphonates, sclerotherapy, and antiangiogenic therapies, as no single therapy is efficacious.[69,73,108]

Conflicts of Interest

The author declares no conflicts of interest.

References

1. SABIN, F. 1901. On the origin of the lymphatic system from the veins and the development of the lymph heart and thoracic duct in the pig. Am. J. Anat. **1:** 367–389.
2. HUNTINGTON, G. & C. McCLURE. 1908. The anatomy and development of the jugular lymph sac in the domestic cat (*Felis domestica*). Anat. Rec. **2:** 1–19.
3. KARKKAINEN, M.J., A. SAARISTO, L. JUSSILA, *et al.* 2001. A model for gene therapy of human hereditary lymphedema. Proc. Natl. Acad. Sci. USA **98:** 12677–12682.
4. KARKKAINEN, M.J., P. HAIKO, K. SAINIO, *et al.* 2004. Vascular endothelial growth factor C is required for sprouting of the first lymphatic vessels from embryonic veins. Nat. Immunol. **5:** 74–80.
5. JELTSCH, M., A. KAIPAINEN, V. JOUKOV, *et al.* 1997. Hyperplasia of lymphatic vessels in VEGF-C transgenic mice. Science **276:** 1423–1425.
6. WIGLE, J.T., N. HARVEY, M. DETMAR, *et al.* 2002. An essential role for Prox1 in the induction of the lymphatic endothelial cell phenotype. EMBO J. **21:** 1505–1513.
7. OLIVER, G. & N. HARVEY. 2002. A stepwise model of the development of lymphatic vasculature. Ann. N.Y. Acad. Sci. **979:** 159–165; discussion, 188–196.
8. YUAN, L., D. MOYON, L. PARDANAUD, *et al.* 2002. Abnormal lymphatic vessel development in neuropilin 2 mutant mice. Development **129:** 4797–4806.
9. HIRAKAWA, S., Y.K. HONG, N. HARVEY, *et al.* 2003. Identification of vascular lineage-specific genes by transcriptional profiling of isolated blood vascular and lymphatic endothelial cells. Am. J. Pathol. **162:** 575–586.
10. MORISADA, T., Y. OIKE, Y. YAMADA, *et al.* 2005. Angiopoietin-1 promotes LYVE-1-positive lymphatic vessel formation. Blood **105:** 4649–4656.
11. TAMMELA, T., T.V. PETROVA & K. ALITALO. 2005. Molecular lymphangiogenesis: new players. Trends Cell Biol. **15:** 434–441.
12. ADAMS, R.H. & K. ALITALO. 2007. Molecular regulation of angiogenesis and lymphangiogenesis. Nat. Rev. Mol. Cell Biol. **8:** 464–478.
13. PODGRABINSKA, S., P. BRAUN, P. VELASCO, *et al.* 2002. Molecular characterization of lymphatic endothelial cells. Proc. Natl. Acad. Sci. USA **99:** 16069–16074.
14. CH'NG, S. & S.T. TAN. 2007. Pr08 lymphatic malformations and the molecular basis of lymphangiogenesis. Aust. N.Z. J. Surg. **77**(Suppl 1): A63.
15. MAKINEN, T. & K. ALITALO. 2002. Molecular mechanisms of lymphangiogenesis. Cold Spring Harb. Symp. Quant. Biol. **67:** 189–196.
16. JELTSCH, M., T. TAMMELA, K. ALITALO & J. WILTING. 2003. Genesis and pathogenesis of lymphatic vessels. Cell Tissue Res. **314:** 69–84.
17. OLIVER, G. 2004. Lymphatic vasculature development. Nat. Rev. Immunol. **4:** 35–45.
18. OLIVER, G. & K. ALITALO. 2005. The lymphatic vasculature: recent progress and paradigms. Annu. Rev. Cell Dev. Biol. **21:** 457–483.
19. STACKER, S.A., R.H. FARNSWORTH, T. KARNEZIS, *et al.* 2007. Molecular pathways for lymphangiogenesis and

their role in human disease. Novartis. Found. Symp. **281:** 38–43; discussion, 44–53, 208–209.

20. WILTING, J., M. PAPOUTSI, B. CHRIST, *et al.* 2002. The transcription factor Prox1 is a marker for lymphatic endothelial cells in normal and diseased human tissues. FASEB J. **16:** 1271–1273.

21. HUANG, H.Y., C.C. HO, P.H. HUANG & S.M. HSU. 2001. Co-expression of VEGF-C and its receptors, VEGFR-2 and VEGFR-3, in endothelial cells of lymphangioma: implication in autocrine or paracrine regulation of lymphangioma. Lab. Invest. **81:** 1729–1734.

22. BANERJI, S., J. NI, S.X. WANG, *et al.* 1999. LYVE-1, a new homologue of the CD44 glycoprotein, is a lymph-specific receptor for hyaluronan. J. Cell Biol. **144:** 789–801.

23. XU, H., J.R. EDWARDS, O. ESPINOSA, *et al.* 2004. Expression of a lymphatic endothelial cell marker in benign and malignant vascular tumors. Hum. Pathol. **35:** 857–861.

24. SHORT, R.F., W.E. SHIELS, 2nd, T.J. SFERRA, *et al.* 2007. Site-specific induction of lymphatic malformations in a rat model for image-guided therapy. Pediatr. Radiol. **37:** 530–534.

25. KASTEN, P., G. SCHNOINK, A. BERGMANN, *et al.* 2007. Similarities and differences of human and experimental mouse lymphangiomas. Dev. Dyn. **236:** 2952–2961.

26. FONKALSRUD, E.W. 1994. Congenital malformations of the lymphatic system. Semin. Pediatr. Surg. **3:** 62–69.

27. KIM, J., D. HAN, C.H. HONG, *et al.* 2005. Colonic lymphangiomatosis associated with protein-losing enteropathy. Dig. Dis. Sci. **50:** 1747–1753.

28. AZIZKHAN, R.G., M.J. RUTTER, R.T. COTTON, *et al.* 2006. Lymphatic malformations of the tongue base. J. Pediatr. Surg. **41:** 1279–1284.

29. LAKSHMAN, R. & A. FINN. 2000. Lymphopenia in lymphatic malformations. Arch. Dis. Child. **83:** 276.

30. DE SERRES, L.M., K.C. SIE & M.A. RICHARDSON. 1995. Lymphatic malformations of the head and neck: a proposal for staging. Arch. Otolaryngol. Head Neck. Surg. **121:** 577–582.

31. HAMOIR, M., I. PLOUIN-GAUDON, P. ROMBAUX, *et al.* 2001. Lymphatic malformations of the head and neck: a retrospective review and a support for staging. Head Neck **23:** 326–337.

32. FAGEEH, N., J. MANOUKIAN, T. TEWFIK, *et al.* 1997. Management of head and neck lymphatic malformations in children. J. Otolaryngol. **26:** 253–258.

33. SMITH, R.J. 2004. Lymphatic malformations. Lymphat. Res. Biol. **2:** 25–31.

34. CHANG, J.C. & T.T. SUN. 1990. Ultrasonographic diagnosis of fetal cystic hygroma. Changgeng Yi Xue Za Zhi. **13:** 214–220.

35. HAEUSLER, M.C., H.M. HOFMANN, W. HOENIGL, *et al.* 1990. Congenital generalized cystic lymphangiomatosis diagnosed by prenatal ultrasound. Prenat. Diagn. **10:** 617–621.

36. DESCAMPS, P., O. JOURDAIN, C. PAILLET, *et al.* 1997. Etiology, prognosis and management of nuchal cystic hygroma: 25 new cases and literature review. Eur. J. Obstet. Gynecol. Reprod. Biol. **71:** 3–10.

37. TANRIVERDI, H.A., A.K. ERTAN, H.J. HENDRIK, *et al.* 2005. Outcome of cystic hygroma in fetuses with normal kary-

otypes depends on associated findings. Eur. J. Obstet. Gynecol. Reprod. Biol. **118:** 40–46.

38. GANAPATHY, R., M. GUVEN, F. SETHNA, *et al.* 2004. Natural history and outcome of prenatally diagnosed cystic hygroma. Prenat. Diagn. **24:** 965–968.

39. HOWARTH, E.S., E.S. DRAPER, J.L. BUDD, *et al.* 2005. Population-based study of the outcome following the prenatal diagnosis of cystic hygroma. Prenat. Diagn. **25:** 286–291.

40. OGITA, K., S. SUITA, T. TAGUCHI, *et al.* 2001. Outcome of fetal cystic hygroma and experience of intrauterine treatment. Fetal Diagn. Ther. **16:** 105–110.

41. KUWABARA, Y., R. SAWA, Y. OTSUBO, *et al.* 2004. Intrauterine therapy for the acutely enlarging fetal cystic hygroma. Fetal Diagn. Ther. **19:** 191–194.

42. CHEN, M., C.P. CHEN, J.C. SHIH, *et al.* 2005. Antenatal treatment of chylothorax and cystic hygroma with OK-432 in nonimmune hydrops fetalis. Fetal Diagn. Ther. **20:** 309–315.

43. STEVENSON, D.A., T.J. PYSHER, R.M. WARD & J.C. CAREY. 2006. Familial congenital non-immune hydrops, chylothorax, and pulmonary lymphangiectasia. Am. J. Med. Genet. A. **140:** 368–372.

44. MOERMAN, P., K. VANDENBERGHE, H. DEVLIEGER, *et al.* 1993. Congenital pulmonary lymphangiectasis with chylothorax: a heterogeneous lymphatic vessel abnormality. Am. J. Med. Genet. **47:** 54–58.

45. BELLINI, C., M. MAZZELLA, C. ARIONI, *et al.* 2003. Hennekam syndrome presenting as nonimmune hydrops fetalis, congenital chylothorax, and congenital pulmonary lymphangiectasia. Am. J. Med. Genet. A **120:** 92–96.

46. BELLINI, C., F. BOCCARDO, C. CAMPISI & E. BONIOLI. 2006. Congenital pulmonary lymphangiectasia. Orphanet. J. Rare Dis. **1:** 43.

47. WILSON, R.D., B. PAWEL, M. BEBBINGTON, *et al.* 2006. Congenital pulmonary lymphangiectasis sequence: a rare, heterogeneous, and lethal etiology for prenatal pleural effusion. Prenat. Diagn. **26:** 1058–1061.

48. FARRELL, P.T. 2004. Prenatal diagnosis and intrapartum management of neck masses causing airway obstruction. Paediatr. Anaesth. **14:** 48–52.

49. RAHBAR, R., A. VOGEL, L.B. MYERS, *et al.* 2005. Fetal surgery in otolaryngology: a new era in the diagnosis and management of fetal airway obstruction because of advances in prenatal imaging. Arch. Otolaryngol. Head Neck. Surg. **131:** 393–398.

50. KUCZKOWSKI, K.M. 2007. Advances in obstetric anesthesia: anesthesia for fetal intrapartum operations on placental support. J. Anesth. **21:** 243–251.

51. HAPPLE, R. 1987. Lethal genes surviving by mosaicism: a possible explanation for sporadic birth defects involving the skin. J. Am. Acad. Dermatol. **16:** 899–906.

52. SCHWARTZ, C.E., A.M. BROWN, V.M. DER KALOUSTIAN, *et al.* 1991. DNA fingerprinting: the utilization of minisatellite probes to detect a somatic mutation in the Proteus syndrome. EXS **58:** 95–105.

53. HAMM, H. 1999. Cutaneous mosaicism of lethal mutations. Am. J. Med. Genet. **85:** 342–345.

54. ERICKSON, R.P. 2003. Somatic gene mutation and human disease other than cancer. Mutat. Res. **543:** 125–136.

55. MAARI, C. & I.J. FRIEDEN. 2004. Klippel-Trenaunay syndrome: the importance of "geographic stains" in identifying lymphatic disease and risk of complications. J. Am. Acad. Dermatol. **51:** 391–398.

56. TIAN, X.L., R. KADABA, S.A. YOU, *et al.* 2004. Identification of an angiogenic factor that when mutated causes susceptibility to Klippel-Trenaunay syndrome. Nature **427:** 640–645.

57. TIMUR, A.A., D.J. DRISCOLL & Q. WANG. 2005. Biomedicine and diseases: the Klippel-Trenaunay syndrome, vascular anomalies and vascular morphogenesis. Cell Mol. Life Sci. **62:** 1434–1447.

58. TIMUR, A.A., A. SADGEPHOUR, M. GRAF, *et al.* 2004. Identification and molecular characterization of a de novo supernumerary ring chromosome 18 in a patient with Klippel-Trenaunay syndrome. Ann. Hum. Genet. **68**(Pt 4): 353–361.

59. WANG, Q., A.A. TIMUR, P. SZAFRANSKI, *et al.* 2001. Identification and molecular characterization of de novo translocation t(8;14)(q22.3;q13) associated with a vascular and tissue overgrowth syndrome. Cytogenet. Cell. Genet. **95:** 183–188.

60. WANG, Q.K. 2005. Update on the molecular genetics of vascular anomalies. Lymphat. Res. Biol. **3:** 226–233.

61. CARDOSO, M.T., T.B. DE CARVALHO, L.A. CASULARI & I. FERRARI. 2003. Proteus syndrome and somatic mosaicism of the chromosome 16. Panminerva Med. **45:** 267–271.

62. THIFFAULT, I., C.E. SCHWARTZ, V. DER KALOUSTIAN & W.D. FOULKES. 2004. Mutation analysis of the tumor suppressor PTEN and the glypican 3 (GPC3) gene in patients diagnosed with Proteus syndrome. Am. J. Med. Genet. A **130:** 123–127.

63. COHEN, M.M., JR. 2005. Proteus syndrome: an update. Am. J. Med. Genet. C Semin. Med. Genet. **137:** 38–52.

64. LOFFELD, A., N.J. MCLELLAN, T. COLE, *et al.* 2006. Epidermal naevus in Proteus syndrome showing loss of heterozygosity for an inherited PTEN mutation. Br. J. Dermatol. **154:** 1194–1198.

65. CAUX, F., H. PLAUCHU, F. CHIBON, *et al.* 2007. Segmental overgrowth, lipomatosis, arteriovenous malformation and epidermal nevus (SOLAMEN) syndrome is related to mosaic PTEN nullizygosity. Eur. J. Hum. Genet. **15:** 767–773.

66. GORHAM, L.W. & A.P. STOUT. 1954. Hemangiomatosis and its relation to massive osteolysis. Trans. Assoc. Am. Physicians. **67:** 302–307.

67. GORHAM, L.W. & A.P. STOUT. 1955. Massive osteolysis (acute spontaneous absorption of bone, phantom bone, disappearing bone); its relation to hemangiomatosis. J. Bone. Joint. Surg. Am. **37-A:** 985–1004.

68. HEYDEN, G., L.G. KINDBLOM & J.M. NIELSEN. 1977. Disappearing bone disease: a clinical and histological study. J. Bone. Joint. Surg. Am. **59:** 57–61.

69. PATEL, D.V. 2005. Gorham's disease or massive osteolysis. Clin. Med. Res. **3:** 65–74.

70. DICKSON, G.R., R.A. MOLLAN & K.E. CARR. 1987. Cytochemical localization of alkaline and acid phosphatase in human vanishing bone disease. Histochemistry **87:** 569–572.

71. DEVLIN, R.D., H.G. BONE, 3RD & G.D. ROODMAN. 1996. Interleukin-**6:** a potential mediator of the massive osteolysis in patients with Gorham-Stout disease. J. Clin. Endocrinol. Metab. **81:** 1893–1897.

72. HIRAYAMA, T., A. SABOKBAR, I. ITONAGA, *et al.* 2001. Cellular and humoral mechanisms of osteoclast formation and bone resorption in Gorham-Stout disease. J. Pathol. **195:** 624–630.

73. HAGENDOORN, J., T.P. PADERA, T.I. YOCK, *et al.* 2006. Platelet-derived growth factor receptor-beta in Gorham's disease. Nat. Clin. Pract. Oncol. **3:** 693–697.

74. LEE, J. & M.S. KONG. 2007. Primary intestinal lymphangiectasia diagnosed by endoscopy following the intake of a high-fat meal. Eur. J. Pediatr. **167:** 237–239.

75. RIVET, C., M.G. LAPALUS, J. DUMORTIER, *et al.* 2006. Use of capsule endoscopy in children with primary intestinal lymphangiectasia. Gastrointest. Endosc. **64:** 649–650.

76. CHAMOUARD, P., H. NEHME-SCHUSTER, J.M. SIMLER, *et al.* 2006. Videocapsule endoscopy is useful for the diagnosis of intestinal lymphangiectasia. Dig. Liver. Dis. **38:** 699–703.

77. DANIELSSON, A., E. TOTH & H. THORLACIUS. 2006. Capsule endoscopy in the management of a patient with a rare syndrome–yellow nail syndrome with intestinal lymphangiectasia. Gut **55:** 196–233.

78. CAMMAROTA, G., R. CIANCI & G. GASBARRINI. 2005. High-resolution magnifying video endoscopy in primary intestinal lymphangiectasia: a new role for endoscopy? Endoscopy **37:** 607.

79. TIMKE, C., M.F. KRAUSE, H.C. OPPERMANN, *et al.* 2007. Interferon alpha 2b treatment in an eleven-year-old boy with disseminated lymphangiomatosis. Pediatr. Blood Can. **48:** 108–111.

80. ROSTOM, A.Y. 2000. Treatment of thoracic lymphangiomatosis. Arch. Dis. Child. **83:** 138–139.

81. KANDIL, A., A.Y. ROSTOM, W.A. MOURAD, *et al.* 1997. Successful control of extensive thoracic lymphangiomatosis by irradiation. Clin. Oncol. (R. Coll. Radiol.) **9:** 407–411.

82. KUROIWA, G., T. TAKAYAMA, Y. SATO, *et al.* 2001. Primary intestinal lymphangiectasia successfully treated with octreotide. J. Gastroenterol. **36:** 129–132.

83. KLINGENBERG, R.D., N. HOMANN & D. LUDWIG. 2003. Type I intestinal lymphangiectasia treated successfully with slow-release octreotide. Dig. Dis. Sci. **48:** 1506–1509.

84. FILIK, L., P. OGUZ, A. KOKSAL, *et al.* 2004. A case with intestinal lymphangiectasia successfully treated with slow-release octreotide. Dig. Liver Dis. **36:** 687–690.

85. TE PAS, A.B., K. VD VEN, M.P. STOKKEL & F.J. WALTHER. 2004. Intractable congenital chylous ascites. Acta Paediatr. **93:** 1403–1405.

86. MAKHIJA, S., P.Y. VON DER WEID, J. MEDDINGS, *et al.* 2004. Octreotide in intestinal lymphangiectasia: lack of a clinical response and failure to alter lymphatic function in a guinea pig model. Can. J. Gastroenterol. **18:** 681–685.

87. RAVEH, E., A.L. DE JONG, G.P. TAYLOR & V. FORTE. 1997. Prognostic factors in the treatment of lymphatic malformations. Arch. Otolaryngol. Head Neck. Surg. **123:** 1061–1065.

88. PADWA, B.L., P.G. HAYWARD, N.F. FERRARO & J.B. MUL-LIKEN. 1995. Cervicofacial lymphatic malformation: clinical course, surgical intervention, and pathogenesis of skeletal hypertrophy. Plast. Reconstr. Surg. **95:** 951–960.

89. LEI, Z.M., X.X. HUANG, Z.J. SUN, *et al.* 2007. Surgery of lymphatic malformations in oral and cervicofacial regions in children. Oral. Surg. Oral. Med. Oral. Pathol. Oral. Radiol. Endod. **104:** 338–344.

90. FISHMAN, S.J., P.E. BURROWS, J. UPTON & W.H. HENDREN. 2001. Life-threatening anomalies of the thoracic duct: anatomic delineation dictates management. J. Pediatr. Surg. **36:** 1269–1272.

91. BARRIER, A., F. LACAINE, P. CALLARD & M. HUGUIER. 2002. Lymphangiomatosis of the spleen and 2 accessory spleens. Surgery **131:** 114–116.

92. SUNG, M.W., S.O. CHANG, J.H. CHOI & J.Y. KIM. 1995. Bleomycin sclerotherapy in patients with congenital lymphatic malformation in the head and neck. Am. J. Otolaryngol. **16:** 236–241.

93. MOLITCH, H.I., E.C. UNGER, C.L. WITTE & E. VAN SONNENBERG. 1995. Percutaneous sclerotherapy of lymphangiomas. Radiology **194:** 343–347.

94. KIM, K.H., M.W. SUNG, J.L. ROH & M.H. HAN. 2004. Sclerotherapy for congenital lesions in the head and neck. Otolaryngol. Head. Neck. Surg. **131:** 307–316.

95. OGITA, S., T. TSUTO, K. TOKIWA & T. TAKAHASHI. 1987. Intracystic injection of OK-432: a new sclerosing therapy for cystic hygroma in children. Br. J. Surg. **74:** 690–691.

96. OGITA, S., T. TSUTO, E. DEGUCHI, *et al.* 1991. OK-432 therapy for unresectable lymphangiomas in children. J. Pediatr. Surg. **26:** 263–268; discussion, 268–270.

97. OGITA, S., T. TSUTO, K. NAKAMURA, *et al.* 1996. OK-432 therapy for lymphangioma in children: why and how does it work? J. Pediatr. Surg. **31:** 477–480.

98. OGITA, S., T. TSUTO, K. NAKAMURA, *et al.* 1994. OK-432 therapy in 64 patients with lymphangioma. J. Pediatr. Surg. **29:** 784–785.

99. GIGUERE, C.M., N.M. BAUMAN, Y. SATO, *et al.* 2002. Treatment of lymphangiomas with OK-432 (Picibanil) sclerotherapy: a prospective multi-institutional trial. Arch. Otolaryngol. Head Neck Surg. **128:** 1137–1144.

100. GREINWALD, J.H., JR., D.K. BURKE, Y. SATO, *et al.* 1999. Treatment of lymphangiomas in children: an update of Picibanil (OK-432) sclerotherapy. Arch. Otolaryngol. Head Neck Surg. **121:** 381–387.

101. SMITH, R.J., D.K. BURKE, Y. SATO, *et al.* 1996. OK-432 therapy for lymphangiomas. Arch. Otolaryngol. Head Neck Surg. **122:** 1195–1199.

102. PETERS, D.A., D.J. COURTEMANCHE, M.K. HERAN, *et al.* 2006. Treatment of cystic lymphatic vascular malformations with OK-432 sclerotherapy. Plast. Reconstr. Surg. **118:** 1441–1446.

103. ALOMARI, A.I., V.E. KARIAN, D.J. LORD, *et al.* 2006. Percutaneous sclerotherapy for lymphatic malformations: a retrospective analysis of patient-evaluated improvement. J. Vasc. Interv. Radiol. **17:** 1639–1648.

104. REINHARDT, M.A., S.C. NELSON, S.F. SENCER, *et al.* 1997. Treatment of childhood lymphangiomas with interferon-alpha. J. Pediatr. Hematol. Oncol. **19:** 232–236.

105. OZEKI, M., M. FUNATO, K. KANDA, *et al.* 2007. Clinical improvement of diffuse lymphangiomatosis with pegylated interferon alfa-2b therapy: case report and review of the literature. Pediatr. Hematol. Oncol. **24:** 513–524.

106. BISDORFF, A., J.B. MULLIKEN, J. CARRICO, *et al.* 2007. Intracranial vascular anomalies in patients with periorbital lymphatic and lymphaticovenous malformations. AJNR Am. J. Neuroradiol. **28:** 335–341.

107. CHEN, Y.L., C.C. LEE, M.L. YEH, *et al.* 2007. Generalized lymphangiomatosis presenting as cardiomegaly. J. Formos. Med. Assoc. **106**(3 Suppl): S10–14.

108. TAMAY, Z., E. SARIBEYOGLU, U. ONES, *et al.* 2005. Diffuse thoracic lymphangiomatosis with disseminated intravascular coagulation in a child. J. Pediatr. Hematol. Oncol. **27:** 685–687.

109. GARZON, M.C., J.T. HUANG, O. ENJOLRAS & I.J. FRIEDEN. 2007. Vascular malformations. Part II. Associated syndromes. J. Am. Acad. Dermatol. **56:** 541–564.

110. VAN BALKOM, I.D., M. ALDERS, J. ALLANSON, *et al.* 2002. Lymphedema-lymphangiectasia-mental retardation (Hennekam) syndrome: a review. Am. J. Med. Genet. **112:** 412–421.

111. VAN DER BURGT, I. 2007. Noonan syndrome. Orphanet. J. Rare. Dis. **2:** 4.

112. KEBERLE, M., H. MORK, M. JENETT, *et al.* 2000. Computed tomography after lymphangiography in the diagnosis of intestinal lymphangiectasia with protein-losing enteropathy in Noonan's syndrome. Eur. Radiol. **10:** 1591–1593.

113. VON KAISENBERG, C.S., K.H. NICOLAIDES & B. BRAND-SABERI. 1999. Lymphatic vessel hypoplasia in fetuses with Turner syndrome. Hum. Reprod. **14:** 823–826.

114. OCHIAI, M., S. HIKINO, H. NAKAYAMA, *et al.* 2006. Nonimmune hydrops fetalis due to generalized lymphatic dysplasia in an infant with Robertsonian trisomy 21. Am. J. Perinatol. **23:** 63–66.

Lymphatics in Lung Disease

Souheil El-Chemaly, Stewart J. Levine, and Joel Moss

Translational Medicine Branch, National Heart, Lung, and Blood Institute, National Institutes of Health, Bethesda, Maryland, USA

The lymphatic circulation appears to be a vital component in lung biology in health and in disease. Animal models have established the role of the lymphatic circulation in neoplastic and inflammatory diseases of the lung, such as asthma and cancer, and allowed for the understanding of the molecular controls of lymphangiogenesis in normal lung development. Understanding the role of lymphatics in human lung disease appears likely to contribute to the understanding of the pathogenesis of disease and the development of novel therapeutic targets.

Key words: lymphangiogenesis; metastasis; lymphangiectasis; lymphangioleiomyomatosis; VEGF-D; VEGF-C

The lymphatic circulation has received little attention in pulmonary research. Fluid homeostasis and host defense are two critical roles for the lung, in addition to gas exchange.[1] The main function of the lymphatic circulation is to drain fluid from tissues and return it to the vascular circulatory system.[2] The lymphatic circulation is also involved in the immune system of the body, since lymphocytes and dendritic cells move through the lymphatic system to reach the lymphoid organs.[2] Therefore, the lymphatic circulation must be a vital component in lung biology in health and in disease.

The lymphatic system comprises a vasculature consisting of thin-walled capillaries as well as larger vessels that are lined by a layer of endothelial cells. Lymphatics structures are distinguished from those of the arterial and venous circulations by the absence of pericytes.[3] Identification of unique lymphatic markers that differentiate its vessels from the arterial and venous circulations enabled a more informed study of the lymphatic circulation.

Among the most important markers are:

1. Prox1, which is found exclusively in the lymphatic endothelium, and is a transcription factor required for programming the phenotype of the lymphatic endothelial cell (LEC)[4];
2. LYVE-1, a CD44 homologue, and an LEC hyaluronan receptor, which is present exclusively on LECs and macrophages[5]; and

3. VEGFR-3 (vascular endothelial growth factor receptor 3), a receptor for VEGF (vascular endothelial growth factor)-C and VEGF-D, which is not detected on blood vascular endothelial cells, and is present on those in the lymphatics.[6]

The discovery of such genes, growth factors, and receptors involved in lymphatic development and function, in addition to the production of transgenic mice and improvement in imaging techniques has enabled a better understanding of the lymphatic circulation. Many of the human diseases associated with lymphatic abnormalities are, however, uncommon and remain poorly characterized.

The best described of numerous pathways that regulate lymphangiogenesis involve VEGF-C and VEGF-D.[7] These proteins promote lymphangiogenesis by activating VEGFR-3, a cell-surface tyrosine kinase receptor, leading to initiation of a downstream signaling cascade. In addition, VEGF-C and VEGF-D bind to neuropilin-2 (Nrp2), a semaphorin receptor, which is also expressed in lymphatic capillaries.[7] Knockout mice for Nrp2 exhibit lymphatic hypoplasia.[8]

During human gestation, lymph sacs appear at 6–7 weeks. Experimental data from mice support the Sabin hypothesis that lymphatics arise by sprouting from embryonic veins.[9] The nuclear transcription factor Prox1[10] and endothelial receptor-ligand VEGF-C are integral to these events. A complete absence of lymphatic vasculature was observed in Prox1-knockout mice, whereas overexpression of Prox1 in human blood vascular endothelial cells suppressed many blood vascular–specific genes and upregulated LEC transcripts.[10] Signals leading to the expression of Prox1 and its target genes in LECs, however, are not known.

Address for correspondence: Joel Moss, M.D., Ph.D., Translational Medicine Branch, National Heart, Lung, and Blood Institute, National Institutes of Health, Bldg. 10, Rm 6D03, MSC 1590, Bethesda, MD 20892-1590. Voice: +1-301-496-1597; fax: +1-301-496-2363.

mossj@nhlbi.nih.gov

VEGF-C deletion leads to a complete absence of lymphatics in mouse embryos.[11] Although LECs initially differentiate, they fail to migrate and form lymphatic sacs. VEGF-D deletion does not affect lymphatic development,[12] although exogenous VEGF-D rescues the phenotype of VEGF-C knockout mice.[11] VEGFR-3 deletion leads to defects in blood vessel remodeling and embryonic death.[13]

Lymphangiogenesis during Lung Organogenesis

VEGF-D mRNA is not detectable during early development of the mouse lung (E12.5). It was first seen at E13.5, the pseudoglandular stage, and by E14.5, it was diffusely abundant throughout the mesenchyme and visible by immunohistochemical study. It remained elevated until birth and was not detected in the postnatal lung.[14] Levels of VEGF-D mRNA were higher in the distal than proximal mesenchyme. VEGF-D was associated with cadherin-11-positive fibroblasts,[15] but not endothelial cells, pericytes, or smooth muscle cells. VEGF-D was not necessary for lymphatic development, as VEGF-D knockout mice did not show evidence of lymphatic abnormalities.[12] Although lungs of VEGF-D-null mice appeared to develop normally, the numbers of VEGFR-3-positive vessels adjacent to the muscular surface of bronchioles in the lungs of knockout mice were significantly lower than those of wild-type mice. This difference did not translate into a difference in the weights of wet and dry lungs, suggesting that knockout mice had a functional lymphatic system.

VEGF-C is critical for the development of the lymphatic system *in utero*, and VEGF-C knockout mice died at E15.5 from accumulation of fluids in their tissues.[11] During normal embryogenesis, there was extensive lymphatic vessel sprouting from embryonic veins and the developing lung was a site of VEGF-C and VEGFR-3 expression.[16]

VEGF (VEGF-A) is a key angiogenic growth factor, with loss of even one allele leading to death *in utero* at E8.5–E9.5 from severe vascular malformations.[17,18] Recently published evidence suggests that the overexpression of VEGF during a late stage of lung organogenesis resulted in increased numbers of lymphatic vessels distally. The increase was mediated by an increase in VEGFR-3 through an unknown mechanism, with associated increased sensitivity to VEGF-C and VEGF-D.[19]

Podoplanin, a transmembrane mucin-type glycoprotein is highly expressed in LECs.[9] In mice, podoplanin deficiency led to abnormal lung develop-

ment and perinatal mortality. Type I cell differentiation was blocked, as indicated by smaller airspaces, fewer type I cells and reduced levels of aquaporin-5, a type I-cell water channel.[20] Moreover, podoplanin knockout mice had defects in lymphatic, but not blood vessel pattern formation.[21] These defects were associated with diminished lymphatic transport, congenital lymphedema, and dilation of lymphatic vessels.[21]

Net, a member of the Ets-domain transcription factors, is a strong inhibitor of transcription when MAP kinases are not activated. Net mutant mice[22] have lymphangiectasia specifically in the thoracic wall, leading to chylothorax in the early postnatal phase, resulting in respiratory failure and death. Lymphangiectasia precedes the appearance of chylothorax, suggesting that lymphatic drainage is impaired. Net is located on chromosome 12q, and there is one case report of partial trisomy of 12q associated with chylothorax in humans.[23]

Mice lacking integrin α9 developed chylous pleural effusions and die shortly after birth (2 days). In α9-null mice, the anatomy of the thoracic duct was normal except for edema and perivascular lymphocytes surrounding the thoracic duct.[24] Possible mechanisms include interaction of the α9 subunit with β1 integrins, which can induce tyrosine phosphorylation of VEGFR-3. Activation of both β1 and VEGFR-3 is necessary for LEC migration.[25] Alternatively, it is now known that Prox-1 induces LEC differentiation through integrin α9.[26]

EphrinB2 is a transmembrane growth factor.[9] Mice with a mutated carboxy-terminal PDZ interaction site of *ephrinB2* die with a chylothorax during the first 3 weeks after birth.[27] Although these mutant mice have a normal blood vasculature, there is an abnormal postnatal remodeling of the lymphatic vasculature, leading to hyperplasia of the collecting lymphatic vessels, lack of luminal valve formation, and a failure to remodel the primary lymphatic plexus.[27]

In humans, the role of VEGF in the maturation and alveolarization of the developing lung has been well established,[28] but the role of lymphatics and the lymphangiogenic growth factors remain poorly understood. Bronchopulmonary dysplasia is a disease characterized by maturation arrest of the lung.[29] A recent study examined the role of lymphangiogenesis and lymphangiogenic growth factors VEGF-D and VEGF-C in the maturation process.[30] Although the role of VEGF-D remains more difficult to determine since VEGF-D levels in tracheal aspirate fluids were too low to detect, VEGF-C seems to play a prominent role in the development of human lung. VEGF-C levels in tracheal fluid declines in the first postnatal week from a high in the first 2 postnatal days. Infants born prematurely may suffer from this decline in VEGF-C with a lack of development of their lymphatic systems, leading to abnormal lung fluid homeostasis. Prenatal injection of corticosteroids increased VEGF-C levels, which may help the maturation process. VEGF-C was present in the bronchial epithelium of fetuses and infants, but only fetuses showed alveolar staining. VEGF-C was seen in alveolar macrophages in infants suffering from respiratory distress syndrome or late bronchopulmonary dysplasia, indicating that inflammatory cells have a major role to play in the lymphangiogenic process.

Lymphangiogenesis and Lung Cancer

Small Cell Lung Cancer (SCLC) and Non–Small Cell Lung Cancer (NSCLC)

It was originally thought that lymphatic spread by tumors occured by means of preexisting peritumoral lymphatics, without intratumoral lymphangiogenesis.[31] In a lung cancer model, there is now direct evidence of the role of high levels of VEGF-C in promoting both tumor lymphangiogenesis and tumor spread to the regional lymph nodes.[32] These effects can be suppressed by blocking VEGFR-3 signaling.[33]

Results of human clinicopathologic studies show that in SCLC, VEGF-C, and VEGF-D transcripts and protein levels are elevated.[34] In one NSCLC study, serum VEGF-C levels were used as markers for lymph node metastasis, and their use with computed tomography scans to stage lymph node invasion was proposed.[35] In NSCLC, a lower survival rate is associated with increased expression of the lymphangiogenic growth factors VEGF-C and VEGF-D and their receptor VEGFR-3.[36] Moreover, angiogenesis in NSCLC is associated with concomitant lymphangiogenesis. In the absence of angiogenesis, NSCLC cells invade host lymphatics during tumor development.[36] Although lymphangiogenesis occurs, the newly formed lymphatic vessels do not survive.[37] Lymph node metastasis occurs via existing lymphatic vessels or newly created shunts by active angiogenesis and lymphangiogenesis. Metastatic tumor spread to lymph nodes plays a major role in staging lung cancer. Strategies that have been found to be useful in animal models to stop lymphatic spread of tumor cells, such as VEGFR-3-blocking antibodies, or small molecule kinase inhibitors, represent new possibilities for the treatment of human cancers.[38]

Kaposi's Sarcoma

Kaposi's sarcoma (KS) is the tumor most often seen in patients suffering from HIV/AIDS. KS most frequently affects the skin,[39] and pulmonary KS carries a poor prognosis.[40]

The causative agent of KS, human herpesvirus-8, was originally thought to infect LECs, since tumor cells express VEGFR-3 and podoplanin.[39] Subsequently, it was shown that blood vascular endothelial cells infected with the virus are re-programmed into LECs that express Prox-1, the master regulator of lymphatic lineage, with downregulation of blood vascular endothelial cell genes.[41,42]

Lymphangiogenesis and Asthma

Patients with chronic asthma can develop chronic airflow obstruction that is fixed or only partially reversible, as well as persistent bronchial hyperreactivity.[43] The pathogenesis of chronic airflow obstruction involves remodeling of the airway wall secondary to inflammatory cell infiltration, myocyte and myofibroblast hyperplasia, mucus metaplasia, subepithelial fibrosis, and edema formation.[43] The lymphatic system, via its ability to facilitate the clearance of edematous fluid, has recently been identified as playing an important role in attenuating airway wall remodeling in asthma.

The pathogenesis of airway lymphatic vessel hyperplasia has been investigated using a murine model of *Mycoplasma pulmonis* pulmonary infection, with evidence of chronic inflammation manifested by angiogenesis and microvascular remodeling.[44,45] This model is relevant for asthma, as mycoplasma infection can initiate or exacerbate disease in humans and animal models.[45–47] In an elegant study by McDonald and colleagues, *M. pulmonis* respiratory infection was shown to induce airway lymphangiogenesis that was dependent upon signaling by VEGF-C and VEGF-D through VEGFR-3.[44] In animals treated with a soluble VEGFR-3-Ig, which binds and sequesters VEGF-C and VEGF-D, lymphatic growth (as measured by the number of lymphatic sprouts) was reduced by 96%, without effects on blood vessel remodeling. Treatment with antibodies directed against VEGFR-1 or VEGFR-2 did not inhibit lymphatic vessel growth, whereas an anti-VEGFR-3 antibody markedly reduced lymphangiogenesis. Although antibiotic treatment of infected animals was associated with a return of remodeled blood vessels to their normal state within 1 month, the extensive network of new lymphatic vessels had not regressed after 12 weeks of treatment. Airway epithelial and inflammatory cells were sources of VEGF-C, whereas VEGF-D was produced by airway neutrophils, macrophages, airway smooth muscle cells, and lymphatics.

To define better the VEGFR-3 ligands that mediate lymphangiogenesis, mice were inoculated with adenoviral constructs that generated VEGF-C or VEGF-D.[45] Overexpression of VEGF-C in airway epithelial cells was associated with lymphangiogenesis similar to that seen in *M. pulmonis* infection, whereas overexpression of VEGF-D induced an exaggerated proliferation of lymphatic vessels. Overexpression of VEGF-C or VEGF-D was specific for lymphatic proliferation, as neither induced angiogenesis. Functional effects of impaired lymphangiogenesis on mucosal edema were also assessed in this model system.[45] Inhibition of lymphangiogenesis by adenoviral delivery of a soluble VEGFR-3-Ig construct prior to *M. pulmonis* infection was associated with significantly increased bronchial lymphedema. In addition, blockade of lymphangiogenesis reduced infection-induced hypertrophy of draining bronchial lymph nodes. This study shows that new lymphatic vessels provide a mechanism for clearance of extravasated fluid from the abnormal leaky blood vessels in inflamed airways.

The murine model of *M. pulmonis* pulmonary infection was also utilized to assess potential roles in lymphangiogenesis of innate and acquired immune responses in host defense. Mast cell–deficient (Kit^{W-sh}/Kit^{W-sh}) mice were used to evaluate the contribution of mast cells, which participate in host defense in septic peritonitis, in contracting *M. pulmonis* infection.[47] Mast cell–deficient mice had a higher mycoplasma burden and mortality rate than wild-type mice, consistent with the conclusion that mast cells play an important role in innate immune responses to pulmonary mycoplasma infection. Furthermore, there was a striking increase in lymphangiogenesis in mast cell–deficient mice that exceeded the angiogenic response at 28 days and persisted despite resolution of the acute infection. This finding supports the notion that lymphangiogenesis reverses more slowly than angiogenic remodeling after *M. pulmonis* infection. In a separate study, lymphatic remodeling after *M. pulmonis* infection was substantially reduced in the airways of lymphocyte-deficient *Rag1*$^{-/-}$ mice and B cell–deficient ($Ig\mu^{MT}$) mice.[48] Remodeling of blood and lymphatic vessels was significantly enhanced by transfer of serum, which contained immune complexes, from infected wild-type mice to B cell–deficient mice. Thus, *M. pulmonis*–induced chronic lymphangiogenesis and angiogenesis were dependent upon immune complex–mediated inflammation.

Taken together, these studies showed that airway inflammation induces a lymphangiogenic response via VEGF-C and VEGF-D, which are produced primarily by inflammatory cells, signaling through VEGFR-3

in LECs. Newly formed lymphatic networks persisted for extended periods of time despite the resolution of inflammation and provided a pathway for removal of extravasated plasma that occurs secondary to endothelial gap formation and microvascular leakage.[49] Thus, with impaired lymphangiogenesis, accumulation of extravasated fluid may result in bronchial lymphedema and contribute to airway remodeling with persistent airflow obstruction in asthmatic lungs.

Lung Transplantation

Obliterative bronchiolitis (OB) is a major cause of late failure of lung transplantation, since it is present in 70% of transplanted lungs by 5 years post transplant.[50] OB is characterized histologically by a fibromyxoid reaction originating from the airway walls that obliterates both small and large airways. This translates physiologically into a progressive decrease in forced expiratory volume in 1 second (FEV$_1$), leading to dyspnea, infections, and loss of the grafted lung. The pathogenesis is not completely understood, but most of the evidence points toward a major role for adaptive immunity in the etiology of OB.[51]

In a rat model of tracheal transplant,[52] transfer of the VEGF gene resulted in the doubling of the number of LYVE-1+ lymphatic vessels in allograft airway walls. Conversely, treatment with PTK 787, a broad VEGFR protein tyrosine kinase inhibitor, resulted not only in a 50% reduction in the number of LYVE-1+ lymphatic vessels, but also in decreased numbers of CD4 and CD8 T cells in airway walls. These observations suggest that lymphangiogenesis enhances antigen presentation to the host immune system, thereby increasing alloimmune sensitization. In agreement with other studies, the effects of VEGF could be due to the recruitment of inflammatory cells such as macrophages that produce VEGF-C and VEGF-D.[45,53] In this model, lymphangiogenesis may lead to continuous presentation of donor tissue to lymph nodes, with persistent alloimmune response, and resulting rejection or graft injury.[52]

The role of lymphangiogenesis in human allograft rejection of solid organs has been well documented in kidney[54,55] and heart transplantation.[56] In human lung transplant recipients, the role of angiogenesis is increasingly recognized[57–60] in the pathogenesis of airway obstruction, leading eventually to OB, but to date, there are no published reports regarding lymphangiogenesis in lung transplantation. Total body lymphoid irradiation has been proven safe and efficacious in slowing the decline of lung function in transplant recipients suffering from OB.[61]

Pulmonary Lymphangiectasia

Lymphangiectasia refers to the dilation of normal lymphatic vessels, most often secondary to more proximal obstruction or to development of lymphatic valvular incompetence. Primary lymphangiectasia is a rare disorder that most commonly affects the pulmonary and the intestinal lymphatic vasculature.[62] Predisposing factors are unknown. Classically, pulmonary lymphangiectasia presents as respiratory distress of the newborn, resulting in death. Historical series had a mortality of 100%, but advances in neonatal respiratory care have improved prognosis.[62] In the prenatal period, the disease often results in stillborn infants. The immediate neonatal presentation involves cyanosis and respiratory distress. Postneonatal or late presentation, which can occur after months without respiratory symptoms, typically includes cough and wheezing. Symptoms in a few cases, did not appear until late childhood or adolescence.[62] Pleural effusions were described in 15% of cases,[63] and typically were chylous exudates, with a high lymphocyte content.[64] In the neonatal presentation, the chest X-ray reveals diffuse interstitial infiltrates. Later in infancy and childhood, reticulonodular infiltrates and hyperinflation were seen,[65] and sometimes pleural effusions. On computerized tomography of the chest there was ground-glass appearance with interstitial densities. Lymphangiography showed an interstitial pattern of pulmonary lymphatics with opacification and formation of collateral lymphatic vessels.[62]

Primary pulmonary lymphangiectasia can be part of generalized lymphangiectasia involving the intestines, or limited to the lungs. It also includes syndromic forms of primary pulmonary lymphangiectasia, where it is associated with unrelated congenital anomalies, including genetic syndromes.[62] Secondary pulmonary lymphangiectasia can be divided into lymphatic and cardiovascular obstructive forms. Thoracic duct agenesis causing lymphatic obstruction has been reported. Among the cardiovascular forms are hypoplastic left heart syndrome, anomalous pulmonary venous return, pulmonary vein atresia, and congenital mitral stenosis.[62]

Autopsy data show that the incidence of pulmonary lymphangiectasia is 0.5–1% of all stillborn or neonatal infant deaths.[62] There was a male predominance of the primary forms. Most cases are sporadic, with a few familial cases of pulmonary lymphangiectasia demonstrating an autosomal-recessive inheritance pattern.[66]

Chylous Pleural Effusions

Chylomicrons and very-low-density lipoproteins produced from dietary fat are delivered to the cisterna chyli, from which one major lymphatic vessel, the thoracic duct, returns them to the vascular circulation.[67] The thoracic duct originates in the cisterna chyli at the level of the second lumbar vertebra, where it crosses into the thoracic cavity through the esophageal hiatus. In the thoracic cavity, the thoracic duct runs extrapleurally in the posterior mediastinum, close to the esophagus and the pericardium, on the right side of the anterior surface of the vertebral column and continues toward the superior mediastinum after crossing to the left at the level of the 4th–6th thoracic vertebrae.[67] The thoracic duct then arches 3–5 cm above the clavicle, passing anterior to the subclavian artery, vertebral artery, and thyrocervical trunk to terminate in the region of left subclavian and jugular veins.[68,69] Depending on diet, 1500–2500 mL of chyle per day are transported through the thoracic duct.[68] The protein content of chyle is about 3g/dL and the electrolyte composition is similar to that of serum. The cellular population is predominantly lymphocytes, (400–6800/mm^3) with most of them T lymphocytes.[70]

A chylothorax is formed when chyle enters the pleural space. Four causes of chylothorax are recognized. In adults, tumors are responsible for more than 50% chylothoraces, and, among those, lymphoma is the cause of about 75% of them.[67] The second most frequent cause of chylothorax is trauma, most often thoracic surgery that requires the mobilization of the left subclavian artery.[71]

Idiopathic chylothorax including congenital chylothorax is the third most common in the neonate, and is the most frequent cause of pleural effusion.[72] Congenital chylothorax is more frequent in males (2:1) and on the right side of the thorax. Thoracotomies have shown that the thoracic duct is normal in most of the afflicted babies, and occasionally generalized oozing is described.[64,73] It may be isolated or part of a larger group of lymphatic abnormalities and it has also been associated with Turner's, Down and the Noonan syndromes.[64] Familial congenital chylothorax has been reported in both males[72] and females.[74] The fourth chylothorax category is "other disorders" (e.g., lymphangioleiomyomatosis, tuberculosis, sarcoidosis, KS).

Idiopathic Pulmonary Fibrosis

Information regarding the role of lymphatics in the pathogenesis of idiopathic pulmonary fibrosis (IPF) is scant. Bleomycin is commonly used in animal models of pulmonary fibrosis.[75] In a rat bleomycin model,[76] deposition of collagen IV and collagen I was observed with time after bleomycin injury. The results showed that collagen type IV accumulation specific for the angiogenic process[77] was followed by collagen type I deposition, indicating the presence of lymphangiogenesis.

There are no studies reported of *de novo* lymphangiogenesis in human IPF or other interstitial pneumonias; however, multiple studies[78,79] have found an increased prevalence (40–60%) of enlarged mediastinal lymph nodes. Although one study showed no correlation with severity of disease,[78] another showed a correlation over time between lymph node enlargement and evolution of the disease[79] in the absence of infection or malignancy.

Tuberculosis

In guinea pigs, inflammation of the pulmonary lymphatics was one of the earliest lung lesions to develop after infection with *M. tuberculosis*.[80] Lymphangitis developed in the stroma of the connective tissue of major lymphatic vessels and in the alveolar spaces 5 – 15 days after infection. Lymphatics surrounding bronchi and vessels were delineated with granulomatous lesions, resulting from inflammation in the lymphatic walls and expanding into the lumen. Lesions in both lung parenchyma and lymphatics progressed similarly, leading to central necrosis and granulocyte infiltration. After the resolution of acute inflammation, chronic stages were characterized by the infiltration of fibroblasts, fibroplasias, and dystrophic mineralization. In humans, the Gohn primary complex represents lung and lymph node lesions. In children and immunocompromised hosts, such as patients with HIV, the lymph nodes represent the first site of extrapulmonary disease.

Despite the increasing evidence from animal models supporting the involvement of lymphatics in lung disease, thus far, studies in humans are few. Understanding the role of lymphatics in human lung disease appears likely to contribute to the understanding of the pathogenesis of disease and the development of novel therapeutic targets.

Acknowledgment

The study was supported, in part, by the Intramural Research Program, NIH/NHLBI.

Conflicts of Interest

The authors declare no conflicts of interest.

References

1. WHITSETT, J.A. 2002. Intrinsic and innate defenses in the lung: intersection of pathways regulating lung morphogenesis, host defense, and repair. J. Clin. Invest. **109:** 565–569.
2. ADAMS, R.H. & K. ALITALO. 2007. Molecular regulation of angiogenesis and lymphangiogenesis. Nat. Rev. Mol. Cell Biol. **8:** 464–478.
3. ARMULIK, A., A. ABRAMSSON & C. BETSHOLTZ. 2005. Endothelial/pericyte interactions. Circ. Res. **97:** 512–523.
4. WIGLE, J.T. *et al.* 2002. An essential role for Prox1 in the induction of the lymphatic endothelial cell phenotype. EMBO J. **21:** 1505–1513.
5. BANERJI, S. *et al.* 1999. LYVE-1, a new homologue of the CD44 glycoprotein, is a lymph-specific receptor for hyaluronan. J. Cell Biol. **144:** 789–801.
6. KAIPAINEN, A. *et al.* 1995. Expression of the fms-like tyrosine kinase 4 gene becomes restricted to lymphatic endothelium during development. Proc. Natl. Acad. Sci. USA **92:** 3566–3570.
7. ROY, H., S. BHARDWAJ & S. YLA-HERTTUALA. 2006. Biology of vascular endothelial growth factors. FEBS Lett. **580:** 2879–2887.
8. YUAN, L. *et al.* 2002. Abnormal lymphatic vessel development in neuropilin 2 mutant mice. Development **129:** 4797–4806.
9. OLIVER, G. & K. ALITALO. 2005. The lymphatic vasculature: recent progress and paradigms. Annu. Rev. Cell Dev. Biol. **21:** 457–483.
10. HONG, Y.K. & M. DETMAR. 2003. Prox1, master regulator of the lymphatic vasculature phenotype. Cell Tissue Res. **314:** 85–92.
11. KARKKAINEN, M.J. *et al.* 2004. Vascular endothelial growth factor C is required for sprouting of the first lymphatic vessels from embryonic veins. Nat. Immunol. **5:** 74–80.
12. BALDWIN, M.E. *et al.* 2005. Vascular endothelial growth factor D is dispensable for development of the lymphatic system. Mol. Cell Biol. **25:** 2441–2449.
13. DUMONT, D.J. *et al.* 1998. Cardiovascular failure in mouse embryos deficient in VEGF receptor-3. Science **282:** 946–949.
14. GREENBERG, J.M. *et al.* 2002. Mesenchymal expression of vascular endothelial growth factors D and A defines vascular patterning in developing lung. Dev. Dyn. **224:** 144–153.
15. ORLANDINI, M. & S. OLIVIERO. 2001. In fibroblasts Vegf-D expression is induced by cell-cell contact mediated by cadherin-11. J. Biol. Chem. **276:** 6576–6581.
16. KUKK, E. *et al.* 1996. VEGF-C receptor binding and pattern of expression with VEGFR-3 suggests a role in lymphatic vascular development. Development **122:** 3829–3837.
17. CARMELIET, P. *et al.* 1996. Abnormal blood vessel development and lethality in embryos lacking a single VEGF allele. Nature **380:** 435–439.
18. FERRARA, N. *et al.* 1996. Heterozygous embryonic lethality induced by targeted inactivation of the VEGF gene. Nature **380:** 439–442.
19. MALLORY, B.P. *et al.* 2006. Lymphangiogenesis in the developing lung promoted by VEGF-A. Microvasc. Res. **72:** 62–73.

20. RAMIREZ, M.I. *et al.* 2003. T1alpha, a lung type I cell differentiation gene, is required for normal lung cell proliferation and alveolus formation at birth. Dev. Biol. **256:** 61–72.

21. SCHACHT, V. *et al.* 2003. T1alpha/podoplanin deficiency disrupts normal lymphatic vasculature formation and causes lymphedema. EMBO J. **22:** 3546–3556.

22. AYADI, A. *et al.* 2001. Net-targeted mutant mice develop a vascular phenotype and up-regulate egr-1. EMBO J. **20:** 5139–5152.

23. HOUFFLIN, V. *et al.* 1993. [Partial 12q trisomy and chylothorax]. J. Gynecol. Obstet. Biol. Reprod. (Paris) **22:** 625–629.

24. HUANG, X.Z. *et al.* 2000. Fatal bilateral chylothorax in mice lacking the integrin alpha9beta1. Mol. Cell Biol. **20:** 5208–5215.

25. WANG, J.F., X.F. ZHANG & J.E. GROOPMAN. 2001. Stimulation of beta 1 integrin induces tyrosine phosphorylation of vascular endothelial growth factor receptor-3 and modulates cell migration. J. Biol. Chem. **276:** 41950–41957.

26. MISHIMA, K. *et al.* 2007. Prox1 induces lymphatic endothelial differentiation via integrin alpha9 and other signaling cascades. Mol. Biol. Cell. **18:** 1421–1429.

27. MAKINEN, T. *et al.* 2005. PDZ interaction site in ephrinB2 is required for the remodeling of lymphatic vasculature. Genes Dev. **19:** 397–410.

28. LASSUS, P. *et al.* 1999. Vascular endothelial growth factor in human preterm lung. Am. J. Respir. Crit. Care Med. **159:** 1429–1433.

29. JOBE, A.J. 1999. The new BPD: an arrest of lung development. Pediatr. Res. **46:** 641–643.

30. JANER, J. *et al.* 2006. Pulmonary vascular endothelial growth factor-C in development and lung injury in preterm infants. Am. J. Respir. Crit. Care Med. **174:** 326–330.

31. PADERA, T.P. *et al.* 2002. Lymphatic metastasis in the absence of functional intratumor lymphatics. Science **296:** 1883–1886.

32. HE, Y. *et al.* 2005. Vascular endothelial cell growth factor receptor 3-mediated activation of lymphatic endothelium is crucial for tumor cell entry and spread via lymphatic vessels. Cancer Res. **65:** 4739–4746.

33. HE, Y. *et al.* 2002. Suppression of tumor lymphangiogenesis and lymph node metastasis by blocking vascular endothelial growth factor receptor 3 signaling. J. Natl. Cancer Inst. **94:** 819–825.

34. TANNO, S. *et al.* 2004. Human small cell lung cancer cells express functional VEGF receptors, VEGFR-2 and VEGFR-3. Lung Cancer **46:** 11–19.

35. TAMURA, M. *et al.* 2004. Chest CT and serum vascular endothelial growth factor-C level to diagnose lymph node metastasis in patients with primary non-small cell lung cancer. Chest **126:** 342–346.

36. RENYI-VAMOS, F. *et al.* 2005. Lymphangiogenesis correlates with lymph node metastasis, prognosis, and angiogenic phenotype in human non-small cell lung cancer. Clin. Cancer Res. **11:** 7344–7353.

37. KOUKOURAKIS, M.I. *et al.* 2005. LYVE-1 immunohistochemical assessment of lymphangiogenesis in endometrial and lung cancer. J. Clin. Pathol. **58:** 202–206.

38. WISSMANN, C. & M. DETMAR. 2006. Pathways targeting tumor lymphangiogenesis. Clin. Cancer Res. **12:** 6865–6868.

39. CUENI, L.N. & M. DETMAR. 2006. New insights into the molecular control of the lymphatic vascular system and its role in disease. J. Invest. Dermatol. **126:** 2167–2177.

40. KANMOGNE, G.D. 2005. Noninfectious pulmonary complications of HIV/AIDS. Curr. Opin. Pulm. Med. **11:** 208–212.

41. WANG, H.W. *et al.* 2004. Kaposi sarcoma herpesvirus-induced cellular reprogramming contributes to the lymphatic endothelial gene expression in Kaposi sarcoma. Nat. Genet. **36:** 687–693.

42. HONG, Y.K. *et al.* 2004. Lymphatic reprogramming of blood vascular endothelium by Kaposi sarcoma-associated herpesvirus. Nat. Genet. **36:** 683–685.

43. ELIAS, J.A. *et al.* 1999. Airway remodeling in asthma. J. Clin. Invest. **104:** 1001–1006.

44. MCDONALD, D.M. 2001. Angiogenesis and remodeling of airway vasculature in chronic inflammation. Am. J. Respir. Crit. Care Med. **164:** S39–45.

45. BALUK, P. *et al.* 2005. Pathogenesis of persistent lymphatic vessel hyperplasia in chronic airway inflammation. J. Clin. Invest. **115:** 247–257.

46. YANO, T. *et al.* 1994. Association of *Mycoplasma pneumoniae* antigen with initial onset of bronchial asthma. Am. J. Respir. Crit. Care Med. **149:** 1348–1353.

47. XU, X. *et al.* 2006. Mast cells protect mice from *Mycoplasma* pneumonia. Am. J. Respir. Crit. Care Med. **173:** 219–225.

48. AURORA, A.B. *et al.* 2005. Immune complex-dependent remodeling of the airway vasculature in response to a chronic bacterial infection. J. Immunol. **175:** 6319–6326.

49. KWAN, M.L. *et al.* 2001. Airway vasculature after mycoplasma infection: chronic leakiness and selective hypersensitivity to substance P. Am. J. Physiol. Lung Cell Mol. Physiol. **280:** L286–297.

50. TRULOCK, E.P. *et al.* 2004. The Registry of the International Society for Heart and Lung Transplantation: twenty-first official adult lung and heart-lung transplant report—2004. J. Heart Lung Transplant. **23:** 804–815.

51. MCDYER, J.F. 2007. Human and murine obliterative bronchiolitis in transplant. Proc. Am. Thorac. Soc. **4:** 37–43.

52. KREBS, R. *et al.* 2005. Dual role of vascular endothelial growth factor in experimental obliterative bronchiolitis. Am. J. Respir. Crit. Care Med. **171:** 1421–1429.

53. CURSIEFEN, C. *et al.* 2004. VEGF-A stimulates lymphangiogenesis and hemangiogenesis in inflammatory neovascularization via macrophage recruitment. J. Clin. Invest. **113:** 1040–1050.

54. KERJASCHKI, D. *et al.* 2006. Lymphatic endothelial progenitor cells contribute to de novo lymphangiogenesis in human renal transplants. Nat. Med. **12:** 230–234.

55. ADAIR, A. *et al.* 2007. Peritubular capillary rarefaction and lymphangiogenesis in chronic allograft failure. Transplantation **83:** 1542–1550.

56. DI CARLO, E. *et al.* 2007. Quilty effect has the features of lymphoid neogenesis and shares CXCL13-CXCR5 pathway with recurrent acute cardiac rejections. Am. J. Transplant. **7:** 201–210.

57. BELPERIO, J.A. *et al.* 2005. Role of CXCR2/CXCR2 ligands in vascular remodeling during bronchiolitis obliterans syndrome. J. Clin. Invest. **115:** 1150–1162.

58. LUCKRAZ, H. *et al.* 2004. Microvascular changes in small airways predispose to obliterative bronchiolitis after lung transplantation. J. Heart Lung Transplant. **23:** 527–531.

59. DUTLY, A.E. *et al.* 2005. A novel model for post-transplant obliterative airway disease reveals angiogenesis from the pulmonary circulation. Am. J. Transplant. **5:** 248–254.

60. LANGENBACH, S.Y. *et al.* 2005. Airway vascular changes after lung transplant: potential contribution to the pathophysiology of bronchiolitis obliterans syndrome. J. Heart Lung Transplant. **24:** 1550–1556.

61. FISHER, A.J. *et al.* 2005. The safety and efficacy of total lymphoid irradiation in progressive bronchiolitis obliterans syndrome after lung transplantation. Am. J. Transplant. **5:** 537–543.

62. ESTHER, C.R., JR. & P.M. BARKER. 2004. Pulmonary lymphangiectasia: diagnosis and clinical course. Pediatr. Pulmonol. **38:** 308–313.

63. GARDNER, T.W. *et al.* 1983. Congenital pulmonary lymphangiectasis: a case complicated by chylothorax. Clin. Pediatr. (Phila.) **22:** 75–78.

64. VAN AERDE, J *et al.* 1984. Spontaneous chylothorax in newborns. Am. J. Dis. Child. **138:** 961–964.

65. BARKER, P.M. *et al.* 2004. Primary pulmonary lymphangiectasia in infancy and childhood. Eur. Respir. J. **24:** 413–419.

66. JACQUEMONT, S. *et al.* 2000. Familial congenital pulmonary lymphangectasia, non-immune hydrops fetalis, facial and lower limb lymphedema: confirmation of Njolstad's report. Am. J. Med. Genet. **93:** 264–268.

67. LIGHT, R.W. 1995. Chylothorax and pseudochylothorax. *In* Pleural Diseases. D.C. Retford, Ed.: 284–298. Williams & Wilkins. Baltimore, MD.

68. BOWER, G.C. 1964. Chylothorax: observations in 20 cases. Dis. Chest. **46:** 464–468.

69. WILLIAMS, K.R. & T.H. BURFORD. 1964. The management of chylothorax. Ann. Surg. **160:** 131–140.

70. TEBA, L. *et al.* 1985. Chylothorax review. Crit. Care Med. **13:** 49–52.

71. STRAUSSER, J.L. & M.W. FLYE. 1981. Management of nontraumatic chylothorax. Ann. Thorac. Surg. **31:** 520–526.

72. ROCHA, G. 2007. Pleural effusions in the neonate. Curr. Opin. Pulm. Med. **13:** 305–311.

73. CHERNICK, V. & M.H. REED. 1970. Pneumothorax and chylothorax in the neonatal period. J. Pediatr. **76:** 624–632.

74. STEVENSON, D.A. *et al.* 2006. Familial congenital nonimmune hydrops, chylothorax, and pulmonary lymphangiectasia. Am. J. Med. Genet. A **140:** 368–372.

75. CHUA, F., J. GAULDIE & G.J. LAURENT. 2005. Pulmonary fibrosis: searching for model answers. Am. J. Respir. Cell Mol. Biol. **33:** 9–13.

76. TELES-GRILO, M.L. *et al.* 2005. Differential expression of collagens type I and type IV in lymphangiogenesis during the angiogenic process associated with bleomycin-induced pulmonary fibrosis in rat. Lymphology **38:** 130–135.

77. POSCHL, E. *et al.* 2004. Collagen IV is essential for basement membrane stability but dispensable for initiation of its assembly during early development. Development **131:** 1619–1628.

78. SOUZA, C.A. *et al.* 2006. Idiopathic interstitial pneumonias: prevalence of mediastinal lymph node enlargement in 206 patients. AJR Am. J. Roentgenol. **186:** 995–999.

79. ATTILI, A.K. *et al.* 2006. Thoracic lymph node enlargement in usual interstitial pneumonitis and nonspecific-interstitial pneumonitis: prevalence, correlation with disease activity and temporal evolution. J. Thorac. Imaging **21:** 288–292.

80. BASARABA, R.J. *et al.* 2006. Pulmonary lymphatics are primary sites of *Mycobacterium tuberculosis* infection in guinea pigs infected by aerosol. Infect. Immun. **74:** 5397–5401.

Gorham's Disease
An Osseous Disease of Lymphangiogenesis?

KAVITA RADHAKRISHNAN AND STANLEY G. ROCKSON

Stanford Center for Lymphatic and Venous Disorders, Division of Cardiovascular Medicine, Stanford University School of Medicine, Stanford, California, USA

Gorham's disease, also known as massive osteolysis, Gorham–Stout disease, vanishing bone disease, or, phantom bone disease is a rare disorder of the musculoskeletal system. The disease is characterized by osteolysis in bony segments, with localized proliferation of lymphatic channels. The presence of abundant, leaky systemic lymphatic vessels is often accompanied by chylous ascites. There is no standardized treatment available for Gorham's disease, and its molecular mechanisms remain unclear. Future strategies for understanding Gorham's disease should emphasize its apparent identity as a disease of disordered lymphangiogenesis. Breakthroughs in lymphatic research have identified several lymphangiogenic pathways that may play a relevant role in Gorham's disease.

Key words: **Gorham's disease; lymphangiogenesis; lymphatic endothelium; platelet-derived growth factor (PDGF)**

Gorham's disease, also known as massive osteolysis, vanishing bone disease, or phantom bone disease[1] is a rare disorder of the musculoskeletal system. The affected bone undergoes osteolysis and is replaced by vascular fibrous connective tissue.[1] Surrounding soft tissue is generally invaded.[2] The disease was first reported by Jackson in 1838[3] in a young male who had a progressively vanishing humerus. The syndrome also became known as Gorham–Stout disease, when these authors published review of 16 cases in 1955.[4]

The disease is characterized by osteolysis in bony segments, with localized proliferation of lymphatic channels.[5] The presence of abundant, leaky systemic lymphatic vessels is often accompanied by chylous ascites.[6] Gorham's disease usually presents in young patients, with no race or gender predilection.[2] It is distinguished from other idiopathic osteolytic disorders by its lack of either a specified mode of transmission or association with a nephropathy. The natural history has no clear pattern.[1] FIGURE 1 shows the clinical features of Gorham's disease.

Johnson and McClure divided Gorham's disease into an early intraosseous stage, with patchy osteoporosis and multiple intramedullary and subcortical radiolucencies, and a later extraosseous stage involving disruption of the cortex with destruction, resorption, and disappearance of the bone.[7]

The etiology and pathophysiology of this disease remain unclear. Historically, following Gorham and Stout's description of hemangiomatosis, the condition was thought to reflect abnormal proliferation of blood vessels. More recently, however, Bruch-Gerharz *et al.* revealed that several types of vascular malformative processes, including capillary, venous, and lymphatic, can accompany the osteolysis characteristic of this condition.[8] These malformations are not neoplastic, but, rather, developmental in nature, as they become aggravated over time and are expressed in numerous tissues.[8] Pathologically, the lesions are distinguished by their lack of endothelial cell proliferation,[2] histologically by the marked proliferation and dilatation of lymphatic vessels,[9] and clinically by the frequent late onset and lack of spontaneous resolution.[8]

Proliferation of abnormal vasculature is evident in numerous cases of Gorham's disease. Recent studies provide insight into a plausible pathophysiology. Histological comparisons of tissue samples derived from patients and controls revealed that more than 90% of the CD-31 (pan-endothelial marker)–expressing endothelial cells in Gorham's disease patients also stained positively for the lymphatic endothelial marker, lymphatic vascular endothelial hyaluronan receptor-1 (LYVE-1).[6] This novel finding suggests that the proliferating vasculature associated with Gorham's disease is derived chiefly from lymphatic endothelium. Further investigation of potential lymphangiogenic or

Address for correspondence: Stanley G. Rockson, M.D., Stanford Center for Lymphatic and Venous Disorders, Division of Cardiovascular Medicine, Stanford University School of Medicine, 300 Pasteur Drive, Stanford, CA 94305. Voice: +1-650-725-7571; fax: +1-650-725-1599.
srockson@cvmed.stanford.edu

Ann. N.Y. Acad. Sci. 1131: 203–205 (2008). © 2008 New York Academy of Sciences.
doi: 10.1196/annals.1413.022

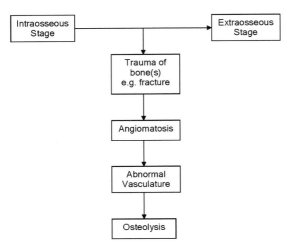

FIGURE 1. Clinical features of Gorham's disease.

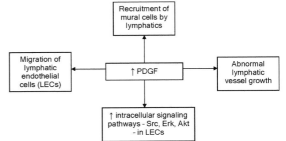

FIGURE 2. Plausible role of PDGF in Gorham's disease.

angiogenic growth factor activation revealed the presence of elevated circulating levels of PDGF-BB in the affected patients, suggesting a potential role of the platelet-derived growth factor (PDGF) signaling pathway in Gorham's disease (FIG. 2).[6]

The PDGF family is involved in angiogenesis, tumor-cell proliferation, and the recruitment of mural cells to tumor blood vessels.[10] Most recently, PDGF-BB overexpression has been implicated in the formation of abnormal lymphatics in the syndrome of lymphedema–distichiasis.[11] Members of the PDGF family have also been demonstrated to stimulate lymphatic vessel growth independently from activation of the vascular endothelial growth factor-C (VEGF-C) pathway, and to induce tumor lymphangiogenesis.[12] The proliferation of abnormal, small lymphatics in Gorham's disease may similarly be driven by lymphangiogenic growth factors, including PDGF-BB. The observation that Gorham's disease is often accompanied by thoracic duct occlusion with chylothorax further supports the notion that it reflects a potential lymphatic pathogenesis.[8]

Future directions for understanding Gorham's disease should emphasize its apparent identity as a disease of disordered lymphangiogenesis. Breakthroughs in lymphatic research have identified several lymphangiogenic pathways, in addition to the PDGF family, that may play a relevant role in Gorham's disease. Early in vascular development, endothelial cells of the cardinal vein express the lymphatic markers vascular endothelial growth factor receptor-3 (VEGFR-3) and LYVE-1.[13] Inductive but unknown mesenchymal signals lead to Prox1 expression and lymphatic commit-

ment.[13] These three pathways are perhaps the most critical in establishing the lymphatic network. A systematic study of these pathways may help in furthering understanding of Gorham's disease.

The first reported lymphatic-specific growth factor, VEGFR-3, was identified by Kaipanen *et al.*[14] A tyrosine kinase receptor, it specifically binds VEGF-C and VEGF-D.[14] Early in development, VEGFR-3 is expressed in developing venous and lymphatic endothelia, while in normal adults it is mainly limited to lymphatic endothelium.[14] Vegfr3 null mice show no lymphatic vasculature, whereas heterozygous mice display delayed lymphatic development, lymphatic hypoplasia, and lymphedema.[14]

LYVE-1 is a marker of lymphatic endothelial competence during development.[13] LYVE-1 expression levels are high in lymphatic capillaries, but down-regulated in collecting lymphatic vessels.[15] The transcription factor Prox1 is the most specific marker for lymphatic endothelium.[13] Prox1 expression leads to lymphatic commitment and specification.[16] Prox1 null mice lack lymphatic vasculature: budding endothelial cells do not express lymphatic endothelial markers.[17] Subsequently, the budding and sprouting of these cells from the veins fails to occur.[17] The disordered lymphangiogenesis seen in Gorham's disease may be a result of disruptions in these pathways.

Since the defining characteristic of Gorham's disease is massive osteolysis, radiographic imaging methods are most useful in establishing the diagnosis.[1] Atrophy, dissolution, fracture, fragmentation, and disappearance of portions of bone are common findings.[1] Blood tests and radiographs are utilized to exclude other causes of osteolysis, including infection, cancer, and inflammatory or endocrine disorders.[1]

As a reflection of the small reported caseload (fewer than 200 cases have been reported), there is no standardized treatment available for patients. Moreover, the molecular mechanisms of disease remain unclear. Treatment modalities (TABLE 1) include surgical and

TABLE 1. Treatment options for Gorham's disease

Treatment/drug	Interventional mechanism	Reference
Bisphosphonate	Anti-osteoclastic	Hammer *et al.*[9]
Alpha-2b interferon	Immune enhancer	Hagendoorn *et al.*[6]
Surgery	Resection of lesion/bone graft	Patel[1]
Radiation	Fractional radiation in 2-Gy fractions for suppression of lesions	Bruch-Gerharz *et al.*[8]
PDGF-BB antagonists	Monoclonal antibodies to PDGF-BB	[speculative/futuristic]
	Anti-PDGF-BB oligo or ribozymes	

radiation therapies, administration of bisphosphonates for their anti-osteoclastic properties, and alpha-2b interferon.[1] Surgical interventions include resection of the lesion, and grafting of bone.[1] Fractionated irradiation (40–45 Gy in 2-Gy fractions) generally results in good clinical outcomes with minimal long-term complications.[18] The presence of extraosseous involvement carries a poor prognosis.[19]

Conflicts of Interest

The authors declare no conflicts of interest.

References

1. PATEL, D.V. 2005. Gorham's disease or massive osteolysis. Clin. Med. Res. **3:** 65–74.
2. DOMINGUEZ, R. & T.L. WASHOWICH. 1994. Gorham's disease or vanishing bone disease: plain film, CT, and MRI findings of two cases. Pediatr. Radiol. **24:** 316–318.
3. JACKSON, J. 1838. A boneless arm. Boston Med. Surg. J. **18:** 368–369.
4. GORHAM, L.W. & A.P. STOUT. 1955. Massive osteolysis (acute spontaneous absorption of bone, phantom bone, disappearing bone); its relation to hemangiomatosis. J. Bone Joint Surg. Am. **37-A:** 985–1004.
5. ABRAHAMS, J. *et al.* 1980. Massive osteolysis in an infant. AJR Am. J. Roentgenol. **135:** 1084–1086.
6. HAGENDOORN, J. *et al.* 2006. Platelet-derived growth factor receptor-beta in Gorham's disease. Nat. Clin. Pract. Oncol. **3:** 693–697.
7. JOHNSON, P.M. & C.J. MC. 1958. Observations on massive osteolysis; a review of the literature and report of a case. Radiology **71:** 28–42.
8. BRUCH-GERHARZ, D. *et al.* 2007. Cutaneous lymphatic malformations in disappearing bone (Gorham-Stout) disease: a novel clue to the pathogenesis of a rare syndrome. J. Am. Acad. Dermatol. **56:** S21–S25.
9. HAMMER, F. *et al.* 2005. Gorham-Stout disease–stabilization during bisphosphonate treatment. J. Bone Miner. Res. **20:** 350–353.
10. JAIN, R.K. 2003. Molecular regulation of vessel maturation. Nat. Med. **9:** 685–693.
11. PETROVA, T.V. *et al.* 2004. Defective valves and abnormal mural cell recruitment underlie lymphatic vascular failure in lymphedema distichiasis. Nat. Med. **10:** 974–981.
12. CAO, R. *et al.* 2004. PDGF-BB induces intratumoral lymphangiogenesis and promotes lymphatic metastasis. Cancer Cell. **6:** 333–345.
13. CUENI, L.N. & M. DETMAR. 2006. New insights into the molecular control of the lymphatic vascular system and its role in disease. J. Invest. Dermatol. **126:** 2167–2177.
14. KAIPAINEN, A. *et al.* 1995. Expression of the fms-like tyrosine kinase 4 gene becomes restricted to lymphatic endothelium during development. Proc. Natl. Acad. Sci. USA **92:** 3566–3570.
15. MAKINEN, T. *et al.* 2005. PDZ interaction site in ephrinB2 is required for the remodeling of lymphatic vasculature. Genes Dev. **19:** 397–410.
16. OLIVER, G. & M. DETMAR. 2002. The rediscovery of the lymphatic system: old and new insights into the development and biological function of the lymphatic vasculature. Genes Dev. **16:** 773–783.
17. WIGLE, J.T. & G. OLIVER. 1999. Prox1 function is required for the development of the murine lymphatic system. Cell **98:** 769–778.
18. DUNBAR, S.F. *et al.* 1993. Gorham's massive osteolysis: the role of radiation therapy and a review of the literature. Int. J. Radiat. Oncol. Biol. Phys. **26:** 491–497.
19. PEDICELLI, G. *et al.* 1984. Gorham syndrome. JAMA **252:** 1449–1451.

Lymphatic Involvement in Lymphangioleiomyomatosis

CONNIE G. GLASGOW,[a] ANGELO M. TAVEIRA-DASILVA,[a]
THOMAS N. DARLING,[b] AND JOEL MOSS[a]

[a]Translational Medicine Branch, National Heart, Lung, and Blood Institute,
National Institutes of Health, Bethesda, Maryland, USA

[b]Department of Dermatology, Uniformed Services University of the Health Sciences,
Bethesda, Maryland, USA

Lymphangioleiomyomatosis (LAM) is a rare, multisystem disease affecting primarily pre-menopausal women. The disease is characterized by cystic lung disease, at times leading to respiratory compromise, abdominal tumors (in particular, renal angiomyolipomas), and involvement of the axial lymphatics (e.g., adenopathy, lymphangioleiomyomas). Disease results from the proliferation of neoplastic cells (LAM cells), which, in many cases, have a smooth muscle cell phenotype, express melanoma antigens, and have mutations in one of the tuberous sclerosis complex genes (*TSC1* or *TSC2*). In the lung, LAM cells found in the vicinity of cysts are, at times, localized in nodules and may be responsible for cyst formation through the production of proteases. Lymphatic channels, expressing characteristic lymphatic endothelial cell markers, are found within the LAM lung nodules. LAM cells may also be localized within the walls of the axial lymphatics, and, in some cases, penetrate the wall and proliferate in the surrounding adipose tissue. Consistent with extensive lymphatic involvement in LAM, the serum concentration of VEGF-D, a lymphangiogenic factor, is higher in LAM patients than in healthy volunteers.

Key words: lymphangiogenesis; metastasis; lymphangioleiomyomatosis; VEGF-D; VEGF-C

Introduction

Lymphangioleiomyomatosis (LAM), a rare multisystem disorder primarily affecting premenopausal women, is characterized by cystic lung destruction, abdominal tumors, in particular, renal angiomyolipomas (AMLs), and involvement of the thoracic and abdominal axial lymphatics (e.g., adenopathy, lymphangioleiomyomas).[1–7] Initial clinical presentation consists of symptoms resulting from lung, kidney, or lymphatic involvement. Disease progression is variable. The course may be insidious: 5 or 6 years may elapse between appearance of symptoms and diagnosis.[5,8] Delayed diagnosis of this rare disease may account for the poor prognosis seen in earlier studies; recent reports indicate a 78% survival in 8.5 years or 90% in 10 years.[5,9]

Cystic lung destruction appears to be responsible for the progressive decline in pulmonary function, which can lead to respiratory failure, dependence on supplemental oxygen, lung transplantation, or death.[8] Pulmonary signs and symptoms include dyspnea, reactive airways disease, and recurrent pneumothoraces. Pulmonary function abnormalities include airflow obstruction (decreased FEV_1) and reduced diffusion capacity (DL_{co}).[1,4,10] High-resolution computed tomography (HRCT) scans reveal thin-walled cysts homogeneously distributed in both lungs and interspersed with normal parenchyma (FIGS. 1A and B).[1,4,5] Ventilation/perfusion scans are consistent with air-trapping in the cystic areas.[4,11]

The hallmark histologic characteristic of LAM is the presence of abnormal-appearing, smooth muscle-like LAM cells.[12] LAM lesions in both pulmonary and extrapulmonary sites consist of LAM cells, in some cases localized in nodules, with slit-like lymphatic channels.[12,13] In the lung, LAM cells are seen in nodular clusters near the vasculature, lymphatics, and bronchioles, and appear to be present in the walls of the cystic lesions; LAM nodules may be responsible for the thin-walled cystic lesions seen on HRCT scans.[1,4,12] LAM cells in lung nodules are morphologically heterogeneous, with both spindle-shaped and epithelioid-like cells observed within the nodule.[12] The LAM nodules

Address for correspondence: Joel Moss, M.D., PhD., Translational Medicine Branch, National Heart, Lung, and Blood Institute, National Institutes of Health, Bldg. 10, Rm. 6D03 MSC 1590, Bethesda, MD 20892-1590. Voice: +1-301-496-1597.

mossj@nhlbi.nih.gov

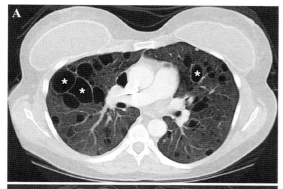

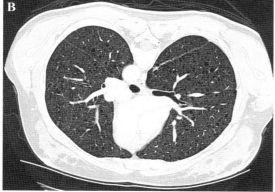

FIGURE 1. Lung CT scan of a patient with LAM **(A)** showing numerous thin-walled cysts (∗) distributed throughout otherwise normal-appearing lungs. **(B)** CT scan of a patient with severe LAM; the lung parenchyma is almost completely replaced by very small cysts.

appear to be covered by hyperplastic type II cells.[14] Spindle-shaped cells are immunoreactive for proliferating cell nuclear antigen, an indicator of mitotic activity and proliferation; the large epithelioid cells are immunoreactive with HMB45, a monoclonal antibody that recognizes gp100, a premelanosomal protein.[12,15] LAM cells, in particular, those with an epithelioid appearance, express progesterone and estrogen receptors, consistent with evidence that hormonal factors may play an important role in disease pathogenesis.[16,17] The LAM nodules may also be regulated by the renal–angiotensin and insulin-like growth factor systems.[18,19]

Extracellular matrix is a structural and functional part of the pulmonary parenchyma.[20] Matrix metalloproteinases (MMPs), a group of Zn^{2+}-dependent endopeptidases, and their regulators, the tissue inhibitors of metalloproteinases (TIMPs), are primary effectors of extracellular matrix turnover and remodeling.[21] LAM lung lesions are immunoreactive for MMPs-1, 2, 9, and 14, and show decreased expression of TIMP-3, an inhibitor of some MMPs.[12,22–25] An imbalance in

the MMP/TIMP ratio caused by the release of MMPs from LAM cells may, in part, result in degradation of the extracellular matrix and thus result in the cystic changes in the LAM lung parenchyma.[12,22,23]

One-third of women with tuberous sclerosis complex (TSC), an autosomal dominant disorder with variable penetrance, characterized by neurologic (tubers, astrocytomas), renal (AMLs), and dermatologic (facial angiofibroma) manifestations and mutations in the *TSC1* or *TSC2* tumor suppressor genes, have cystic lung lesions seen on HR CT scans.[26–28] LAM cells from the kidney (AMLs), lung, and lymph nodes of patients with sporadic LAM (no evidence of germline TSC mutations) may possess identical *TSC2* mutations and/or loss of heterozygosity for the allele.[29–31] *TSC2* mutations result in activation of the guanine nucleotide-binding protein Rheb, and its downstream target, mammalian target of rapamycin (*mTOR*), leading to increases in cell size and cell number.[32–34] These studies are consistent with the conclusion that the *TSC* genes are involved in the susceptibility to LAM.

Extrapulmonary manifestations, which occur in more than 70% of patients with LAM, include AMLs, lymphadenopathy, lymphangioleiomyomas, chylous ascites, and pleural effusions.[6,8,12,13] AMLs are benign hamartomatous lesions, with abnormal, heterogenous smooth muscle-like cells similar to those in pulmonary and other extrapulmonary LAM lesions. Unlike these other LAM lesions, AMLs may be associated with adipose tissue and underdeveloped thick-walled blood vessels[12,13,35–38] (FIG. 2). The adipose tissues and blood vessels may possess *TSC* gene mutations.[39]

Patients with LAM frequently present with pulmonary symptoms, but initial symptoms may originate from extrapulmonary LAM lesions.[12,13,40,41] Roentgenographic appearance of diffusely distributed pulmonary cysts, coupled with CT evidence of extrathoracic disease (e.g., AMLs, lymphatic involvement, chylous effusions), and/or manifestations of are sufficient for a diagnosis of pulmonary LAM. If these extrapulmonary manifestations of disease are not evident, a lung biopsy may be necessary to confirm the diagnosis.[4,8,37,42]

Improved understanding of prognostic indicators would facilitate selection of subjects for clinical trials and, perhaps, provide new insight into the etiology of LAM. To this end, Matsui *et al.* correlated histopathologic features and predictors of survival by developing a grading system, the LAM histologic score (LHS), based on a review of surgical biopsy specimens of lung from 105 LAM patients.[43] Scores based on the percentage of lung tissue involved with both cystic lesions and areas of proliferation in the lung tissue sample, were: LHS-1

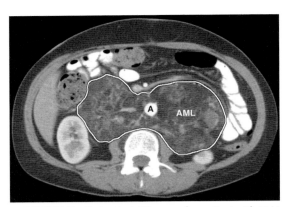

FIGURE 2. Abdominal CT scan of a patient with LAM showing a very large angiomyolipoma (AML, outlined area) completely involving the retroperitoneal area, resulting in anterior displacement of the aorta (A).

<25% lung involvement, LHS-2 = 25–50% involvement, and LHS-3 >50% involvement. LHS-3 was correlated with poor prognosis and more rapid progression to lung transplantation or death. In an analysis of pulmonary function tests data from 143 LAM patients, Taveira-DaSilva et al. concluded that DL_{co} correlated with the LHS and could also, potentially, be a predictor of survival or time to lung transplantation.[10]

Avila et al. proposed a grading system using HRCT lung scans of LAM patients.[11] Lungs were divided into three equal zones for assessment of the extent of cystic involvement of parenchyma: grade 0 = none, grade 1 <30%, grade 2 = 30%–60%, grade 3 >60% judged abnormal. In 39 LAM patients, comparison of these grades with pulmonary function tests data revealed an inverse correlation between CT grade and both FEV_1 and DL_{co}.[11] In another study of 80 patients with LAM, a correlation was observed between severity of lung disease, assessed by HRCT scans (CT grade), and the presence of enlarged abdominal lymph nodes.[44]

There is no effective treatment for preventing the progression of pulmonary LAM. Hormonal manipulations were, in the past, a frequent choice of treatment, but no study has confirmed their efficacy.[5,8,45] One serious complication of extrapulmonary LAM is AML hemorrhage.[46] The initial symptom is abdominal pain, with hematuria, and shock caused by blood loss. Treatment may involve embolization of the artery responsible for the bleeding or surgical resection of the tumor.[46] There is no effective treatment for lymphangioleiomyomas. Chylothorax, an accumulation of lymph or chyle in the pleural space, is a complication, which occurs in fewer than 20% of LAM patients.[1,4,5,47] Depending on the size and clinical effects of the effusion, thoracente-

sis, pleurodesis, parietal pleurectomy, and thoracic duct ligation are possible options.[47] Pneumothorax can be treated by thoracostomy, but because of its frequent recurrence, pleurodesis may be necessary, even though the procedure may complicate subsequent lung transplantation.[48] Evidence of pleural repair may be seen on HRCT after pleurodesis.[49] The presence of pleurodesis has variable influence on the correlation of quantitative thin-section CT with pulmonary function tests.[50] The occurrence of pneumothoraces may have genetic and morphologic determinants.[51]

There is genetic evidence that cells in the recurrent LAM lesions in transplanted lungs originate in the recipient.[52,53] These observations are consistent with a model of metastatic disease in which cells move from AML, lymphatics, or a presently unknown site of origin, into the lung. Isolation of LAM cells from blood and other body fluids (e.g., urine, chylous pleural effusions, ascites) is additional support for the metastatic model.[54]

Lymphatic Studies in LAM

Extrapulmonary LAM

Lymphatic involvement in LAM, which may include the posterior mediastinum, and upper retroperitoneal, or pelvic regions, appears to be related to the distribution of lymphatic vessels.[4,13,40,41] Lymphangioleiomyomas on CT scans are well-circumscribed lesions of variable dimensions that are seen as thick- or thin-walled lobulated masses with a central fluid-rich region (FIGS. 3A and B). These lesions are most frequently found in the retroperitoneal and mediastinal regions.[13,44,55] Abdominal lymphangioleiomyomas, formed from a collection of chyle in lymphatic vessels, when overdistended, can rupture or leak, thus leading to chylous ascites.[13,44] Enlarged lymph nodes appear round or oval and are most often found in the retroperitoneal area.[4,40,44] They have also been observed in the mediastinum and pelvic regions, and are commonly associated with other extrapulmonary lesions.[44,56,57] Unlike pulmonary LAM cells, most extrapulmonary lymphatic LAM cells are found in lymph nodes along the lymphatic vessels of the mediastinum and retroperitoneum, where they form fascicles and papillary patterns, rather than nodules, and have been reported to extend beyond the connective tissue capsule of the lymphatics.[4,12,13,40,41] Mutations in the *TSC* genes have been observed in lymphatic LAM cells and in the constituent tissues of the AMLs.[29,39]

With the exception of AMLs, the different manifestations of extrapulmonary LAM appear to arise from a common pathology. Proliferating LAM cells

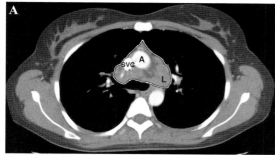

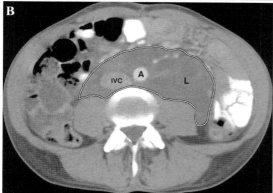

FIGURE 3. Chest CT scan of a patient with LAM **(A)** showing a large lymphangioleiomyoma (L, outlined area) involving the pre-tracheal and left lung hilar area, and an abdominal CT scan **(B)** of a patient with LAM showing a very large lymphangioleiomyoma (L, outlined area) located in the retroperitoneal area surrounding the aorta and inferior vena cava. SVC, superior vena cava; IVC, inferior vena cava; A, aorta.

appear to compress or obstruct lymphatic vessels, leading to obstruction of the flow of chyle, and the observed lymphangioleiomyomas or dilatation of lymphatic vessels seen on abdominal CT scans of patients with LAM.[13,44,58] Biopsies of involved lymph nodes, with characteristic adenopathy, reveal replacement of normal components by LAM cells.[45]

In two separate studies, lymphangioleiomyomas in retroperitoneal, pelvic, thoracic, and/or cervical regions were found to have a diurnal variation in size in patients with LAM. Patients underwent CT or sonogram examinations in the morning and afternoon of the same day. The studies showed that the lesions in 12/13 patients and 12/21 patients, respectively, exhibited an increase between morning and afternoon.[59,60] The median percent changes in volume were 140% and 38%, respectively.[59,60] Diurnal variation may be due to the greater lymph flow resulting from food intake and exercise, and/or the effect of gravity on lymph flow. Lymphangioleiomyomas, viewed by CT or sonog-

raphy, are sometimes difficult to distinguish from neoplasms such as lymphoma or sarcoma. Documentation of diurnal variation may aid in diagnosis and obviate the need for biopsy.

Lymphatic Involvement in TSC

Hamartomas in patients with TSC are most frequently found in the skin, brain (tubers, subependymal lesions), kidneys (AMLs), and heart (rhabdomyomas).[61] Angiogenesis in subependymal giant cell astrocytoma and AML lesions has been demonstrated by immunoreactivity with anti-CD31 antibody.[62] In agreement, AMLs have been shown to produce VEGF *in vitro*.[62]

Lymphatic involvement (thoracic duct dilation, chylous pleural effusion, ascites, and lymphangioleiomyomas) is less common in patients with TSC/LAM than in those with sporadic LAM.[63] Patients with TSC develop multiple different types of skin lesions, including facial angiofibromas, forehead plaques, and ungual fibromas. The papules and plaques are typically pink to red, and contain accumulated collagen plus increased numbers of ectatic vessels in the dermis.[64] Sections of TSC skin tumors have been studied using immunohistochemical markers for angiogenesis. Skin tumor lesions (angiofibromas) of TSC patients express VEGF.[65] Many dilated tubular structures are blood vessels that contain immunoreactive CD31 or CD34.[62,66] Not all of the endothelial-lined spaces in angiofibromas and periungual fibromas react with anti-CD34 antibodies (FIGS. 4A and B). The vessels in TSC skin tumors have not been closely examined, and although blood vessels are often assumed to be present, some reports mentioned increased numbers of lymphatic as well as blood vessels.[67,68] These data are consistent with the presence of dilated lymphatic vessels, since anti-CD34 antibodies do not react with lymphatic endothelial cells in skin.[69]

Histopathologic Studies of LAM Cells and Lymphatics in LAM

Genetic and pathologic evidence suggests a metastatic potential for LAM cells. Angiogenesis and lymphangiogenesis have been postulated as mediators of metastatic spread and tumor growth by providing new vessels for the invading tumor cells.[70,71] The role of angiogenesis in tumor spread is well documented, but research regarding lymphangiogenesis in tumor progression presents a new avenue of approach and has been facilitated by the identification of specific lymphatic endothelial cell markers, such as LYVE-1 (transports hyaluronan by lymphatic endothelial cells to lymph), podoplanin (marker for lymphatic endothelial cells), VEGF-R3 (growth factor receptor for ligands

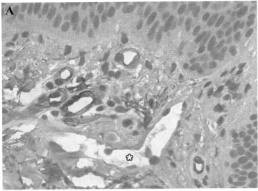

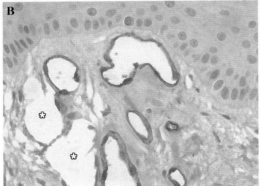

FIGURE 4. Lymphangiogenesis in TSC skin lesions. Sections of angiofibromas **(A)** and periungual fibromas **(B)** were stained for CD34, a marker for blood vessel endothelial cells. Vessels negative for CD34 (∗), presumed to be lymphatic vessels, are interspersed with blood vessels positive for CD34 (darkly stained endothelial cells). Original magnification ×400.

VEGF-C and VEGF-D), and PROX-1 (transcription factor required for lymphatic vessel growth).[70]

Lymphatic studies in LAM focus on defining the role of lymphangiogenesis and angiogenesis in the dissemination of LAM cells. Among three recent LAM studies that focus on lymphatic involvement, the first used immunohistochemistry to analyze 21 samples of LAM tissue (lung, lymph nodes, uterus, ovary) obtained from autopsy or lung explants.[72] Angiogenesis and lymphangiogenesis were evaluated, respectively, with antibodies against CD31, a vascular endothelial cell marker and VEGFR-3 (Flt-4), a specific lymphatic endothelial cell marker. LAM lesions exhibited minimal cellular reactive CD31, but were reactive with anti-VEGFR-3 (Flt-4) antibodies in both pulmonary and extrapulmonary tissue samples of the lung, lymph nodes (cervical, axillary, supraclavicular, mediastinal, retroperitoneal, and mesenteric regions), uterus and ovary. The slit-like spaces surrounding and infiltrat-ing LAM cell foci and nodules were identified in this way as lymphatic tracts. Lymphatic endothelial cells, immunoreactive with Flt-4 antibodies, were observed in retroperitoneal and mediastinal lymph nodes to surround LAM cell foci, which were formed in bundle-like structures. LAM cell clusters (LCCs), or LAM cell nests covered by lymphatic endothelial clusters appeared to be fragments of LAM foci and were observed in lymphatic vessels, some near lymph nodes, indicating a region of potential interaction of LAM cells and lymphatic endothelial clusters. Lymphangiogenesis was given a semiquantitative assessment based on the extent of each lesion occupied by lymphatic endothelial cells.[72]

In the same study, LAM tissue was analyzed immunohistochemically for VEGF-C as a potential mediator of lymphangiogenesis. Diffuse immunoreactivity with antibodies against VEGF-C and with monoclonal antibody HMB45 was observed in the cytoplasm of pulmonary and extrapulmonary LAM tissue, including lung samples from biopsy and explanted lung, as well as lymph nodes, uterus, and lung cells in primary cultures. Relationships between the intensity of VEGF-C immunostaining in lung tissue, the extent of lymphangiogenesis in lung lesions (as reflected by estimates of relative amounts of Flt-4 immunostaining), and the LHS score developed by Matsui *et al.*[43] were evaluated. A statistically significant correlation between greater lymphatic involvement and higher VEGF-C content was demonstrated, and both were associated with worse prognosis and time to lung transplantation as expressed by LHS.[72] The results of this study would support a lymphangiogenic model for the progression of LAM, in which VEGF-C produced by LAM cells enhances lymphangiogenesis via the VEGFR-3 receptor.

Immunologic examination of chylous fluid and retroperitoneal lymphangioleiomyoma samples from six patients with LAM, and examination of tissue from five autopsy cases that included samples of the thoracic duct, and left jugular subclavian angle, and cervical, axillary, supraclavicular, mediastinal, retroperitoneal, and mesenteric region lymph nodes were employed to assess the presence of LAM cells in lymphatic structures containing chyle and to evaluate a potential route for LAM cell entry into the lymphatic system.[73] LAM cells, immunoreactive with both HMB45 and anti-α-SMA (smooth muscle actin) antibodies, were observed, in all cases, as clusters covered by lymphatic endothelial clusters in chylous effusions and within lymphatic vessels, and were also detected floating in the extra-lymphatic spaces. The clusters showed an outer layer of lymphatic endothelial cells containing immunoreactive VEGFR-3. In addition, in the lymphangioleiomyoma sample, LAM cell lesions appeared as LAM cell clusters surrounded by lymphatic endothelial clusters. Retrospective examination of the autopsy samples confirmed the involvement of LAM cells extending along the axial lymphatics, including the region of thoracic duct drainage into the venous circulation (left jugular and subclavian junction). LAM cell–positive lymph nodes were most frequently identified in the retroperitoneal region. After LCCs were incubated *in vitro* on collagen-coated slides for several days, clusters were no longer visible, but two types of cells were seen, spindle-shaped LAM cells (containing immunoreactive α-SMA) and lymphatic endothelial cells (immunopositive for VEGFR-3).[73]

Based on the results of these two studies, a model for the dissemination of AM cells was formulated in which lymphatic endothelial cells lining lymphatic channels in LAM cell nodules, cystic lesions, and lymph nodes eventually envelop LAM cells, and the resulting clusters may be shed into the lymphatic circulation or extrapulmonary space. In the lymphatic channels, interaction of the lymphatic endothelial cells on the cluster and the lymphatic endothelial cells lining lymphatic vessels, initiate fragmentation, separating LCCs into LAM cells and lymphatic endothelial cells. LAM cells are able then to invade the extracellular matrix, which is degraded by LAM cell–derived MMPs with facilitation by downregulation of TIMPs. LAM cell invasion and proliferation results in the formation of a new LAM lesion.[73]

In the most recent report of LAM-associated lymphangiogenesis, it was proposed that since the pathogenesis and progression of LAM (evaluated by LHS) is associated with extent of lymphatic involvement, serum levels of lymphatic growth factors might be elevated, and correlated with clinical characteristics of LAM.[74] Such a growth factor could be considered a biological marker for lymphangiogenesis and disease severity. Serum levels of VEGF-A, an angiogenic factor, and VEGF-C and VEGF-D, described lymphangiogenic factors, in 44 patients with LAM were quantified by serum enzyme-linked immunosorbent assay. Serum concentrations of VEGF-A in patients with LAM were similar to those of the control group, whereas the levels of VEGF-C were significantly lower in LAM than in control groups. Serum VEGF-D was, however, significantly higher in patients with LAM than in controls. There was a significant negative correlation between serum levels of VEGF-D and pulmonary function as measured by FEV_1/FVC (forced expiratory volume in 1 second/forced vital capacity) or percent predicted DL_{co}/VA (diffusing capacity for carbon monoxide/alveolar volume).

A subset of LAM patients was divided into those who had been given hormonal treatment and those who had not.[74] The group of patients who had undergone hormone therapy had more restricted air flow (lower FEV_1 percent predicted, FEV_1/FVC), lower diffusing capacity (percent predicted lower DL_{co}/VA), and higher serum VEGF-D concentration than those patients with no history of recent hormone therapy. This study established an association between serum levels of VEGF-D, a lymphangiogenic growth factor, and clinical characteristics of patients with LAM, suggesting that VEGF-D could be a marker for disease severity.

Conclusion and Future Directions

Extensive association of lymphangiogenesis was found with pulmonary and extrapulmonary lesions in LAM. The production of VEGF-C and VEGF-D by pulmonary and extrapulmonary LAM lesions might reflect LAM cell–induction of lymphangiogenesis through VEGFR-3 signaling. The interesting observation of LCCs covered with lymphatic endothelial cells located in lymphatic vessels suggests a potential mechanism for a lymphangiogenesis-driven metastatic process. *In vitro* evidence of LCC fragmentation suggests a mechanism whereby LAM cells have the potential to interact with the extracellular matrix to form new lesions and to reach the vascular circulation via the thoracic duct. Future studies may reveal histopathologic evidence of anatomic connections between blood and lymphatic vessels in LAM lesions.

Findings from the first two reports, based on immunohistochemical evidence of protein expression, implicated the growth factor VEGF-C as a driving force for LAM-directed lymphangiogenesis.[72,73] In contrast, on the basis of serum levels of VEGF growth factors in the latest study, VEGF-D appears much more likely to be a biomarker that correlates with lymphangiogenesis and disease progression.[74] Protein expression studies and studies correlating clinical phenotypes to serum levels of VEGF-D and VEGF-C would help to define more clearly the differential roles of VEGF-C and VEGF-D in LAM. The study of lymphangiogenesis and LAM presents a promising avenue to elucidate a model for dissemination of LAM cells.

Acknowledgments

This study was supported by the Intramural Research Program of the National Institutes of Health,

NHLBI. We thank Dr. Martha Vaughan for helpful discussions and critical review of the manuscript. We thank the LAM Foundation and the Tuberous Sclerosis Alliance for patient referrals, and we thank the patients with LAM for their commitment and inspiration.

Conflicts of Interest

The authors declare no conflicts of interest.

References

1. KITAICHI, M., K. NISHIMURA, *et al.* 1995. Pulmonary lymphangioleiomyomatosis: a report of 46 patients including a clinicopathologic study of prognostic factors. Am. J. Respir. Crit. Care Med. **151:** 527–533.

2. URBAN, T., R. LAZOR, *et al.* 1999. Pulmonary lymphangioleiomyomatosis: a study of 69 patients. Medicine **78:** 321–337.

3. RYU, J.H., J. MOSS, *et al.* 2006. The NHLBI lymphangioleiomyomatosis registry: characteristics of 230 patients at enrollment. Am. J. Respir. Crit. Care Med. **173:** 105–111.

4. CHU, S.C., K. HORIBA, *et al.* 1999. Comprehensive evaluation of 35 patients with lymphangioleiomyomatosis. Chest **115:** 1041–1052.

5. TAYLOR, J.R., J. RYU, *et al.* 1990. Lymphangioleiomyomatosis: clinical course in 32 patients. N. Engl. J. Med. **323:** 1254–1260.

6. JOHNSON, S.R. & A.E. TATTERSFIELD. 2002. Lymphangioleiomyomatosis. Semin. Respir. Crit. Care Med. **23:** 85–92.

7. HAYASHIDA, M., K. SEYAMA, *et al.* 2007. The epidemiology of lymphangioleiomyomatosis in Japan: a nationwide cross-sectional study of presenting features and prognostic factors. Respirology **12:** 523–530.

8. TAVEIRA-DASILVA, A.M., W.K. STEAGALL, *et al.* 2006. Lymphangioleiomyomatosis. Cancer Control **13:** 276–285.

9. JOHNSON, S.R., C.I. WHALE, *et al.* 2004. Survival and disease progression in UK patients with lymphangioleiomyomatosis. Thorax **59:** 800–803.

10. TAVEIRA-DASILVA, A.M., C. HEDIN, *et al.* 2001. Reversible airflow obstruction: proliferation of abnormal smooth muscle cells, and impairment of gas exchange as predictors of outcome in lymphangioleiomyomatosis. Am. J. Respir. Crit. Care Med. **164:** 1072–1076.

11. AVILA, N.A., C.C. CHEN, *et al.* 2000. Pulmonary lymphangioleiomyomatosis: correlation of ventilation-perfusion scintigraphy, chest radiography, and CT with pulmonary function tests. Radiology **214:** 441–446.

12. FERRANS, V.J., Z.-X. YU, *et al.* 2000. Lymphangioleiomyomatosis (LAM): a review of clinical and morphological features. J. Nippon Med. Sch. **67:** 311–329.

13. MATSUI, K., A. TATSUGUCHI, *et al.* 2000. Extrapulmonary lymphangioleiomyomatosis (LAM): clinicopathologic features in 22 cases. Hum. Pathol. **31:** 1242–1248.

14. MATSUI, K., W.K. RIEMENSCNEIDER, *et al.* 2000. Hyperplasia of type II pneumocytes in pulmonary lymphangioleiomyomatosis: immunohistochemical and electron microscopic study. Arch. Pathol. Lab. Med. **124:** 1642–1648.

15. MATSUMOTO, Y., K. HORIBA, *et al.* 1999. Markers of cell proliferation and expression of melanosomal antigen in lymphangioleiomyomatosis. Am. J. Respir. Cell Mol. Biol. **21:** 327–336.

16. OHORI, N.P., S.A. YOUSEM, *et al.* 1991. Estrogen and progesterone receptors in lymphangioleiomyomatosis, epithelioid hemangioendothelioma, and sclerosing hemangioma of the lung. Am. J. Clin. Pathol. **96:** 529–535.

17. MATSUI, K., K. TAKEDA, *et al.* 2000. Downregulation of estrogen and progesterone receptors in the abnormal smooth muscle cells in pulmonary lymphangioleiomyomatosis following therapy: an immunohistochemical study. Am. J. Respir. Crit. Care Med. **161:** 1002–1009.

18. VALENCIA, J.C., G. PACHECO-RODRIGUEZ, *et al.* 2006. Tissue-specific renin-angiotensin system in pulmonary lymphangioleiomyomatosis. Am. J. Respir. Cell Mol. Biol. **35:** 40–47.

19. VALENCIA, J.C., K. MATSUI, *et al.* 2001. Distribution and mRNA expression of insulin-like growth factor system in pulmonary lymphangioleiomyomatosis. J. Invest. Med. **49:** 421–433.

20. SHAPIRO, S.D. 2006. Chairman's summary. Proc. Am. Thorac. Soc. **3:** 397–400.

21. PARKS, W.C. & S. SHAPIRO. 2001. Review: matrix metalloproteinases in lung biology. Respir. Res. **2:** 10–19.

22. MATSUI, K., K. TAKEDA, *et al.* 2000. Role for activation of matrix metalloproteinases in the pathogenesis of pulmonary lymphangioleiomyomatosis. Arch. Pathol. Lab. Med. **124:** 267–275.

23. HAYASHI, T., M.V. FLEMING, *et al.* 1997. Immunohistochemical study of matrix metalloproteinases (MMPs) and their tissue inhibitors (TIMPs) in pulmonary lymphangioleiomyomatosis (LAM). Hum. Pathol. **28:** 1071–1078.

24. ZHE, X., Y. YANG, *et al.* 2003. Tissue inhibitor of metalloproteinase-3 downregulation in lymphangioleiomyomatosis: potential consequence of abnormal serum response factor expression. Am. J. Respir. Cell Mol. Biol. **28:** 504–511.

25. KRYMSKAYA, V.P. & J.M. SHIPLEY. 2003. Lymphangioleiomyomatosis: a complex tale of serum response factor-mediated tissue inhibitor of metalloproteinase-3 regulation. Am. J. Respir. Cell Mol. Biol. **28:** 546–550.

26. MOSS, J., N.A. AVILA, *et al.* 2001. Prevalence and clinical characteristics of lymphangioleiomyomatosis (LAM) in patients with tuberous sclerosis complex. Am. J. Respir. Crit. Care Med. **163:** 669–671.

27. COSTELLO, L.C., T.E. HARTMAN, *et al.* 2000. High frequency of pulmonary lymphangioleiomyomatosis in women with tuberous sclerosis complex. Mayo Clin. Proc. **75:** 591–594.

28. FRANZ, D.N., A. BRODY, *et al.* 2001. Mutational and radiographic analysis of pulmonary disease consistent with lymphangioleiomyomatosis and micronodular pneumocyte hyperplasia in women with tuberous sclerosis. Am. J. Respir. Crit. Care Med. **164:** 661–668.

29. SMOLAREK, T.A., L.L. WESSNER, *et al.* 1998. Evidence that lymphangioleiomyomatosis is caused by *TSC2* mutations: chromosome 16p13 loss of heterozygosity in

angiomyolipomas and lymph nodes from women with lymphangiomyomatosis. Am. J. Hum. Genet. **62:** 810–815.

30. CARSILLO, T., A. ASTRINIDIS, *et al.* 2000. Mutations in the tuberous sclerosis complex gene *TSC2* are a cause of sporadic pulmonary lymphangioleiomyomatosis. Proc. Natl. Acad. Sci. USA **97:** 6085–6090.

31. SATO, T., K. SEYAMA, *et al.* 2002. Mutation analysis of the *TSC1* and *TSC2* genes in Japanese patients with pulmonary lymphangioleiomyomatosis. J. Hum. Genet. **47:** 20–28.

32. GONCHAROVA, E.A., D.A. GONCHAROV, *et al.* 2002. Tuberin regulates p70 S6 kinase activation and ribosomal protein S6 phosphorylation: a role for the TSC2 tumor suppressor gene in pulmonary lymphangioleiomyomatosis (LAM). J. Biol. Chem. **277:** 30958–30967.

33. TEE, A.R., D.C. FINGAR, *et al.* 2002. Tuberous sclerosis complex-1 and -2 gene products function together to inhibit mammalian target of rapamycin (mTOR)-mediated downstream signaling. Proc. Natl. Acad. Sci. USA **99:** 13571–13576.

34. TEE, A.R., B.D. MANNING, *et al.* 2003. Tuberous sclerosis complex gene products, Tuberin and Hamartin, control mTOR signaling by acting as a GTPase-activating protein complex toward Rheb. Curr. Biol. **13:** 1259–1268.

35. MAZIAK, D.E., S. KESTON, *et al.* 1996. Extrathoracic angiomyolipomas in lymphangioleiomyomatosis. Eur. Respir. J. **9:** 402–405.

36. KERR, L.A., M.L. BLUTE, *et al.* 1993. Renal angiomyolipoma in association with pulmonary lymphangioleiomyomatosis: forme fruste of tuberous sclerosis? Urology **41:** 440–444.

37. KELLY, J. & J. MOSS. 2001. Lymphangioleiomyomatosis. Am. J. Med. Sci. **321:** 17–25.

38. MCINTOSH, G.S., S.H. DUTOIT, *et al.* 1989. Multiple unilateral renal angiomyolipomas with regional lymphangioleiomyomatosis. J. Urol. **142:** 1305–1307.

39. HENSKE, E.P., H.P. NEUMANN, *et al.* 1995. Loss of heterozygosity in the tuberous sclerosis (*TSC2*) region of chromosome band 16p13 occurs in sporadic as well as TSC-associated renal angiomyolipomas. Genes Chromosomes Cancer **13:** 295–298.

40. KEBRIA, M., D. BLACK, *et al.* 2007. Primary retroperitoneal lymphangioleiomyomatosis in a postmenopausal woman: a case report and review of the literature. Int. J. Gynecol. Cancer **17:** 528–532.

41. LAM, B., G.C. OOI, *et al.* 2003. Extrapulmonary presentation of asymptomatic pulmonary lymphangioleiomyomatosis. Respirology **8:** 544–547.

42. JOHNSON, S.R. 2006. Lymphangioleiomyomatosis. Eur. Respir. J. **27:** 1056–1065.

43. MATSUI, K., M.B. BEASLEY, *et al.* 2001. Prognostic significance of pulmonary lymphangioleiomyomatosis histologic score. Am. J. Surg. Pathol. **29:** 1356–1366.

44. AVILA, N.A., J.A. KELLY, *et al.* 2000. Lymphangioleiomyomatosis: abdominopelvic CT and US findings. Radiology **216:** 147–153.

45. TAVEIRA-DASILVA, A.M., M.P. STYLIANOU, *et al.* 2004. Decline in lung function in patients with lymphangioleiomyomatosis treated with or without progesterone. Chest **126:** 1867–1874.

46. BISSLER, J.J. & J.C. KINGSWOOD. 2004. Renal angiomyolipomata. Kidney Int. **66:** 924–934.

47. RYU, J.H., C.H. DOERR, *et al.* 2003. Chylothorax in lymphangioleiomyomatosis. Chest **123:** 623–627.

48. ALMOOSA, K.F., J.H. RYU, *et al.* 2006. Management of pneumothorax in lymphangioleiomyomatosis: effects on recurrence and lung transplantation complications. Chest **129:** 1274–1281.

49. AVILA, N.A., A.J. DWYER, *et al.* 2006. CT of pleural abnormalities in lymphangioleiomyomatosis and comparison of pleural findings after different types of pleurodesis. AJR Am. J. Roentgenol. **186:** 1007–1012.

50. AVILA, N.A., J.A. KELLY, *et al.* 2002. Lymphangioleiomyomatosis: correlation of qualitative and quantitative thin-section CT with pulmonary function tests and assessment of dependence on pleurodesis. Radiology **223:** 189–197.

51. STEAGALL, W.K., C.G. GLASGOW, *et al.* 2007. Genetic and morphologic determinants of pneumothorax in lymphangioleiomyomatosis. Am. J. Physiol. Lung Cell. Mol. Physiol. **293:** L800–L808.

52. BITTMAN, I., B. ROLF, *et al.* 2003. Recurrence of lymphangioleiomyomatosis after single lung transplantation: new insights into pathogenesis. Hum. Pathol. **34:** 95–98.

53. KARBOWNICZEK, M., A. ASTRINIDIS, *et al.* 2003. Recurrent lymphangiomyomatosis after transplantation: genetic analyses reveal a metastatic mechanism. Am. J. Respir. Crit. Care Med. **167:** 976–982.

54. CROOKS, D.M., G. PACHECO-RODRIGUEZ, *et al.* 2004. Molecular and genetic analysis of disseminated neoplastic cells in lymphangioleiomyomatosis. Proc. Natl. Acad. Sci. USA **101:** 17462–17467.

55. TAI, H.-C., J.-W. LIN, *et al.* 2006. Pulmonary lymphangioleiomyomatosis followed by a localized retroperitoneal lymphangioleiomyoma. A.P.M.I.S. **114:** 821–824.

56. CORNOG, J.L., JR. & H.T. ENTERLINE. 1966. Lymphangiomyoma, a benign lesion of chyliferous lymphatics synonymous with lymphangiopericytoma. Cancer **19:** 1909–1930.

57. WOODRING, J.H., R.S. HOWARD II, *et al.* 1994. Massive low-attenuation mediastinal, retroperitoneal, and pelvic lymphadenopathy on CT from lymphangioleiomyomatosis case report. Clinical Imaging **18:** 7–11.

58. PALLISA, E., P. SANZ, *et al.* 2002. Lymphangioleiomyomatosis: pulmonary and abdominal findings with pathologic correlation. RadioGraphics **22:** S185–S198.

59. AVILA, N.A., J. BECHTLE, *et al.* 2001. Lymphangioleiomyomatosis: CT of diurnal variation of lymphangioleiomyomas. Radiology **221:** 415–421.

60. AVILA, N.A., A.J. DWYER, *et al.* 2005. Sonography of lymphangioleiomyoma in lymphangioleiomyomatosis: demonstration of diurnal variation in lesion size. AJR Am. J. Roentgenol. **184:** 459–464.

61. LEUNG, A.K.C., W.L.M. ROBSON. 2007. Tuberous sclerosis complex: a review. J. Pediatr Health Care **21:** 108–114.

62. ARBISER, J.L., D. BRAT, *et al.* 2002. Tuberous sclerosis-associated lesions of the kidney, brain, and skin are angiogenic neoplasms. J. Am. Acad. Dermatol. **46:** 376–380.

63. Avila, N.A., A.J. Dwyer, *et al.* 2007. Sporadic lymphangioleiomyomatosis and tuberous sclerosis complex with lymphangioleiomyomatosis: comparison of CT features. Radiology **242:** 277–285.

64. Webb, D.W., A. Clarke, *et al.* 1996. The cutaneous features of tuberous sclerosis: a population study. Br. J. Dermatol. **135:** 1–5.

65. Nguyen-Vu, P.-A., I. Facker *et al.* 2001. Loss of tuberin, the tuberous-sclerosis-complex-2 gene product is associated with angiogenesis. J. Cutan. Pathol. **28:** 470–475.

66. Li, S., F. Takeuchi, *et al.* 2005. MCP-1 overexpressed in tuberous sclerosis lesions acts as a paracrine factor for tumor development. J. Exp. Med. **202:** 617–624.

67. Nickel, W.R. & W.B. Reed. 1962. Tuberous sclerosis: special reference to the microscopic alterations in the cutaneous hamartomas. Arch. Dermatol. **85:** 89–106.

68. Sanchez, N.P., M.R. Wick, *et al.* 1981. Adenoma sebaceum of Pringle: a clinicopathologic review, with a discussion of related pathologic entities. J. Cutan. Pathol. **8:** 395–403.

69. Hirakawa, S., Y.-K. Hong, *et al.* 2003. Identification of vascular lineage-specific genes by transcriptional profiling of isolated blood vascular and lymphatic endothelial cells. Am. J. Pathol. **162:** 575–586.

70. Stacker, S.A., M.E. Baldwin, *et al.* 2002. The role of tumor lymphangiogenesis in metastatic spread. FASEB J. **16:** 922–934.

71. Sleeman, J.P. 2000. The lymph node as a bridgehead in the metastatic dissemination of tumors. Rec. Results Cancer Res. **157:** 55–81.

72. Kumasaka, T., K. Seyama, *et al.* 2004. Lymphangiogenesis in lymphangioleiomyomatosis: its implication in the progression of lymphangioleiomyomatosis. Am. J. Surg Pathol. **8:** 1007–1015.

73. Kumasaka, T., K. Seyama, *et al.* 2005. Lymphangiogenesis-mediated shedding of LAM cell clusters as a mechanism for dissemination in lymphangioleiomyomatosis. Am. J. Surg. Pathol. **29:** 1356–1366.

74. Seyama, K., T. Kumasaka, *et al.* 2006. Vascular endothelial growth factor-D is increased in serum of patients with lymphangioleiomyomatosis. Lymph. Res. Biol. **3:** 143–152.

Targeted Treatment for Lymphedema and Lymphatic Metastasis

TOMI TERVALA,[a] ERKKI SUOMINEN,[a] AND ANNE SAARISTO[a,b]

[a]*Department of Plastic Surgery, Turku University Central Hospital, Turku, Finland*

[b]*Molecular/Cancer Biology Laboratory and Ludwig Institute for Cancer Research, Biomedicum Helsinki and Helsinki University Central Hospital, University of Helsinki, Helsinki, Finland*

The presence of lymphatic vessels has been known for centuries, but the key players regulating the lymphatic vessel growth and function have only been discovered during the recent decade. The lymphatic vasculature is essential for maintenance of normal fluid balance and for the immune response. Hypoplasia or dysfunction of the lymphatic vessels can lead to lymphedema. Currently, lymphedema is treated primarily by physiotherapy, compression garments, and occasionally by surgery, but the means to reconstitute the collecting lymphatic vessels and cure the condition are limited. Specific growth factor therapy has been used in experimental models to regenerate lymphatic capillaries and collecting vessels after surgical damage. Recent results provide a new concept of combining growth factor therapy with lymph node transplantation as a rationale for treating secondary lymphedema. Lymphatic vessels are also involved in lymph node and systemic metastasis of cancer cells; our understanding of mechanisms of lymphatic metastasis has increased remarkably.

Key words: lymphangiogenesis; VEGF-C; growth factor therapy

Regulation of Lymphangiogenesis

The development of lymphatic vessels begins in the embryo at midgestation, at a time when the cardiovascular system is established and fully functional. There are two different theories for the origin of the lymphatic vessels. Sabin in 1902 proposed that primitive lymph sacs originate by endothelial cell budding from embryonic veins.[1] Later the peripheral lymphatic system grows from these primary lymph sacs by endothelial sprouting into the surrounding tissues and organs. Huntington and McClure, on the other hand, proposed that primordial lymph sacs arise in the mesenchyme from precursor cells ("lymphangioblasts"), independently of the veins and that the venous connections are established after this.[2,3]

The development and growth of the lymphatic vessels are regulated by a complex array of growth factors and cytokines secreted by stromal or inflammatory cells in a temporally and spatially organized manner. In the past decade specific growth factors capable of directly inducing the growth of lymphatic vessels *in vivo* have been discovered, some of which are summarized in TABLE 1.

Prox1

The homeobox transcription factor Prox1 is essential for the establishment of the identity of the lymphatic endothelial cells.[4] In mice, the first Prox1-positive endothelial cells emerge on embryonic day 10 in the jugular vein, from where they migrate to form the first lymphatic sprout.[4] In the absence of Prox1, endothelial cells bud from the cardinal vein, but cannot migrate further; they fail to establish a lymphatic endothelial cell–specific gene expression profile.[5] Accordingly, overexpression of Prox1 in human blood vascular endothelial cells suppresses many blood vascular-specific genes and upregulates lymphatic endothelial cell-specific transcripts.[6,7] Prox1 is also important for maintenance of the lymphatic vasculature during later stages of development and in adulthood, as Prox+/− mice develop chylous ascites, abnormal lymphatic vessels, and adult-onset obesity.[8] The signals leading to polarized expression of Prox1 in differentiating lymphatic endothelial cells are not known. It is, however, known that Prox1 expression is severely reduced in the intestinal lymphatic vessels of postnatal mice lacking

Address for correspondence: Dr. Anne Saaristo, Molecular/Cancer Biology Laboratory, Biomedicum Helsinki, P.O.B. 63 (Haartmaninkatu 8), University of Helsinki, 00014 Finland. Voice: +358-2-3130257; fax: +358-9-1912 5510.

Anne.Saaristo@helsinki.fi

Ann. N.Y. Acad. Sci. 1131: 215–224 (2008). © 2008 New York Academy of Sciences.
doi: 10.1196/annals.1413.019

TABLE 1. Genes that mediate lymphatic vessel formation

Gene	Function
VEGF-C/D[11–14]	Lymphangiogenic growth factors, ligands for VEGFR-3
Angs[40–43]	Growth factors regulating patterning of lymphatic vessels, ligands for Tie-1 and Tie-2
LYVE-1[25–30]	Widely used marker for lymphatic endothelial cells; specific function unknown
Prox1[4–7]	Transcription factor, essential for determination of lymphatic endothelial cell identity
Foxc2[36–39]	Transcription factor regulating later development of lymphatic vessels and valves; mutations cause a disease called lymphedema–distichiasis (LD)
EphrinB2[31,45,46]	Ligand for EphB4 receptor; remodeling of the lymphatic vasculature
Podoplanin[33–35]	Membrane glycoprotein; knockout mice have deficient lymphatic vessel formation and patterning
Syk and Slp76[47,48]	Involved in controlling the separation of lymphatic and blood vascular systems
Integrins, neuropilins[20–24]	Coreceptors modulating VEGF receptor signaling

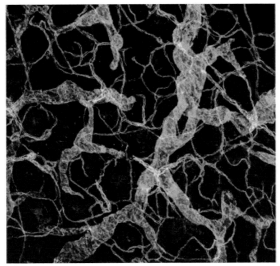

FIGURE 1. Blood (thin) and lymphatic (thick) vessel capillaries in the skin visualized by immunohistochemical staining. (In color in *Annals* online.)

fasting-induced adipose factor/angiopoietin-like protein 4 (Fiaf/Angptl4).[9] This leads to dilated and blood-filled lymphatic vessels with abnormal connections to blood vasculature.

VEGF-C/D

The vascular endothelial growth factor (VEGF) family is a group of ligands for the endothelial cell–specific VEGF tyrosine kinase receptors. These growth factors play pivotal roles in the regulation of vascular and lymphatic growth, vascular permeability, and influx of inflammatory cells (for a review see Tammela *et al.*[10]). VEGF-C and VEGF-D can stimulate the growth of lymphatic vessels, a process called lymphangiogenesis.[11,12] Although not required for lymphatic endothelial commitment, VEGF-C is essential for the initial sprouting and directed migration and for the subsequent survival of lymphatic endothelial cells.[13]

Deletion of both *Vegfc* alleles is lethal to the murine embryo, and heterozygous *Vegf-c*$^{+/-}$ mice display defects in lymphatic vascular development, indicating the importance of tightly regulated VEGF-C concentrations in the embryo.[13] VEGF-D, on the other hand, is dispensable for the embryonic development of blood vascular and lymphatic systems; the only defect caused by deletion of *Vegf-d* in mice is a slight reduction in the number of peribronchiolar lymphatic vessels; the primary site of VEGF-D expression.[14] VEGF-C and VEGF-D are ligands for the endothelial cell–specific tyrosine kinase receptors VEGFR-2 and VEGFR-3. In normal adult tissues their main target receptor is VEGFR-3, which is expressed predominantly in the lymphatic endothelial cells.[15] However, in chronic wounds VEGFR-3 is abnormally expressed on the blood vessel endothelium.[16,17] Interestingly, inactivating mutations of the *VEGFR-3* gene are a known cause of hereditary lymphedema.[18,19]

Both neuropilins and integrins modulate VEGF receptor signaling. *In vitro*, neuropilin-2 is co-internalized along with VEGFR-3 upon VEGF-C or VEGF-D binding, suggesting that neuropilin-2 is involved in modulating VEGFR-3 signaling.[20] Neuropilin-2 mutant mice are born without small lymphatic vessels and capillaries, while the blood vasculature and larger lymphatic vessels develop normally.[21] Integrin β_1 can directly interact with and induce tyrosine phosphorylation of VEGFR-3, and stimulate migration of endothelial cells to some extent, even in the absence of a ligand.[22] Integrin $\alpha_5\beta_1$ participates in the activation of VEGFR-3, which is essential for fibronectin-mediated LEC survival and proliferation.[23] Integrin α_9 chain–deficient mice die during the early postnatal period of chylothorax, which suggests an underlying function for integrin $\alpha_9\beta_1$ in lymphatic development.[24]

LYVE-1

LYVE-1 (lymphatic vessel hyaluronan resceptor-1) is expressed predominantly in lymphatic vessels and it is one of the most widely used markers of lymphatic endothelial cells in normal and tumor tissues (FIG. 1).[25] However, LYVE-1 expression is also detected in blood vessel endothelial cells of liver sinusoids, in the lungs, in high endothelial venules, and in activated tissue macrophages.[26–29] In mice LYVE-1 is expressed in a polarized manner in venous endothelium starting from E9.[30] The expression of LYVE-1 in adults is downregulated in collecting lymphatic vessels, but remains high in the lymphatic capillaries.[31] LYVE-1 is thought to contribute to hyaluronan turnover or leukocyte trafficking. Of interest, mice deficient for LYVE-1 have a normal lymphatic vasculature and secondary lymphoid tissue, normal trafficking of cells to draining lymph nodes, and a normal skin inflammation response.[32]

Podoplanin

Podoplanin is a small transmembrane mucin-like protein that is highly expressed in lymphatic endothelial cells.[33] Targeted inactivation of podoplanin leads to abnormal lung development and perinatal death in an experimental setting.[34] However, podoplanin-deficient mice have also abnormally dilated and nonfunctioning superficial lymphatic vessels in the skin.[34] In cultured endothelial cells, podoplanin promotes cell adhesion, migration, and tube formation.[34] Podoplanin is also upregulated in the invasive front of many human carcinomas.[35]

Foxc2

The forkhead transcription factor Foxc2 is highly expressed in the developing lymphatic vessels and also in lymphatic valves in adults.[36] Mutation in Foxc2 causes the hereditary lymphedema–distichiasis (LD) syndrome.[37] Foxc2 deficiency in mice leads to perinatal death due to aortic arch malformations.[38] The initial development of lymphatic vasculature proceeds normally in the absence of Foxc2, but at later stages of embryogenesis lymphatic vessels are severely affected.[36] Lymphatic vessels in Foxc2−/− mice fail to develop valves, and lymphatic capillaries acquire ectopic coverage by basal lamina components and an abnormal covering of smooth muscle cells.[36] Mutations in Foxc2 have been strongly associated with primary venous valve failure in both the superficial and deep veins in the lower limb.[39]

Angiopoietins and Ephrins

Angiopoietin 1 (Ang-1) and 2 (Ang-2) modify VEGF responses by affecting vessel maturation and stabilization via the Tie receptors.[40,41] Angiopoietins seems to play a role in the patterning of the lymphatic vessels,

and overexpression of Ang1 was shown to induce lymphangiogenesis.[42,43] The primordial lymphatic vascular plexus needs to undergo extensive remodeling and maturation in order to establish a functional lymphatic vasculature (reviewed in Karpanen and Makinen[44]). EphrinB receptor tyrosine kinases and their transmembrane-type ephrin-B2 ligands play essential roles in the remodeling of the arterial–venous capillary plexus and in the postnatal maturation of lymphatic vasculature.[31,45,46] Mice expressing a mutated form of the transmembrane growth factor ephrinB2 have normal blood vasculature, but the maturation of lymphatic vessels and development of valves is disturbed, which lead to abnormal lymphatic functions, such as retrograde lymph flow.[31] In these mice the lymphatic capillaries were covered by smooth muscle cells and also exhibited a hyperplasia of the collecting lymphatic vessels, lack of luminal valve formation, and a failure to remodel the primary lymphatic capillary plexus.[31]

SYK and SLP76

The tyrosine kinase Syk and adaptor protein SLP76 are involved in controlling the separation of the lymphatic and blood vascular systems.[47] Mice deficient of the intracellular signaling molecules *Syk* or *Slp76*, expressed almost exclusively in hematopoietic but not in endothelial cells, develop arteriovenous shunts and abnormal lymphaticovenous connections, leading to blood-filled lymphatic vessels.[47] Circulating hematopoietic cells, presumably endothelial progenitors, are necessary for controlling the separation of the lymphatic and blood vascular systems.[48]

Other Growth Factors

In addition to factors listed above, several growth factors, including VEGF, hepatocyte growth factor, insulin-like growth factors 1 and 2, platelet-derived growth factor B (PDGF-B), and fibroblast growth factor have been described to induce lymphangiogenesis either in a VEGFR-3-independent or -dependent manner (reviewed in Tammela *et al.*[10] and Cueni and Detmar[49]). These growth factors may primarily stimulate other biological processes and induce lymphangiogenesis indirectly.

Lymphedema

Impaired lymphatic function results in inadequate transport of extracellular fluid, macromolecules, and cells from the interstitium, and is associated with several diseases characterized by tissue edema, impaired immunity, and fibrosis.[50] Lymphedema may be hereditary or idiopathic, in which case it is called primary lymphedema. Lymphedema may also result from some disease, trauma, surgery or radiotherapy, and this type is called secondary or acquired lymphedema.

Infection, especially filariasis, is the most common cause of secondary lymphedema worldwide (reviewed in Dreyer *et al.*[51]). Cancer treatment is the largest single cause of secondary lymphedema in developed countries. Surgical treatments (e.g., lymph node dissection to treat malignant disease) may cause secondary lymphedema. Around 20–30% of patients who undergo axillary dissection as a part of the treatment of breast cancer will develop lymphedema of variable severity.[52] The combination of radiation therapy and lymphadenectomy even more often causes lymphedema, especially of the upper extremity. Common complications of lymphedema include progressive dermal fibrosis, accumulation of adipose and connective tissue, impaired wound healing, decreased immune defense, and an increased susceptibility to infections and subsequent cellulitis.[53] Long-term lymphedema may also place the patient at risk for developing lymphangiosarcoma.

Primary lymphedemas are rare (1:6000) inherited conditions caused by aberrant prenatal development of lymph vessels. Many of these syndromes have been recently linked with a specific growth factor gene mutation.[54] Milroy's disease is an inherited disorder that begins in infancy. It affects primarily the legs and feet. Lymphangiography has revealed that the lymphatic vessels of these patients are absent or extremely hypoplastic in affected but not in unaffected areas.[55] Milroy's syndrome is an autosomal dominant disease and inheritance of this disease has been linked to the VEGFR-3 locus in chromosome 5q35.3.[56–58] Most of the disease-associated alleles contain missense mutations encoding tyrosine-kinase-negative VEGFR-3 proteins with extended cellular half-lives, which function as dominant negative receptors that reduce downstream signaling.[19,59] A novel VEGFR-3 mutation in the exon 22 of chromosome 5 has recently been described in four generations of a family.[60]

Meige's disease or lymphedema praecox (hereditary lymphedema type II, OMIM 153200) causes lymphedema in childhood or around puberty. It is the most common type of primary lymphedema, accounting for approximately 90% of all cases and is often unilateral. This hereditary disorder causes the lymph vessels to form without the valves. The inheritance of this disease has been linked to an inactivating mutation in the FOXC2 gene, at least in one family.[61] There may be more than one gene causing Meige's disease because of genetic heterogeneity within this group of patients.[62]

The LD syndrome (OMIM 153400) is an autosomal dominant condition that present by late-onset lymphedema of the legs and double rows of eyelashes (distichiasis), which is often the diagnostic feature of this disease. Lymphedema usually develops after or at puberty. Incompetent venous valves and venous insufficiency are frequent.[39] The complications may also include cardiac defects, extradural cysts, photophobia, and cleft palate.[37] LD is caused by heterozygous loss-of-function mutations of FOXC2.[37,63] The lymphatic vasculature in LD is normal or hyperplastic, but there is backflow of lymph because of abnormal lymphatic valves, defective vessel patterning, and the presence of ectopic smooth muscle cells surrounding abnormal lymph vessels.[36,63]

An unusual association of lymphedema with hypotrichosis and telangiectasia (OMIM 607823) was linked to mutations of the gene encoding for the transcription factor SOX18.[64] In addition, the locus responsible for the variable symptoms of Turner's syndrome, frequently associated with lymphedema, has been mapped to region Xp11.2-p22.1, which, of interest, also harbors the *VEGF-D* gene.[65]

Treatment of Lymphedema

Reconstructive Lymphatic Microsurgery: Current Approach to Curative Treatment

The treatment of lymphedema is currently based on physiotherapy, compression garments, and occasionally surgery.[53] Liposuction and different modifications of reducing operations are often used to treat chronic lymphedema.[66–68] They often provide good cosmetic or functional result, but have no effect on lymphatic dysfunction. A durable result requires regular use of compression garments.[67]

Reconstructive lymphatic microsurgery aims at curative treatment of lymphedema. However, lymphatic vessels are a surgical challenge—they are difficult to prepare and it may be difficult to identify the optimal strategy for reconstruction. Lymphavenous anastomoses have been the most common strategy for lymphatic bypass surgery,[69,70] but the late patency of these anastomoses in humans is considered uncertain.[71] Backflow of venous blood and coagulation in the lymphatic vessels may cause problems in these anastomoses. Lymphatic endothelial cells express a small transmembrane mucin-like protein called podoplanin (or platelet aggregation–inducing factor) that directly induces blood coagulation.[33] The concept of lymphatic grafting is attractive because it avoids problems of thromboses in the anastomoses. In the lymphatic bypass surgery pioneered by Prof. Baumeister, lymphatic grafts are harvested from the medial aspect of patient's thigh.[72] Long-term patency of lymphatic grafts has been documented by lymphography.[73] Nevertheless, lymphatic grafting is not widely used in the clinical work. The most recent reconstructive approach for the treatment of lymphedema is autotransplantation of lymphatic tissue in a free flap. Lymph nodes and vessels have been harvested from the patient's inguinal or axillary region and lymphatic anastomoses should form spontaneously.[74,75]

Growth Factor Therapy: Curative Treatment of Lymphedema in the Future

Were it possible to anastomose lymphatic vessels in the area of lymphatic vasculature defect without surgery and donor side defects, such a therapeutic strategy would be certainly useful. A combination of reconstructive surgery and therapeutic growth factors may also open new possibilities.

In a previous volume of the *Annals* entitled *The Lymphatic Continuum* Prof. Alitalo's group reported a new experimental growth factor therapy for Milroy's primary lymphedema.[76] VEGF-C therapy has been used in preclinical lymphedema models and the results have been encouraging. VEGF-C gene transfer to the skin of mice with primary lymphedema (Milroy's disease) induced regeneration of the cutaneous lymphatic vessel network.[77,78] Recombinant human VEGF-C, naked plasmids encoding VEGF-C, and adenoviral VEGF-C have also reportedly induced lymphatic capillary growth and ameliorated the symptoms of secondary lymphedema in rabbit and mouse models.[79–82] However, these studies have limited themselves to analyses of lymphatic capillaries, whereas clinical lymphedema usually occurs after damage to the collecting lymphatic vessels. Recently Tammela *et al.* showed significant improvement in lymphatic draining function by viral VEGF-C or VEGF-D in an orthotopic model that is clinically relevant, since it involved treatment of damage to the axillary lymph nodes[83] (FIG. 2). The result demonstrated for the first time that growth factor therapy can indeed be used to generate functional and mature collecting lymphatic vessels. Interestingly, intraluminal valves were also formed in the growth factor–induced vessels, suggesting that the FOXC2 pathway and other developmental mechanisms are reactivated in the lymphatic vasculature generated by VEGF-C or VEGF-D.[83]

Lymph node transplantation has been used to treat of lymphedema without growth factor therapy.[74,75] The expectation is that the lymphatic vessels anastomose spontaneously. However, lymphoscintigraphic data imply that autologous lymph nodes do not always incorporate optimally into existing lymphatic vasculature.[75,84] On the other hand, connection with lymphatic vessels is required for maintenance and function of lymph nodes.[85] Recent experimental results show that transplantation of VEGF-C transfected lymph nodes results in restoration of normal anatomy of the lymphatic network in the defect area, and this includes both the lymphatic vessels and the nodes[83] (FIG. 2). Combined growth factor treatment and lymph node transplantation was used to incorporate lymph nodes with the pre-existing lymphatic vessel network in the defect area in the axilla.[83] The advantage of this rationale compared to lymph node transfer alone is increased incorporation efficiency of the transplanted nodes. The advantage of combination therapy in comparison to growth factor therapy alone is increased safety of patients with recurrent malignancies, as lymph nodes provide an immunologic barrier against systemic dissemination of cancer cells, as well as other pathogens (FIG. 2).

In secondary lymphedema, the lymphatic defect area is usually restricted (for example, to the axillary or inguinal area) and in this setting brief gene expression or even protein therapy may be sufficient. Thus, secondary lymphedema seems to be the most attractive target to start with as human prolymphangiogenic therapy trials are concerned. Results from experimental models also suggest that if lymphatic neovasculature becomes functional, it remains stable and mature in the target tissue even without growth factor support.[81,83] Vectors that provide transient gene expression are likely to be useful, at least in the treatment of secondary lymphedema. Combining autologous lymph node transfer and therapeutic growth factors also provides a promising tool for curative treatment of secondary lymphedema.

Mechanisms of Lymphatic Metastasis

Cancer kills mainly through metastatic spread. Tumor cells can spread by several mechanisms: local invasion, through the blood vascular system and through lymphatic vasculature. Several clinical and pathologic observations suggest that the most common way of initial metastasis of many types of solid human tumors is through the lymphatic system to the regional lymph nodes (reviewed in Sleeman[86] and Stacker *et al.*[87]). Dissemination of tumor cells to the lymphatic system requires physical contacts between the tumor cells and lymphatic vessels.[88–90] The size of the peritumoral lymphatic vessels may be the most important factor

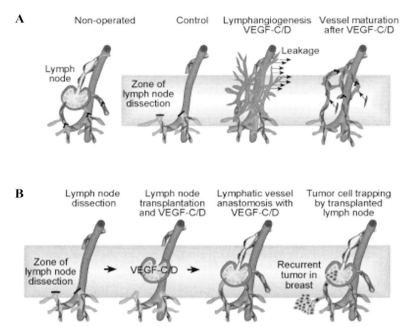

FIGURE 2. Schematic illustration of lymphatic growth factor growth factor therapy. **(A)** Differentiation and remodeling of the VEGF-C/D-induced lymphatic vessels into functional collecting vessels in the defect area after lymph node dissection. The lymphatic vessels gradually stabilize, and acquire SMC coverage and intraluminal valves. **(B)** Normal lymphatic vessel anatomy is restored after growth factor–treated lymph node transplantation. The transplanted lymph nodes incorporate into the host's lymphatic vasculature and can function as sentinel lymph nodes in the event of a recurrent tumor. (From Tammela et al.[83] Modified by permission.)

contributing to lymph node metastasis in human malignant melanoma.[90] Dissemination of tumor cells from the primary sites to the lymphatic system occurs either by invasion into pre-existing lymphatic vessels in surrounding tissues or by invasion into intratumoral lymphatic networks.[87,91] Tumor, stromal, and inflammatory cells that are present in the tumor microenvironment produce multiple lymphangiogenic factors that may stimulate intratumoral lymphangiogenesis.[11,12,42,92–96] Several factors, including VEGF-A, VEGF-C, VEGF-D, and PDGF-BB, have been reported to induce lymphatic metastasis.[87,91,92,97–100] Experimental studies show that VEGF-C overexpression in tumors potently increases intratumoral lymphangiogenesis, and that this significantly enhances metastasis to regional lymph nodes and to distant sites.[97,100,101] Experimental studies also indicate that the degree of tumor lymphangiogenesis correlates strongly with the tendency of tumor to metastases.[88,97–102] Of interest, recent experimental studies indicated that tumors already induce lymphangiogenesis in the sentinel lymph nodes before tumor cells disseminate.[103,104] Identical changes were found in the axillary lymph nodes

of breast cancer patients who had no evidence of metastasis.[104]

The lymphatic endothelial–specific receptor VEGFR-3 may also contribute to tumor angiogenesis or to the integrity of tumor blood vessels.[105–107] A recent study shows that by blocking VEGFR-3, tumor angiogenesis and tumor growth can be reduced; this observation confirms the involvement of VEGFR-3 in tumor angiogenesis.[108]

Lymphatic vessels have a discontinuous basement membrane and lack tight interendothelial junctions, in contrast to blood vessels.[109] It is therefore believed that it would be easier for tumor cells to get into the lymphatic vessels than blood circulation. This kind of mechanism is thought to be passive. However, different tumor types show different patterns of spread to distant organs, suggesting that metastasis to a specific organ could be a guided process, possibly as an active event.[110] VEGF-C-stimulated growth of new tumor-associated lymphatic vessels may eliminate one rate-limiting step of the metastatic process by simply increasing the contact area of malignant cells with the lymphatic vasculature.[88,97,100,101,111–115] Experimental evidence also suggests that lymphatic endothelial cells could attract tumor cells by secreting chemokines, thus actively promoting lymphatic metastasis. The observation that adjacent lymphatic endothelium forms extensive filopodia toward VEGF-C producing tumor cells, suggests that tumor emboli could be actively entrapped inside the sprouting lymphatic vessels.[54,116]

Secondary lymphoid chemokine (SLC/CCL21) is one of the chemokines that is highly expressed in lymph nodes and in the lymphatic endothelium of many organs.[117] CCR7 and CXCR4, receptors for SLC/CCL21 and CXCL12, respectively, are highly expressed in human breast cancer cells, and their ligands exhibit peak levels of expression in regional lymph nodes, lung, bone marrow, and liver, which represent the first destinations of breast cancer metastasis.[118] Inhibition of the CXCL12/CXCR4 interaction significantly reduces breast cancer metastases to the lymph nodes and the lung.[118] Similarly, expression of CCR7 in murine melanoma cells promoted metastasis to lymph nodes, which was blocked by neutralizing SLC/CCL21 antibodies.[119] This information suggests that there are active interactions between endothelial cells and tumor cells.

Conclusions

Lymphangiogenic growth factor therapy has enormous therapeutic potential in the treatment of lymphedema. Angiogenic growth factor therapy has been extensively studied in the cardiovascular ischemic diseases and several clinical studies are ongoing. Data from these studies will benefit the planning of clinical trials on lymphedema patients. Angiogenic gene therapy has raised some concerns regarding the possibility of stimulation of dormant tumor cells as a consequence of increased angiogenesis. Many human tumors use lymphatic vessels for metastatic dissemination and therefore induction of lymphangiogenesis in a patient with malignant disease may raise safety concerns. By combining growth factor therapy with lymph node transfer, we may increase patient safety in instances of recurrent malignancies, as lymph nodes function as immunologic barriers against systemic dissemination of cancer cells and other pathogens. Results from animal studies have also shown that prolonged gene therapy is not required, at least in the treatment of secondary lymphedema. Once the functional lymphatic vessels are formed, they stabilize and mature and no longer need growth factor support.[81,83] Data from preclinical studies imply that therapeutic lymphangiogenesis provides a new treatment option for lymphedema patients. Formal clinical trials are now needed. Detailed understanding of the mechanisms regulating lymphatic metastasis may also result in the development of new therapies. For example, tumor lymphangiogenesis and lymphatic metastasis could efficiently be inhibited by interfering with the VEGFR-3 signaling by neutralizing antibodies, ligand traps, or RNA interference. Understanding the general mechanisms guiding tumor cell homing to specific destinations may also provide therapeutic means of inhibiting tumor progression to a lethal metastatic disease.[120]

Acknowledgments

Our studies were supported by the Special Government Funding (EVO) allocated to Turku University Central Hospital, the Turku University Foundation, the Finnish Diabetes Foundation, the Paavo and Eila Salonen Foundation, and the Novo Nordisk Foundation. We thank Tuomas Tammela for preparing the figures and Robert Paul for preparing the manuscript.

Conflicts of Interest

The authors declare no conflicts of interest.

References

1. SABIN, F.R. 1902. On the origin of the lymphatic system from the veins and the development of the lymph hearts and thoracic duct in the pig. Am. J. Anat. **1:** 367–391.
2. HUNTINGTON, G.S. & C.F.W. MCCLURE. 1908. The anatomy and development of the jugular lymph sac in the domestic cat (*Felis domestica*). Anat. Rec. **2:** 1–19.
3. KAMPMEIER, O.F. 1912. The value of the injection method in the study of lymphatic development. Anat. Rec. **6:** 223–232.
4. WIGLE, J.T. & G. OLIVER. 1999. Prox1 function is required for the development of the murine lymphatic system. Cell **98:** 769–778.
5. WIGLE, J.T. *et al.* 2002. An essential role for Prox1 in the induction of the lymphatic endothelial cell phenotype. EMBO J. **21:** 1505–1513.
6. HONG, Y.K. *et al.* 2002. Prox1 is a master control gene in the program specifying lymphatic endothelial cell fate. Dev. Dyn. **225:** 351–357.
7. PETROVA, T.V. *et al.* 2002. Lymphatic endothelial reprogramming of vascular endothelial cells by the Prox-1 homeobox transcription factor. EMBO J. **21:** 4593–4599.
8. HARVEY, N.L. *et al.* 2005. Lymphatic vascular defects promoted by Prox1 haploinsufficiency cause adult-onset obesity. Nat. Genet. **37:** 1072–1081.
9. BACKHED, F. *et al.* 2007. Postnatal lymphatic partitioning from the blood vasculature in the small intestine requires fasting-induced adipose factor. Proc. Natl. Acad. Sci. USA **104:** 606–611.

10. TAMMELA, T. *et al.* 2005. The biology of vascular endothelial growth factors. Cardiovasc. Res. **65:** 550–563.

11. JOUKOV, V. *et al.* 1996. A novel vascular endothelial growth factor, VEGF-C, is a ligand for the Flt4 (VEGFR-3) and KDR (VEGFR-2) receptor tyrosine kinases. EMBO J. **15:** 290–298.

12. ACHEN, M.G. *et al.* 1998. Vascular endothelial growth factor D (VEGF-D) is a ligand for the tyrosine kinases VEGF receptor 2 (Flk1) and VEGF receptor 3 (Flt4). Proc. Natl. Acad. Sci. USA **95:** 548–553.

13. KARKKAINEN, M.J. *et al.* 2004. Vascular endothelial growth factor C is required for sprouting of the first lymphatic vessels from embryonic veins. Nat. Immunol. **5:** 74–80.

14. BALDWIN, M.E. *et al.* 2005. Vascular endothelial growth factor D is dispensable for development of the lymphatic system. Mol. Cell Biol. **25:** 2441–2449.

15. PARTANEN, T.A. *et al.* 2000. VEGF-C and VEGF-D expression in neuroendocrine cells and their receptor, VEGFR-3, in fenestrated blood vessels in human tissues. FASEB J. **14:** 2087–2096.

16. PAAVONEN, K. *et al.* 2000. Vascular endothelial growth factor receptor-3 in lymphangiogenesis in wound healing. Am. J. Pathol. **156:** 1499–1504.

17. WITMER, A.N. *et al.* 2001. VEGFR-3 in adult angiogenesis. J Pathol. **195:** 490–497.

18. KARKKAINEN, M.J. & T.V. PETROVA. 2000. Vascular endothelial growth factor receptors in the regulation of angiogenesis and lymphangiogenesis. Oncogene **19:** 5598–5605.

19. KARKKAINEN, M.J. *et al.* 2000. Missense mutations interfere with VEGFR-3 signalling in primary lymphoedema. Nat. Genet. **25:** 153–159.

20. KARPANEN, T. *et al.* 2006. Functional interaction of VEGF-C and VEGF-D with neuropilin receptors. FASEB J. **20:** 1462–1472.

21. YUAN, L. *et al.* 2002. Abnormal lymphatic vessel development in neuropilin 2 mutant mice. Development **129:** 4797–4806.

22. WANG, J.F., X.F. ZHANG & J.E. GROOPMAN. 2001. Stimulation of beta 1 integrin induces tyrosine phosphorylation of vascular endothelial growth factor receptor-3 and modulates cell migration. J. Biol. Chem. **276:** 41950–41957.

23. ZHANG, X., J.E. GROOPMAN & J.F. WANG. 2005. Extracellular matrix regulates endothelial functions through interaction of VEGFR-3 and integrin alpha5beta1. J. Cell Physiol. **202:** 205–214.

24. HUANG, X.Z. *et al.* 2000. Fatal bilateral chylothorax in mice lacking the integrin a9b1. Mol. Cell. Biol. **20:** 5208–5215.

25. JACKSON, D.G. 2004. Biology of the lymphatic marker LYVE-1 and applications in research into lymphatic trafficking and lymphangiogenesis. APMIS **112:** 526–538.

26. WROBEL, T. *et al.* 2005. LYVE-1 expression on high endothelial venules (HEVs) of lymph nodes. Lymphology **38:** 107–110.

27. BANERJI, S. *et al.* 1999. LYVE-1, a new homologue of the CD44 glycoprotein, is a lymph-specific receptor for hyaluronan. J. Cell Biol. **144:** 789–801.

28. MOUTA CARREIRA, C. *et al.* 2001. LYVE-1 is not restricted to the lymph vessels: expression in normal liver blood sinusoids and down-regulation in human liver cancer and cirrhosis. Cancer Res. **61:** 8079–8084.

29. PREVO, R. *et al.* 2001. Mouse LYVE-1 is an endocytic receptor for hyaluronan in lymphatic endothelium. J. Biol. Chem. **276:** 19420–19430.

30. OLIVER, G. 2004. Lymphatic vasculature development. Nat. Rev. Immunol. **4:** 35–45.

31. MAKINEN, T. *et al.* 2005. PDZ interaction site in ephrinB2 is required for the remodeling of lymphatic vasculature. Genes Dev. **19:** 397–410.

32. GALE, N.W. *et al.* 2007. Normal lymphatic development and function in mice deficient for the lymphatic hyaluronan receptor LYVE-1. Mol. Cell Biol. **27:** 595–604.

33. BREITENEDER-GELEFF, S. *et al.* 1999. Angiosarcomas express mixed endothelial phenotypes of blood and lymphatic capillaries: podoplanin as a specific marker for lymphatic endothelium. Am. J. Pathol. **154:** 385–394.

34. SCHACHT, V. *et al.* 2003. T1alpha/podoplanin deficiency disrupts normal lymphatic vasculature formation and causes lymphedema. EMBO J. **22:** 3546–3556.

35. WICKI, A. *et al.* 2006. Tumor invasion in the absense of epithelial-mesenchymal transition: podoplanin-mediated remodeling of actin cytoskeleton. Cancer Cell **9:** 261–272.

36. PETROVA, T.V. *et al.* 2004. Defective valves and abnormal mural cell recruitment underlie lymphatic vascular failure in lymphedema distichiasis. Nat. Med. **10:** 974–981.

37. FANG, J. *et al.* 2000. Mutations in FOXC2 (MFH-1), a forkhead family transcription factor, are responsible for the hereditary lymphedema-distichiasis syndrome. Am. J. Hum. Genet. **67:** 1382–1388.

38. IIDA, K. *et al.* 1997. Essential roles of the winged helix transcription factor MFH-1 in aortic arch patterning and skeletogenesis. Development **124:** 4627–4638.

39. MELLOR, R.H. *et al.* 2007. Mutations in FOXC2 are strongly associated with primary valve failure in veins of the lower limb. Circulation **115:** 1912–1920.

40. DAVIS, S. *et al.* 1996. Isolation of angiopoietin-1, a ligand for the TIE2 receptor, by secretion-trap expression cloning. Cell **87:** 1161–1169.

41. MAISONPIERRE, P.C. *et al.* 1997. Angiopoietin-2, a natural antagonist for Tie2 that disrupts in vivo angiogenesis. Science **277:** 55–60.

42. GALE, N.W. *et al.* 2002. Angiopoietin-2 is required for postnatal angiogenesis and lymphatic patterning, and only the latter role is rescued by angiopoietin-1. Dev. Cell. **3:** 411–423.

43. TAMMELA, T. *et al.* 2005. Angiopoietin-1 promotes lymphatic sprouting and hyperplasia. Blood **105:** 4642–4648.

44. KARPANEN, T. & T. MAKINEN. 2006. Regulation of lymphangiogenesis–from cell fate determination to vessel remodeling. Exp. Cell Res. **312:** 575–583.

45. WANG, H.U., Z.F. CHEN & D.J. ANDERSON. 1998. Molecular distinction and angiogenic interaction between embryonic arteries and veins revealed by ephrin-B2 and its receptor Eph-B4. Cell **93:** 741–753.

46. ADAMS, R.H. *et al.* 1999. Roles of ephrinB ligands and EphB receptors in cardiovascular development: demarcation of arterial/venous domains, vascular morphogenesis, and sprouting angiogenesis. Genes Dev. **13:** 295–306.

47. ABTAHIAN, F. *et al.* 2003. Regulation of blood and lymphatic vascular separation by signaling proteins SLP-76 and Syk. Science **299:** 247–251.

48. SEBZDA, E. *et al.* 2006. Syk and Slp-76 mutant mice reveal a cell-autonomous hematopoietic cell contribution to vascular development. Dev. Cell. **11:** 349–361.

49. CUENI, L.N., & M. DETMAR 2006. New insights into the molecular control of the lymphatic vascular system and its role in disease. J. Invest. Dermatol. **126:** 2167–2177

50. WITTE, M.H. & C.L. WITTE. 1986. Lymphangiogenesis and lymphologic syndromes. Lymphology **19:** 21–28.

51. DREYER, G. *et al.* 2000. Pathogenesis of lymphatic disease in bancroftian filariasis: a clinical perspective. Parasitol. Today **16:** 544–548.

52. MOFFATT, C.J. *et al.* 2003. Lymphoedema: an underestimated health problem. Q. J. Med. **96:** 731–738.

53. ROCKSON, S.G. 2001. Lymphedema. Am. J. Med. **110:** 288–295.

54. ALITALO, K., T. TAMMELA & T.V. PETROVA. 2005. Lymphangiogenesis in development and human disease. Nature **438:** 946–953.

55. BRICE, G. *et al.* 2005. Milroy disease and the VEGFR-3 mutation phenotype. J Med. Genet. **42:** 98–102.

56. FERRELL, R.E. *et al.* 1998. Hereditary lymphedema: evidence for linkage and genetic heterogeneity. Hum. Mol. Genet. **7:** 2073–2078.

57. WITTE, M.H. *et al.* 1998. Phenotypic and genotypic heterogeneity in familial Milroy lymphedema. Lymphology **31:** 145–155.

58. EVANS, A.L. *et al.* 1999. Mapping of primary congenital lymphedema to the 5q35.3 region. Am. J. Hum. Genet. **64:** 547–555.

59. IRRTHUM, A. *et al.* 2000. Congenital hereditary lymphedema caused by a mutation that inactivates VEGFR3 tyrosine kinase. Am. J. Hum. Genet. **67:** 295–301.

60. BUTLER, M.G. *et al.* 2007. A novel VEGFR3 mutation causes Milroy disease. Am. J. Med. Genet. A **143:** 1212–1217.

61. FINEGOLD, D.N. *et al.* 2001. Truncating mutations in FOXC2 cause multiple lymphedema syndromes. Hum. Mol. Genet. **10:** 1185–1189.

62. WHEELER, E.S. *et al.* 1981. Familial lymphedema praecox: Meige's disease. Plast. Reconstr. Surg. **67:** 362–364.

63. BRICE, G. *et al.* 2002. Analysis of the phenotypic abnormalities in lymphoedema-distichiasis syndrome in 74 patients with FOXC2 mutations or linkage to 16q24. J. Med. Genet. **39:** 478–483.

64. IRRTHUM, A. *et al.* 2003. Mutations in the transcription factor gene SOX18 underlie recessive and dominant forms of hypotrichosis-lymphedema-telangiectasia. Am. J. Hum. Genet. **72:** 1470–1478.

65. ZINN, A.R. *et al.* 1998. Evidence for a Turner syndrome locus or loci at Xp11.2-p22.1. Am. J. Hum. Genet. **63:** 1757–1766.

66. NAVA, V.M. & W.T. LAWRENCE. 1988. Liposuction on a lymphedematous arm. Ann. Plast. Surg. **21:** 366–368.

67. BRORSON, H. & H. SVENSSON. 1998. Liposuction combined with controlled compression therapy reduces arm lymphedema more effectively than controlled compression therapy alone. Plast. Reconstr. Surg. **102:** 1058–1067; discussion 1068.

68. BROWSE, N.L. 1986. A Colour Atlas of Reducing Operations for Lymphoedema of the Lower Limb. Wolfe. London.

69. NIELUBOWICZ, J. & W. OLSZEWSKI. 1968. Experimental lymphovenous anastomosis. Br. J. Surg. **55:** 449–451.

70. CAMPISI, C. & F. BOCCARDO. 2004. Microsurgical techniques for lymphedema treatment: derivative lymphatic-venous microsurgery. World J. Surg. **28:** 609–613.

71. GLOVICZKI, P. *et al.* 1988. Microsurgical lymphovenous anastomosis for treatment of lymphedema: a critical review. J. Vasc. Surg. **7:** 647–652.

72. BAUMEISTER, R.G., J. SEIFERT & D. HAHN. 1981. Autotransplantation of lymphatic vessels. Lancet **1:** 147.

73. BAUMEISTER, R.G. 1994. Microsurgical lymphatic grafting: first demonstration of patent grafts by indirect lymphography and long term follow-up studies. Progr. Lymphol. **27:** 787.

74. TREVIDIC, P. 1992. Free axillary lymph node transfer. Progr. Lymphol. **13:** 415–420.

75. BECKER, C. *et al.* 2006. Postmastectomy lymphedema: long-term results following microsurgical lymph node transplantation. Ann. Surg. **243:** 313–315.

76. SAARISTO, A., M.J. KARKKAINEN & K. ALITALO. 2002. Insights into the molecular pathogenesis and targeted treatment of lymphedema. Ann. N.Y. Acad. Sci. **979:** 94–110.

77. KARKKAINEN, M.J. *et al.* 2001. A model for gene therapy of human hereditary lymphedema. Proc. Natl. Acad. Sci. USA **98:** 12677–12682.

78. SAARISTO, A. *et al.* 2002. Lymphangiogenic gene therapy with minimal blood vascular side effects. J. Exp. Med. **196:** 719–730.

79. SZUBA, A. *et al.* 2002. Therapeutic lymphangiogenesis with VEGF-C. FASEB J. **16:** 1985–1987.

80. YOON, Y.S. *et al.* 2003. VEGF-C gene therapy augments postnatal lymphangiogenesis and ameliorates secondary lymphedema. J. Clin. Invest. **111:** 717–725.

81. SAARISTO, A. *et al.* 2004. Vascular endothelial growth factor-C gene therapy restores lymphatic flow across incision wounds. FASEB J. **18:** 1707–1709.

82. SAARISTO, A. *et al.* 2006. Vascular endothelial growth factor-C accelerates diabetic wound healing. Am. J. Pathol. **169:** 1080–1087.

83. TAMMELA, T. 2007. Therapeutic differentiation and maturation of lymphatic vessels after lymph node dissection and transplantation. Nat. Med. **13:** 1458–1466.

84. RABSON, J.A. *et al.* 1982. Tumor immunity in rat lymph nodes following transplantation. Ann. Surg. **196:** 92–99.

85. MEBIUS, R.E. *et al.* 1991. The influence of afferent lymphatic vessel interruption on vascular addressin expression. J. Cell Biol. **115:** 85–95.

86. SLEEMAN, J.P. 2000. The lymph node as a bridgehead in the metastatic dissemination of tumors. Rec. Results Cancer Res. **157:** 55–81.

87. STACKER, S.A. *et al.* 2002. Lymphangiogenesis and cancer metastasis. Nat. Rev. Cancer **2:** 573–583.

88. PADERA, T.P. 2002. Lymphatic metastasis in the absence of functional intratumor lymphatics. Science **296:** 1883–1886.

89. ALITALO, K. 2004. Lymphangiogenesis and cancer: meeting report. Cancer Res. **64:** 9225–9229.

90. DADRAS, S.S. 2003. Tumor lymphangiogenesis: a novel prognostic indicator for cutaneus melanoma metastasis and survival. Am. J. Pathol. **162:** 1951–1960.

91. ACHEN, M.G. 2005. Focus on lymphangiogenesis in tumor metastasis. Cancer Cell **7:** 121–127.

92. CAO, R. 2004. PDGF-BB induces intratumoral lymphangiogenesis and promotes lymphatic metastasis. Cancer Cell **6:** 333–345.

93. CHANG, L.K. 2004. Dose-dependent response of FGF-2 for lymphangiogenesis. Proc. Natl. Acad. Sci. USA **101:** 11658–11663.

94. KUBO, H. *et al.* 2002. Blockade of vascular endothelial growth factor receptor-3 signaling inhibits fibroblast growth factor-2-induced lymphangiogenesis in mouse cornea. Proc. Natl. Acad. Sci. USA **99:** 8868–8873.

95. NAGY, J.A. *et al.* 2002. Vascular permeability factor/vascular endothelial growth factor induces lymphangiogenesis as well as lngiogenesis. J. Exp. Med. **196:** 1497–1506.

96. CURSIEFEN, C. *et al.* 2004. Inhibition of hemangiogenesis and lymphangiogenesis after normal-risk corneal transplantation by neutralizing VEGF promotes graft survival. Invest. Ophthalmol. Vis. Sci. **45:** 2666–2673.

97. SKOBE, M. *et al.* 2001. Induction of tumor lymphangiogenesis by VEGF-C promotes breast cancer metastasis. Nat. Med. **7:** 192–198.

98. STACKER, S.A. *et al.* 2001. VEGF-D promotes the metastatic spread of tumor cells via the lymphatics. Nat. Med. **7:** 186–191.

99. HIRAKAWA, S. 2005. VEGF-A induces tumor and sentinel lymph node lymphangiogenesis and promotes lymphatic metastasis. J. Exp. Med. **201:** 1089–1099.

100. KARPANEN, T. *et al.* 2001. Vascular endothelial growth factor C promotes tumor lymphangiogenesis and intralymphatic tumor growth. Cancer Res. **61:** 1786–1790.

101. MANDRIOTA, S.J. *et al.* 2001. Vascular endothelial growth factor-C-mediated lymphangiogenesis promotes tumour metastasis. EMBO J. **20:** 672–682.

102. SCHOPPMANN, S.F. *et al.* 2001. Lymphatic microvessel density and lymphovascular invasion assessed by anti-podoplanin immunostaining in human breast cancer. Anticancer Res. **21:** 2351–2355.

103. HARRELL, M.I., B.M. IRITANI & A. RUDDELL. 2007. Tumor-induced sentinel lymph node lymphangiogenesis and increased lymph flow precede melanoma metastasis. Am. J. Pathol. **170:** 774–786.

104. QIAN, C.N. *et al.* 2006. Preparing the "soil": the primary tumor induces vasculature reorganization in the sentinel lymph node before the arrival of metastatic cancer cells. Cancer Res. **66:** 10365–10376.

105. VALTOLA, R. *et al.* 1999. VEGFR-3 and its ligand VEGF-C are associated with angiogenesis in breast cancer. Am. J. Pathol. **154:** 1381–1390.

106. CLARIJS, R. 2002. Induction of vascular endothelial growth factor receptor-3 expression on tumor microvasculature as a new progression marker in human cutaneus melanoma. Cancer Res. **62:** 7059–7065.

107. KUBO, H. *et al.* 2000. Involvement of vascular endothelial growth factor receptor-3 in maintenance of integrity of endothelial cell lining during tumor angiogenesis. Blood **96:** 546–553.

108. LAAKKONEN, P. 2007. Vascular endothelial growth factor receptor 3 is involvent in tumor angiogenesis and growth. Cancer Res. **67:** 593–599.

109. LEAK, L.V. 1976. The structure of lymphatic capillaries in lymph formation. Fed. Proc. **35:** 1863–1871.

110. HE, Y., T. KARPANEN & K. ALITALO. 2004. Role of lymphangiogenic factors in tumor metastasis. Biochim. Biophys. Acta **1654:** 3–12.

111. HE, Y. *et al.* 2002. Suppression of tumor lymphangiogenesis and lymph node metastasis by blocking vascular endothelial growth factor receptor 3 signaling. J. Natl. Cancer Inst. **94:** 819–825.

112. HE, Y. *et al.* 2005. Vascular endothelial cell growth factor receptor 3-mediated activation of lymphatic endothelium is crucial for tumor cell entry and spread via lymphatic vessels. Cancer Res. **65:** 4739–4746.

113. SKOBE, M. *et al.* 2001. Concurrent induction of lymphangiogenesis, angiogenesis, and macrophage recruitment by vascular endothelial growth factor-C in melanoma. Am. J. Pathol. **159:** 893–903.

114. YANAI, Y. *et al.* 2001. Vascular endothelial growth factor C promotes human gastric carcinoma lymph node metastasis in mice. J. Exp. Clin. Cancer Res. **20:** 419–428.

115. KAWAKAMI, M. *et al.* 2005. Vascular endothelial growth factor C promotes lymph node metastasis in a rectal cancer orthotopic model. Surg. Today **35:** 131–138.

116. HE, Y. *et al.* 2005. Vascular endothelial cell growth factor receptor 3-mediated activation of lymphatic endothelium is crucial for tumor cell entry and spread via lymphatic vessels. Cancer Res. **65:** 4739–4746.

117. GUNN, M.D. *et al.* 1998. A chemokine expressed in lymphoid high endothelial venules promotes the adhesion and chemotaxis of naive T lymphocytes. Proc. Natl. Acad. Sci. USA **95:** 258–263.

118. MULLER, A. *et al.* 2001. Involvement of chemokine receptors in breast cancer metastasis. Nature **410:** 50–56.

119. WILEY, H.E. *et al.* 2001. Expression of CC chemokine receptor-7 and regional lymph node metastasis of B16 murine melanoma. J. Natl. Cancer Inst. **93:** 1638–1643.

120. ZLOTNIK, A. 2004. Chemokines in neoplastic progression. Semin. Cancer Biol. **14:** 181–185.

Molecular Control of Lymphatic Metastasis

MARC G. ACHEN AND STEVEN A. STACKER

*Ludwig Institute for Cancer Research, Royal Melbourne Hospital,
Melbourne, Victoria 3050, Australia*

The metastatic spread of tumor cells is the most lethal aspect of cancer and often occurs via the lymphatic vasculature. Both experimental tumor models and human clinicopathologic data indicate that growth of lymphatic vessels (lymphangiogenesis) near solid tumors is often associated with lymph node metastasis. Changes in the adhesive properties of lymphatic endothelium near tumors may also facilitate metastatic spread via the lymphatics. Lymphangiogenic growth factors have been identified that promote formation of tumor lymphatics and metastatic spread of tumor cells to lymph nodes. These include the secreted glycoproteins vascular endothelial growth factor-C (VEGF-C) and VEGF-D, which act via their cognate receptor tyrosine kinase VEGF receptor-3 (VEGFR-3) located on lymphatic endothelial cells. Other signaling molecules that have been reported to promote lymphangiogenesis and/or lymphatic metastasis in cancer include VEGF-A, platelet-derived growth factor-BB, and hepatocyte growth factor. However, the quantitative contribution of these proteins to tumor lymphangiogenesis and lymphatic metastasis in different tumor types requires further investigation. In addition, chemokines are thought to play a role in attracting tumor cells and lymphatic vessels to each other. Moreover, it has recently been shown that lymphangiogenic growth factors secreted from a primary tumor can induce lymphangiogenesis in nearby lymph nodes, even before arrival of tumor cells, which may facilitate further metastasis. This article provides an overview of the molecular mechanisms that control lymphatic metastasis and discusses potential therapeutic approaches for inhibiting this process in human cancer.

Key words: angiogenesis; cancer; lymphangiogenesis; lymphatic vessel; lymph node

Introduction

The spread of tumor cells to lymph nodes is an early and common event in human cancer, and the lymphatic vasculature is therefore considered an important route of metastatic spread. Lymph node metastasis can be an early event in metastatic disease and is useful for the staging of cancer in the clinic.[1] The presence of tumor foci in lymph nodes is considered an adverse prognostic factor in almost all carcinomas, in what otherwise might have been considered localized disease. Further, the detection of tumor cells in lymphatic vessels and regional lymph nodes can lead to treatment of regional lymph nodes by surgery and radiation therapy.[2–4] Importantly, inhibition of lymph node metastasis was associated with reduced distant organ metastasis in rodent models of cancer, suggesting there may be a pathway facilitating distant organ metastasis via the lymph nodes.[5,6]

Lymphatic metastasis was previously thought to be a passive process by which detached tumor cells enter lymph nodes via pre-existing lymphatic vessels in the vicinity of a primary tumor[7]—entry to the lymphatics would be facilitated by the thin walls and incomplete basement membrane of lymphatics[8–10] which would not provide a significant barrier to entry of tumor cells. However, studies in animal models of cancer, employing lymphangiogenic growth factors and histochemical markers that discriminate between blood vessels and lymphatics, indicated that lymphangiogenesis is associated with, and may facilitate, metastasis (for review see Saharinen *et al.*[11] and Stacker *et al.*[12]). Furthermore, clinicopathologic studies indicated that lymphangiogenesis can occur near or in human tumors (for review of methodology for quantitating lymphangiogenesis in human tumors see Van der Auwera *et al.*[13]), and this may correlate with metastasis to lymph nodes in some tumors, such as cutaneous melanoma,[14] inflammatory breast carcinoma,[15] non–small cell lung cancer,[16] muscle-invasive transitional cell carcinoma of the bladder,[17] and head and neck cancer.[18] The location of lymphatic vessels relative to a primary tumor may be a determinant of metastatic spread as studies in animal models have indicated that intratumoral lymphatics are nonfunctional, but peritumoral lymphatics

Address for correspondence: Marc G. Achen and Steven A. Stacker, Ludwig Institute for Cancer Research, Post Office Box 2008, Royal Melbourne Hospital, Victoria 3050, Melbourne, Australia. Voice: +61-3-9341-3155; fax: +61-3-9341-3107.

Marc.achen@ludwig.edu.au

Steven.stacker@ludwig.edu.au

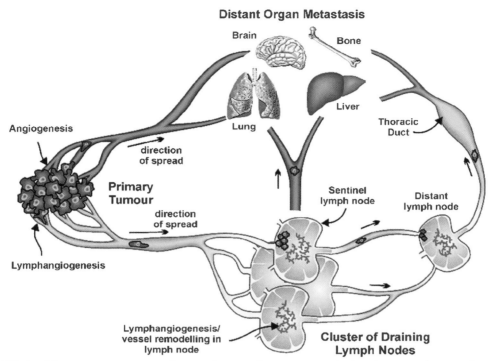

FIGURE 1. Schematic representation of potential routes of metastasis via the lymphatic vasculature (yellow), blood vessels (red) and lymph nodes. Lymphangiogenesis in the vicinity of the primary tumor (denoted by lymphatics around the tumor) and metastatic spread of tumor cells to lymph nodes can be promoted by lymphangiogenic growth factors, such as VEGF-C and VEGF-D, and chemokines may contribute by attracting tumor cells toward lymphatic vessels. These factors may be secreted directly by the tumor cells or by stromal cells, such as macrophages/monocytes. Larger lymphatic vessels may undergo an increase in their diameter, which can also encourage the spread of tumor cells. The presence of tumor cells can also induce lymphangiogenesis in the lymph nodes (denoted by extra lymphatics in the nodes) and potentially other changes to the morphology and function of various vessel types in the lymph node, providing a site for rearrangement of the vessel structures and abnormal connections between the vascular and lymphatic networks. These changes to vessels can occur before (due to the release of soluble factors) or after arrival of tumor cells at the node, and this may facilitate further spread to other distant lymph nodes or major organs, such as the lungs, liver, brain, or bone. Tumor cells could spread from lymph nodes to distant organs via blood vessels associated with the nodes or by entering the venous system via the major lymphatic ducts (e.g., the thoracic duct) and then spreading via blood vessels. Alternatively, tumor cells could spread directly from the primary tumor to distant organs via blood vessels, a process that may be facilitated by tumor angiogenesis stimulated by such growth factors as VEGF-A, VEGF-C, and VEGF-D. (Adapted from Tobler and Detmar.[136])

are capable of draining fluid and cells from a tumor so they may be more important for the metastatic process.[19] Despite the studies of animal tumor models indicating that tumor lymphangiogenesis was associated with metastasis, metastatic spread to lymph nodes in some animal models occurred in the absence of tumor lymphangiogenesis, presumably via pre-existing lymphatic vessels.[20] This suggests that the importance of lymphangiogenesis in metastasis may vary, depending on parameters, such as the tumor type, and the position of the primary tumor relative to the lymphatic network.

An alternative site at which lymphangiogenesis may promote metastatic spread is within the sentinel lymph nodes (FIG. 1). It has been observed in animal tumor models that this can begin even before tumor cells arrive in the nodes, presumably promoted by lymphangiogenic growth factors derived from the primary tumor, and it has been proposed that this process may facilitate further metastatic spread to distant organs.[6,21–23] In a clinical setting, lymphangiogenesis has been detected in lymph node metastases in breast cancer.[24]

The experimental and clinicopathologic studies of tumor lymphangiogenesis suggest that the molecular signaling pathways controlling this process may be targets for therapeutics designed to restrict the metastatic spread of cancer.[25,26] Given that in many cases the primary tumor is surgically excised from the patient, an antimetastatic therapeutic approach could be useful for restricting the spread of tumor cells from remnants of primary tumor that were not successfully removed during surgery. Alternatively, such an approach may be beneficial for inhibiting spread of tumor cells from inoperable primary tumors or from pre-existing cancer metastases in lymph nodes or distant organs.[27,28]

Lymphangiogenic Signaling in the Primary Tumor

The VEGF-C/VEGF-D/VEGFR-3 Signaling Axis

The most extensively studied molecular system that signals for tumor lymphangiogenesis and is associated with lymphatic spread from primary cancers is the VEGF-C/VEGF-D/VEGFR-3 signaling axis. The secreted glycoproteins VEGF-C or VEGF-D activate VEGFR-3,[29,30] a cell surface receptor tyrosine kinase on lymphatic endothelium,[31] leading to growth of lymphatic vessels.[32–36] Activation of VEGFR-3 is sufficient to protect cultured lymphatic endothelial cells from apoptosis and to induce proliferation and migration of these cells—some of these signals are transmitted via activation of the p42/p44 MAPK signaling cascade in a protein kinase C–dependent fashion and via induction of Akt phosphorylation, signaling systems that are associated with cell growth and survival.[37] Further aspects of VEGFR-3 signaling may be complicated by the fact that VEGFR-3 can form heterodimers with the structurally related receptor VEGFR-2,[38] which can also be localized on lymphatic endothelial cells.[39]

Expression of VEGF-C or VEGF-D by tumor cells in xenograft or transgenic animal models of cancer led to an increase in the abundance of lymphatic vessels at the periphery of and sometimes inside the primary tumor, promoted spread of tumor cells to lymph nodes and, in some models, facilitated distant organ metastasis.[34,40–46] Inhibition of this signaling pathway, using a soluble form of VEGFR-3 to sequester VEGF-C and VEGF-D, reduced both tumor lymphangiogenesis and lymphatic metastasis in some animal models of cancer.[47–49] Similar results were observed with a neutralizing VEGF-D monoclonal antibody (mAb)[34] or a neutralizing VEGFR-3 mAb.[50,51]

Despite the abundance of data from animal tumor models showing that the VEGF-C/VEGF-D/VEGFR-3 axis promotes tumor lymphangiogenesis and metastatic spread, the question arises as to whether these proteins exert similar effects in human cancer. Importantly, the expression of VEGF-C, VEGF-D, and VEGFR-3 correlates with lymphatic

metastasis in some prevalent forms of human cancer, a topic which has been reviewed on multiple occasions in the past.[12,52,53] For example, VEGF-C levels in primary tumors correlate with lymph node metastasis in prostate,[54] gastric,[55] thyroid,[56] colorectal,[57] esophageal,[58] and lung carcinomas.[59] In addition, VEGF-C was reported to be a potential prognostic factor in cervical and ovarian carcinomas.[60,61] VEGF-D was reported to be an independent prognostic marker for disease-free and overall survival in colorectal carcinoma that correlates with lymphatic involvement.[62] Furthermore, it has been reported that the presence of both VEGF-D and VEGFR-3 in endometrial carcinoma may predict myometrial invasion and lymph node metastasis and may prospectively identify patients at increased risk for poor outcome,[63] and that VEGF-D and VEGFR-3 are independent prognostic markers in gastric adenocarcinoma.[64] In addition, VEGF-D was observed to be expressed in glioblastoma multiforme[65] and was found to be an independent predictor of poor outcome in epithelial ovarian carcinoma,[66] and expression of VEGFR-3 by lymphatic endothelial cells was reported to be associated with lymph node metastasis in prostate cancer.[67] Although not all studies of VEGF-C or VEGF-D expression in human cancer support a role for these molecules in tumor lymphangiogenesis and metastatic spread (see, for example, the report from Gisterek and colleagues[68]), it is fair to conclude that the majority of the many studies conducted on this topic support this hypothesis.

Of importance, VEGF-C and VEGF-D can be proteolytically processed[69–71] by enzymes, such as plasmin[72] and proprotein convertases[73,74]—this process increases the affinity of these growth factors for both VEGFR-3 and VEGFR-2. Processing of these ligands appears to be particularly important for the capacity to activate VEGFR-2,[69,70,74] a receptor that is also activated by VEGF-A and signals for tumor angiogenesis.[75,76] VEGF-C and VEGF-D can promote angiogenesis in different biological settings,[77–82] including cancer,[34,42,46] and the proteolytically processed forms of these growth factors may contribute to signaling for tumor angiogenesis by binding VEGFR-2.[83] Clinical studies demonstrated that a neutralizing mAb to VEGF-A (known as Avastin[TM] or bevacizumab), which prevents binding to VEGFR-2, is beneficial for patients with metastatic colorectal cancer (for review see Ferrara *et al.*[84]). It is possible that human tumors could "escape" therapies targeting VEGF-A by employing other angiogenic signaling molecules,[85] such as the alternative VEGFR-2 ligands VEGF-C or VEGF-D. Moreover, VEGFR-3 can be expressed on blood vessels in cancer[86,87] and may contribute to tumor angiogenesis and solid tumor growth: This hypothesis is supported by the finding that treatment of tumor xenograft models with neutralizing VEGFR-3 mAbs restricted both angiogenesis and tumor growth.[88] Therefore, therapeutics that target VEGF-C and VEGF-D may augment the anti-angiogenic effects of targeting VEGF-A, and thereby further restrict solid tumor growth, in addition to restricting tumor lymphangiogenesis and metastasis via the lymphatic vasculature.

VEGF-A/VEGFR-2

VEGF-A is an angiogenic protein that can be expressed in hypoxic regions of tumors[89,90] and that activates VEGFR-1 and VEGFR-2 on the endothelial cells lining blood vessels (for review, see Ferrara *et al.*[91]). VEGFR-2 can also be expressed on the endothelium of lymphatic vessels,[39] suggesting that its activation by VEGF-A could contribute to lymphangiogenic signaling. Expression of VEGF-A in a chemically induced skin carcinogenesis model[21] and in a tumor xenograft model utilizing fibrosarcoma cells[92] led to enhanced lymphangiogenesis and lymph node metastasis. Furthermore, treatment of an orthotopic breast carcinoma model with a neutralizing VEGF-A antibody reduced both lymphatic vessel density and lymph node metastasis.[93] However, VEGF-A did not promote lymphangiogenesis or lymph node metastasis in some animal tumor models[34,94] and the variables determining whether or not it exerts these effects in cancer are not well understood. These variables could include (*a*) the recruitment by VEGF-A of monocytes or other cell types expressing VEGF-C and/or VEGF-D to the tumor; (*b*) the isoform(s) of VEGF-A being expressed by tumor cells or tumor stroma; and (*c*) the absolute levels of VEGFR-2 on the endothelium of lymphatics near the tumor.

Other Secreted Growth Factors

A range of other growth factors has been reported to exhibit lymphangiogenic activity and promote lymph node metastasis in cancer. For example, expression of platelet-derived growth factor-BB in murine fibrosarcoma cells induced tumor lymphangiogenesis in syngeneic mice, causing formation of intratumoral lymphatics that led to enhanced metastasis to lymph nodes.[95] Hepatocyte growth factor (HGF) promoted lymphangiogenesis *in vivo* when overexpressed in transgenic mice or delivered subcutaneously[96] and appeared to promote peritumoral lymphangiogenesis in a transgenic model of mammary tumorigenesis.[97] However, the role of HGF in lymph node metastasis is yet to be established, although a correlation has been reported between expression or mutation of its receptor,

known as HGF-R or MET/c-met, and the metastatic spread of cancer.[98] The angiopoietins (Angs) are ligands for the endothelial cell surface receptor Tie2 (for review, see Thurston[99]): Ang-1 activates Tie2 and positively regulates remodeling of blood vessels,[100] whereas Ang-2 promotes blood vessel regression by blocking Tie2 activation.[101] Loss of Ang-2 in mice causes defects in patterning and function of the lymphatic vasculature that can be rescued by Ang-1, suggesting a role for these proteins in lymphangiogenesis,[102] which has been supported by studies in other animal models.[103,104] The VEGF-C/VEGF-D/VEGFR-3 system is thought to be involved in Ang-1-mediated lymphangiogenic signaling.[103] However, there have not yet been any reports showing that Ang-1 or Ang-2 promote tumor lymphangiogenesis or lymphatic metastasis in animal models. The insulin-like growth factors (IGFs), IGF-1 and IGF-2, exhibited lymphangiogenic activity in mouse cornea,[105] are expressed in many human tumors and have been associated with malignant progression and poor prognosis.[106] Of interest, IGF-1 receptor promoted expression of VEGF-C and lymph node metastasis in a Lewis lung carcinoma model, suggesting that it could act as a positive regulator of VEGF-C in cancer and thereby promote lymphatic metastasis.[107] However, analyses of IGF effects on tumor lymphatics and lymph node metastasis in animal models have not been reported.

It will be important to assess the role of these growth factors in lymphatic metastasis by (*a*) specifically targeting them, or their receptors, in a range of animal tumor models that exhibit metastatic spread via the lymphatics; (*b*) determining whether their effects in cancer are mediated directly or indirectly via the VEGF-C/VEGF-D/VEGFR-3 signaling axis; and (*c*) establishing whether their expression in a range of prevalent forms of human cancer correlates with lymph node and/or distant metastasis.

Lymph Node Lymphangiogenesis in Cancer

It has been observed that lymphangiogenesis in cancer is not restricted to regions within or immediately adjacent to a primary tumor, but can also occur in the sentinel lymph nodes (FIG. 1). For example, lymphangiogenesis was reported in lymph nodes that contained tumor cells expressing VEGF-D in the setting of a mouse xenograft model,[34] and Hirakawa and colleagues employed a chemically induced skin carcinogenesis model overexpressing VEGF-A to demonstrate that VEGF-A promoted lymphangiogenesis in metastasis-containing lymph nodes.[21] Intriguingly, the primary tumors expressing VEGF-A in this model induced sentinel lymph node lymphangiogenesis even before metastasizing tumor cells had arrived at these nodes. In fact, it has been known for many years that regional lymph nodes draining tumors can be enlarged without evidence of metastasis, a condition known as tumor-reactive lymphadenopathy (for review, see Ioachim[108]). VEGF-C also promoted expansion of the lymphatic networks in sentinel lymph nodes prior to the onset of metastasis in a chemically induced skin carcinogenesis model.[6] Similarly, a study of B16 melanoma cells implanted in syngeneic mice demonstrated that tumor-draining lymph nodes featured greatly increased lymphatic sinuses and that lymphangiogenesis began in these nodes before the arrival of melanoma cells, indicating that primary tumors induce these alterations at a distance.[22] Further, lymph flow was greatly increased through tumor-draining relative to non-draining lymph nodes. This study indicated that lymph node lymphangiogenesis can be dependent on the accumulation of B lymphocytes in the draining lymph nodes. In a another study utilizing animal models it was observed that before the establishment of metastasis in the sentinel lymph node, there is reorganization of the lymphatic channels, such that the lymph node became a "lymph vessel/sinus enriched organ."[23] Lymphangiogenesis has also been identified in the lymph nodes of mice developing lymphomas.[109] Human lymph nodes infiltrated with metastatic melanoma or breast cancer cells also exhibited lymphatic vessel growth, indicating that lymph node lymphangiogenesis can be a feature of human cancer.[24,110]

It has been proposed that the lymphangiogenesis in lymph nodes containing tumor cells might facilitate further metastatic spread throughout the lymphatic system. Interestingly, in the chemically induced skin carcinogenesis model, mice with metastasis-containing sentinel lymph nodes that expressed VEGF-C were more likely to have metastasis to additional organs including distal lymph nodes and lungs, and no metastases were observed in distant organs in the absence of lymph node metastases.[6] These findings suggest an important role for both lymph node lymphangiogenesis and tumor lymphangiogenic growth factors in cancer metastasis beyond the sentinel lymph nodes. Further, the occurrence of lymphangiogenesis in sentinel lymph nodes prior to arrival of tumor cells indicates that signals derived from the primary tumor are transported to the draining lymph nodes, where they induce localized lymphatic vessel growth.[21] This is a new variant of the "seed and soil" hypothesis by which organ-specific characteristics are responsible for the preferential patterns of metastasis exhibited by distinct tumor types.[111] In this scenario, primary tumors might modify the "soil" in the lymph node to make it better suited for supporting further metastatic spread. The analysis of more murine and human cancers is required to establish whether lymphangiogenesis in tumor-draining lymph nodes and increased lymph flow are features of cancer in general, and if they might allow identification of those lesions more likely to metastasize. The assessment of the propensity for metastasis could be based on monitoring the abundance of lymphatics in or adjacent to a primary tumor or sentinel lymph node, or analyzing expression of lymphangiogenic growth factors and their cognate receptors at these sites.

Chemokines and Lymphatic Metastasis

Chemokines are a family of more than 40 small chemoattractant cytokines that bind to G protein–coupled receptors expressed on target cells, allowing these cells to follow chemokine concentration gradients into selected tissues.[112,113] Chemokines and their receptors are essential for leukocyte trafficking and have also been implicated in cancer metastasis to specific organs. For example, transduction of the B16 murine melanoma cell line with a retroviral vector encoding the chemokine receptor CCR7 resulted in greatly enhanced spread of tumor cells to draining lymph nodes which could be blocked by neutralizing antibodies to CCL21, a chemokine that binds CCR7 and is constitutively produced by lymphatic endothelial cells in the skin.[114] This study suggests that chemokines can attract tumor cells to lymphatic vessels in the vicinity of a tumor to promote lymph node metastasis, which has also been observed in another animal model of melanoma.[115] Furthermore, overexpression of the chemokine receptor CXCR4 in oral squamous cell carcinoma promoted lymph node metastasis in an orthotopic mouse model,[116] and neutralizing the interaction of the chemokine CXCL12 with CXCR4 expressed on tumor cells significantly impaired metastasis of breast cancer cells to regional lymph nodes and lung in a mouse model of tumor development.[117] A role for the chemokine receptor CXCR3 in metastasis of colon cancer and melanoma cells to lymph nodes was also demonstrated in animal models.[118,119]

The involvement of chemokines in lymphatic metastasis is also supported by clinicopathologic data: For example, it has been reported that expression of CCR7 is associated with lymph node metastasis of gastric carcinoma,[120] colorectal carcinoma[121] and breast cancer,[122] and that expression of CXCR4 significantly correlated with lymphatic metastasis as well as distant dissemination in hepatocellular carcinoma[123] and is associated with lymph node metastasis in oral squamous cell carcinoma[124] and with lymph node status in breast cancer.[125]

In summary, studies over the past 10 years suggest that while production of lymphangiogenic growth factors, such as VEGF-C or VEGF-D, by tumor cells or the tumor stroma attract growth of lymphatic vessels expressing VEGFR-3 toward the primary tumor, chemokines secreted by lymphatic vessels or lymph nodes may attract the movement of tumor cells expressing cognate chemokine receptors toward these structures.[115] These signal transduction systems could thereby constitute a two-way signaling mechanism, promoting lymph node and possibly distant organ metastases, which offers attractive therapeutic targets for novel drugs designed to restrict the metastatic spread of cancer.

Therapeutic Opportunities

The VEGF-C/VEGF-D/VEGFR-3 axis is the best validated signaling system for promoting lymphangiogenesis associated with solid tumors and the metastatic spread of tumor cells to lymph nodes. A role for VEGF-A is also supported by studies in multiple animal models of cancer. Given that these growth factors can promote lymphangiogenesis in the setting of solid tumors, they are also likely candidates for driving lymph node lymphangiogenesis that appears to promote metastatic spread to sentinel lymph nodes and perhaps to more distant sites. In fact, VEGF-C, VEGF-D, and VEGF-A have already been shown to promote lymph node lymphangiogenesis, at least in animal tumor models overexpressing these growth factors.[6,21,34] Hence it would be attractive to target VEGF-C, VEGF-D, and VEGF-A signaling in order to block lymphangiogenesis, both in the vicinity of the primary tumor and in draining lymph nodes, and thereby restrict the metastatic spread of cancer. This may have the attraction of restricting metastasis via the blood vessels by decreased tumor angiogenesis, as well as restricting metastasis via the lymphatic vasculature. VEGF-A can be specifically targeted with Avastin, which could be combined with reagents targeting VEGF-C and VEGF-D, such as a soluble version of VEGFR-3,[5,47–49] or neutralizing mAbs to these growth factors.[34,126] Alternatively, Avastin could be combined with a neutralizing mAb to VEGFR-3,[88,127] although this would

not block the effects of VEGF-C and VEGF-D mediated by VEGFR-2. (It would be attractive to block the signaling via VEGFR-2 to restrict tumor angiogenesis and solid tumor growth.) Another approach would be to make use of small-molecule tyrosine kinase inhibitors that block the enzymatic activity of both VEGFR-2 and VEGFR-3, such as sorafenib[128] or sunitinib (also known as SU-11248)[129,130] and thus inhibit the contribution of both receptors to tumor angiogenesis and lymphangiogenesis. However, such small-molecule inhibitors usually inhibit a range of kinases in the cell so side-effects, particularly when these agents are combined with chemotherapy, must be carefully examined.[131] A range of studies utilizing a soluble form of VEGFR-3,[132] a neutralizing VEGFR-3 mAb,[127,132] or mice in which the *Vegf-d* gene has been inactivated[133] suggest that specifically targeting the VEGF-C/VEGF-D/VEGFR-3 signaling axis in cancer patients is unlikely to be problematic in terms of perturbing the function of the adult lymphatic system.

Similar strategies could be employed to target the chemokine signaling systems in cancer. Neutralizing antibodies to specific chemokines or their cognate receptors have been employed in animal models to explore the significance of these molecules in metastasis,[114,115,117] and may also be useful in a clinical setting. Further, small-molecule orally active inhibitors of chemokine receptors could also be employed—such an inhibitor for CCR1, designated MLN3897, has been characterized and is already under evaluation in Phase II clinical trials for rheumatoid arthritis and multiple sclerosis.[134] Further, the bicyclam AMD3100 is an inhibitor of CXCR4 which inhibited growth of tumors in an animal model.[135]

Concluding Remarks

Our knowledge of the molecular mechanisms controlling lymphatic metastasis has advanced greatly over the past 10 years, and this progress has provided opportunities for therapeutic interventions designed to restrict the metastatic spread of cancer. The VEGF-C/VEGF-D/VEGFR-3 lymphangiogenic signaling axis is an attractive target for this purpose, but an approach whereby signaling of VEGF-C and VEGF-D via VEGFR-2 is also blocked might provide the extra benefit of restricting tumor angiogenesis and solid tumor growth. The chemokine system is another potential target for restricting cancer metastasis and could conceivably be targeted simultaneously with the lymphangiogenic signaling systems to achieve the most complete inhibition of lymphatic metastasis. A range of potential therapeutic agents has been generated to each of these signaling systems, some of which are already approved for use in the clinic, whereas others await clinical trials to establish their utility as cancer therapeutics. Testing these novel agents in carefully designed clinical trials to assess effects on metastatic disease is now a high priority.

Acknowledgments

M.G.A. and S.A.S. are supported by Senior Research Fellowships from the National Health and Medical Research Council of Australia (NHMRC) and the Pfizer Foundation, respectively, and by a Program Grant from the NHMRC. We thank Tony Burgess for helpful comments.

Conflicts of Interest

M.G.A. and S.A.S. are consultants to Vegenics Ltd., and S.A.S. holds a Pfizer Australia Fellowship.

References

1. TUTTLE, T.M. 2004. Technical advances in sentinel lymph node biopsy for breast cancer. Am. Surg. **70:** 407–413.
2. DUKES, C.E. 1932. The classification of cancer of the rectum. J. Pathol. **35:** 323–332.
3. FISHER, B. *et al.* 1983. Relation of number of positive axillary nodes to the prognosis of patients with primary breast cancer: an NSABP update. Cancer **52:** 1551–1557.
4. TAIPALE, J. *et al.* 1999. Vascular endothelial growth factor receptor-3. Curr. Top. Microbiol. Immunol. **237:** 85–96.
5. KRISHNAN, J. *et al.* 2003. Differential in vivo and in vitro expression of vascular endothelial growth factor (VEGF)-C and VEGF-D in tumors and its relationship to lymphatic metastasis in immunocompetent rats. Cancer Res. **63:** 713–722.
6. HIRAKAWA, S. *et al.* 2007. VEGF-C-induced lymphangiogenesis in sentinel lymph nodes promotes tumor metastasis to distant sites. Blood **109:** 1010–1017.
7. PEPPER, M.S. 2001. Lymphangiogenesis and tumor metastasis: myth or reality? Clin. Cancer Res. **7:** 462–468.
8. LEAK, L.V. 1970. Electron microscopic observations on lymphatic capillaries and the structural components of the connective tissue-lymph interface. Microvasc. Res. **2:** 361–391.
9. CASLEY-SMITH, J.R. & H.W. FLOREY. 1961. The structure of normal small lymphatics. Quart. J. Expt. Physiol. **46:** 101–116.
10. SWARTZ, M.A. & M. SKOBE. 2001. Lymphatic function, lymphangiogenesis, and cancer metastasis. Microsc. Res. Tech. **55:** 92–99.
11. SAHARINEN, P. *et al.* 2004. Lymphatic vasculature: development, molecular regulation and role in tumor metastasis and inflammation. Trends Immunol. **25:** 387–395.
12. STACKER, S.A. *et al.* 2002. Lymphangiogenesis and cancer metastasis. Nat. Rev. Cancer **2:** 573–583.
13. VAN DER AUWERA, I. *et al.* 2006. First international consensus on the methodology of lymphangiogenesis quantification in solid human tumours. Br. J. Cancer **95:** 1611–1625.
14. DADRAS, S.S. *et al.* 2003. Tumor lymphangiogenesis: a novel prognostic indicator for cutaneous melanoma metastasis and survival. Am. J. Pathol. **162:** 1951–1960.
15. VAN DER AUWERA, L. *et al.* 2005. Tumor lymphangiogenesis in inflammatory breast carcinoma: a histomorphometric study. Clin Cancer Res. **11:** 7637–7642.
16. RENYI-VAMOS, F. *et al.* 2005. Lymphangiogenesis correlates with lymph node metastasis, prognosis, and angiogenic phenotype in human non-small cell lung cancer. Clin. Cancer Res. **11:** 7344–7353.
17. FERNANDEZ, M.I. *et al.* 2007. Prognostic implications of lymphangiogenesis in muscle-invasive transitional cell carcinoma of the bladder. Eur. Urol. **53:**571–580.
18. BEASLEY, N.J. P. *et al.* 2002. Intratumoral lymphangiogenesis and lymph node metastasis in head and neck cancer. Cancer Res. **62:** 1315–1320.
19. PADERA, T.P. *et al.* 2002. Lymphatic metastasis in the absence of functional intratumor lymphatics. Science **296:** 1883–1886.
20. WONG, S.Y. *et al.* 2005. Tumor-secreted vascular endothelial growth factor-C is necessary for prostate cancer lymphangiogenesis, but lymphangiogenesis is unnecessary for lymph node metastasis. Cancer Res. **65:** 9789–9798.
21. HIRAKAWA, S. *et al.* 2005. VEGF-A induces tumor and sentinel lymph node lymphangiogenesis and promotes lymphatic metastasis. J. Exp. Med. **201:** 1089–1099.
22. HARRELL, M.I., B.M. IRITANI & A. RUDDELL. 2007. Tumor-induced sentinel lymph node lymphangiogenesis and increased lymph flow precede melanoma metastasis. Am. J. Pathol. **170:** 774–786.
23. QIAN, C.N. *et al.* 2006. Preparing the "soil": the primary tumor induces vasculature reorganization in the sentinel lymph node before the arrival of metastatic cancer cells. Cancer Res. **66:** 10365–10376.
24. VAN DEN EYNDEN, G.G. *et al.* 2006. Induction of lymphangiogenesis in and around axillary lymph node metastases of patients with breast cancer. Br. J. Cancer **95:** 1362–1366.
25. JAIN, R.K. & T.P. PADERA. 2002. Prevention and treatment of lymphatic metastasis by antilymphangiogenic therapy. J. Natl. Cancer Inst. **94:** 785–787.
26. STACKER, S.A., R.A. HUGHES & M.G. ACHEN. 2004. Molecular targeting of lymphatics for therapy. Curr. Pharm. Des. **10:** 65–74.
27. THIELE, W. & J.P. SLEEMAN. 2006. Tumor-induced lymphangiogenesis: a target for cancer therapy? J. Biotechnol. **124:** 224–241.
28. ACHEN, M.G., G.B. MANN & S.A. STACKER. 2006. Targeting lymphangiogenesis to prevent tumour metastasis. Br. J. Cancer **94:** 1355–1360.
29. JOUKOV, V. *et al.* 1996. A novel vascular endothelial growth factor, VEGF-C, is a ligand for the Flt-4 (VEGFR-3) and KDR (VEGFR-2) receptor tyrosine kinases. EMBO J. **15:** 290–298.
30. ACHEN, M.G. *et al.* 1998. Vascular endothelial growth factor D (VEGF-D) is a ligand for the tyrosine kinases VEGF

receptor 2 (Flk-1) and VEGF receptor 3 (Flt-4). Proc. Natl. Acad. Sci. USA **95:** 548–553.

31. KAIPAINEN, A. *et al.* 1995. Expression of the fms-like tyrosine kinase 4 gene becomes restricted to lymphatic endothelium during development. Proc. Natl. Acad. Sci. USA **92:** 3566–3570.

32. JELTSCH, M. *et al.* 1997. Hyperplasia of lymphatic vessels in VEGF-C transgenic mice. Science **276:** 1423–1425.

33. OH, S.J. *et al.* 1997. VEGF and VEGF-C: specific induction of angiogenesis and lymphangiogenesis in the differentiated avian chorioallantoic membrane. Dev. Biol. **188:** 96–109.

34. STACKER, S.A. *et al.* 2001. VEGF-D promotes the metastatic spread of tumor cells via the lymphatics. Nat. Med. **7:** 186–191.

35. VEIKKOLA, T. *et al.* 2001. Signalling via vascular endothelial growth factor receptor-3 is sufficient for lymphangiogenesis in transgenic mice. EMBO J. **20:** 1223–1231.

36. BALDWIN, M.E., S.A. STACKER & M.G. ACHEN. 2002. Molecular control of lymphangiogenesis. Bioessays **24:** 1030–1040.

37. MAKINEN, T. *et al.* 2001. Isolated lymphatic endothelial cells transduce growth, survival and migratory signals via the VEGF-C/D receptor VEGFR-3. EMBO J. **20:** 4762–4773.

38. DIXELIUS, J. *et al.* 2003. Ligand-induced vascular endothelial growth factor receptor-3 (VEGFR-3) heterodimerization with VEGFR-2 in primary lymphatic endothelial cells regulates tyrosine phosphorylation sites. J. Biol. Chem. **278:** 40973–40979.

39. SAARISTO, A. *et al.* 2002. Adenoviral VEGF-C overexpression induces blood vessel enlargement, tortuosity, and leakiness but no sprouting angiogenesis in the skin or mucous membranes. FASEB J. **16:** 1041–1049.

40. SKOBE, M. *et al.* 2001. Induction of tumor lymphangiogenesis by VEGF-C promotes breast cancer metastasis. Nat. Med. **7:** 192–198.

41. MANDRIOTA, S.J. *et al.* 2001. Vascular endothelial growth factor-C-mediated lymphangiogenesis promotes tumour metastasis. EMBO J. **20:** 672–682.

42. SKOBE, M. *et al.* 2001. Concurrent induction of lymphangiogenesis, angiogenesis, and macrophage recruitment by vascular endothelial growth factor-C in melanoma. Am. J. Pathol. **159:** 893–903.

43. KARPANEN, T. *et al.* 2001. Vascular endothelial growth factor C promotes tumor lymphangiogenesis and intralymphatic tumor growth. Cancer Res. **61:** 1786–1790.

44. ACHEN, M.G., B.K. McCOLL & S.A. STACKER. 2005. Focus on lymphangiogenesis in tumor metastasis. Cancer Cell **7:** 121–127.

45. KOPFSTEIN, L. *et al.* 2007. Distinct roles of vascular endothelial growth factor-d in lymphangiogenesis and metastasis. Am. J. Pathol. **170:** 1348–1361.

46. VON MARSCHALL, Z. *et al.* 2005. Vascular endothelial growth factor-D induces lymphangiogenesis and lymphatic metastasis in models of ductal pancreatic cancer. Int. J. Oncol. **27:** 669–679.

47. HE, Y. *et al.* 2002. Suppression of tumor lymphangiogenesis and lymph node metastasis by blocking vascular endothelial growth factor receptor 3 signaling. J. Natl. Cancer Inst. **94:** 819–825.

48. HE, Y. *et al.* 2005. Vascular endothelial cell growth factor receptor 3-mediated activation of lymphatic endothelium is crucial for tumor cell entry and spread via lymphatic vessels. Cancer Res. **65:** 4739–4746.

49. LIN, J. *et al.* 2005. Inhibition of lymphogenous metastasis using adeno-associated virus-mediated gene transfer of a soluble VEGFR-3 decoy receptor. Cancer Res. **65:** 6901–6909.

50. HOSHIDA, T. *et al.* 2006. Imaging steps of lymphatic metastasis reveals that vascular endothelial growth factor-C increases metastasis by increasing delivery of cancer cells to lymph nodes: therapeutic implications. Cancer Res. **66:** 8065–8075.

51. ROBERTS, N. *et al.* 2006. Inhibition of VEGFR-3 activation with the antagonistic antibody more potently suppresses lymph node and distant metastases than inactivation of VEGFR-2. Cancer Res. **66:** 2650–2657.

52. STACKER, S.A., M.E. BALDWIN & M.G. ACHEN. 2002. The role of tumor lymphangiogenesis in metastatic spread. FASEB J. **16:** 922–934.

53. HE, Y., T. KARPANEN & K. ALITALO. 2004. Role of lymphangiogenic factors in tumor metastasis. Biochim.Biophys. Acta **1654:** 3–12.

54. TSURUSAKI, T. *et al.* 1999. Vascular endothelial growth factor-C expression in human prostatic carcinoma and its relationship to lymph node metastasis. Br. J. Cancer **80:** 309–313.

55. YONEMURA, Y. *et al.* 1999. Role of vascular endothelial growth factor C expression in the development of lymph node metastasis in gastric cancer. Clin. Cancer Res. **5:** 1823–1829.

56. BUNONE, G. *et al.* 1999. Expression of angiogenesis stimulators and inhibitors in human thyroid tumors and correlation with clinical pathological features. Am. J. Pathol. **155:** 1967–1976.

57. AKAGI, K. *et al.* 2000. Vascular endothelial growth factor-C (VEGF-C) expression in human colorectal cancer tissues. Br. J. Cancer **83:** 887–891.

58. KITADAI, Y. *et al.* 2001. Clinicopathological significance of vascular endothelial growth factor (VEGF)-C in human esophageal squamous cell carcinomas. Int. J. Cancer **93:** 662–666.

59. KAJITA, T. *et al.* 2001. The expression of vascular endothelial growth factor C and its receptors in non-small cell lung cancer. Br. J. Cancer **85:** 255–260.

60. UEDA, M. *et al.* 2002. Correlation between vascular endothelial growth factor-C expression and invasion phenotype in cervical carcinomas. Int. J. Cancer **98:** 335–343.

61. NISHIDA, N. *et al.* 2004. Vascular endothelial growth factor C and vascular endothelial growth factor receptor 2 are related closely to the prognosis of patients with ovarian carcinoma. Cancer **101:** 1364–1374.

62. WHITE, J.D. *et al.* 2002. Vascular endothelial growth factor-D expression is an independent prognostic marker for survival in colorectal carcinoma. Cancer Res. **62:** 1669–1675.

63. YOKOYAMA, Y. *et al.* 2003. Expression of vascular endothelial growth factor (VEGF)-D and its receptor, VEGF receptor 3, as a prognostic factor in endometrial carcinoma. Clin. Cancer Res. **9:** 1361–1369.

64. JUTTNER, S. *et al.* 2006. Vascular endothelial growth factor-D and its receptor VEGFR-3: two novel independent prognostic markers in gastric adenocarcinoma. J. Clin. Oncol. **24:** 228–240.

65. DEBINSKI, W. *et al.* 2001. VEGF-D is an X-linked/AP-1 regulated putative onco-angiogen in human glioblastoma multiforme. Mol. Med. **7:** 598–608.

66. YOKOYAMA, Y. *et al.* 2003. Vascular endothelial growth factor-D is an independent prognostic factor in epithelial ovarian carcinoma. Br. J. Cancer **88:** 237–244.

67. ZENG, Y. *et al.* 2004. Expression of vascular endothelial growth factor receptor-3 by lymphatic endothelial cells is associated with lymph node metastasis in prostate cancer. Clin. Cancer Res. **10:** 5137–5144.

68. GISTEREK, I. *et al.* 2007. Evaluation of prognostic value of VEGF-C and VEGF-D in breast cancer—10 years follow-up analysis. AntiCancer Res. **27:** 2797–802.

69. JOUKOV, V. *et al.* 1997. Proteolytic processing regulates receptor specificity and activity of VEGF-C. EMBO J. **16:** 3898–3911.

70. STACKER, S.A. *et al.* 1999. Biosynthesis of vascular endothelial growth factor-D involves proteolytic processing which generates non-covalent homodimers. J. Biol. Chem. **274:** 32127–32136.

71. BALDWIN, M.E. *et al.* 2001. Multiple forms of mouse vascular endothelial growth factor-D are generated by RNA splicing and proteolysis. J. Biol. Chem. **276:** 44307–44314.

72. MCCOLL, B.K. *et al.* 2003. Plasmin activates the lymphangiogenic growth factors VEGF-C and VEGF-D. J. Exp. Med. **198:** 863–868.

73. SIEGFRIED, G. *et al.* 2003. The secretory proprotein convertases furin, PC5, and PC7 activate VEGF-C to induce tumorigenesis. J. Clin. Invest. **111:** 1723–1732

74. MCCOLL, B.K. *et al.* 2007. Proprotein convertases promote processing of VEGF-D, a critical step for binding the angiogenic receptor VEGFR-2. FASEB J. **21:** 1088–1098.

75. PLATE, K.H. *et al.* 1992. Vascular endothelial growth factor is a potential tumour angiogenesis factor in human gliomas in vivo. Nature **359:** 845–848.

76. MILLAUER, B. *et al.* 1993. High affinity VEGF binding and developmental expression suggest Flk-1 as a major regulator of vasculogenesis and angiogenesis. Cell **72:** 835–846.

77. WITZENBICHLER, B. *et al.* 1998. Vascular endothelial growth factor-C (VEGF-C/VEGF-2) promotes angiogenesis in the setting of tissue ischemia. Am. J. Pathol. **153:** 381–394.

78. CAO, Y. *et al.* 1998. Vascular endothelial growth factor C induces angiogenesis in vivo. Proc. Natl. Acad. Sci. USA **95:** 14389–14394.

79. BYZOVA, T.V. *et al.* 2002. Adenovirus encoding vascular endothelial growth factor-D induces tissue-specific vascular patterns in vivo. Blood **99:** 4434–4442.

80. RISSANEN, T.T. *et al.* 2003. VEGF-D is the strongest angiogenic and lymphangiogenic effector among VEGFs delivered into skeletal muscle via adenoviruses. Circ. Res. **92:** 1098–1106.

81. RUTANEN, J. *et al.* 2004. Adenoviral catheter-mediated intramyocardial gene transfer using the mature form of vascular endothelial growth factor-D induces transmural angiogenesis in porcine heart. Circulation **109:** 1029–1035.

82. BHARDWAJ, S. *et al.* 2003. Angiogenic responses of vascular endothelial growth factors in periadventitial tissue. Hum. Gene Ther. **14:** 1451–1562.

83. ACHEN, M.G. *et al.* 2002. The angiogenic and lymphangiogenic factor vascular endothelial growth factor-D exhibits a paracrine mode of action in cancer. Growth Factors **20:** 99–107.

84. FERRARA, N. *et al.* 2004. Discovery and development of bevacizumab, an anti-VEGF antibody for treating cancer. Nat. Rev. Drug Discov. **3:** 391–400.

85. SHOJAEI, F. *et al.* 2007. Tumor refractoriness to anti-VEGF treatment is mediated by CD11b(+)Gr1(+) myeloid cells. Nat. Biotechnol. **25:** 911–920.

86. VALTOLA, R. *et al.* 1999. VEGFR-3 and its ligand VEGF-C are associated with angiogenesis in breast cancer. Am. J. Pathol. **154:** 1381–1390.

87. KUBO, H. *et al.* 2000. Involvement of vascular endothelial growth factor receptor-3 in maintenance of integrity of endothelial cell lining during tumor angiogenesis. Blood **96:** 546–553.

88. LAAKKONEN, P. *et al.* 2007. Vascular endothelial growth factor receptor 3 is involved in tumor angiogenesis and growth. Cancer Res. **67:** 593–599.

89. PLATE, K.H. *et al.* 1993. Up-regulation of vascular endothelial growth factor and its cognate receptors in a rat glioma model of tumor angiogenesis. Cancer Res. **53:** 5822–5827.

90. IKEDA, E. *et al.* 1995. Hypoxia-induced transcriptional activation and increased mRNA stability of vascular endothelial growth factor in C6 glioma cells. J. Biol. Chem. **270:** 19761–19766.

91. FERRARA, N., H.P. GERBER & J. LECOUTER. 2003. The biology of VEGF and its receptors. Nat. Med. **9:** 669–676.

92. BJORNDAHL, M.A. *et al.* 2005. Vascular endothelial growth factor-a promotes peritumoral lymphangiogenesis and lymphatic metastasis. Cancer Res. **65:** 9261–9268.

93. WHITEHURST, B. *et al.* 2007. Anti-VEGF-A therapy reduces lymphatic vessel density and expression of VEGFR-3 in an orthotopic breast tumor model. Int. J. Cancer **121:** 2181–2191.

94. GANNON, G. *et al.* 2002. Overexpression of vascular endothelial growth factor-A165 enhances tumor angiogenesis but not metastasis during á-cell carcinogenesis. Cancer Res. **62:** 603–608.

95. CAO, R. *et al.* 2004. PDGF-BB induces intratumoral lymphangiogenesis and promotes lymphatic metastasis. Cancer Cell **6:** 333–345.

96. KAJIYA, K. *et al.* 2005. Hepatocyte growth factor promotes lymphatic vessel formation and function. EMBO J. **24:** 2885–2895.

97. CAO, R. *et al.* 2006. Hepatocyte growth factor is a novel lymphangiogenic factor with an indirect mechanism of action. Blood **107:** 3531–3536.

98. DANILKOVITCH-MIAGKOVA, A. & B. ZBAR. 2002. Dysregulation of Met receptor tyrosine kinase activity in invasive tumors. J. Clin. Invest. **109:** 863–867.

99. THURSTON, G. 2003. Role of angiopoietins and Tie receptor tyrosine kinases in angiogenesis and lymphangiogenesis. Cell Tissue Res. **314:** 61–68.

100. SURI, C. *et al.* 1996. Requisite role of angiopoietin-1, a ligand for the TIE2 receptor, during embryonic angiogenesis. Cell **87:** 1171–1180.

101. MAISONPIERRE, P.C. *et al.* 1997. Angiopoietin-2, a natural antagonist for Tie2 that disrupts in vivo angiogenesis. Science **277:** 55–60.

102. GALE, N.W. *et al.* 2002. Angiopoietin-2 is required for postnatal angiogenesis and lymphatic patterning, and only the latter role is rescued by angiopoietin-1. Dev. Cell. **3:** 411–423.

103. TAMMELA, T. *et al.* 2005. Angiopoietin-1 promotes lymphatic sprouting and hyperplasia. Blood **105:** 4642–4648.

104. MORISADA, T. *et al.* 2005. Angiopoietin-1 promotes LYVE-1-positive lymphatic vessel formation. Blood **105:** 4649–4656.

105. BJORNDAHL, M. *et al.* 2005. Insulin-like growth factors 1 and 2 induce lymphangiogenesis in vivo. Proc. Natl. Acad. Sci. USA **102:** 15593–15598.

106. MITA, K., M. NAKAHARA & T. USUI. 2000. Expression of the insulin-like growth factor system and cancer progression in hormone-treated prostate cancer patients. Int. J. Urol. **7:** 321–329.

107. TANG, Y. *et al.* 2003. Vascular endothelial growth factor C expression and lymph node metastasis are regulated by the type I insulin-like growth factor receptor. Cancer Res. **63:** 1166–1171.

108. IOACHIM, H. 2002. Tumor-Reactive Lymphadenopathy. Lippincott, Williams and Wilkins. Philadelphia, PA.

109. RUDDELL, A. *et al.* 2003. B lymphocyte-specific c-Myc expression stimulates early and functional expansion of the vasculature and lymphatics during lymphomagenesis. Am. J. Pathol. **163:** 2233–2245.

110. DADRAS, S.S. *et al.* 2005. Tumor lymphangiogenesis predicts melanoma metastasis to sentinel lymph nodes. Mod. Pathol. **18:** 1232–1242.

111. PAGET, S. 1889. The distribution of secondary growths in cancer of the breast. Lancet **1:** 571–573.

112. MACKAY, C.R. 2001. Chemokines: immunology's high impact factors. Nat. Immunol. **2:** 95–101.

113. LAURENCE, A.D. 2006. Location, movement and survival: the role of chemokines in haematopoiesis and malignancy. Br. J. Haematol. **132:** 255–267.

114. WILEY, H.E. *et al.* 2001. Expression of CC chemokine receptor-7 and regional lymph node metastasis of B16 murine melanoma. J. Natl. Cancer Inst. **93:** 1638–1643.

115. SHIELDS, J.D. *et al.* 2007. Chemokine-mediated migration of melanoma cells towards lymphatics—a mechanism contributing to metastasis. Oncogene **26:** 2997–3005.

116. UCHIDA, D. *et al.* 2004. Acquisition of lymph node, but not distant metastatic potentials, by the overexpression of CXCR4 in human oral squamous cell carcinoma. Lab. Invest. **84:** 1538–1546.

117. MÜLLER, A. *et al.* 2001. Involvement of chemokine receptors in breast cancer metastasis. Nature **410:** 50–56.

118. KAWADA, K. *et al.* 2007. Chemokine receptor CXCR3 promotes colon cancer metastasis to lymph nodes. Oncogene **26:** 4679–4688.

119. KAWADA, K. *et al.* 2004. Pivotal role of CXCR3 in melanoma cell metastasis to lymph nodes. Cancer Res. **64:** 4010–4017.

120. MASHINO, K. *et al.* 2002. Expression of chemokine receptor CCR7 is associated with lymph node metastasis of gastric carcinoma. Cancer Res. **62:** 2937–2941.

121. GUNTHER, K. *et al.* 2005. Prediction of lymph node metastasis in colorectal carcinoma by expression of chemokine receptor CCR7. Int. J. Cancer **116:** 726–733.

122. CABIOGLU, N. *et al.* 2005. CCR7 and CXCR4 as novel biomarkers predicting axillary lymph node metastasis in T1 breast cancer. Clin. Cancer Res. **11:** 5686–5693.

123. SCHIMANSKI, C.C. *et al.* 2006. Dissemination of hepatocellular carcinoma is mediated via chemokine receptor CXCR4. Br. J. Cancer **95:** 210–217.

124. ISHIKAWA, T. *et al.* 2006. CXCR4 expression is associated with lymph-node metastasis of oral squamous cell carcinoma. Int. J. Oncol. **28:** 61–66.

125. SALVUCCI, O. *et al.* 2006. The role of CXCR4 receptor expression in breast cancer: a large tissue microarray study. Breast Cancer Res. Treat. **97:** 275–283.

126. ACHEN, M.G. *et al.* 2000. Monoclonal antibodies to vascular endothelial growth factor-D block interactions with both VEGF receptor-2 and VEGF receptor-3. Eur. J. Biochem. **267:** 2505–2515.

127. PYTOWSKI, B. *et al.* 2005. Complete and specific inhibition of adult lymphatic regeneration by a novel VEGFR-3 neutralizing antibody. J. Natl. Cancer Inst. **97:** 14–21.

128. GRIDELLI, C. *et al.* 2007. Sorafenib and sunitinib in the treatment of advanced non-small cell lung cancer. Oncologist **12:** 191–200.

129. MANLEY, P.W. *et al.* 2004. Advances in the structural biology, design and clinical development of VEGF-R kinase inhibitors for the treatment of angiogenesis. Biochim. Biophys. Acta **1697:** 17–27.

130. ROSKOSKI, R., JR. 2007. Sunitinib: a VEGF and PDGF receptor protein kinase and angiogenesis inhibitor. Biochem. Biophys. Res. Commun. **356:** 323–328.

131. VERHEUL, H.M. & H.M. PINEDO. 2007. Possible molecular mechanisms involved in the toxicity of angiogenesis inhibition. Nat. Rev. Cancer **7:** 475–485.

132. KARPANEN, T. *et al.* 2006. Lymphangiogenic growth factor responsiveness is modulated by postnatal lymphatic vessel maturation. Am. J. Pathol. **169:** 708–718.

133. BALDWIN, M.E. *et al.* 2005. Vascular endothelial growth factor d is dispensable for development of the lymphatic system. Mol. Cell. Biol. **25:** 2441–2449.

134. VALLET, S. *et al.* 2007. MLN3897, a novel CCR1 inhibitor, impairs osteoclastogenesis and inhibits the interaction of multiple myeloma cells and osteoclasts. Blood **110:**3744–3752.

135. SMITH, M.C. *et al.* 2004. CXCR4 regulates growth of both primary and metastatic breast cancer. Cancer Res. **64:** 8604–8612.

136. TOBLER, N.E. & M. DETMAR. 2006. Tumor and lymphnode lymphangiogenesis—impact on cancer metastasis. J. Leuk. Biol. **80:** 692.

Lymphatic Vessel Activation in Cancer

SUVENDU DAS AND MIHAELA SKOBE

Department of Oncological Sciences, Mount Sinai School of Medicine, New York, New York, USA

Most cancerous lesions metastasize through the lymphatic system and the status of regional lymph nodes is the most important indicator of a patient's prognosis. The extent of lymph node involvement with cancer is also an important parameter used for determining treatment options. Although the importance of the lymphatic system for metastasis has been well recognized, traditionally, the lymphatic vessels have not been considered actively involved in the metastatic process. Recent evidence, however, indicates that the activation of the lymphatic system is an important factor in tumor progression to metastasis. Tumor lymphangiogenesis has been associated with increased propensity for metastasis, and lymphatic vessel density has emerged as another promising prognostic indicator. More recently, lymphangiogenesis in the sentinel lymph nodes has been shown to contribute to malignant progression. In addition to its role as a transport system for tumor cells, the lymphatic system may also be more actively involved in metastases by directly facilitating tumor cell recruitment into the lymphatic vessels. This review highlights recent advances in our understanding of the mechanisms by which lymphatic vessels participate in metastasis.

Key words: lymphangiogenesis; lymphatic endothelium; lymphatic metastasis; sentinel lymph node biopsy; VEGF; VEGFR

Henry LeDran (1684–1770), a French surgeon, first proposed the theory that cancer begins in its earliest stages as a local disease, which initially spreads to the lymph nodes and subsequently enters the blood circulation. LeDran also observed that the cure was much less likely when lymph nodes were involved. This period was referred to as an optimistic period in the history of breast cancer therapy, because his theory offered the hope that surgery might cure the disease if performed sufficiently early.[1] Until then, the predominant model, which originated in Greek philosophy, implied that cancer is the local manifestation of a systemic disease.[2] In about 1840, Virchow proposed that lymph nodes filter particles from lymph, and that they act as a first-line of defense filtering out the cancer cells. Once these filters became saturated, lymph nodes themselves acted as a nidus for spread to the next line of defense, and ultimately to the skeleton and vital organs. Based on these arguments, radical mastectomy was introduced in the 19th century and adopted as a default therapy for breast cancer worldwide until the late 20th century.[1] The failures of radical operations to cure patients of breast cancer, however, prompted

Fisher to propose a new hypothesis that rejected the mechanistic models of the past.[1,2] He postulated that cancer spreads via the bloodstream even before its clinical detection, with the outcome determined by the biology of tumor–host interactions. Although the importance of the lymphatic system in cancer metastasis is well appreciated, even today the field continues to be plagued with unresolved issues and controversies. As illustrated in this brief historical overview, such fundamental issues, as to whether cancer cells within the lymph node partake in the formation of distant metastases or whether they are only an early marker of distant disease, remain unresolved.

Clinical Implications of Lymphatic Vessel Involvement in Metastasis

Malignant cells commonly disseminate first through the lymphatic system into the regional lymph node basin. The status of regional lymph nodes is an important parameter used for determining the stage of disease progression for different types of cancer, and it is a powerful predictor of patient survival and future treatment options. While lymph node metastases are not directly responsible for cancer-related death, cancer cells could spread from the lymph nodes to distant organs, where they can develop a secondary tumor and perturb critical function of that organ. To date, however, this concept has not been directly

Address for correspondence: Mihaela Skobe, Department of Oncological Sciences, Mount Sinai School of Medicine, One Gustave L. Levy Place, Box 1130, New York, NY 10029. Voice: +1-212-659-5570; fax: +1-212-987-2240.

mihaela.skobe@mssm.edu

Ann. N.Y. Acad. Sci. 1131: 235–241 (2008). © 2008 New York Academy of Sciences.
doi: 10.1196/annals.1413.021

proven. Clinical data demonstrating improved patient survival upon removal of the regional lymph nodes involved with cancer, however, indirectly support this model. Sentinel node biopsy and lymphadenectomy have been shown to reduce lymph node relapse and to prolong disease-free survival, and today are the standard of care for several types of cancer. The first lymph node which receives lymphatic drainage from the primary tumor site is defined as a sentinel lymph node. A sentinel lymph node biopsy procedure involves removal and examination of sentinel nodes, usually identified by preoperative lymphoscintigraphy and by intraoperative blue staining after blue dye injection at the primary tumor site.[3,4] The sentinel lymph node biopsy procedure is based on the assumption that if the sentinel lymph nodes are free of metastases, other lymph nodes will also be free of metastases and their removal can be avoided. If the biopsy shows a presence of lymph node metastases, lymphadenectomy is performed. Ultimately, the extent of lymph node involvement will influence decisions about the choice of adjuvant therapy. While the use of sentinel lymph node biopsy clearly demonstrates the value in improving the staging and treatment in many types of cancer, such as melanoma, breast, prostate, and colon cancer, whether the complete regional lymphadenectomy in patients with metastatic disease in a sentinel node confers benefit in terms of patient survival remains controversial. Many clinical studies indicate that patients who undergo extensive lymph node dissection have higher survival rates in melanoma,[5] gastric cancer,[6] colon[7] and prostate cancer[8] compared to those who received limited or no lymph node dissection. However, other clinical studies contradict these results.[9–11]

In melanoma, the number of micrometastases is a criterion for staging and the sentinel node status is the most significant predictor of survival. Recently, Morton *et al.* reported the largest clinical trial to date of sentinel lymph node biopsy for melanoma.[5] In the trial, 1269 patients with intermediate-thickness melanoma were randomly assigned to immediate sentinel node biopsy or to observation for clinically detectable (palpable) lymph nodes. If the biopsy showed microscopic metastases, patients underwent immediate radical lymphadenectomy. In the observation group, lymphadenectomy was performed only after palpable lymph nodes were detected. The trial clearly demonstrates that sentinel node biopsy provides important prognostic information and identifies patients with nodal metastases whose survival can be prolonged by lymphadenectomy. Among patients with nodal metastases, the 5-year survival rate was significantly greater in patients who had undergone lym-

phadenectomy than in those who did not. Sentinel node biopsy is also the standard of care for breast cancer, where axillary lymph node status continues to be the single most important prognostic indicator of patient survival. The axillary lymph nodes receive 85 % of the lymphatic drainage from all quadrants of the breast, and axillary lymph node dissection has been associated with reduced frequency of visceral and supraclavicular metastases, and lymph node recurrences. An increasing body of evidence indicates that improved local control in breast cancer indeed results in a significantly better 5-year survival rate.[12,13]

Lymphovascular invasion (LVI) is another important parameter in assessing the risk for tumor metastasis. Vascular involvement denotes invasion of tumor cells into the microvasculature either by abutting the endothelium or by penetrating the endothelium and lodging within the vessel lumen.[14] A number of reports have indicated that LVI significantly increased the risk of lymph node involvement, relapse, distant metastases, and death. In melanoma, LVI was present in 57% of nodular melanomas with lymph node metastasis at the time of diagnosis compared with only 12% of those confined to the skin.[14] Distant metastases were observed in 74% of patients with LVI and in 22% of patients without lymphovascular involvement.[15] In thick melanomas, the presence of LVI is associated with a 5-year survival rate of 25%, compared to 50% in patients without vascular involvement.[14] LVI is also indicative of an unfavorable prognosis in breast cancer,[16,17] gastric cancer,[18,19] bladder cancer,[20] prostate cancer,[21] colorectal cancer,[22] rectal carcinoma,[23] and endometrial carcinoma.[24] In node-negative cancer, LVI is an independent adverse prognostic factor for metastases to regional and distant sites. Most importantly, it was shown to be negatively correlated with survival in breast, gastric, bladder, and prostate cancer.[18,25–27]

More recently, lymphatic vessel density (LVD) has emerged as another promising indicator of patient prognosis.[28,29] An increasing number of studies is showing a relationship between patient survival and lymphatic density in different tumor types. High LVD has been associated with a high incidence of regional and distant metastases as well as poor survival in melanoma, breast, lung, colorectal, gastric, and endometrial cancer.[28,29] Quantification of tumor lymphangiogenesis has a potential of being an early prognostic marker and would be particularly beneficial for patients presenting primary tumors without lymph node involvement. While lymphangiogenesis has been correlated with poor prognosis in humans and has been shown to promote metastases in many animal models,[29–31] the mechanisms by which newly formed lymphatics promote cancer progression are not well understood. The infiltration of lymphatic vessels into the tumor may increase the probability of metastasis by creating an increased opportunity for metastatic cancer cells to leave the primary tumor site. Tumor lymphatics may also aid in boosting the transport of toxic metabolites from the tumor, produced by rapidly proliferating tumor cells. Some studies have suggested that tumor lymphatics have impaired transport function for solutes and macromolecules.[32] In this case, change of lymphatic function may result in alterations of tumor microenvironment in the vicinity of lymphatic vessels, which may in turn promote metastasis. Furthermore, lymphatic endothelial cells (LECs) may facilitate tumor spread more directly, by promoting tumor cell migration and invasion into the vessels.

VEGFs and the Regulation of Tumor Lymphangiogenesis

Studies demonstrating occurrence of tumor lymphangiogenesis were the first line of evidence for lymphatic vessel activation in cancer. Traditionally, the lymphatic system has not been considered to be actively involved in the process of metastasis. Tumor cells were believed to be passively carried into the lymphatic vessels with the interstitial fluid and proteins[33] and the prevailing view had been that lymphangiogenesis is not a part of tumorigenesis.[34,35] During the past several years, two members of the VEGF family, VEGF-C and VEGF-D, have been demonstrated to play an important role in tumor lymphangiogenesis via activation of VEGFR-3, which is expressed in normal adult tissues mainly by LECs.[36–38] VEGF-C and VEGF-D, when fully proteolytically processed, can also activate VEGFR-2[39] but whether VEGFR-2 plays a direct role in lymphangiogenesis is less clear. VEGF-C also binds to a non-kinase receptor neuropilin-2 (NRP2),[40] which may cooperate with VEGFR-3 to mediate VEGF-C-dependent lymphangiogenesis.[41]

Direct evidence for the role of VEGF-C and VEGF-D in lymphangiogenesis and metastasis was obtained by using mouse models. In a xenograft model of human breast cancer transplanted into immunodeficient mice, VEGF-C overexpression induced tumor lymphangiogenesis, and drastically increased metastasis to regional lymph nodes and to lungs.[42] Similarly, VEGF-D-overexpressing epitheloid tumors induced the formation of intratumoral lymphatic vessels and promoted lymph node metastases.[43] While VEGF-D promoted tumor dissemination to lymph nodes, VEGF overexpression in the same experimental model

did not, implying differential roles of the VEGF family members in determining the route of metastases. Furthermore, double-transgenic mice, generated by crossing Rip1-Tag2 and Rip1-VEGF-C mice, formed pancreatic tumors surrounded by an extensive lymphatic network. Although Rip1-Tag2 mice developed pancreatic β cell tumors that are not metastatic, metastases to lymph nodes were frequently observed in double-transgenic VEGF-C-overexpressing mice.[44] These data demonstrate that VEGF-C-mediated activation of the lymphatic system is sufficient to induce metastasis to lymph nodes in previously non-metastatic tumor cells. An important role of VEGF-C in lymphangiogenesis and metastasis was also shown in a mouse model of B16 melanoma, T-241 fibrosarcoma, MCF-7 breast carcinoma, AZ521 gastric carcinoma, and NCI-H460-LNM35 lung carcinoma.[35,45–47] Contribution of VEGF-D to lymph node metastasis was also shown in a model of breast and pancreatic cancer.[48,49] VEGF-C also has been reported to induce hyperplasia and dilation of the tumor lymphatic vessels[42,44] as well as to increase lymph flow rate,[50,51] which may have an impact on lymph node metastases by enhancing tumor cell delivery to the lymph nodes.

Many human tumors and cell lines express high levels of VEGF-C, the expression of which is in most cases correlated with metastases.[29] What are the factors regulating VEGF-C expression? Notably, hypoxia, Ras oncoprotein, and mutant p53 tumor suppressor, which are potent inducers of VEGF-A mRNA, do not increase VEGF-C mRNA levels, whereas serum and its component growth factors, PDGF, EGF and TGF-β, stimulate VEGF-C expression.[52] Androgen ablation was shown to increase VEGF-C mRNA expression in prostate cancer cells.[53] VEGF-C expression can also be upregulated in tumors upon interaction with growth factors, or inflammatory molecules secreted from the host cells. Activation of type I insulin-like growth factor receptor IGF-IR, expressed on the tumor cell surface, induces the production of VEGF-C by the tumor cells and thereby influences lymphatic metastasis.[54] In human lung adenocarcinoma cells, VEGF-C is one of the major downstream genes of COX-2, which catalyzes synthesis of prostaglandins and plays a vital role in inflammation, tissue damage, and tumorigenesis. COX-2-mediated VEGF-C upregulation was found to correlate with LVD and lymph node metastasis in a mouse model of lung cancer.[55] Neural cell adhesion molecule (NCAM)–deficient Rip1Tag2 transgenic mice showed substantial increase in lymphangiogenesis and lymph node metastasis compared to Rip1Tag2 transgenic mice, which was correlated with an increase in VEGF-C and VEGF-D expression.[56] Thus, deletion of a gene may result in a bystander effect and induce expression of a lymphangiogenic factor.

Macrophages are another important source of lymphangiogenic factors.[57] Tumor-associated macrophages were shown to express large amounts of VEGF-C and VEGF-D in cervical and in gastric cancer.[58,59] In breast cancer, VEGF-C-expressing tumor-associated macrophages were positively correlated with lymphangiogenesis.[60] Macrophages express VEGFR-3, and VEGF-C was shown to promote macrophage chemotaxis and increase macrophage densities in a melanoma model.[61] Thus, VEGF-C-expressing tumor cells may attract macrophages, which in turn also secrete lymphangiogenic factors and contribute to the process of lymphangiogenesis. In addition to VEGFs, several pleiotropic growth factors have also been implicated in lymphangiogenesis, including bFGF, PDGF, IGF, and HGF.[62]

Lymphatic Vessel Activation in the Lymph Nodes

Recent studies demonstrated that lymphangiogenesis occurs not only at the primary tumor site, but also in the sentinel lymph nodes. Transgenic mice overexpressing VEGF-A or VEGF-C in mouse keratinocytes and subjected to the chemical carcinogenesis protocol show expansion of the lymphatic network in the tumor-draining lymph nodes.[63,64] Notably, lymph node lymphangiogenesis was observed even before the onset of metastasis. Similarly, tumor-induced sentinel lymph node lymphangiogenesis was shown to precede metastasis of nasopharyngeal carcinoma and melanoma.[51,65] P19/Arf and p53 deficiency were recently associated with increased lymph node lymphangiogenesis.[66] In papilloma-bearing p19/Arf or p53-deficient mice, lymph node lymphangiogenesis was greatly accelerated, which coincided with the greater propensity of these tumors to progress to carcinomas and metastasize. In breast cancer patients with a positive sentinel lymph node, increased sentinel node lymphangiogenesis was associated with nonsentinel lymph node involvement.[67,68] While lymph node lymphangiogenesis has been positively correlated with metastasis in several tumor models, it is not an event specific for malignancy. Lymphangiogenesis has been observed also in the lymph nodes draining benign tumors in the context of inflammation, when induced with the tetradecanoyl phorbol acetate in chemical carcinogenesis protocols.[63,64,66] Expansion of the lymphatic network has been observed within the immunized lymph nodes in which B cells were shown to be a requirement

for the process.[69] Hence, lymph node lymphangiogenesis appears to be a constituent of the normal host immune response, which, paradoxically, in a tumor setting enhances tumor progression to metastasis.

Role of Inflamed Lymphatic Endothelium in Cancer

In addition to its role as a transport system, the lymphatic system may play a more active role in metastases, by directly facilitating tumor cell recruitment into the lymphatic vessels. Alterations of the lymphatic endothelial junctions and changes in expression of cell adhesion molecules by LECs in inflammation may promote tumor metastasis via lymphatics. However, while tumor cells have been shown to adhere to LECs, there is no direct experimental evidence to date demonstrating that adhesive interactions with LECs are indeed required for tumor cell entry into the lymphatics. Several studies suggest that macrophage mannose receptor I (MR) and CLEVER-1 may be important mediators of cancer cell adhesion to the lymphatic endothelium. MR is highly expressed by cultured LECs[70,71] and it has been shown to mediate adhesion of lymphocytes to the lymphatics in lymph nodes.[72] CLEVER-1 is an inducible vascular adhesion molecule which is implicated in binding of lymphocytes to LECs in the lymph nodes during inflammation.[73] Notably, MR and CLEVER-1 expression has been detected on peri- and intratumoral lymphatic vessels in human head and neck and breast carcinomas. In addition, expression of MR on intratumoral lymphatic vessels was associated with increased lymph node metastases in breast cancer.[74]

Soluble factors constitutively expressed or induced in LECs may also facilitate tumor cell invasion of lymphatic vessels. Activation of LECs in inflammation could promote release of chemokines, which may attract tumor cells into the lymphatics. As the migration of cancer cells to regional lymph nodes resembles physiological migration of leukocytes, it is conceivable that chemokine-mediated mechanisms for the mobilization of lymphocytes may also be utilized for metastatic dissemination. LECs constitutively express chemokines CCL19 and CCL21, ligands for CCR7, which is critical for the migration of lymphocytes and dendritic cells to the lymph nodes. CCL21 has been shown to promote melanoma chemotaxis toward lymphatics[75] and to enhance the metastatic colonization of CCR7-expressing mouse melanoma cells to regional lymph nodes.[76] However, there is little information about the chemokines induced in LECs upon inflammatory stimuli. *In vitro*, human LECs release macrophage inflammatory protein (MIP)-3alpha/CCL20 upon activation.[77] CCL1, the sole ligand for chemokine receptor CCR8, is strongly upregulated in human LECs upon TNF=α stimulation *in vitro*, and the inhibition of CCR8 signaling in human melanoma cells decreased the incidence of lymph node metastases (Das and Skobe, unpublished data). The lymphatic system is believed to play a role in resolution of inflammation by removing extravasated fluids, inflammatory mediators, and cells.[78] Recent evidence, however, demonstrates that LECs are more directly involved in regulating inflammation. LECs constitutively express atypical chemokine receptor D6, which acts as a decoy receptor, which sequesters and destroys many proinflammatory CC chemokines.[79] D6 function on lymphatic vessels appears to be critical for effective resolution of inflammation *in vivo*.[80] Moreover, *D6*-deficient mice have increased susceptibility to cutaneous tumor development in response to a chemical carcinogenesis protocol and, remarkably, *D6* deletion is sufficient to make resistant mouse strains susceptible to invasive squamous cell carcinoma.[81] Conversely, transgenic D6 expression in keratinocytes dampens cutaneous inflammation and can confer considerable protection from tumor formation.[81] Thus, somewhat surprisingly, in addition to affecting metastatic potential, the lymphatic endothelium also has the capacity to have a direct impact on tumorigenesis.

Conflicts of Interest

The authors declare no conflicts of interest.

References

1. RAYTER, Z. 2003. Medical Therapy of Breast Cancer. Cambridge University Press. Cambridge, UK.
2. TANIS, P.J. *et al.* 2001. History of sentinel node and validation of the technique. Breast Cancer Res. **3:** 109–112.
3. MORTON, D.L. *et al.* 1992. Technical details of intraoperative lymphatic mapping for early stage melanoma. Arch. Surg. **127:** 392–399.
4. COCHRAN, A.J. *et al.* 2000. The Augsburg Consensus: techniques of lymphatic mapping, sentinel lymphadenectomy, and completion lymphadenectomy in cutaneous malignancies. Cancer **89:** 236–241.
5. MORTON, D.L. *et al.* 2006. Sentinel-node biopsy or nodal observation in melanoma. N. Engl. J. Med. **355:** 1307–1317.
6. WU, C.W. *et al.* 2006. Nodal dissection for patients with gastric cancer: a randomised controlled trial. Lancet Oncol. **7:** 309–315.
7. RICCIARDI, R. & N.N. BAXTER. 2007. Association versus causation versus quality improvement: setting benchmarks for lymph node evaluation in colon cancer. J. Natl. Cancer Inst. **99:** 414–415.
8. SWANSON, G.P., I.M. THOMPSON & J. BASLER. 2006. Current status of lymph node-positive prostate cancer: incidence and predictors of outcome. Cancer **107:** 439–450.
9. BALCH, C.M. 1988. The role of elective lymph node dissection in melanoma: rationale, results, and controversies. J. Clin. Oncol. **6:** 163–172.
10. BEMBENEK, A.E. *et al.* 2007. Sentinel lymph node biopsy in colon cancer: a prospective multicenter trial. Ann. Surg. **245:** 858–863.
11. HARTGRINK, H.H. *et al.* 2004. Extended lymph node dissection for gastric cancer: who may benefit? Final results of the randomized Dutch gastric cancer group trial. J. Clin. Oncol. **22:** 2069–2077.
12. TRUONG, P.T. *et al.* 2002. Age-related variations in the use of axillary dissection: a survival analysis of 8038 women with T1-ST2 breast cancer. Int. J. Radiat. Oncol. Biol. Phys. **54:** 794–803.
13. JOSLYN, S.A. & B.R. KONETY. 2005. Effect of axillary lymphadenectomy on breast carcinoma survival. Breast Cancer Res. Treat. **91:** 11–18.
14. KASHANI-SABET, M. *et al.* 2001. Vascular involvement in the prognosis of primary cutaneous melanoma. Arch. Dermatol. **137:** 1169–1173.
15. STRAUME, O. & L.A. AKSLEN. 1996. Independent prognostic importance of vascular invasion in nodular melanomas. Cancer **78:** 1211–1219.
16. HODA, S.A. *et al.* 2006. Issues relating to lymphovascular invasion in breast carcinoma. Adv. Anat. Pathol. **13:** 308–315.
17. KIRUPARAN, P. & L. FORREST. 2007. Prediction in breast cancer of the extent of axillary node involvement from the size and lymphovascular invasion status of the primary tumour: medico-legal considerations. Eur. J. Surg. Oncol. **33:** 435–437.
18. LEE, C.C. *et al.* 2007. Survival predictors in patients with node-negative gastric carcinoma. J. Gastroenterol. Hepatol. **22:** 1014–1018.
19. DICKEN, B.J. *et al.* 2006. Lymphovascular invasion is associated with poor survival in gastric cancer: an application of gene-expression and tissue array techniques. Ann. Surg. **243:** 64–73.
20. QUEK, M.L. *et al.* 2005. Prognostic significance of lymphovascular invasion of bladder cancer treated with radical cystectomy. J. Urol. **174:** 103–106.
21. CHENG, L. *et al.* 2005. Lymphovascular invasion is an independent prognostic factor in prostatic adenocarcinoma. J. Urol. **174:** 2181–2185.
22. KIM, J.C. *et al.* 2005. Genetic and pathologic changes associated with lymphovascular invasion of colorectal adenocarcinoma. Clin. Exp. Metastasis **22:** 421–428.
23. BENSON, R. *et al.* 2001. Local excision and postoperative radiotherapy for distal rectal cancer. Int. J. Radiat. Oncol. Biol. Phys. **50:** 1309–1316.
24. HACHISUGA, T. *et al.* 1999. The grading of lymphovascular space invasion in endometrial carcinoma. Cancer **86:** 2090–2097.
25. LOTAN, Y. *et al.* 2005. Lymphovascular invasion is independently associated with overall survival, cause-specific

survival, and local and distant recurrence in patients with negative lymph nodes at radical cystectomy. J. Clin. Oncol. **23:** 6533–6539.

26. MAY, M. *et al.* 2007. Prognostic impact of lymphovascular invasion in radical prostatectomy specimens. BJU Int. **99:** 539–544.

27. LEE, A.H. *et al.* 2006. Prognostic value of lymphovascular invasion in women with lymph node negative invasive breast carcinoma. Eur. J. Cancer **42:** 357–362.

28. VAN DER AUWERA, I. *et al.* 2006. First international consensus on the methodology of lymphangiogenesis quantification in solid human tumours. Br. J. Cancer **95:** 1611–1625.

29. PEPPER, M.S. *et al.* 2003. Lymphangiogenesis and tumor metastasis. Cell Tissue Res. **314:** 167–177.

30. SWARTZ, M.A. & M. SKOBE. 2001. Lymphatic function, lymphangiogenesis, and cancer metastasis. Microsc. Res. Tech. **55:** 92–99.

31. CASSELLA, M. & M. SKOBE. 2002. Lymphatic vessel activation in cancer. Ann. N.Y. Acad. Sci. **979:** 120–130.

32. ISAKA, N. *et al.* 2004. Peritumor lymphatics induced by vascular endothelial growth factor-C exhibit abnormal function. Cancer Res. **64:** 4400–4404.

33. HARTVEIT, E. 1990. Attenuated cells in breast stroma: the missing lymphatic system of the breast. Histopathology **16:** 533–543.

34. CARMELIET, P. & R.K. JAIN. 2000. Angiogenesis in cancer and other diseases. Nature **407:** 249–257.

35. PADERA, T.P. *et al.* 2002. Lymphatic metastasis in the absence of functional intratumor lymphatics. Science **296:** 1883–1886.

36. ACHEN, M.G. *et al.* 1998. Vascular endothelial growth factor D (VEGF-D) is a ligand for the tyrosine kinases VEGF receptor 2 (Flk1) and VEGF receptor 3 (Flt4). Proc. Natl. Acad. Sci. USA **95:** 548–553.

37. JOUKOV, V. *et al.* 1996. A novel vascular endothelial growth factor, VEGF-C, is a ligand for the Flt4 (VEGFR-3) and KDR (VEGFR-2) receptor tyrosine kinases. EMBO J. **15:** 290–298.

38. LEE, J. *et al.* 1996. Vascular endothelial growth factor-related protein: a ligand and specific activator of the tyrosine kinase receptor Flt4. Proc. Natl. Acad. Sci. USA **93:** 1988–1992.

39. JOUKOV, V. *et al.* 1997. Proteolytic processing regulates receptor specificity and activity of VEGF-C. EMBO J. **16:** 3898–3911.

40. KARKKAINEN, M.J. *et al.* 2001. A model for gene therapy of human hereditary lymphedema. Proc. Natl. Acad. Sci. USA **98:** 12677–12682.

41. YUAN, L. *et al.* 2002. Abnormal lymphatic vessel development in neuropilin 2 mutant mice. Development **129:** 4797–4806.

42. SKOBE, M. *et al.* 2001. Induction of tumor lymphangiogenesis by VEGF-C promotes breast cancer metastasis. Nat. Med. **7:** 192–198.

43. STACKER, S.A. *et al.* 2001. VEGF-D promotes the metastatic spread of tumor cells via the lymphatics. Nat. Med. **7:** 186–191.

44. MANDRIOTA, S.J. *et al.* 2001. Vascular endothelial growth factor-C-mediated lymphangiogenesis promotes tumour metastasis. EMBO J. **20:** 672–682.

45. KARPANEN, T. *et al.* 2001. Vascular endothelial growth factor C promotes tumor lymphangiogenesis and intralymphatic tumor growth. Cancer Res. **61:** 1786–1790.

46. HE, Y. *et al.* 2002. Suppression of tumor lymphangiogenesis and lymph node metastasis by blocking vascular endothelial growth factor receptor 3 signaling. J. Natl. Cancer Inst. **94:** 819–825.

47. YANAI, Y. *et al.* 2001. Vascular endothelial growth factor C promotes human gastric carcinoma lymph node metastasis in mice. J. Exp. Clin. Cancer Res. **20:** 419–428.

48. KOPFSTEIN, L. *et al.* 2007. Distinct roles of vascular endothelial growth factor-D in lymphangiogenesis and metastasis. Am. J. Pathol. **170:** 1348–1361.

49. KRISHNAN, J. *et al.* 2003. Differential in vivo and in vitro expression of vascular endothelial growth factor (VEGF)-C and VEGF-D in tumors and its relationship to lymphatic metastasis in immunocompetent rats. Cancer Res. **63:** 713–722.

50. HOSHIDA, T. *et al.* 2006. Imaging steps of lymphatic metastasis reveals that vascular endothelial growth factor-C increases metastasis by increasing delivery of cancer cells to lymph nodes: therapeutic implications. Cancer Res. **66:** 8065–8075.

51. HARRELL, M.I., B.M. IRITANI & A. RUDDELL. 2007. Tumor-induced sentinel lymph node lymphangiogenesis and increased lymph flow precede melanoma metastasis. Am. J. Pathol. **170:** 774–786.

52. ENHOLM, B. *et al.* 1997. Comparison of VEGF, VEGF-B, VEGF-C and Ang-1 mRNA regulation by serum, growth factors, oncoproteins and hypoxia. Oncogene **14:** 2475–2483.

53. LI, J. *et al.* 2005. Upregulation of VEGF-C by androgen depletion: the involvement of IGF-IR-FOXO pathway. Oncogene **24:** 5510–5520.

54. TANG, Y. *et al.* 2003. Vascular endothelial growth factor C expression and lymph node metastasis are regulated by the type I insulin-like growth factor receptor. Cancer Res. **63:** 1166–1171.

55. SU, J.L. *et al.* 2004. Cyclooxygenase-2 induces EP1- and HER-2/Neu-dependent vascular endothelial growth factor-C up-regulation: a novel mechanism of lymphangiogenesis in lung adenocarcinoma. Cancer Res. **64:** 554–564.

56. CRNIC, I. *et al.* 2004. Loss of neural cell adhesion molecule induces tumor metastasis by up-regulating lymphangiogenesis. Cancer Res. **64:** 8630–8638.

57. KERJASCHKI, D. 2005. The crucial role of macrophages in lymphangiogenesis. J. Clin. Invest. **115:** 2316–2319.

58. SCHOPPMANN, S.F. *et al.* 2002. Tumor-associated macrophages express lymphatic endothelial growth factors and are related to peritumoral lymphangiogenesis. Am. J. Pathol. **161:** 947–956.

59. IWATA, C. *et al.* 2007. Inhibition of cyclooxygenase-2 suppresses lymph node metastasis via reduction of lymphangiogenesis. Cancer Res. **67:** 10181–10189.

60. SCHOPPMANN, S.F. *et al.* 2006. VEGF-C expressing tumor-associated macrophages in lymph node positive breast cancer: impact on lymphangiogenesis and survival. Surgery **139:** 839–846.

61. Skobe, M. *et al.* 2001. Concurrent induction of lymphangiogenesis, angiogenesis, and macrophage recruitment by vascular endothelial growth factor-C in melanoma. Am. J. Pathol. **159:** 893–903.

62. Cao, Y. 2005. Opinion: emerging mechanisms of tumour lymphangiogenesis and lymphatic metastasis. Nat. Rev. Cancer **5:** 735–743.

63. Hirakawa, S. *et al.* 2005. VEGF-A induces tumor and sentinel lymph node lymphangiogenesis and promotes lymphatic metastasis. J. Exp. Med. **201:** 1089–1099.

64. Hirakawa, S. *et al.* 2007. VEGF-C-induced lymphangiogenesis in sentinel lymph nodes promotes tumor metastasis to distant sites. Blood **109:** 1010–1017.

65. Qian, C.N. *et al.* 2006. Preparing the "soil": the primary tumor induces vasculature reorganization in the sentinel lymph node before the arrival of metastatic cancer cells. Cancer Res. **66:** 10365–10376.

66. Ruddell, A. *et al.* 2007. p19/Arf and p53 suppress sentinel lymph node lymphangiogenesis and carcinoma metastasis. Oncogene [Epub ahead of print].

67. Van Den Eynden, G.G. *et al.* 2006. Induction of lymphangiogenesis in and around axillary lymph node metastases of patients with breast cancer. Br. J. Cancer **95:** 1362–1366.

68. Van Den Eynden, G.G. *et al.* 2007. Increased sentinel lymph node lymphangiogenesis is associated with nonsentinel axillary lymph node involvement in breast cancer patients with a positive sentinel node. Clin. Cancer Res. **13:** 5391–5397.

69. Angeli, V. *et al.* 2006. B cell-driven lymphangiogenesis in inflamed lymph nodes enhances dendritic cell mobilization. Immunity **24:** 203–215.

70. Podgrabinska, S. *et al.* 2002. Molecular characterization of lymphatic endothelial cells. Proc. Natl. Acad. Sci. USA **99:** 16069–10674.

71. Petrova, T.V. *et al.* 2002. Lymphatic endothelial reprogramming of vascular endothelial cells by the Prox-1 homeobox transcription factor. EMBO J. **21:** 4593–4599.

72. Irjala, H. *et al.* 2001. Mannose receptor is a novel ligand for L-selectin and mediates lymphocyte binding to lymphatic endothelium. J. Exp. Med. **194:** 1033–1042.

73. Irjala, H. *et al.* 2003. The same endothelial receptor controls lymphocyte traffic both in vascular and lymphatic vessels. Eur. J. Immunol. **33:** 815–824.

74. Irjala, H. *et al.* 2003. Mannose receptor (MR) and common lymphatic endothelial and vascular endothelial receptor (CLEVER)-1 direct the binding of cancer cells to the lymph vessel endothelium. Cancer Res. **63:** 4671–4676.

75. Shields, J.D. *et al.* 2007. Chemokine-mediated migration of melanoma cells towards lymphatics—a mechanism contributing to metastasis. Oncogene **26:** 2997–3005.

76. Wiley, H.E. *et al.* 2001. Expression of CC chemokine receptor-7 and regional lymph node metastasis of B16 murine melanoma. J. Natl. Cancer Inst. **93:** 1638–1643.

77. Kriehuber, E. *et al.* 2001. Isolation and characterization of dermal lymphatic and blood endothelial cells reveal stable and functionally specialized cell lineages. J. Exp. Med. **194:** 797–808.

78. Szuba, A. *et al.* 2002. Therapeutic lymphangiogenesis with human recombinant VEGF-C. FASEB J. **16:** 1985–1987.

79. Nibbs, R.J. *et al.* 2001. The beta-chemokine receptor D6 is expressed by lymphatic endothelium and a subset of vascular tumors. Am. J. Pathol. **158:** 867–877.

80. Jamieson, T. *et al.* 2005. The chemokine receptor D6 limits the inflammatory response in vivo. Nat. Immunol. **6:** 403–411.

81. Nibbs, R.J. *et al.* 2007. The atypical chemokine receptor D6 suppresses the development of chemically induced skin tumors. J. Clin. Invest. **117:** 1884–1892.

Index of Contributors

Swarm Intelligence
Methods for
Statistical
Regression

Swarm Intelligence Methods for Statistical Regression

Soumya D. Mohanty

CRC Press
Taylor & Francis Group
Boca Raton London New York

CRC Press is an imprint of the
Taylor & Francis Group, an **informa** business

CRC Press
Taylor & Francis Group
6000 Broken Sound Parkway NW, Suite 300
Boca Raton, FL 33487-2742

© 2019 by Taylor & Francis Group, LLC
CRC Press is an imprint of Taylor & Francis Group, an Informa business

No claim to original U.S. Government works

Printed on acid-free paper
Version Date: 20181128

International Standard Book Number-13: 978-1-138-55818 (Hardback)

Visit the Taylor & Francis Web site at
http://www.taylorandfrancis.com

and the CRC Press Web site at
http://www.crcpress.com

Dedication
To my parents, Rama Ranjan and Krishna

Contents

Preface

This book is based on a set of lectures on big data analysis delivered at the BigDat International Winter School held at Bari, Italy, in 2017. The lectures focused on a very practical issue encountered in the statistical regression of non-linear models, namely, the numerical optimization of the fitting function. The optimization problem in statistical analysis, especially in big data applications, is often a bottleneck that forces either the adoption of simpler models or a shift to linear models even where non-linearity is known to be a better option. The goal of the lectures was to introduce the audience to a set of relatively recent biology inspired stochastic optimization methods, collectively called swarm intelligence (SI) methods, that are proving quite effective in tackling the optimization challenge in statistical analysis.

It was clear from the audience response at these lectures that, despite their collective background in very diverse areas of data analysis ranging from the natural sciences to the social media industry, many had not heard of and none had seriously explored SI methods. The root causes behind this lacuna seem to be (a) a lack of familiarity, within the data analysis community, of the latest literature in stochastic optimization, and (b) lack of experience and guidance in tuning these methods to work well in real-world problems. Matters are not helped by the fact that there are not a whole lot of papers in the optimization community examining the role of SI methods in statistical analysis, most of the focus in that field being on optimization problems in engineering.

I hope this small book helps in bridging the current divide between the two communities. As I have seen within my own research, the statistical data analyst will find that success in solving the optimization challenge spurs the contemplation of better, more sophisticated models for data. Students and researchers in the SI community reading this book will find that statistical data analysis offers a rich source of challenging test beds for their methods.

The aim of the book is to arm the reader with practical tips and rules of thumb that are observed to work well, not to provide a deeper than necessary theoretical background. In particular, the book does not delve deep into the huge range of SI methods out there, concentrating rather on one particular method, namely particle swarm optimization (PSO). My experience in teaching SI methods to students has shown that it is best to learn these methods by starting with one and understanding it well. For this purpose, PSO provides the simplest entry point. Similarly, this book does not provide a more than superficial background in optimization theory, choosing to only highlight some its important results.

It is assumed that the reader of this book has a basic background in probability theory and function approximation at the level of undergraduate or graduate courses. Nonetheless, appendices are provided that cover the required material succinctly. Instead of a problem set, two realistic statistical regression problems covering both parametric and nonparametric approaches form the workhorse exercises in this book. The reader is highly encouraged to independently implement these examples and reproduce the associated results provided in the book.

References are primarily provided in the "Notes" section at the end of each chapter. While it was tempting to include in the book a more complete review of the technical literature than what is currently provided, I decided to point the reader to mostly textbooks or review articles. This was done keeping in mind the expected readership of this book, which I assume would be similar to the student-heavy makeup of the BigDat participants. This inevitably means that many key references have been left out but I hope that the ones included will give readers a good start in seeking out the technical material that is appropriate for their application areas.

Acknowledgements: It is a pleasure to acknowledge colleagues who supported the inception and development of this book. I thank Carlos Martin-Vide, Donato Malerba, and other organizers of the BigDat schools for inviting me as a lecturer and for their hospitality. This annual school provides a wonderful and refreshing opportunity to interact with data scientists and students from a broad spectrum of fields, and I hope that it continues to have several future editions.

Before embarking on the writing of this book, I had the pleasure of discussing the core content of the lectures with Innocenzo Pinto and Luigi Troiano at the University of Sannio at Benevento, Italy. I had similar opportunities, thanks to Runqiu Liu and Zong-kuan Guo at the Chinese Academy of Sciences, Beijing, to present the material to undergraduate and graduate students during my lectures there on gravitational wave data analysis. I am greatly indebted to Yan Wang at Huazhong University of Science and Technology, Wuhan, for a thorough reading of the draft and many invaluable comments and suggestions. Several useful comments from Ram Valluri at the University of Western Ontario are appreciated as well. Finally, I thank Randi Cohen at CRC press for contacting me and initiating this project.

Conventions and Notation

Technical terms are italicized when they are introduced but appear thereafter in normal font. Italics are also used for emphasis where needed.

Much of the notation used in this book is collected here for a quick reference.

$\forall$	For all.
$A \propto B$	A is proportional to B.
$\sum_{k=i}^{j}$	Summation of quantities indexed by integer k.
$\prod_{k=i}^{j}$	Product of quantities indexed by integer k.
$\mathbb{B}$	A set. Small sets will be shown explicitly, when needed, as $\{a, b, \ldots\}$. Otherwise, $\{a \,|\, C\}$ will denote the set of elements for which the statement C is true. A set of indexed elements, such as $\{a_0, a_1, \ldots\}$, will be denoted by $\{a_i\}$, $i = 0, 1, \ldots$, where needed.
$\alpha \in \mathbb{A}$	α is an element of $\mathbb{A}$.
N	An integer.
$\mathbb{A} \times \mathbb{B}$	Direct product of $\mathbb{A}$ and $\mathbb{B}$. It is the set $\{(x, y) \,|\, x \in \mathbb{A}, y \in \mathbb{Y}\}$.
$\mathbb{A}^2$	The set $\mathbb{A} \times \mathbb{A}$ with each element, called a 2-tuple, of the form $(\alpha \in \mathbb{A}, \beta \in \mathbb{A})$.
$\mathbb{A}^N$	For $N \geq 2$, the set $\mathbb{A} \times \mathbb{A}^{N-1}$ with $\mathbb{A}^1 = \mathbb{A}$. Each element, called an N-tuple, is of the form $(\alpha_0, \alpha_1, \ldots, \alpha_{N-1})$ with $\alpha_i \in \mathbb{A} \; \forall i$.

$\mathbb{Z}^N$	The set of positive integer N-tuples. Each element is a list of N positive integers.
$\mathbb{R}^N$	The set of real N-tuples. Each element is a list of N real numbers.
$\mathbb{R}^+$	The set of non-negative real numbers .
x	A scalar (element of $\mathbb{R}^1$ or $\mathbb{Z}^1$).
$\overline{x}$	A row vector $\overline{x} = (x_0, x_1, \ldots, x_{N-1})$ with $x_i \in \mathbb{R}^1$ or $x_i \in \mathbb{Z}^1$. If $x_i \in \mathbb{R}^1$ $\forall i$, $\overline{x} \in \mathbb{R}^N$ and if $x_i \in \mathbb{Z}^1$ $\forall i$, $\overline{x} \in \mathbb{Z}^N$.
$\|\overline{x}\|$	The norm of a vector.
$\mathbf{A}$	A matrix. The element of $\mathbf{A}$ in the ith row and j^{th} column is denoted by A_{ij}.
$\mathbf{A}^T, \overline{x}^T$	The transpose of $\mathbf{A}$ and $\overline{x}$ respectively.
X	A scalar random variable. It can be discrete or continuous.
$X \in \mathbb{A}$	The outcome of a trial gives a value of X in a set $\mathbb{A}$.
$\Pr(X \in \mathbb{A})$ or $\Pr(\mathbb{A})$	The probability of $X \in \mathbb{A}$.
$\overline{X}$	A vector random variable $\overline{X} = (X_0, X_1, \ldots, X_{N-1})$.
$P_X(x)$	The probability of a discrete random variable X having value x in a trial.
$p_X(x)$	The probability density function (pdf) of a continuous random variable X.
$p_{\overline{X}}(\overline{x})$	The joint pdf of $\overline{X}$.
$p_{\overline{X}|\overline{Y}}(\overline{x}|\overline{y})$	The conditional pdf of $\overline{X}$ given the trial value $\overline{y}$ of $\overline{Y}$.
$\min(a, b)$	The smaller of the two scalars a and b.
$\min_{\overline{x}} f(\overline{x})$	The minimum value of $f(\overline{x})$ over $\overline{x}$.
$\max_{\overline{x}} f(\overline{x})$	The maximum value of $f(\overline{x})$ over $\overline{x}$.
$\arg\min_{\overline{x}} f(\overline{x})$	The value of $\overline{x}$ at which $f(\overline{x})$ is minimum.

Introduction

CONTENTS

This chapter introduces the important role of optimization in statistical data analysis and provides an overview of the latter that, although not comprehensive, is adequate for the purpose of this book. Two concrete data analysis problems are described that constitute the testbeds used in this book for illustrating the application of swarm intelligence methods.

1.1 OPTIMIZATION IN STATISTICAL ANALYSIS

The objective of any kind of statistical analysis of given data is to consider a set of models for the process or phenomenon associated with the data and find the model that is best supported by it. This so-called *best-fit model* can then be used to make predictions about the phenomenon. The same idea figures prominently in the drive to develop computer algorithms in *machine learning* that can learn and generalize from examples.

Obtaining the best-fit model becomes a non-trivial task due to the presence of *randomness* (also called *noise* or *error*) – an unpredictable contaminant in the data. (When noise can be ignored, the analysis problem becomes one of *interpolation* or *function approximation*.) Due to this unpredictability, the theory of probability becomes the natural framework within which the methods of statistical analysis are derived. (See App. A for a brief review of some of the concepts in probability theory that are relevant to this book.)

The task of scanning across a given set of models to find the best-fit one is an optimization problem. Hence, optimization is at the core of statistical analysis as much as probability theory. Often, our choice of models is itself limited by the difficulty of the associated optimization problem. As our technology for handling optimization has evolved, mainly in terms of advances in computing and numerical algorithms, so has the type of models we are able to infer from data.

Another evolution in parallel, again enabled by advances in computing, has been in the technology for data collection that has resulted in very high data volumes and the era of *big data*. A large amount of data that is rich in information demands the use of models in its statistical analysis that are more flexible and, hence, have greater complexity. This, in turn, increases the difficulty of the optimization step.

Optimization problems arise in a wide array of fields, of course, and not just in statistical analysis. As such, a large variety of approaches have been and continue to be developed in diverse application domains. For the same reason, however, not all of them have found their way into the standard toolkit used in statistical analysis. Among the more recent ones are those based on emulating the behavior of biological swarms such as a flock of birds or an ant colony. It turns out that nature has figured out, through the blind process of evolution, that a swarm of cooperating entities, with each following a fairly simple set of behavioral rules, can solve difficult optimization problems. Optimization methods based on such approaches fall under the rubric of *swarm intelligence* (SI).

In this book, we will explore the use of SI methods in solving the optimization bottleneck that is often encountered in statistical analyses. Optimization methods that can handle difficult problems cannot, in general, be used as black-boxes: Some tuning of these methods is always needed in order to extract good performance from them. In the experience of the author, the best way to become familiar with this process is to implement SI methods on some concrete but realistic statistical analyses. This is the path followed in this book, where we focus on getting the reader started with some worked out examples that are simple enough to implement fairly easily but sufficiently complicated to mimic realistic problems.

1.2 STATISTICAL ANALYSIS: BRIEF OVERVIEW

A formal description of a statistical analysis problem is as follows. One is given trial values $\overline{z}_i$, $i = 0, 1, \ldots, N - 1$ of a vector random variable $\overline{Z}$ – the set $\mathbb{T} = \{\overline{z}_0, \overline{z}_1, \ldots, \overline{z}_{N-1}\}$ of trial values[1] is called a *data realization* (or just *data*) – and the goal is to consider a set of models for the joint probability density function (pdf), $p_{\overline{Z}}(\overline{z})$, of $\overline{Z}$ and find the one that is best supported by the data.

The task of obtaining this best-fit model is called *density estimation* . With the best-fit $p_{\overline{Z}}(\overline{z})$ in hand, one can make useful predictions or inferences about the phenomenon that generated the data. (While $\overline{z}_i$ can be a mixed combination of integers and real numbers in general, we restrict attention in the following to the case $\overline{z}_i \in \mathbb{R}^N$.)

Besides density estimation, statistical analysis also includes problems of *hypothesis testing*. In hypothesis testing, one has two competing sets of models for $p_{\overline{Z}}(\overline{z})$. The objective is to decide which of the two sets of models is best supported by the data. It is possible that each set itself contains more

---

[1]Alternatively denoted by $\{\overline{z}_i\}$, $i = 0, 1, \ldots, N - 1$.

than one model. So, the decision must take into account the variation of models within each set. We discuss hypothesis testing in further detail in Sec. 1.4.

Methods for density estimation can be divided, rather loosely, into *parametric* and *non-parametric* ones. This division is essentially based on how strong the assumptions embodied in the models of $p_{\overline{Z}}$ are *before* the data is obtained. If the set of models is prescribed independently of the data, we get a purely parametric method. If $p_{\overline{Z}}(\overline{z})$ is determined from the data itself without strong prior assumptions, we get a non-parametric method.

For example, deciding ahead of obtaining the data that $p_{\overline{Z}}(\overline{z})$ is a multivariate normal pdf (see App. A), with only its parameters, namely the mean vector and/or the covariance matrix, being unknown, yields an example of a parametric method. In contrast, inferring $p_{\overline{Z}}(\overline{z})$ by making a (possibly multi-dimensional) *histogram* is an example of a non-parametric method.

Strictly speaking, non-parametric methods are not parameter-free since they do contain parameters, such as the size and number of bins in the histogram example above. The main distinction is that the assumptions made about $p_{\overline{Z}}(\overline{z})$ are much weaker in the non-parametric approach. It is best to see parametric and non-parametric methods as lying at the extreme ends of a continuum of approaches in statistical analysis.

The appearance of the parameters is more explicit in parametric analysis. This is often shown by using $p_{\overline{Z}}(\overline{z}; \overline{\theta})$, where $\overline{\theta} = (\theta_0, \theta_1, \ldots, \theta_{P-1})$ is the set of parameters defining the models. For example, in the case of the multivariate normal pdf model (see Sec. A.6), $\overline{\theta}$ may include the vector of mean values, or the covariance matrix, or both.

A real-world example of a density estimation problem is finding groups in data consisting of the luminosity and temperature of some set of stars in our Galaxy. (We are talking here about the famous *Hertzsprung-Russell diagram* [50].) A parametric approach to the above problem could be based on using a *mixture model* where $p_{\overline{Z}}(\overline{z})$ is given by a superposition of prescribed (such as multivariate normal) pdfs. The fitting of mixture models to data is called *clustering* analysis. Clustering analysis can also be formulated in a non-parametric way.

In a subset of density estimation problems, the trial values are of a random vector of the form $\overline{Z} = (\overline{Y}, \overline{X})$, and the goal is to obtain the best-fit model for the conditional pdf $p_{\overline{Y}|\overline{X}}(\overline{y}|\overline{x})$. Such a problem is called a *statistical regression* problem and a model of $p_{\overline{Y}|\overline{X}}(\overline{y}|\overline{x})$ is called a *regression model*. $\overline{X}$ is called the *independent* and $\overline{Y}$ is called the *dependent* variable. The terminology in the literature on statistical analysis can vary a lot and, depending on the application area, the independent variable is also called *predictor, input,* or *feature*. Similarly, the dependent variable is also called *response, output,* or *outcome*. The set of regression models and the method used to find the best fit one among them constitute a *regression method*.

A simple example of a statistical regression problem is the prediction of temperature variation in a room during the course of a day. In this example, $z_i = (y_i, x_i)$, $i = 0, 1, \ldots, N-1$, with $y_i \in \mathbb{R}^1$ being temperature and $x_i \in \mathbb{R}^1$ being the time at which y_i is measured. Statistical analysis provides a best fit model for $p_{Y|X}(y|x)$ and this model can be used to predict the temperature at a time instant x that does not belong to the set $\{x_0, x_1, \ldots, x_{N-1}\}$.

Another example is that of an image where $\overline{x}_i \in \mathbb{Z}^2$ is the location of a pixel in the image, while $y_i \in \mathbb{R}^1$ is the intensity of the recorded image at that pixel. For a colored digital image, $\overline{y}_i = (r_i, g_i, b_i) \in \mathbb{Z}^3$, where r_i, g_i, and b_i are the red, green, and blue intensities of the recorded image. The goal of statistical regression here could be to *denoise* the image: obtain a new image given by $E[\overline{Y}|\overline{X}]$, where the expectation is taken with respect to the best fit model of $p_{\overline{Y}|\overline{X}}(\overline{y}|\overline{x})$.

Statistical regression, being a subset of density estimation, also inherits parametric and non-parametric approaches. However, there is considerably more ambiguity in the literature on the distinction between the two as far as regression problems go. Non-parametric regression can also include the case of models that are parametric but have a large number of parameters that allow greater flexibility in capturing complexity in the data. An important issue that arises in both parametric and non-parametric regression is that of *overfitting* where too many parameters can cause a model to adapt to every feature of the data, including the noise. Such a model then loses all predictive power because a small perturbation in the data can cause a significant change in the best fit model. Overfitting is an especially important issue for non-parametric regression since the models used are designed to be more adaptable to the data. The mitigation of overfitting requires the use of *model selection* and *regularization* techniques. Instances of these advanced topics will appear later on but they will not be covered in great depth in this book.

1.3 STATISTICAL REGRESSION

1.3.1 Parametric regression

In parametric regression, the set of models for $p_{\overline{Y}|\overline{X}}(\overline{y}|\overline{x})$ consists of functions that are labeled by a fixed number of parameters $\overline{\theta} = (\theta_0, \theta_1, \ldots, \theta_{P-1})$, where $\theta_i \in \mathbb{R}^1$ (or $\theta_i \in \mathbb{Z}^1$). As such, we can denote the conditional probability as $p_{\overline{Y}|\overline{X}}(\overline{y}|\overline{x}; \overline{\theta})$.

Consider the well-known example of fitting a straight line to a set $\mathbb{T} = \{(x_i \in \mathbb{R}^1, y_i \in \mathbb{R}^1)\}$, $i = 0, 1, \ldots, N - 1$, of data points. In the simplest case, the regression model is $Y = f(X; \overline{\theta}) + E$, where $f(X; \overline{\theta}) = aX + b$, $\overline{\theta} = (a, b)$, and the error E is a normal random variable (see Sec. A.6) with zero mean and variance σ^2. (Error arising from X is a subtle issue that is briefly discussed in Sec. 1.5.1.) The conditional probability model is then given by

$$p_{Y|X}(y|x; \overline{\theta}) = p_E\left(y - f(x; \overline{\theta})\right) = N(y; ax + b, \sigma^2). \quad (1.1)$$

Now, let the conditional probability of obtaining the given data, $\mathbb{T}$, for a given $\overline{\theta}$ be denoted by $L(\mathbb{T}; \overline{\theta})$. For a fixed $\mathbb{T}$,

$L(\mathbb{T}; \overline{\theta})$ considered as a function $\overline{\theta}$ is called the *likelihood function* of the data[2]. In the present case,

$$L(\mathbb{T}; \overline{\theta}) = \Pi_{i=0}^{N-1} p_{\mathrm{E}} \left(y_i - f(x_i; \overline{\theta}) \right) , \qquad (1.2)$$

$$\propto \exp \left(-\frac{1}{2\sigma^2} \sum_{i=0}^{N-1} (y_i - ax_i - b)^2 \right) . \qquad (1.3)$$

In the *Maximum Likelihood Estimation* (MLE) method, the best-fit model is obtained by maximizing the likelihood function over the model parameters $\overline{\theta}$. Since the exponential function depends monotonically on its argument, maximimization of $L(\mathbb{T}; \overline{\theta})$ is equivalent to maximizing the exponent (i.e., its natural logarithm), or (since the exponent is negative) minimizing

$$L_S(\mathbb{T}; \overline{\theta}) = \frac{1}{2\sigma^2} \sum_{i=0}^{N-1} \left(y_i - f(x_i; \overline{\theta}) \right)^2 . \qquad (1.4)$$

over $\overline{\theta}$. Most readers will recognize $L_S(\mathbb{T}; \overline{\theta})$ – barring the unimportant constant factor $1/(2\sigma^2)$ – as the *sum of squared residuals* that is used in *least squares* fitting of a straight line to a set of points. From here on, therefore, we will call L_S the *least squares function*.

The minimization problem above illustrates the role of optimization in statistical regression. It is straightforward to do the optimization analytically in the straight line fitting problem because the parameters appear linearly in the regression model, making $L_S(\mathbb{T}; \overline{\theta})$ a simple quadratic function of the parameters. In general, however, the optimization problem is not trivial if the parameters appear non-linearly in the regression model. The only option then is to optimize the likelihood numerically.

In this book, we will use the following non-linear model when discussing the application of SI methods to parametric regression.

---

[2]Note that the likelihood function could have been introduced in the discussion of density estimation itself. However, we prefer to do so here because our focus in this book is only on statistical regression.

Quadratic chirp in iid noise:

$$Y = q_c(X; \overline{\theta}) + E , \qquad (1.5)$$

where

$$q_c(X) = A\sin(2\pi\phi(X)) , \qquad (1.6)$$
$$\phi(X) = a_1 X + a_2 X^2 + a_3 X^3 . \qquad (1.7)$$

There are 4 parameters in this model $\overline{\theta} = (A, a_1, a_2, a_3)$. (For clarity in notation, $\overline{\theta}$ is dropped when there is no scope for confusion.) Fig. 1.1 shows an example of $q_c(X)$ and a data realization for regularly spaced values of X.

As before (c.f., Eq. 1.4), the maximization of the likelihood reduces to the minimization of

$$L_S(\mathbb{T}; \overline{\theta}) = \frac{1}{2\sigma^2} \sum_{i=0}^{N-1} (y_i - q_c(x_i))^2 . \qquad (1.8)$$

As we will see later in Sec. 5.1.1, the minimization of $L_S(\mathbb{T}; \overline{\theta})$ over A can be performed analytically, leaving behind a function that must be minimized numerically over the remaining three parameters.

1.3.2 Non-parametric regression

In non-parametric regression problems, no explicit assumption is made about the functional form of models for $p_{\overline{Y}|\overline{X}}(\overline{y}|\overline{x})$. Instead, the assumptions made are about some global properties of the models. For example, one may assume that $f(\overline{X}) = E[\overline{Y}|\overline{X}]$, the conditional expectation of $\overline{Y}$ given $\overline{X}$, is a "smooth" function in the sense that its derivatives exist up to a prescribed order. In other cases, one may forgo smoothness but impose a property such as positivity on $f(\overline{X})$.

Another approach in non-parametric regression problems is to adopt a functional form for $p_{\overline{Y}|\overline{X}}(\overline{y}|\overline{x})$ but keep the set of parameters sufficiently large such that the functional form is very flexible. Since having too much flexibility can lead to

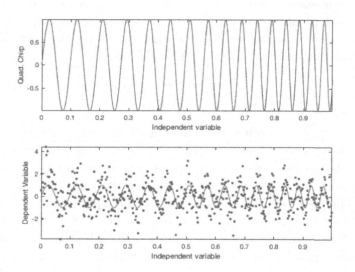

Figure 1.1 The top panel shows a quadratic chirp $q_c(X)$, defined in Eqs. 1.6 and 1.7, for equally spaced values of X with $x_i = i\Delta$, $\Delta = 1/512$, and $i = 0, 1, \ldots, 511$. The parameter values are $A = 1$, $a_1 = 10$, and $a_2 = a_3 = 3$. The bottom panel shows a data realization $\{(x_i, y_i)\}$, $i = 0, 1, \ldots, 511$, following the model given in Eq. 1.5.

a model that offers a very good fit but only to the given data, namely an overfitted model, additional constraints are imposed that restrict the choice of parameter values.

We will illustrate the application of SI methods to non-parametric regression problems through the following example.

Spline-based smoothing:

The regression model is $Y = f(X) + E$, where $f(X)$ is only assumed to be a smooth, but an otherwise unknown, function. One way to implement smoothness is to require that the average squared curvature of $f(X)$, defined as

$$\frac{1}{(b-a)} \int_a^b dx \left(\frac{d^2 f}{dx^2}\right)^2 , \qquad (1.9)$$

for $x \in [a, b]$ be sufficiently small. It can be shown that the best least-squares estimate of $f(X)$ under this requirement must be a *cubic spline*. (See App. B for a bare-bones review of splines and the associated terminology.)

Thus, we will assume $f(X)$ to be a cubic spline defined by a set of M *breakpoints* denoted by $\overline{b} = (b_0, b_1, \ldots, b_{M-1})$, where $b_{i+1} > b_i$. The set of all cubic splines defined by the same $\overline{b}$ is a linear vector space. One of the basis sets of this vector space is that of *B-spline* functions $B_{j,4}(x; \overline{b})$, $j = 0, 1, \ldots, M-1$. (It is assumed here that the cubic splines decay to zero at $X = b_0$ and $X = b_{M-1}$.) It follows that $f(X)$ is a linear combination of the B-splines given by

$$f(X) = \sum_{j=0}^{M-1} \alpha_j B_{j,4}(X; \overline{b}) , \qquad (1.10)$$

where $\overline{\alpha} = (\alpha_0, \alpha_1, \ldots, \alpha_{M-1})$ is the set of coefficients in the linear combination.

Finding the best-fit model here involves finding the values of $\overline{\theta} = (\overline{\alpha}, \overline{b})$ for which the least-squares function

$$L_S(\mathbb{T}; \overline{\theta}) = \frac{1}{2\sigma^2} \sum_{i=0}^{N-1} \left(y_i - \sum_{j=0}^{M-1} \alpha_j B_{j,4}(X; \overline{b}) \right)^2 , \quad (1.11)$$

is minimized. If $\bar{b}$ is fixed, the coefficients $\bar{\alpha}$ can be found easily since, as in the case of the straight line fitting problem, they appear quadratically in the least-squares function. The regression method based on this approach is called *spline smoothing*.

However, if $\bar{b}$ is allowed to vary, then the minimization problem cannot be solved analytically alone and numerical methods must be used. Varying the breakpoints in order to find the best fit spline is called *knot optimization* (See App. B for the distinction between knots and breakpoints) and the corresponding regression method is called *regression spline*. It is known [24] to provide a better fitting model than spline smoothing but it is also an extremely challenging optimization problem that has prevented its use from becoming more widespread. Various approaches have been devised to tackle this problem[3] but it is only recently that SI methods have begun to be explored (e.g., [17]) for this interesting problem.

To test the performance of the regression spline method, we will use data where

$$f(X) \propto B_{0,4}(X; \bar{c}) , \qquad (1.12)$$

namely, a single B-spline function defined by breakpoints $\bar{c}$. The values of X are confined to $[0, 1]$ and $\bar{c} = (0.3, 0.4, 0.45, 0.5, 0.55)$. Fig. 1.2 shows examples of both the $f(X)$ defined above and the data.

1.4 HYPOTHESES TESTING

As discussed above, statistical regression is a special case of the more general density estimation problem where one has to obtain the best fit among a set of models for the joint pdf $p_{\bar{Z}}(\bar{z})$ of data $\bar{z}$.

A different statistical analysis problem, that of hypotheses testing, is concerned with deciding which among two sets of

---

[3] A good entry point for the literature on this subject is the journal *Computer-Aided Design*.

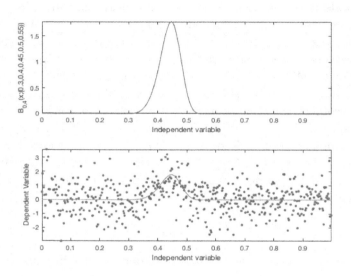

Figure 1.2 The top panel shows the function $f(X)$ in Eq. 1.12 that is used for generating test data for the regression spline method. Here, X is equally spaced with $x_i = i\Delta$, $\Delta = 1/512$, and $i = 0, 1, \ldots, 511$, and $f(X) = 10B_{0,4}(x; \bar{c})$, with $\bar{c} = (0.3, 0.4, 0.45, 0.5, 0.55)$. The bottom panel shows a data realization.

models best explains the data. The new element of making a decision in the presence of noise leads to two types of errors, called *false alarm* and *false dismissal*, that are absent in density estimation. The theory behind hypotheses testing is an elaborate subject in itself and lies outside the scope of this book. However, it will suffice here to discuss just a few of the main results from this theory.

One is that it is possible to come up with a "best" test in the special case of *binary hypotheses* where there are only two competing models to choose from. An example is when we have two models, $Y = E$ and $Y = f(X) + E$, with $f(X)$ a known function, and a choice needs to be made between the two models. Traditionally, the two models are respectively called the H_0 or *null hypothesis*, and the H_1 or *alternative hypothesis*. The joint pdf of the data under H_i, $i = 0, 1$, is denoted by $p_{\overline{Z}}(\overline{z}|H_i)$.

Deciding on H_1 (H_0) when the true hypothesis is H_0 (H_1) leads to a false alarm (false dismissal) error. Under the *Neyman-Pearson* criterion [31], the best test for a binary hypotheses is the one that minimizes the false dismissal probability for a given probability of false alarm. The test consists of computing the *log-likelihood ratio* (L_R) of the data,

$$L_R = \ln \left(\frac{p_{\overline{Z}}(\overline{z}|H_1)}{p_{\overline{Z}}(\overline{z}|H_0)} \right) , \qquad (1.13)$$

and comparing L_R to a threshold value η (called the *detection threshold*): H_1 is selected if $L_R \geq \eta$, and H_0 otherwise.

Note that L_R is a random variable because of noise in the data $\overline{z}$. Depending on which hypothesis is the true one, its pdf is $p_{L_R}(x|H_i)$. The detection threshold is fixed once a false alarm probability, P_{FP}, given by

$$P_{\text{FP}} = \int_\eta^\infty dx\, p_{L_R}(x|H_0) , \qquad (1.14)$$

for the test is specified.

A scenario that is closer to real-world problems than the binary hypothesis test, is that of a *composite hypotheses* test. An example is when we have two sets of models, H_0 : Y = E and H_1 : Y = $f(X; \overline{\theta})$ + E, where $f(X; \overline{\theta})$ is a function parameterized by $\overline{\theta}$ (such as the quadratic chirp with $\overline{\theta}$ = (A, a_1, a_2, a_3)) but the true value of $\overline{\theta}$ is unknown. Here, the set of models under H_1 has many models in it while the set of models under H_0 has just one. In the general case, the joint pdf of the data under H_1 is $p_{\overline{Z}}(\overline{z}|H_1; \overline{\theta})$, where $\overline{\theta}$ is unknown. (Here, $\overline{\theta}$ should be seen as a label for models and, hence, includes the case of non-parametric regression also.)

The theory of hypotheses testing tells us that there is no best test, except in some trivial cases, that can minimize the false dismissal probability across all values of $\overline{\theta}$ for a given P_{FP}. However, it is found quite often that a straightforward generalization of the log-likelihood ratio test, called the *Generalized Likelihood Ratio Test* (GLRT), gives a good performance. The GLRT requires that we compute a quantity L_G, given by

$$L_G = \max_{\overline{\theta}} \ln \left(\frac{p_{\overline{Z}}(\overline{z}|H_1; \overline{\theta})}{p_{\overline{Z}}(\overline{z}|H_0)} \right), \qquad (1.15)$$

and compare it to a detection threshold. As before, H_1 is selected if L_G exceeds the threshold.

For a regression problem where $\overline{Z} = (\overline{Y}, \overline{X})$, $\overline{X}$ and $\overline{Y}$ being the independent and dependent variables respectively, we get

$$L_G = \max_{\overline{\theta}} \ln \left(\frac{p_{\overline{Y}|\overline{X}}(\overline{y}|\overline{x}; H_1, \overline{\theta})}{p_{\overline{Y}|\overline{X}}(\overline{y}|\overline{x}; H_0)} \right). \qquad (1.16)$$

Note that, since $\overline{\theta}$ is absent in $p_{\overline{Y}|\overline{X}}(\overline{y}|\overline{x}; H_0)$, the maximimization over $\overline{\theta}$ in Eq. 1.16 is only over $p_{\overline{Y}|\overline{X}}(\overline{y}|\overline{x}; H_1, \overline{\theta})$. This is nothing but the likelihood function introduced in Sec. 1.3.1. Thus, GLRT is closely linked to statistical regression.

The general form of the GLRT for the examples in Secs. 1.3.1 and 1.3.2 is,

$$L_G = \min_{\overline{\theta}} L_S(\mathbb{T}; \overline{\theta}) - \frac{1}{2\sigma^2} \sum_{i=0}^{N-1} y_i^2, \qquad (1.17)$$

where $\overline{\theta}$ denotes the respective parameters. Thus, L_G in these examples is obtained in a straightforward manner from the global minimum of the least squares function.

1.5 NOTES

The brevity of the overview of statistical analysis presented in this chapter demands incompleteness in many important respects. Topics such as the specification of errors in parameter estimation and obtaining confidence intervals instead of point estimates are some of the important omissions. Also omitted are some of the important results in the theoretical foundations of statistics, such as the Cramer-Rao lower bound on estimation errors and the RBLS theorem.

While the overview, despite these omissions, is adequate for the rest of the book, the reader will greatly benefit from perusing textbooks such as [26, 25, 15] for a firmer foundation in statistical analysis. Much of the material in this chapter is culled from these books. Non-parametric smoothing methods, including spline smoothing, are covered in detail in [20].

Some important topics that were touched upon only briefly in this chapter are elaborated upon a little more below.

1.5.1 Noise in the independent variable

In the examples illustrating statistical regression in Sec. 1.2, one may object to treating the independent variable, $\overline{X}$, as random. After all, we think of a quantity such as the time instant at which the temperature of a room is measured as being completely under our control and not the trial outcome of a random variable.

The nature of the independent variable – random or not – is actually immaterial to a statistical regression problem since one is only interested in the conditional probability $p_{\overline{Y}|\overline{X}}$, where the value of $\overline{X}$ is a given.

That said, one must remember that even the independent variable in real data is the outcome of some measurement and, hence, has error in it. For example, time is measured using a clock and no clock is perfect. Usually, this error is negligible and can be ignored compared to the error in the measurement of the dependent variable. Problems where the error in the independent variable is not negligible are called *errors-in-variables* regression problems. The Wikipedia entry [49] on this topic is a good starting point for further exploration.

1.5.2   Statistical analysis and machine learning

The advent of computers has led to a dramatic evolution of statistical methodology. Some methods, such as *Bootstrapping* [13], are purely a result of merging statistics with computing. At the same time, it is now widely accepted that the probabilistic framework underlying statistical analysis is also fruitful in advancing a holy grail of computing, namely the creation of machines that can learn like humans from experience and apply this learning to novel situations. Thus, there is a strong overlap between the concepts and methods in the fields of statistical analysis and *machine learning*. However, due to their different origins and historical development, there is a considerable divergence in the terminology used in the two fields for essentially the same concepts.

In the machine learning literature, the task of obtaining the best-fit model is called *learning*, density estimation is called *unsupervised learning*, and statistical regression is called *supervised learning*. The data provided in a supervised learning problem is called *training data*.

The main difference between statistical regression and machine learning is that the element of randomness is mainly localized in the dependent variable, $\overline{Y}$, in the former while it is predominantly in the independent variable, $\overline{X}$, for the latter. This is because instances of X, such as images, are obtained from generic sources, such as the web, without a fine degree of control.

The emphasis in machine learning is on obtaining a best fit model for the joint pdf of data from a training data set such that the model generalizes well to new data. Since the typical applications of machine learning, such as object recognition in images or the recognition of spoken words, involve inherently complicated models for the joint pdf, a large amount of training data is required for successful learning. This requirement has dovetailed nicely with the emergence of big data, which has led to the availability of massive amounts of training data related to real world tasks. It is no surprise, therefore, that the field of machine learning has seen breathtaking advances recently.

Deeper discussions of the overlap between statistical and machine learning concepts can be found in [15, 19].

Stochastic Optimization Theory

CONTENTS

This chapter touches upon some of the principal results from optimization theory that have a bearing on swarm intelligence methods. After establishing the terminology used to describe optimization problems in general, the two main classes of these problems are discussed that determine the types of optimization methods – deterministic or stochastic – that can be used. Some general results pertaining to stochastic optimization methods are outlined that set the boundaries for their performance. This is followed by a description of the metrics used in the characterization and comparison of stochastic optimization methods. A simple strategy is described for extracting better performance from stochastic optimization methods that is well-suited to parallel computing.

2.1 TERMINOLOGY

The objective in an optimization problem is to find the optimum (maximum or minimum) value of a *fitness function* $f(\overline{x})$, $\overline{x} \in \mathbb{D} \subseteq \mathbb{R}^D$. The subset $\mathbb{D}$ is called the *search space* for the optimization problem and D is the *dimensionality* of the search space. Alternative terms used in the literature are *objective function* and *constraint space* (or *feasible set*) for the fitness function and search space respectively. (Strictly speaking, the above definition is that of a *continuous optimization* problem. We do not consider *discrete* or combinatorial optizimation problems, where $\overline{x} \in \mathbb{Z}^D$, in this book.)

The location, denoted as $\overline{x}^*$ below, of the optimum is called the *optimizer* (*maximizer* or *minimizer* depending on the type of optimization problem). Since finding the maximizer of a fitness function $f(\overline{x})$ is completely equivalent to finding the minimizer of $-f(\overline{x})$, we only consider minimization problems in the following.

Formally stated, a minimization problem consists of solving for $\overline{x}^*$ such that

$$f(\overline{x}^*) \leq f(\overline{x}), \forall \overline{x} \in \mathbb{D}. \tag{2.1}$$

Equivalently,

$$f(\overline{x}^*) = \min_{\overline{x} \in \mathbb{D}} f(\overline{x}). \tag{2.2}$$

This is stated more directly as,

$$\overline{x}^* = \arg\min_{\overline{x} \in \mathbb{D}} f(\overline{x}). \tag{2.3}$$

Note that the minimizer of a function need not be a unique point. A simple counterexample is a fitness function that has a constant value over $\mathbb{D}$. Then all $\overline{x} \in \mathbb{D}$ are minimizers of the fitness function.

One can classify a minimizer further as a *local minimizer* or a *global minimizer*. The corresponding values of the fitness function are the *local minimum* and *global minimum* respectively. The definition of a minimizer given above is that of the global minimizer over $\mathbb{D}$. If $f(\overline{x})$ is differentiable and its gradient

$$\nabla f(\overline{x}) = \left(\frac{\partial f}{\partial x_0}, \frac{\partial f}{\partial x_1}, \ldots, \frac{\partial f}{\partial x_{N-1}} \right), \quad (2.4)$$

vanishes $(\nabla f = (0, 0, \ldots, 0))$ at $\overline{x}^*$, then $\overline{x}^*$ is a local minimizer. Formally, a local minimizer is the global minimizer but over an open subset of the search space. A local minimizer may also be the global minimizer over the whole search space but this need not be true in general.

In this chapter, the following functions serve as concrete examples of fitness functions.

Generalized Rastrigin:

$$f(\overline{x}) = \sum_{i=1}^{D} \left[x_i^2 - 10\cos(2\pi x_i) + 10 \right]. \quad (2.5)$$

Generalized Griewank:

$$f(\overline{x}) = \frac{1}{4000} \sum_{i=1}^{D} x_i^2 - \prod_{i=1}^{D} \cos\left(\frac{x_i}{\sqrt{i}} \right) + 1. \quad (2.6)$$

Fig. 2.1 shows the generalized Rastrigin function for $D = 2$. The global minimizer of this function is the origin $\overline{x}^* = (0, 0, \ldots, 0)$. The function also exhibits numerous local minima. Here, the global minimizer is also a local one.

2.2 CONVEX AND NON-CONVEX OPTIMIZATION PROBLEMS

A set $\mathbb{D} \subset \mathbb{R}^D$ is said to be a *convex set* if for any two elements $\overline{a}, \overline{b} \in \mathbb{D}$ and $0 < \lambda < 1$, the element $\overline{c} = (1 - \lambda)\overline{a} + \lambda\overline{b}$ also belongs to $\mathbb{D}$. In other words, all the points on the straight line joining any two points in $\mathbb{D}$ also lie within $\mathbb{D}$. As an example, the points constituting an elliptical area in a two dimensional plane form a convex set.

Consider a function $f(\overline{x})$ with $\overline{x} \in \mathbb{D} \subset \mathbb{R}^D$. As illustrated in Fig. 2.1 for the $D = 2$ case, the function defines a "surface"

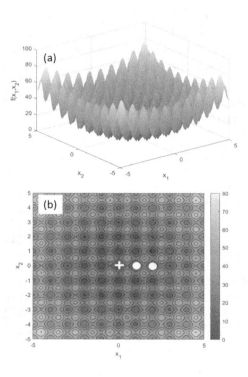

Figure 2.1 (a) The generalized Rastrigin function for a 2-dimensional search space, and (b) its contour plot showing the global ('+') and some local ('○') minimizers.

over $\mathbb{D}$. Taking all the points above this surface yields a "volume" in $D + 1$ dimensional space. If this volume is convex, then the function surface is said to be convex. The formal definition goes as follows: Take the subset $\mathbb{D}_f$ of the Cartesian product set $\mathbb{D} \times \mathbb{R}$ such that for each element $(\overline{x}, y) \in \mathbb{D}_f$, $\overline{x} \in \mathbb{D}$ and $y \geq f(\overline{x})$. If $\mathbb{D}_f$ is convex, then $f(\overline{x})$ is a convex function. While we skip the formal proof here, it is easy to convince oneself that the definition of convexity of a function implies that $f((1 - \lambda)\overline{a} + \lambda\overline{b}) \leq (1 - \lambda)f(\overline{a}) + \lambda f(\overline{b})$. This inequality is an alternative definition of a convex function.

One of the principal theorems in optimization theory states that the global minimum of a convex function $f(\overline{x})$ over a convex set $\mathbb{D}$ is also a local minimum and that there is only one local minimum. A *non-convex* function, on the other hand, can have multiple local minima in the search space and the global minimum need not coincide with any of the local minima. The Rastrigin function is an example of a non-convex fitness function. The optimization of a non-convex function is a *non-convex optimization* problem.

The local minimizer of a convex function can be found using the *method of steepest descent*: step along a sequence of points in the search space such that at each point, the step to the next point is directed opposite to the gradient of the fitness function at that point. Since the gradient at the local minimizer vanishes, the method will terminate once it hits the local minimizer. Steepest descent is an example of a *deterministic optimization* method since the same starting point and step size always results in the same path to the local minimizer.

For a non-convex optimization problem, the method of steepest descent will only take us to some local minimizer. When the local minima are spaced closely, steepest descent will not move very far from its starting point before it hits a local minimizer and terminates. Thus, steepest descent or any similar deterministic optimization method is not useful for locating the global minimizer in a non-convex optimization problem.

Before proceeding further, we caution the reader that the description of steepest descent just presented is very simplistic. In reality, much more sophistication is required in the choice of step sizes and other parameters. However, since we do not dwell on local minimization further in this book, the simple description above is appropriate for our purpose.

2.3 STOCHASTIC OPTIMIZATION

The only deterministic method to find the global optimum in a non-convex optimization problem is to set up a grid of points in the search space, evaluate the fitness function at each point, and interpolate these values to find the minimum fitness value. While this is a reasonable approach in low-dimensional search spaces (2 to 3), it is easy to see that it is a poor strategy as the dimensionality grows.

Take the simple case of a fixed number, N_G, of grid locations along each dimension. Then the total number of grid points throughout a search space grows exponentially fast, as N_G^D, with the dimensionality, D, of the search space. As an example, with $N_G = 100$, the total number of points goes up to 10^{10} even with a modest dimensionality of $D = 5$.

Even with smarter grid-based search strategies, it is practically impossible to avoid the exponential explosion in the number of points with the dimensionality of a search space. In general, all deterministic global minimization methods, which are all variations of a grid-based search, become computationally infeasible quite rapidly even in problems with moderate search space dimensionalities.

Given the infeasibility of deterministic methods for non-convex optimization, almost every method proposed to solve this problem involves some element of randomness. Such a method, which we call a *stochastic optimization* method, evaluates the fitness at points that are trial outcomes of a random vector. The element of randomness in the search strategy is a critical ingredient because it allows a stochastic optimization method to avoid getting trapped by local minimizers. (There also exist stochastic local minimization methods, but we do not consider them here.) It is best to clarify here that, in the literature, stochastic optimization also refers to the optimization of stochastic functions using methods such as *stochastic approximation*. While the fitness functions in statistical regression are indeed stochastic, since the underlying data is a realization from a joint pdf, this book only considers the more restricted definition where a stochastic optimization method is applied to a deterministic function.

There exists a mathematical framework [44] that includes most stochastic optimization methods as special instances of a general method. The reader may wish to peruse App. A for a brief review of probability theory and associated notation before proceeding further.

General stochastic optimization method: An iterative algorithm where $\overline{x}[k]$ is the candidate position of the minimizer at step $k \in \mathbb{Z}^1$.

- *Initialization*: Set $k = 0$ and draw $\overline{x}[0] \in \mathbb{D} \subset \mathbb{R}^D$, where $\mathbb{D}$ is the search space. Normally, $\overline{x}[0]$ is drawn from a specified joint pdf, an example of which is provided later.

- Repeat

  - *Randomization*: Draw a trial value $\overline{v}[k] \in \mathbb{R}^D$ of a vector random variable $\overline{V} = (V_0, V_1, \ldots, V_{D-1})$ with a joint pdf $p_{\overline{V}}(\overline{x}; k)$. (The dependence on k means that the pdf can change form as the iterations progress.)

  - *Update position*: $\overline{x}[k+1] = A(\overline{x}[k], \overline{v}[k])$, which indicates that the new position is some prescribed function of the old position and $\overline{v}[k]$. The range of the function should be $\mathbb{D}$, ensuring that the new position stays within the search space.

  - *Update pdf*: $p_{\overline{V}}(\overline{x}; k) \rightarrow p_{\overline{V}}(\overline{x}; k+1)$.

  - Increment k to $k + 1$.

- Until a *termination* condition becomes true. The simplest termination condition is to stop after a preset number of iterations are completed, but more sophisticated alternatives are also possible.

Different stochastic optimization algorithms correspond to different choices for $p_{\overline{V}}(\overline{x}; k)$, $A(\overline{x}, \overline{y})$ and the update rule for $p_{\overline{V}}(\overline{x}; k)$. It is important to emphasize here that each of these components can contain a set of more elaborate operations. For example, the trial value of $\overline{V}$ could itself be the result of combining the trial values of a larger set of random vectors and the set of fitness values found in past iterations.

Since a stochastic optimization algorithm, by definition, does not have a predictable terminal location in the search space, success for such algorithms must be defined in probabilistic terms. One cannot define success as the probability of $\overline{x}[k] = \overline{x}^*$ because $\overline{x}[k]$ is the trial value of a continuous vector random variable and the probability that it takes an exact value is zero (see App. A). However, it is legitimate to ask if $\overline{x}[k]$ falls in some volume, no matter how small, containing $\overline{x}^*$.

Formally, one defines an *optimality region*, for a given $\epsilon > 0$, as the set $\mathbb{R}_\epsilon = \{\overline{x} \in \mathbb{D} | f(\overline{x}) \leq f(\overline{x}^*) + \epsilon\}$. A stochastic algorithm is said to converge to the global minimum if

$$\lim_{k \to \infty} P(\overline{x}[k] \in \mathbb{R}_\epsilon) = 1 \,. \tag{2.7}$$

That is, the probability of the final solution landing in the optimality region goes to unity asymptotically with the number of iterations.

The following conditions are sufficient for the general stochastic optimization method to converge.

- **Algorithm condition**: The fitness value always improves; $f(\overline{x}[k+1]) \leq f(\overline{x}[k])$ and if $\overline{v}[k] \in \mathbb{D}$, $f(\overline{x}[k+1]) \leq f(\overline{v}[k])$.

- **Convergence condition**: There is no region of the search space that the random vector, $\overline{V}$, does not visit at least once. Equivalently, for any given region $\mathbb{S} \subset \mathbb{D}$, the probability of not getting a trial value $\overline{v}[k]$ in $\mathbb{S}$ diminishes to zero as the iterations progress; $\prod_{k=0}^{\infty} (1 - P(\overline{v}[k] \in \mathbb{S})) = 0$.

Unfortunately, these two conditions work against each other: On one hand, the method must keep descending to lower fitness values but on the other, it should also examine every little region of the search space no matter how small. It is remarkable, therefore, that it is at all possible to come up with stochastic optimization methods that satisfy these conditions.

One such algorithm is "Algorithm 3" of [44]) that is defined as follows. (a) The position update rule is simply

$$
A(\overline{x}, \overline{y}) = \begin{cases} \overline{y}, & f(\overline{y}) < f(\overline{x}), \\ \overline{x}, & \text{otherwise} \end{cases} . \tag{2.8}
$$

(b) In the randomization step, the trial value $\overline{v}[k]$ of the vector random variable $\overline{V}$ is obtained as follows. First, a trial value $\overline{w}$ is drawn for a vector random variable $\overline{W}$ that has a uniform joint pdf over $\mathbb{D}$: This means that $P(\overline{w} \in \mathbb{S} \subset \mathbb{D}) = \mu(\mathbb{S})/\mu(\mathbb{D})$ for any $\mathbb{S} \subset \mathbb{D}$, where $\mu(\mathbb{A})$ denotes the volume of $\mathbb{A}$. Then, a local minimization method is started at $\overline{w}$. The location returned by the local minimization is $\overline{v}[k]$. (c) The pdf $p_{\overline{V}}$ from which $\overline{v}[k]$ is drawn is determined implicitly by the procedure in (b) above. (c) The algorithm is initialized by drawing $\overline{x}[0]$ from the same joint pdf as that of $\overline{W}$.

While guaranteed to converge in the limit of infinite iterations, the above algorithm is not a practical one. First, since the number of iterations must always be finite in practice, convergence need not happen. Secondly, the computational cost of evaluating the fitness function is an important consideration in the total number of iterations that can be performed. This often requires moving the number of iterations towards as few as possible, a diametrically opposite condition to the one above for convergence. Shortening the number of iterations necessarily means that an algorithm will not be able to visit every region of the search space.

Thus, practical stochastic optimization algorithms can never be guaranteed to converge in a finite number of iterations. Different methods will have different probabilities of landing in the optimality region, and this probability, which depends on a complicated interaction between the algorithm and the fitness function, is usually difficult to compute. In fact, the *no free lunch* (NFL) theorem in optimization theory [52] tells us that there is no magic stochastic optimization method that has the best performance across all optimization problems. More precisely, the NFL theorem states that, averaged across all fitness functions, the performance of one stochastic optimization method is as good as any other. In particular, this means that there is an infinity of fitness functions over which a pure random search is better than any other more sophisticated optimization method (and vice versa)! That said, the NFL theorem does not prevent one stochastic optimization method from being the best over a suitably restricted class of fitness functions.

2.4 EXPLORATION AND EXPLOITATION

While the sufficiency conditions for convergence are not satisfied by most practical stochastic optimization methods, they serve to lay out the type of behavior a method must have in order to be useful. The convergence condition entails that a stochastic optimization method should spend some fraction of iterations in exploring the search space before finding a candidate optimality region for a deeper search using the remaining iterations. These two phases are generally called *exploration* and *exploitation*. During both of these phases, the algorithm must continue to descend the fitness function in keeping with the algorithm condition.

The relative lengths of the exploration and exploitation phases that an optimization method should have depend, among other factors, on the degree to which the local minima of a fitness function are crowded together, a property that is often called *ruggedness*, and the dimensionality of the search space. Too short an exploration phase can lead to convergence to a local minima rather than the global minimum. Too long an exploration phase, on the other hand, increases the computational cost of the optimization.

The effect of the dimensionality of the search space on the exploration behavior of an optimization method can be particularly counter-intuitive. Consider the initialization step for a stochastic optimization method and let us assume that the initial location is obtained from a uniform pdf over the search space. For simplicity, let the search space be a cube in D-dimensions (a *hypercube*) with side length b in each dimension. This means that each component x_i, $i = 0, 1, \ldots, D$, of the position vector $\overline{x} \in \mathbb{D}$ can vary independently in an interval $[a_i, a_i + b]$. (A more general definition of a hypercube is that $x_i \in [a_i, b_i]$, $b_i - a_i > 0$.)

Given that the pdf is uniform over the search space, our intuition built on low-dimensional examples, such as the uniform pdf over the real line, would suggest that the fraction of trials in which the initial location falls around the center of the cube is the same or more than that around a point near the boundary. However, this is not so: the volume contained in an inner cube of side length $a < b$ around the center, which is a^D, decreases exponentially relative to the volume $b^D - a^D$ in the rest of the cube. Hence, most of the initial locations will actually be quite far from the center.

For example, for $a = 0.9b$, the ratio of the inner and outer volumes for $D = 2$ is 2.7, showing that most of the volume is occupied by the inner cube, but it is only 0.14 when $D = 20$! Thus, even though the inner cube sides are nearly as big as those of the outer cube, most of the volume in the latter is near its walls. Counter-intuitive effects like this in higher dimensions are clubbed together under the term *curse of dimensionality*. Due to such effects, stochastic optimization methods that shorten exploration relative to exploitation may work well and converge quickly in low dimensional search spaces but fail miserably as the dimensionality grows.

In order to be a practical and useful tool, it should be easy for the user of a stochastic optimization method to control the relative lengths of the exploration and exploitation phases. If the algorithm has a large number of parameters whose influence on these two phases is obscure and difficult to disentangle, it becomes very difficult for a non-expert user to set these parameters for achieving success in a real-world problem.

2.5 BENCHMARKING

Since stochastic optimization methods are neither guaranteed to converge nor perform equally well across all possible fitness functions, it is extremely important to test a given stochastic optimization method across a sufficiently diverse suite of fitness functions to gauge the limits of its performance. This process is called *benchmarking*.

Any measure of the performance of a stochastic optimization method on a given fitness function is bound to be a random variable. Hence, benchmarking and the comparison of different methods must employ a statistical approach. This requires running a given method on not only a sufficiently diverse set of fitness functions but also running it multiple times on each one. The initialization of the method and the sequence of random numbers should be statistically independent across the runs.

To keep the computational cost of doing the resulting large number of fitness function evaluations manageable, the fitness functions used for benchmarking must be computationally inexpensive to evaluate. At the same time, these functions should provide enough of a challenge to the optimization methods if benchmarking is to be useful. In particular, the fitness functions should be extensible to an arbitrary number of search space dimensions. Several such benchmark fitness functions have been developed in the literature and an extensive list can be found in Table I of [4]. The Rastrigin and Griewank functions introduced earlier in Eq. 2.5 and Eq. 2.6 are in fact two such benchmark fitness functions.

Comparisons of optimization methods based on benchmark fitness functions should not only look at the final fitness value found but also their exploration-exploitation trade-off. The following is a standard method that is often adopted in the literature in this context.

For each method, multiple runs are carried out on a given fitness function and the fitness value at each iteration for each run are recorded. Let the sequence of fitness values obtained at each iteration step be denoted by $\overline{f}^{(i)}$ for the ith independent run. Then

$$\overline{f}^{(\text{av})} = \frac{1}{N_{\text{runs}}} \sum_{i=1}^{N_{\text{runs}}} \overline{f}^{(i)}, \qquad (2.9)$$

is the sequence of average fitness values found by the optimization method as a function of the number of elapsed iterations. Since the actual number of fitness function evaluations made at each iteration step depends on the details of the optimization algorithm, $\overline{f}^{(\text{av})}$ is often shown as a function of the number of fitness evaluations rather than the number of iterations. This leads to a fairer comparison of optimization methods because the computational cost of algorithms is often one of the prime considerations when deciding which one to use.

Fig. 2.2 shows examples of $\overline{f}^{(\text{av})}$ and illustrates several of the concepts covered earlier. In this figure, two different stochastic optimization methods (described later in Sec. 4.7.2) are used for minimizing the generalized Griewank function over a 30 dimensional search space. The curves clearly show the exploration and exploitation phases for each of the methods: The fitness value drops rapidly during the exploration phase, followed by a much slower decay during the exploitation phase. As is evident, the method with the longer exploration phase finds a better fitness value. Note that in both cases, the methods do not converge to the true global minimum, which has a value of zero. The leveling off of the curves indicates that, due to the finite length of the exploration phase and consequent failure to maintain the convergence condition, convergence may not occur even if the number of iterations increases without limit.

In addition to the $\overline{f}^{(\text{av})}$ plot, another common approach to comparing the performance of two methods is to perform a two-sample hypotheses test, such as Student's t-test or the Kolmogorov-Smirnov test [31], on the sample of final fitness values from multiple independent runs.

2.6 TUNING

While benchmarking is essential in the study and objective comparison of stochastic optimization methods, it does not guarantee, thanks to the NFL theorem, that the best performer on benchmark fitness functions will also work well in the intended application area. At best, benchmarking establishes the order in which one should investigate the effectiveness of a set of methods. Having arrived at an adequately performing method, one must undertake further statistical characterization in order to understand and improve its performance on the problem of interest. We call this process, which is carried out with the fitness functions pertaining to the actual application area, *tuning*.

To summarize, benchmarking is used for comparing optimization methods or delimiting the class of fitness functions that a method is good for. It is carried out with fitness functions that are crafted to be computationally cheap but sufficiently challenging. Tuning is the process of improving the performance of a method on the actual problem of interest and involves computing the actual fitness function (or surrogates).

A pitfall in tuning is *over-tuning* the performance of an optimization method on one specific fitness function. This is especially important for statistical regression problems because the fitness function, such as the least squares function in Eq. 1.4, depends on the data realization and, hence, changes every time the data realization changes. Thus, over-tuning the performance for a particular data realization runs the danger of worsening the performance of the method on other real-

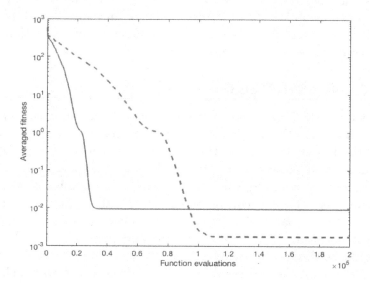

Figure 2.2 Fitness value averaged over independent trials for two different stochastic optimization methods. The X-axis shows the number of fitness evaluations. The fitness function is the 30-dimensional generalized Griewank function (Eq. 2.6) with each dimension of the search space confined to the interval $[-600, 600]$. For each curve, the averaging was performed over 30 independent runs. The two methods have different durations for the exploration phase, which extends out to roughly 3×10^4 and 1×10^5 function evaluations for the solid and dashed curves respectively. Details of the methods used are provided in Sec. 4.7.2.

izations. In fact, this situation is pretty much guaranteed to occur due to the NFL theorem. Tuning of optimization methods for statistical regression problems must be carried out using metrics that are based on a sufficiently large set of data realizations. This almost always requires using *simulated data* since there may just be one realization of real data that is available. The use of simulations in tuning is covered in more detail in Ch. 5.

2.7 BMR STRATEGY

A simple strategy to safeguard against over-tuning is *best-of-M-runs* (BMR). In this strategy, a stochastic optimization method is tuned only to the point where a moderate probability, p_{hit}, of success in any single run is achieved. (Here, success is to be interpreted in the sense of landing in an optimality region but it can be redefined, as will be discussed in Sec 5.2.1, for parametric statistical regression problems.) One then performs M independent runs and picks the one that produces the best fitness value.

There are several notable features of the BMR approach. One is that the probability of failure across all the runs decreases exponentially as $(1 - p_{hit})^M$. Equivalently, the probability of success in at least one run increases rapidly with M. For example, if $p_{hit} = 0.5$, the probability of failure in $M = 10$ runs is only $\approx 10^{-3}$. In other words, success in finding the global minimum is practically assured.

The second feature is that the effort needed to tune a stochastic optimization method to perform moderately well is often significantly smaller than getting it to perform very well. This is especially true as the dimensionality of the search space increases. Hence, the BMR approach reduces the burden on the user in terms of tuning.

The third feature of BMR is that it falls in the class of so-called *embarrassingly parallel problems* that are the easiest to implement in a parallel computing environment. This is because each run can be executed by a parallel worker completely independently of the other runs, and does not involve the often tricky part of managing communications between processes in a parallel program. Since different runs occupy different processors, the execution time of the BMR strategy is essentially independent of M. Thus, it is as computationally cheap as a single run provided one has access to a sufficient number of processors.

Another factor in favor of BMR is that, having reached a plateau in Moore's law for the packing density of transistors on a chip, the chip industry is moving towards more processing cores to sustain future growth in computing power. This will drive down the dollar cost of the BMR strategy for optimization applications even further.

Finally, since the stochastic optimization method is only moderately tuned for any one run, the BMR strategy prevents overtuning on any particular fitness function.

2.8 PSEUDO-RANDOM NUMBERS AND STOCHASTIC OPTIMIZATION

An issue of practical importance when using stochastic optimization methods is that of random number generation. On computers, random numbers are generated using deterministic algorithms. In this sense, they are *pseudo-random* numbers that are not the trial values of a true random variable. However, if one does not know which algorithm is generating these numbers and its initial state, they satisfy the criteria of unpredictability of the next trial value based on past values. Once the initial state of the algorithm is specified, by specifying an integer called a *seed*, the sequence of numbers produced by it is predictable and reproducible.

All PRNGs have a *cycle length* (or period) for the sequences they produce. This is the number of trial values one can produce using a given seed before the sequence starts to repeat. In practice, most of the standard PRNGs have very long periods and one does not worry about exceeding them in typical applications. However, stochastic optimization methods are random number hungry algorithms and more so if coupled with the BMR strategy. Moreover, a long cycle length does not mean that the quality of the random number stream remains the same throughout. Therefore, it is best to err on the side of conserving the number of draws from a PRNG in stochastic optimization. In the context of BMR, this can be done by fixing the seeds for the PRNGs used by the independent parallel runs. While the fitness function may change, and itself may be a major consumer of PRNGs in the case of statistical regression applications, this ensures that the BMR consumes only a limited number of random numbers.

The reproducibility of a PRNG actually has some advantages when it comes to stochastic optimization. Fixing the seed allows the entire search sequence of a stochastic optimization method on a given fitness function to be reproduced exactly. This is usually invaluable in understanding its behavior in response to changes in its parameters or changes in the fitness function. Thus, the reader is encouraged to look carefully at the use of PRNGs when benchmarking or tuning a given method.

2.9 NOTES

The mathematically rigorous approach to optimization theory is well described in [46]. A comprehensive list of fitness functions commonly used in the benchmarking of stochastic optimization methods is provided in Table I of [4]. Deterministic optimization methods are described in great detail in [34]. The curse of dimensionality is discussed in the context of statistical analysis and machine learning in Ch. 2 of [16] and Ch. 5 of [19] respectively. Exploration and exploitation have been encountered in many different optimization scenarios. A good review can be found in [5]. The methodology for comparative assessment of optimization algorithms (i.e., benchmarking) is discussed critically in [2], which also includes references for the BMR strategy. For an insightful study of PRNGs, the reader may consult the classic text [29] by Knuth.

CHAPTER 3

Evolutionary Computation and Swarm Intelligence

CONTENTS

An overview is provided of stochastic optimization methods with a focus on the ones, generically called *evolutionary computation* (EC) or SI algorithms, that are derived from models of biological systems. The main goal of this chapter is to ease the novice user of EC and SI methods into this dauntingly diverse field by providing some guideposts.

3.1   OVERVIEW

The universe of stochastic optimization methods is vast and continues to inflate all the time. However, it is possible to discern some well-established approaches in the literature, called *metaheuristics*, that most stochastic optimization methods can be grouped under. This classification is only useful as an organizing principle that helps in comprehending the diversity of methods. There is a continuous flow of ideas between the metaheuristics and the boundaries are not at all impermeable. In fact, there is no universal agreement on the list of metaheuristics themselves. Therefore, our partial list of metaheuristics below should by no means be seen as either exhaustive or definitive.

- **Random sampling**: Methods in this group treat the fitness function as a joint pdf over the search space (normalization of the pdf is unnecessary) and draw trial values from it. The frequency with which any region of the search space hosts a trial outcome then depends on the fitness value in that region, allowing the global maximizer to be identified as the point around which the frequency of trial outcomes is the highest. (Recall that any minimization problem is equivalent to a maximization problem and vice versa.) The random sampling metaheuristic is derived from the theory of *Markov chains* [41], a class of memory-less stochastic processes.

- **Randomized local optimization**: The main idea here is to inject some randomness in the way deterministic optimization methods, such as steepest descent, work so that they can extricate themselves from local minima. The *stochastic gradient descent* method, which is used widely in the training of neural networks in machine learning, is an example under this metaheuristic.

- **Nature-inspired optimization**:

  - **Physics based**: *Simulated Annealing* [28] is a well-known method that is based on a model of how a molten metal turns into a crystal as its temperature is reduced. As it cools, the system of atoms searches stochastically through different possible spatial configurations, which correspond to different internal energies, until the crystalline configuration with the lowest energy is found.

This process is mimicked in simulated annealing by treating points in the search space as different configurations of a system and the fitness function as the energy associated with each configuration. A random walk is performed in the space, with the probability of transitioning to a point with a worse fitness than the current point decreasing as the iteration progresses. The change in the transition probability is controlled by a parameter called the temperature that is decreased following some user-specified schedule.

– **Biology based**: The hallmark of the methods in this class is the use of a population of *agents* in the search space with some means of communication between them. An agent is nothing but a location in the search space at which the fitness function is evaluated. However, the agents and their means of communication are modeled in a rudimentary way after actual biological organisms and systems. Within biology based optimization methods, one can further distinguish two metaheuristics, namely EC and SI that are discussed in more detail below.

3.2   EVOLUTIONARY COMPUTATION

Biological evolution of species is nothing but an optimization algorithm that moves the genetic makeup of a population of organisms towards higher fitness in their environment. The elementary picture of evolution, which is more than sufficient for our purpose, is that individuals in a species have variations in their genetic code that is brought about by mutation and gene mixing through sexual reproduction. Over multiple generations, the process of natural selection operates on the population to weed out individuals with genetic traits that result in lower fitness. Individuals with higher fitness in any given generation have a better chance at surviving and go on to produce offsprings that share the beneficial genetic trait. This results in the advantageous genetic trait gradually spreading throughout the population.

The idea of evolution as an optimization algorithm forms the foundation of the EC metaheuristic. A method under this metaheuristic is also known as a *Genetic Algorithm* (GA). The origin of GA methods dates back to early attempts at modeling biological evolution on computers.

A typical GA is an iterative algorithm that has the following components. (An iteration is more commonly called a *generation* in the GA literature.)

- A population of *genomes*, where a genome is simply a representation of a point in the search space. GAs were initially designed to handle discrete optimization problems. For such problems, one of the commonly used genomic representations encodes the possible integer values of each coordinate as a binary string with a specified number of bits.

- The fitness of a genome, which is simply the value of the fitness function at the location encoded in the genome.

- A *Crossover operation* that takes two or more genomes, called *parents*, as input and produces one or more new genomes, called *children*, as an output.

- A *Mutation operation* that creates random changes in a genome.

- A *Selection operation* that picks the members of the current generation (parents and children) who will survive to the next generation. Following natural evolution, the fitness of a genome only governs the probability of its survival. The number of genomes left after selection is kept constant in each generation.

The normal sequence in which the above operations are applied in any one generation is: Crossover $\rightarrow$ Mutation $\rightarrow$ Selection. The initialization of the iterations consists of picking the locations of the genomes randomly, such as by drawing a trial value for each bit with equal probabilities for getting 0 or 1. A variety of termination conditions have been proposed in the GA literature, the simplest being the production of a user-specified number of generations.

The limitation imposed by a binary genomic representation for optimization in continuous search spaces is overcome in a GA called *differential evolution* (DE) [45]. In DE, the genomic representation of a location in the search space is simply its position vector. For each parent, the crossover and mutation operations involve mixing its genes with those obtained by linearly combining three other randomly picked individuals. The selection operation picks the parent or the child based on their fitness values.

3.3 SWARM INTELLIGENCE

SI methods are based on modeling the cooperative behavior often seen in groups (swarms) or societies of biological organisms. The main difference between the SI and EC metaheuristics is that agents in SI are not born and do not die from one generation to the next. Agents in an SI method also move around in the search space like in EC methods, but the rules governing the change in location are modeled after how organisms in a collective communicate. SI methods are the newest entrants in the field of biology inspired optimization methods and have proven to be competitive with EC methods in most application areas.

The reasons behind the tendency of certain organisms, such as birds, fish and insects, to form collectives are quite varied but generally center around the search for food and avoiding predators. The behavior of swarms, such as a flock of birds or a school of fish, can be quite complex as a whole but the rules governing the behavior of individuals in a swarm are likely quite simple. This is simply a byproduct of the fact that an individual inside a dense swarm has limited visibility of what the rest of the swarm is doing. It can only process the detailed behavior of just a few individuals in its neighborhood while inferring some sort of an average for the behavior of more distant individuals. Numerical simulations of swarms support this hypothesis regarding local rules of behavior. The same principle operates in large animal or insect societies. In an ant colony, for instance, a single ant is not capable of making sophisticated decisions on its own. However, through simple rules governing the behavior of each ant, the ant society as a whole is able to construct remarkably sophisticated structures, find large enough food caches to support itself, and mount effective defenses against much larger predators. Unlike swarms such as bird flocks that rely mostly on vision, individual ants react to chemical signaling that has contributions from not just its nearest neighbors but potentially the entire colony. This is reflected in models of the foraging behavior of ants, where each ant not only follows the chemical tracers left behind by other ants but also adds to the tracer to increase its concentration.

Both types of collective behaviors seen in swarms and colonies are modeled in SI methods. The prime example of the former type is *particle swarm optimization* (PSO), which will be the focus of the next chapter and the remainder of the book. The main example of the latter is *Ant Colony Optimization* (ACO). While PSO is better suited to optimization problems in continuous search spaces, ACO is aimed at discrete optimization.

3.4 NOTES

Methods under the random sampling metaheuristic are popular in Bayesian statistical analysis and a good description of the latter can be found in [18, 38]. Randomized local optimization methods play a key role in the training of artificial neural networks: See [19]. A comprehensive review of EC and SI algorithms, along with references to the original papers, is provided in [14]. The journal *IEEE Transactions on Evolutionary Computation* provides an excellent portal into the literature on EC and SI methods. Many of the seminal works in both fields have been reported in this journal. The biological models behind EC and SI algorithms are discussed in [12].

A promising new physics-based approach, called *quantum annealing*, is emerging on the horizon that has the potential to revolutionize how optimization problems are solved [3]. This approach is based on programming an optimization problem onto a new type of computing hardware [22], called a quantum annealer, that exploits the laws of quantum mechanics. While major corporations are already working on testing this approach, it remains to be seen whether and when it becomes mainstream.

Particle Swarm Optimization

CONTENTS

This chapter contains detailed descriptions of some of the principal variants under the PSO metaheuristic. The emphasis in this chapter is on getting the reader up-and-running with a successful implementation of PSO rather than providing a deep theoretical analysis. As such, the chapter contains some practical advice that can help the beginner in writing a correct and efficient PSO code.

In common with all EC and SI algorithms (see Ch. 3), the PSO metaheuristic consists of a population (*swarm*) of agents (*particles*) that move iteratively in the search space of an optimization problem. (Throughout this chapter, the search space will be assumed to be a D-dimensional hypercube.) The movement of each particle is determined by a vector, called its *velocity*[1], that provides its future location. The heart of any PSO algorithm is in the rules, called the *dynamical equations*, that govern the velocity update from one iteration to the next. Variants of PSO differ mostly in terms of the dynamical equations that they use.

4.1   KINEMATICS: GLOBAL-BEST PSO

The essence of the velocity update rules in global-best PSO is that a particle explores the search space randomly but constantly feels an attractive force towards the best location, in terms of fitness function value, it has found so far and the best location found by the swarm so far. These two locations are respectively called the particle-best (*pbest*) and global-best (*gbest*). As such, global-best PSO is also commonly called *gbest PSO*.

The mathematical notation for the quantities defined above, which establish the kinematics of the gbest PSO algorithm, are summarized in Table 4.1. The same quantities are shown pictorially in Fig. 4.1. As we will see later, these kinematical quantities remain essentially the same across the different variants of PSO but may be substituted with alternatives in some cases. Let $f(\overline{x})$ denote the fitness function that is being minimized. Then,

$$f\left(\overline{p}^{(i)}[k]\right) = \min_{j \leq k} f\left(\overline{x}^{(i)}[j]\right), \qquad (4.1)$$

$$f\left(\overline{g}[k]\right) = \min_{j} f\left(\overline{p}^{(j)}[k]\right), \qquad (4.2)$$

define pbest and gbest.

---

[1]If we adhere strictly to physics terminology, a more appropriate name would be the displacement vector.

Symbol	Description
N_{part}	The number of particles in the swarm.
$\overline{x}^{(i)}[k]$	Position of the ith particle at the kth iteration.
$\overline{v}^{(i)}[k]$	Velocity of the ith particle at the kth iteration.
$\overline{p}^{(i)}[k]$	Best location found by the ith particle (pbest).
$\overline{g}[k]$	Best location found by the swarm (gbest).

Table 4.1 Symbols used for describing kinematical quantities in the gbest PSO algorithm.

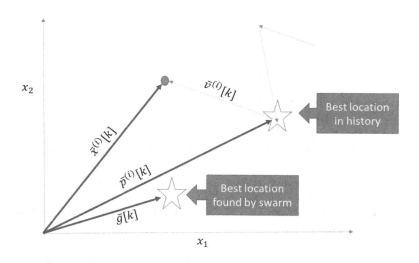

Figure 4.1 Kinematical quantities in the gbest PSO algorithm illustrated for a 2-dimensional search space. The dotted line shows the trajectory of the a particle over a few iterations before the current (k^{th}) one.

4.2 DYNAMICS: GLOBAL-BEST PSO

The dynamical rules for updating the velocity of particles in gbest PSO are given below.

Velocity update:

$$\overline{v}^{(i)}[k+1] \quad = \quad w\overline{v}^{(i)}[k] + \mathbf{R}_1 c_1 \overline{F}_c^{(i)}[k] + \mathbf{R}_2 c_2 \overline{F}_s^{(i)}[k], \quad (4.3)$$

$$\overline{F}_c^{(i)}[k] \quad = \quad \overline{p}^{(i)}[k] - \overline{x}^{(i)}[k], \quad (4.4)$$

$$\overline{F}_s^{(i)}[k] \quad = \quad \overline{g}[k] - \overline{x}^{(i)}[k]. \quad (4.5)$$

Velocity clamping: For each component $v_j^{(i)}[k+1]$, $j = 0, 1, \ldots, D-1$, of $\overline{v}^{(i)}[k+1]$,

$$v_j^{(i)}[k+1] \quad = \quad \begin{cases} v_{\max} & \text{if } v_j^{(i)}[k+1] > v_{\max} \\ -v_{\max} & \text{if } v_j^{(i)}[k+1] < -v_{\max} \end{cases}, \quad (4.6)$$

where $v_{\max}$ is a positive number.

Position update:

$$\overline{x}^{(i)}[k+1] \quad = \quad \overline{x}^{(i)}[k] + \overline{v}^{(i)}[k+1]. \quad (4.7)$$

In the velocity update equation:

- $\mathbf{R}_i$, $i = 1, 2$, is a diagonal matrix with each diagonal element being an independently drawn trial value, at each iteration, of a random variable R having a uniform pdf, $p_R(x) = 1$ for $x \in [0, 1]$ and zero outside this interval.

- The parameter $w \in \mathbb{R}^+$ is a positive number called the *inertia weight*.

- $c_i \in \mathbb{R}^+$, $i = 1, 2$, is called an *acceleration constant*.

- $\overline{F}_c^{(i)}[k]$ is a vector pointing from the current location of the particle towards its current pbest.

- $\overline{F}_s^{(i)}[k]$ is a vector pointing from the current location of the particle towards the current gbest.

4.2.1 Initialization and termination

To complete the description of the algorithm, we need to specify its initialization and termination conditions.

Initialization in gbest PSO is fairly straightforward: (i) Each initial position, $\overline{x}^{(i)}[0]$, $0 \leq i \leq N_{\text{part}} - 1$, is assigned an independently drawn trial value from a joint uniform pdf over the search space. This simply means that each component $x_j^{(i)}[0]$, $0 \leq j \leq D-1$, is assigned an independently drawn trial value from the uniform pdf $U(x; a_j, b_j)$ (see App. A), with a_j and b_j being the minimum and maximum values along the jth dimension of the search space. (ii) Each initial velocity vector $\overline{v}^{(i)}[0]$ is first drawn independently from a joint uniform pdf, but one that is centered on $x^{(i)}[0]$ and confined to lie within the search space boundary. That is, each component $\overline{v}_j^{(i)}[0]$, $0 \leq j \leq D - 1$, is assigned an independently drawn trial value from the uniform pdf $U(x; a_j - x_j^{(i)}, b_j - x_j^{(i)})$. The initial velocity is then subjected to velocity clamping before starting the subsequent iterations.

The simplest termination condition is to stop the algorithm once a maximum number of iterations, N_{iter}, is reached. However, a number of more sophisticated termination conditions have been proposed in the literature. Some of these are enumerated in Sec. 4.7.

With the initialization, termination and dynamical equations defined, we can present an outline of the whole algorithm. This is displayed in Fig 4.2.

4.2.2 Interpreting the velocity update rule

The velocity update equation is quite straightforward to understand if one focuses on each of its terms in isolation. The *inertia term* $w\overline{v}^{(i)}[k]$ alone makes the next velocity vector point in the same direction as the current velocity. Hence, if this were the only term, a particle would simply move along a straight line. This makes the name of this term quite appropriate because the role of the inertia of a real object is indeed

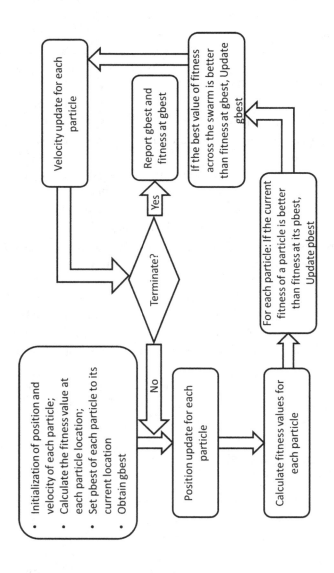

Figure 4.2 Schematic illustrating the PSO algorithm.

to maintain its state of motion and offer resistance to external forces that seek to change it.

The next term in the equation is called the *cognitive* term. If one considers $\overline{F}_c^{(i)}[k]$ alone, it is a force vector that wants to pull the particle towards pbest. The role of $\mathbf{R}_1$ in this term is to randomize this pull such that it is not always exactly attracting the particle towards pbest. In the same way, if one considers $\overline{F}_s^{(i)}[k]$ alone in the last term, which is called the *social* term, its role is to attract the particle towards gbest. Again the presence of $\mathbf{R}_2$ randomizes this pull so that it does not always point towards gbest. For each of the forces, the corresponding acceleration constant determines the importance (on average) of that force in the motion of a particle. Fig. 4.3 shows a cartoon to illustrate the meaning of cognitive and social forces using an analogy with a swarm of birds.

4.2.3   Importance of limiting particle velocity

During the course of their movement, some particles can acquire a velocity vector that, combined with their current location, can take them out of the boundary of the search space. In the early versions of PSO, it was found that for some settings of the algorithm parameters, most particles would leave the search space – a phenomenon called *particle explosion*. The occurrence of instabilities in iterative equations (e.g., $x[k + 1] = 5x[k]$ leads to an exponential increase in x) is well known and it is not surprising that something similar can happen in PSO.

In the early versions of PSO, particle explosion was tackled by velocity clamping as shown in Eq. 4.6. This was subsequently replaced by the inertia term with a weight $w < 1$. (An iterative equation of the form $x[k + 1] = wx[k]$ is then guaranteed to converge since the value of x keeps decreasing.) Present day versions of PSO use both velocity clamping and the inertia term but with a very generous clamping.

Another approach to limiting particle explosion is called *velocity constriction*, where the velocity update equation is

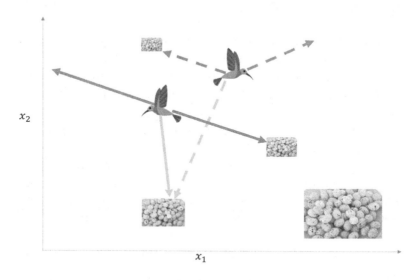

Figure 4.3 A cartoon that illustrates the nature of the inertia (red), cognitive (green) and social (yellow) terms. A bird swarm is searching for the best food source (i.e., best fitness value). The size of the picture of the food indicates how good it is. Each bird remembers the best food location it found in its own searching and has some desire to return back to it (cognitive term). At the same time it knows that the rest of the swarm has found a better food source and would like to move towards it (social term). All the while each bird also knows that there may be an even better food source that no member of the swarm has found. Hence, each one wants to continue searching (inertia term) and not be attracted completely towards either of the food sources found so far. Not shown here is the randomization of the cognitive and social forces, which is the key ingredient that makes PSO work.

altered as follows.

Velocity constriction:

$$\overline{v}^{(i)}[k+1] = K\left(\overline{v}_i[k] + \mathbf{R}_1 c_1 \overline{F}_c^{(i)} + \mathbf{R}_2 c_2 \overline{F}_s^{(i)}\right) . \quad (4.8)$$

The factor K, called the *constriction factor*, is prescribed to be [11]

$$K = \frac{2}{\left|2 - \phi - \sqrt{\phi^2 - 4\phi}\right|} , \quad (4.9)$$

$$\phi = c_1 + c_2 > 4 . \quad (4.10)$$

Note that the velocity constriction expression on the RHS of Eq. 4.8 is the same as Eq. 4.3 if the inertia weight is pulled out as a common factor. Even when velocity constriction is not used, the acceleration constants c_1 and c_2 are, in most cases, set to $c_1 = c_2 = 2$ (or just around this value) in keeping with Eq. 4.10 above.

Whether one uses clamping or constriction, it is important to somehow limit the velocities of particles in order to keep them in the search space for a sufficiently large number of iterations.

4.2.4 Importance of proper randomization

The stochastic nature of the PSO algorithm arises from the random matrices $\mathbf{R}_1$ and $\mathbf{R}_2$ in Eq. 4.3. The most common mistake observed by the author in implementations of PSO by beginners is an improper use of randomization: It is very tempting for beginners to replace these matrices by scalar random variables. After all, it looks very natural that the particle experiences attractive forces that point towards pbest and gbest, which would be the case for scalar weights multiplying the vectors $\overline{F}_c^{(i)}$ and $\overline{F}_s^{(i)}$, but that have random strengths. However, this is a serious error!

Consider a particle at some iteration step k in a $D = 3$ dimensional search space for which $\overline{v}^{(i)}[k] = 0$. Then, if R_1 and R_2 are scalars, the velocity vector will be updated to some random vector lying in the plane formed by $\overline{F}_c^{(i)}$ and $\overline{F}_s^{(i)}$. If $\overline{F}_s^{(i)}$ does not change in the next update, the particle position as well as the new $\overline{F}_{c,s}^{(i)}$ vectors will continue to lie in the same plane. Hence, the next updated velocity vector will continue to lie in this plane, and so on. In fact, as long as $\overline{F}_s^{(i)}$ stays the same, all future velocity vectors will only cause the particle to move in the initial plane. (When $D > 3$ and the starting velocity is not zero, the motion will be confined to a 3-dimensional subspace of the D-dimensional space.)

This confinement to a restricted region of the full search space leads to inefficient exploration and, hence, loss of performance. On the other hand, if one uses the matrices $\mathbf{R}_1$ and $\mathbf{R}_2$, the particle moves off the initial plane even when the starting velocity vector is zero. Fig. 4.4 illustrates the difference between using scalar and matrix random variables, clearly showing that proper randomization is the most important ingredient of the velocity update equation.

4.2.5 Role of inertia

The role of the inertia term is to prevent the particle from being attracted too strongly by the cognitive and social forces. In this way, it encourages exploration in the algorithm. The inertia weight w is usually kept between 0 and 1 since higher values can quickly lead to runaway growth in the velocity, thereby forcing it to stay at the clamped value throughout most of the iterations.

Keeping the inertia weight at a value close to unity promotes exploration but also prevents the particles from responding to the attractive forces. This keeps them from entering the exploitation phase in which they should start converging towards a promising region of the search space that is likely to contain the global minimum. Hence, the standard prescription is to allow the inertia weight to start out near unity and decay to a small value as the iterations progress. Typically, one starts with $w = 0.9$ and allows it to decay linearly to a more moderate value, such as 0.4, as the algorithm

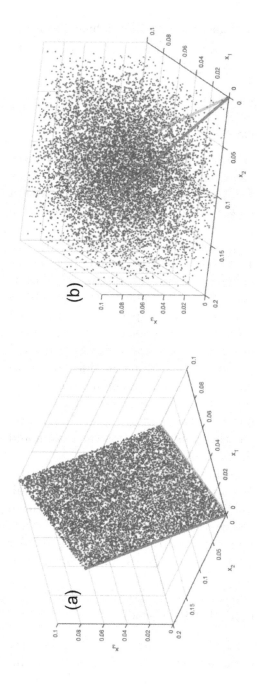

Figure 4.4 A large number of trial outcomes from the probability distribution of $\overline{v}^{(i)}[k+1]$ when $\overline{v}^{(i)}[k] = (0,0,0)$ and the random factors multiplying the social and cognitive forces in the velocity update equation are (a) scalars (i.e., same for all the velocity components), and (b) matrices (i.e., different random numbers for different velocity components). The search space in this illustration is 3-dimensional. The particle position $\overline{x}^{(i)}[k]$ is the origin and the red (green) line points from the origin to the pbest (gbest). (For this illustration, we set $w = 0.9$, $c_1 = c_2 = 1.0$.)

approaches termination. Thus, the inertia weight is given by

$$w[k] = -\frac{0.5}{N_{\text{iter}} - 1}(k - 1) + 0.9 , \qquad (4.11)$$

for termination based on a maximum number of iterations.

4.2.6 Boundary condition

It is quite possible that the velocity update equation generates a velocity vector that, despite clamping, moves a particle out of the search space. An important part of the PSO algorithm is the specification of the *boundary conditions* to use when this happens.

A number of boundary conditions have been explored in the literature. However, it has often been found that the so-called *let them fly* boundary condition tends to work best. Under this condition, no change is made to the dynamical equations if a particle exits the search space. The only thing that happens is that its fitness value is set to $+\infty$ if the problem is that of minimization (or $-\infty$ if maximizing). Since the particle has the worst possible fitness value outside the search space, its pbest does not change and, consequently, neither is the gbest affected. Eventually, the attractive forces from these two locations, which are always within the search space, pull the particle back. At this point, it rejoins the swarm and searches the space in the normal way.

Other boundary conditions that have been proposed include the so-called *reflecting walls* and *absorbing walls* conditions. These require overriding the velocity update equation when a boundary crossing is detected for a particle. In the former case, the sign of the component of velocity perpendicular to the boundary being breached is flipped. In the latter, this component is set to zero.

4.3 KINEMATICS: LOCAL-BEST PSO

The local-best variant of PSO, also called *lbest PSO*, involves a single change to the dynamical equations of gbest PSO. This one change has been observed across a large range of statistical regression problems to make lbest PSO much more effective at locating the global optimum. However, we remind the reader that, following the NFL theorem (see Sec. 2.3), there will always exist optimization problems where lbest PSO performs worse than other variants.

For lbest PSO, we first need to define the concept of a particle *neighborhood*. The neighborhood $\mathbb{N}^{(i)}$ of the ith particle is simply a subset, including i, of the set of indices labeling the particles in the swarm.

$$\mathbb{N}^{(i)} = \mathbb{I}^{(i)} \cup \{i\}, \tag{4.12}$$
$$\mathbb{I}^{(i)} \subseteq \{1, 2, \ldots, N_{\text{part}}\} \setminus \{i\}. \tag{4.13}$$

($\mathbb{A} \setminus \mathbb{B}$, called the *absolute complement* of $\mathbb{B} \subset \mathbb{A}$ in $\mathbb{A}$, is the set of elements in $\mathbb{A}$ that are not in $\mathbb{B}$.) The number of elements in $\mathbb{N}^{(i)}$ is called *neighborhood size* for the ith particle.

For a given particle and iteration k, the local-best particle is the particle in its neighborhood whose pbest has the best fitness value. The location of the local-best particle, denoted by $\bar{l}^{(i)}[k]$, is called the *lbest* for the ith particle.

$$f(\bar{l}^{(i)}[k]) = \min_{j \in \mathbb{N}^{(i)}} f(\bar{p}^{(j)}[k]). \tag{4.14}$$

The definition of lbest is a generalization of gbest. The gbest location is simply the lbest when the neighborhood of each particle consists of all the particles in the swarm.

A particular specification of $\mathbb{N}^{(i)}$ for each i is said to define a particular *topology*. The topology described above, where each particle has the same lbest (i.e., the gbest), is called the *gbest topology*. A less trivial topology is the so-called *ring topology*. Here, each particle has the same neighborhood size, given by $2m + 1$ for some specified m, and

$$\mathbb{N}^{(i)} = \{r(i - k); -m \leq k \leq m, \, m \geq 1\}, \tag{4.15}$$

$$r(j) = \begin{cases} j & 1 \leq j \leq N_{\text{part}}, \\ j + N_{\text{part}} & j < 1, \\ j - N_{\text{part}} & j > N_{\text{part}} \end{cases} \tag{4.16}$$

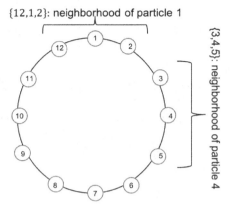

Figure 4.5 Graph representation of the ring topology with a neighborhood size of 3 for a swarm consisting of 12 particles. Each node in this graph indicates a particle and an edge connecting two nodes indicates that they belong to the same neighborhood. Note that every particle in this topology belongs to three different neighborhoods (one of which is its own).

Fig. 4.5 illustrates the the ring topology using a graph.

4.4 DYNAMICS: LOCAL-BEST PSO

The dynamical rules for lbest PSO remain the same as those of gbest PSO except for the change $\overline{F}_s^{(i)}[k] \to \bar{l}^{(i)}[k] - \bar{x}^{(i)}[k]$. (The initialization and termination conditions for lbest PSO are the same as those in Sec. 4.2.1.)

Velocity update (lbest PSO):

$$\bar{v}^{(i)}[k+1] = w\bar{v}^{(i)}[k] + \mathbf{R}_1 c_1 \overline{F}_c^{(i)}[k] + \mathbf{R}_2 c_2 \overline{F}_s^{(i)}[k] \quad (4.17)$$

$$\overline{F}_c^{(i)}[k] = \bar{p}^{(i)}[k] - \bar{x}^{(i)}[k] , \quad (4.18)$$

$$\overline{F}_s^{(i)}[k] = \bar{l}^{(i)}[k] - \bar{x}^{(i)}[k] . \quad (4.19)$$

The main significance of replacing gbest with lbest is that any given particle is not immediately aware of the best fitness found by the whole swarm. It only knows about the best fitness in its own neighborhood at any given iteration step. However, since every particle is part of multiple neighborhoods (see Fig. 4.5), the information about gbest gradually percolates down to every particle. This happens at a much slower rate (in terms of iteration number) compared to gbest PSO. The end result is that particles spend more time in the exploration phase since they are not all attracted towards the same current best fitness location. For the same fitness function, this generally increases the number of iterations required for lbest PSO to converge to the global optimum, but the probability of convergence goes up significantly.

4.5 STANDARDIZED COORDINATES

Consider a search space $\mathbb{D} \subset \mathbb{R}^D$ that is in the form of a D-dimensional hypercube with each component x_i, $i = 0, 1, \ldots, D$, of the position vector $\overline{x} \in \mathbb{D}$ obeying $x_i \in [a_i, b_i]$, $b_i - a_i > 0$. One can reparametrize the search space by shifting to translated and normalized coordinates $s_i = (x_i - a_i)/(b_i - a_i)$ such that $s_i \in [0, 1]$. We call s_i a *standardized coordinate*.

Coordinate standardization is a great convenience when writing a code for PSO since the information about the actual interval $[a_i, b_i]$, and the inversion of s_i to the actual x_i, can be hidden in the code for the fitness function. Alternatively, the interval $[a_i, b_i]$ can be passed to the PSO code, which in turn simply passes it to the fitness function [2]. With standardization, the PSO code is written such that it always searches a D-dimensional hypercube with side lengths of unity irrespective of the actual size of the search hypercube.

Standardization only works if the search space is a hypercube. But this is actually a tacit assumption embedded in the design of PSO itself. In fact, the performance of PSO suffers

---

[2]Further simplification can be achieved by using a C-style pointer to the fitness function as an input argument, making the PSO code completely independent of the fitness function it is applied to.

if it runs in a search space with too complicated a shape. This mainly happens due to the leakage of an excessive number of particles from the search space, reducing the effectiveness of the exploration phase of the search.

4.6 RECOMMENDED SETTINGS FOR REGRESSION PROBLEMS

In the author's experience with regression problems, both gbest and lbest PSO perform well but the performance of the latter becomes better as the number of parameters increases. For non-parametric problems, which tend to have higher search space dimensionalities, lbest PSO is certainly the better option. The choice between the two often boils down to the computational cost of the methods. As mentioned above, lbest PSO generally takes more iterations to converge. This entails a larger number of fitness evaluations, which increases the overall computational cost of the method since the cost of fitness evaluation in regression problems always dominates over that of the PSO dynamical equations and book-keeping.

Regarding the settings for the parameters in both gbest and lbest PSO, an in-depth study on benchmark fitness functions is reported in [4]. Based on [4] and our own experience, the recommended settings for non-linear regression problems involving more than ≈ 2 parameters (or almost any non-parametric regression problem) are summarized in Table 4.2.

4.7 NOTES

PSO was proposed by Kennedy and Eberhart in [27]. The important idea of an inertia weight was introduced in [43]. Velocity constriction was obtained from an analytical study of a simplified PSO in [8]. A text that goes deeper into the theoretical foundations of PSO is [7]. Performance under different boundary conditions is studied in [39].

Setting Name	Setting Value
Position initialization	$x_j^{(i)}[0]$ drawn from $U(x; 0, 1)$
Velocity initialization	$v_j^{(i)}[0]$ drawn from $U(x; 0, 1) - x_j^{(i)}[0]$
v_{max}	0.5
N_{part}	40
$c_1 = c_2$	2.0
$w[k]$	Linear decay from 0.9 to 0.4
Boundary condition	Let them fly
Termination condition	Fixed number of iterations
lbest PSO	Ring topology; Neighborhood size = 3

Table 4.2 Recommended settings for gbest and lbest PSO in the case of regression problems. The search space coordinates are assumed to be standardized.

While we have cited some of the important original papers above, the PSO literature is too vast to be reviewed properly in this book. The reader interested in delving deeper into the PSO literature should consult books such as [14] or can easily find up to date reviews through web searches. We will now discuss some of the PSO variations that have been explored in the literature. Again, this is by no means intended to be a comprehensive review and we do not attempt to reference all the original papers.

4.7.1 Additional PSO variants

In this chapter, we have discussed two out of the many variants of PSO that have been proposed in the literature. The main reason for exploring variations in algorithms under any particular metaheuristic is that there is still no comprehensive theoretical understanding, if something like this is at all possible, of what makes these algorithms succeed in any given instance. In fact the NFL theorem promises us that every variant will have an associated set of optimization problems where it will be the best performer. That said, it is important to understand that creating a new variant should not be taken up lightly because our intuitive prediction of its behavior in a high-dimensional search space may be very far from its actual behavior.

One set of variations is in the termination condition. While termination based on a fixed number of iterations is the simplest, one does not have an intuitive way of fixing this parameter. Some alternatives are listed below.

- In some optimization problems, we know the minimum value f_{min} of the fitness function but we need the location of the minimum. In such problems, the natural termination criterion is to stop when $f(\overline{g}[k]) - f_{min} \leq \epsilon$, where ϵ is a sufficiently small threshold set by the user.

- Stop at step k_{max} if the change in gbest averaged over some number, K, of past iterations is small: $(1/K) \sum_{i=0}^{K-1} \|\overline{g}[k_{max} - i] - \overline{g}[k_{max} - i - 1]\| < \epsilon$. Here, the user needs to fix K and ϵ.

- Stop at step k_{max} if the velocity averaged over some number, K, of past iterations and over the swarm is small: $(1/K) \sum_{j=0}^{K-1} \sum_{i=1}^{N_{part}} \|\overline{v}^{(i)}[k_{max} - j]\| < \epsilon$. Here, the user needs to fix K and ϵ.

- Stop at step k_{max} if the change in the fitness value averaged over some number, K, of past iterations is small:

$$\frac{1}{K} \sum_{i=0}^{K-1} \left(f\left(\overline{g}[k_{max} - i]\right) - f\left(\overline{g}[k_{max} - i - 1]\right) \right) < \epsilon \ .$$

- Define the *swarm radius*, $R[k] = \max_i \|\overline{x}^{(i)}[k] - \overline{g}[k]\|$ and the normalized swarm radius, $r[k] = R[k]/R[1]$. Stop at step k_{max} if the normalized swarm radius is small: $r[k_{max}] \leq \epsilon$. Here, the user needs to fix ϵ.

Some alternatives to the linear inertia decay law are as follows.

- The inertia $w[k]$ is a random variable with a uniform pdf over some interval $[w_1, w_2]$ with $w_2 < 1.0$. This tends to lengthen the exploration phase of PSO.

- Non-linear decrease in inertia.

Several variations in the dynamical equations of PSO have been proposed. In one class of variations, called *memetic searches*, a local minimzation method (e.g., steepest descent) is incorporated into PSO [36]. For some fitness functions, memetic searches are more efficient since the local minimization can move the search more efficiently to the global minimum. However, memetic searches also tend to have a shortened exploration phase that can be detrimental for very rugged fitness functions.

An interesting variant of PSO is *Standard PSO* (SPSO) [53]. This is an algorithm that has gone through several versions and seeks to capture the latest advances in PSO research. The motivation behind SPSO is to present a baseline, well-characterized algorithm against which new variants can be compared. One of the key features of SPSO is that it changes the distribution of the velocity vector compared to the distribution seen in gbest or lbest PSO (see Fig. 4.4). For the latter two, the distribution is aligned along the coordinate axes of the search space. This can be disadvantageous if the fitness function is rotated while keeping the same coordinates since the distribution, which will remain aligned along the axes, will not be well-adapted to the rotated fitness function. In SPSO, the distribution becomes spherical in shape making it insensitive to a rotation of the fitness function.

4.7.2 Performance example

Previously, in Fig. 2.2, we showed the performance of two stochastic optimization methods on the generalized Griewank benchmark function. The two methods, corresponding to the solid and dashed curves in Fig. 2.2, are gbest and lbest PSO respectively. The parameters of both the methods were set to the prescribed values in Table 4.2 except for the linear decay of inertia. For gbest PSO, the inertia decayed from 0.9 to 0.4 over 500 iterations, while the same drop in inertia happened over 2000 iterations for lbest PSO. In both cases, the total number of iterations at termination was 5000 and the inertia remains constant over the remaining iterations once it reaches its lowest value.

PSO Applications

CONTENTS

This chapter describes the application of PSO to concrete parametric and non-parametric regression problems. Some common issues of practical importance in these problems are identified and possible ways to handle them are presented.

The regression problems that will be considered here are the examples described in Sec. 1.3.1 and Sec. 1.3.2 for parametric and non-parametric regression respectively.

5.1   GENERAL REMARKS

5.1.1   Fitness function

Statistical regression involves the optimization of a fitness function over a set of regression models. We have encountered two such fitness functions earlier, namely the likelihood and the least squares function. In the following, we will concentrate exclusively on the latter in which the optimization task is that of minimization.

In most regression problems, it is possible to organize the search for the global minimum of the fitness function as a set of nested minimizations. Consider the parametric regression example where we have 4 parameters defining the quadratic chirp: $\overline{\theta} = (A, a_1, a_2, a_3)$. Let θ' denote the subset (a_1, a_2, a_3). We can organize the minimization of the least squares function in Eq. 1.8 over these parameters as follows.

$$\min_{\overline{\theta}} L_S(\mathbb{T}; \overline{\theta}) = \min_{\overline{\theta}'} \left(\min_A L_S(\mathbb{T}; \overline{\theta}) \right) , \qquad (5.1)$$

$$= \min_{\overline{\theta}'} l_S(\overline{\theta}') , \qquad (5.2)$$

$$l_S(\overline{\theta}') = \frac{1}{2\sigma^2} \min_A \sum_{i=0}^{N} (y_i - A \sin(\phi(x_i)))^2 . (5.3)$$

It is straightforward to carry out the minimization in Eq. 5.3 analytically since one simply has a quadratic polynomial in A.

Similarly, the minimization of $L_S(\mathbb{T}; \overline{\theta})$ in the non-parametric regression example, where $\overline{\theta} = (\overline{\alpha}, \overline{b})$ (see Eq. 1.11), can be organized as an inner minimization over $\overline{\alpha}$ followed by an outer minimization over $\overline{b}$. Since $\overline{\alpha}$ appears linearly in the regression model, it can be minimized over analytically.

In the following, we will deal only with the function l_S that is left over after the inner minimizations in both the parametric and non-parametric examples. Therefore, l_S constitutes the fitness function for PSO and the minimization is over the set of parameters left over in the outer minimization step. The explicit forms of the function l_S for the parametric and non-parametric regression examples are derived in App. C.

In the general case, analytical minimizations should be carried out first before using a numerical minimization method if a regression model admits this possibility. Doing so significantly reduces the burden on the numerical method due to both (i) a reduction in the dimensionality of the search space, and (ii) a reduction in the ruggedness of the leftover fitness function.

5.1.2 Data simulation

In both the parametric and non-parametric examples, a data realization is of the form $\mathbb{T} = \{(y_i, x_i)\}$, $i = 0, 1, \ldots, N - 1$, where y_i is a trial value of $Y = f(X) + E$. In each data realization $y_i = f(x_i) + \epsilon_i$, $i = 0, 1, \ldots, N - 1$, where ϵ_i is a trial value of E, a normally distributed random variable with zero mean and unit variance.

For a data realization such as $\mathbb{T}$, let $\overline{x}$ denote $(x_0, x_1, \ldots, x_{N-1})$ and $\overline{y} = (y_0, y_1, \ldots, y_{N-1})$. It is convenient to refer to the sequence $(f(x_0), f(x_1), \ldots, f(x_{N-1}))$ present in a given $\overline{y}$ as the *signal*. The sequence $(\epsilon_0, \epsilon_1, \ldots, \epsilon_{N-1})$ in a data realization is called a *noise realization*. In both the examples, the elements of $\overline{x}$ are fixed and equally spaced: $x_i - x_{i-1}$, $i = 1, 2, \ldots, N - 1$, is constant. Thus, $\overline{x}$ stays the same in the different data realizations but $\overline{y}$ changes.

For the parametric example, we call the values of the parameters (A, a_1, a_2, a_3) that define the quadratic chirp present in the data as the *true signal parameters*. The quadratic chirp corresponding to the true signal parameters is called the *true signal*. While the true signal sequence used in the non-parametric example, $(f(x_0), f(x_1), \ldots, f(x_{N-1}))$, where $f(X) \propto B_{0,4}(X; \overline{b})$, is also defined by a set of parameters, we deliberately avoid references to its parameters because they are not directly estimated in non-parametric regression.

We will illustrate the application of PSO and measure its performance using multiple data realizations. Data realizations will be generated under both the null (H_0) and alternative (H_1) hypotheses (see Sec. 1.4). A data realization under H_0 consists of only a noise realization, while a realization under H_1 has a true signal added to a noise realization. In the H_1 data realizations, the true signal will be held fixed and only the noise realization will change from one realization to another. Since the noise realizations are obtained using a PRNG, the data realizations constitute *simulated* data.

As a measure of the overall strength of the signal in both the parametric and non-parametric examples, we use the *signal to noise ratio* (SNR) defined as,

$$ \text{SNR} \;=\; \frac{1}{\sigma} \left[\sum_{i=0}^{N-1} f(x_i)^2 \right]^{\frac{1}{2}}, \tag{5.4} $$

where σ^2 is the variance of the noise. Fixing the strength of a signal in terms of its SNR, instead of a parameter such as A in the case of the quadratic chirp, ensures that the effect of the signal on inferences drawn from the data is the same no matter the variance of the noise.

5.1.3 Parametric degeneracy and noise

Multiple local minima in statistical regression fitness functions arise from two sources. This is illustrated in Fig. 5.1 where the fitness function of the parametric regression example is shown across a 2-dimensional cross-section of the 3-dimensional search space.

We see that even when a data realization has no noise in it, the fitness function has multiple local minima. These local minima arise because the parameters used in the fitness function evaluation can conspire to make the associated quadratic chirp resemble the true signal. We call this source of local minima *parametric degeneracy* and it is a hallmark of non-linear regression models. When the data has noise in it, we see that additional local minima appear in the fitness function while the ones due to parametric degeneracy are shifted, suppressed, or enhanced.

A stochastic optimization method must escape these local minima while searching for the global minimum. While we

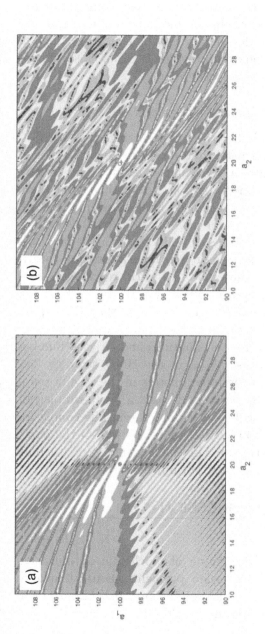

Figure 5.1 Contour plot of the fitness function for the quadratic chirp regression problem in Sec. 1.3.1. Shown here is a 2-dimensional cross-section of the 3-dimensional search space. In panel (a), the data, $\bar{y}$, is just a quadratic chirp, with $a_1 = 100$, $a_2 = 20$, $a_3 = 10$, without any added noise. In panel (b), there is noise present in $\bar{y}$ along with the quadratic chirp from (a). Areas with white color contain local minima. The location of the true parameters is marked by '+', while 'o' marks the global minimizer. The global minimizer coincides with the location of the true parameters in (a) but not in (b). The fitness function is shown on a logarithmic scale because the local minima are not as prominent as the global minimum on a linear scale.

have no control over the noise induced local minima, every effort should be made to help the method by mitigating the ones due to parametric degeneracy. Not much can be done in the parametric regression example given here but this becomes a critical topic in non-parametric regression.

5.1.4 PSO variant and parameter settings

We use lbest PSO and the parameter settings listed in Table 4.2 for both the parametric and non-parametric regression examples. This illustrates a particularly advantageous feature of PSO in general compared to other stochastic optimization methods: its parameter settings need little tweaking, if at all, across a wide variety of problems. This remarkable robustness of the parameter settings is well-known in the PSO literature from empirical testing on benchmark functions but it also seems to carry over into the real-world regression problems that the author has dealt with.

From practical experience, the settings that do require tuning to achieve good performance with PSO are (i) the number of iterations, N_{iter}, for termination, and (ii) the number of independent PSO runs, N_{runs}, in the BMR strategy (see Sec. 2.7). The tuning of these parameters in the case of parametric regression can be performed using a fairly simple but effective strategy that will be described later. The tuning for non-parametric regression involves more of a seat-of-the-pants approach but it is greatly aided by the fact that there are only two such parameters that require tuning in most cases.

5.2 PARAMETRIC REGRESSION

5.2.1 Tuning

As discussed in Sec. 2.6, tuning of stochastic optimization methods in statistical regression problems should take into account a large number of data realizations. It is important to avoid the pitfall of over-tuning a method on a single or a few data realizations because, thanks to the NFL theorem, this is likely to result in a poor overall performance.

While a typical statistical regression problem involves just one data realization, one can always simulate multiple data realizations, as described in Sec. 5.1.2, for the purpose of tuning. These simulated data realizations need only be derived from a reasonably good approximation of the regression model for the tuning to work well on the actual data.

Since a regression problem that needs a stochastic optimization method is also often the one where a deterministic method is computationally infeasible, it is generally not possible to ascertain if the global minimum of the fitness function has been found successfully. Fortunately, in the case of parametric regression, there exists a surrogate condition that at least tells us if a stochastic optimization method is worth using or not. The same condition can also be used to set up a general purpose strategy for tuning the method.

Minimal performance condition: For a given simulated data realization, let f^{opt} be the fitness found by the stochastic optimization method and f^{true} be the fitness at the point in the search space corresponding to the true signal in the data. If the condition $f^{\mathrm{opt}} < f^{\mathrm{true}}$ is satisfied, the method can be said to have met the minimal performance expected of any statistical regression method.

The minimal performance condition is based on the fact that the global minimizer is always located away from the true values of the parameters when there is noise in the data. This is evident in the example shown in Fig. 5.1. Equivalently, $f^{\mathrm{opt}} < f^{\mathrm{true}}$. Hence, any stochastic optimization method should, at the very least, satisfy this condition even if we can never ascertain its success in finding the actual global minimum.

Ideally, the minimal performance condition should hold for any data realization for a well-performing stochastic optimization method. However, in practice, we can only ensure that this happens with a sufficiently high probability. The higher this probability, the lower is the effect of failure in localizing the global minimum on the statistical inference drawn from given data. This observation leads to the definition of a performance metric that can be used to tune any stochastic optimization method for parametric regression.

For concreteness, let us now focus on PSO where we need to tune only two parameters, namely N_{runs} and N_{iter}. Let $\mathcal{M}(N_{\mathrm{runs}}, N_{\mathrm{iter}})$ denote the performance metric. Then, for a set of N_{tune} data realizations, we define

$$\mathcal{M}(N_{\mathrm{runs}}, N_{\mathrm{iter}}) = \frac{N_{\mathrm{mpc}}}{N_{\mathrm{tune}}}, \qquad (5.5)$$

where N_{mpc} is the number of data realizations where the minimal performance condition was satisfied. In the next stage, one goes through different combinations of N_{runs} and N_{iter} until $\mathcal{M}(N_{\mathrm{runs}}, N_{\mathrm{iter}})$ reaches a sufficiently low value. This value depends on the requirements the user has on the quality of the solutions found by PSO as well as the available computational resources.

Fig. 5.2 illustrates the above process by taking two extreme combinations of N_{runs} and N_{iter}. Shown is the scatterplot of f^{opt} and f^{true} for $N_{\mathrm{tune}} = 100$ realizations of data. As expected for these extreme combinations, $\mathcal{M}(N_{\mathrm{runs}}, N_{\mathrm{iter}})$ goes from being nearly zero to being unity. For the parametric regression example, a reasonable setting for N_{runs} and N_{iter} lies somewhere in between these extremes.

Recalling that the tuning process only ensures a certain probability for success, obtaining $\mathcal{M}(N_{\mathrm{runs}}, N_{\mathrm{iter}}) = 1$ for the set of tuning data realizations does not mean that it will remain unity for a much larger number of realizations. Therefore, once a satisfactory setting for N_{runs} and N_{iter} has been found, it is always a good idea to validate the performance of the tuned PSO by running it on a much larger and independent set of *validation data* realizations. The validation should be done before analyzing the actual data with the tuned PSO.

Once PSO has been tuned using simulated data as described above and its performance validated to be acceptable, it can be applied to the actual data that needs to be ana-

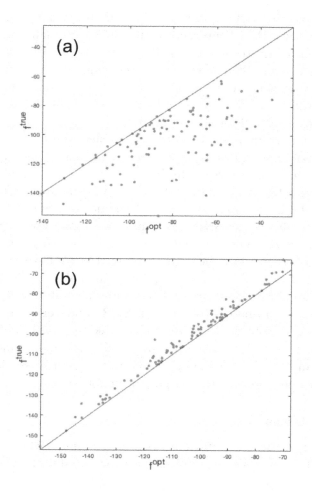

Figure 5.2 Scatterplot of f^{opt} and f^{true} for $N_{\mathrm{tune}} = 100$ realizations of data containing the quadratic chirp. The panels correspond to PSO with (a) $N_{\mathrm{runs}} = 2$, $N_{\mathrm{iter}} = 50$, and (b) $N_{\mathrm{runs}} = 8$, $N_{\mathrm{iter}} = 1000$. The minimal performance condition is satisfied when a point lies above the straight line representing $f^{\mathrm{opt}} = f^{\mathrm{true}}$. For (a), $\mathcal{M}(2, 50) = 0.07$, while $\mathcal{M}(8, 1000) = 1.0$ for (b). The same data realizations are used in both (a) and (b). The true signal parameters are $a_1 = 100$, $a_2 = 20$, $a_3 = 10$ and the SNR is 10.

lyzed. Since it would, in general, be computationally infeasible to check if PSO converged to the global minimum for the actual data, the result obtained with the tuned PSO is the best one can get. The more faithful the simulations are to the process that generated the real data, the more confident one can be that the PSO tuned and validated on simulated data has performed well for the actual data.

An important point to note here is that the tuning procedure described above is for a fixed true signal. This implies that all the parameters, including the SNR of the signal, are fixed across the N_{tune} data realizations. In general, this is not a major limitation because the performance of PSO typically depends only on the SNR and tuning it for the lowest SNR value of interest ensures that it will perform well for higher values as well. It does, however, require that we have an idea of how low in SNR we wish to go in a particular application. It does no harm also to repeat the tuning for different sets of true signal parameters (other than SNR) and verifying that the tuning is robust. A more sophisticated approach, not explored here, could involve using an overall performance metric that is an average over the individual metric values associated with a discrete set of true signal parameter values.

5.2.2 Results

Given that the computational cost of running PSO on the parametric regression example is quite low, we skip the metric-based tuning and validation step, leaving it as an exercise for the reader, and simply adopt the highest settings, $N_{\text{runs}} = 8$ and $N_{\text{iter}} = 1000$ (see Fig. 5.2). The results obtained with this setting and presented below serve as a reference for comparison when reproducing the performance of PSO on this example.

For each data realization, the minimizer found by PSO gives us the estimated values of the true signal parameters in that realization. Due to the presence of noise, the estimated value of a parameter is a random variable. Fig. 5.3 shows the marginal and pair-wise bivariate distributions of the estimated values of a_1, a_2, and a_3.

Fig. 5.3 also allows us to make a crude estimate of the computational cost of a grid-based search. A reasonable choice is to let the grid spacing along a given parameter be some fraction of the standard deviation of its estimated value. For a fractional spacing between 1 and 1/5 times the standard deviation, and using the search range and standard deviation of each parameter as shown in Fig. 5.3, we get the total number of grid points to be between $\approx 3.8 \times 10^5$ to $\approx 4.7 \times 10^7$ respectively. In contrast, the number of fitness evaluations in PSO is fixed at $40 \times 8 \times 1000 = 3.2 \times 10^5$, which is lower than the lowest estimate above for the grid-based search[1].

Fig. 5.4 shows the distribution of the GLRT (Eq. 1.17) under H_0 and H_1. The lack of any overlap between the two estimated distributions indicates that the signal can be detected almost all the time even when the false alarm probability is very low. (Setting the detection threshold at the maximum observed value of the GLRT under H_0 gives a false alarm probability of $\approx 1/500$ given the number of H_0 data realizations used here.)

Both the parameter estimation and detection results tell us that PSO is able to get us a performance that is quite satisfactory given that its computational cost is lower than any reasonable grid-based search. (Note that with proper tuning, one can bring the cost of PSO down further.)

It should be noted here that each fitness evaluation in PSO incurs the extra computational cost of generating a quadratic chirp on the fly, which can be avoided for a grid-based search with a fixed grid by generating and storing the quadratic chirps in advance. However, storing and efficiently retrieving such a large number of vectors from memory or disc, not to mention the time taken to initialize them, can lead to inefficiencies. A true comparison of computational costs, therefore,

---

[1]Of course, smarter strategies for placing the grid are conceivable that can reduce the number of points. However, these would rapidly fail as the number of parameters increases.

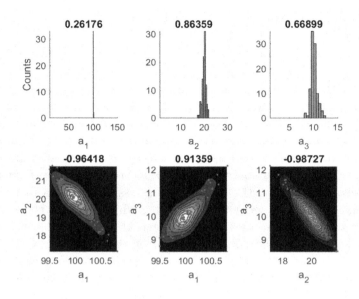

Figure 5.3 (Top row) Histograms of estimated values of a_1, a_2, and a_3 obtained from 100 data realizations under H_1. The red line in each plot indicates the true parameter value. The range of the horizontal axis in each histogram is the search range used in PSO for the corresponding parameter: $[10, 150]$, $[1, 30]$, and $[1, 15]$ for a_1, a_2, and a_3 respectively. The error in a_1 is so small that the width of the histogram for a_1 is nearly equal to that of the line showing the true parameter value, making it difficult to resolve the details. The number at the top of each panel is the standard deviation of the estimated values. (Bottom row) Scatterplots of a_i and a_j, for all combinations of i and $j \neq i$. The number at the top of each panel is the Pearson Correlation Coefficient [51] for the scatterplot. The high values of correlation coefficient (which is always limited to $[-1, 1]$) show that the estimation errors in the different parameters are not independent but highly correlated for the quadratic chirp. The background of each scatterplot is a contour map of the bivariate pdf estimated from the data points. The true signal in the simulated data under H_1 has SNR $= 10$, $a_1 = 100$, $a_2 = 20$, and $a_3 = 10$.

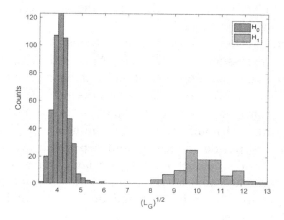

Figure 5.4 Histograms of the square root of the GLRT (L_G) for the quadratic chirp . The H_0 histogram was obtained with 500 and the H_1 histogram with 100 data realizations.

requires a measurement of the *computation time* of a given method. The curious reader may wish to undertake this as an exercise!

5.3 NON-PARAMETRIC REGRESSION

The minimal performance condition (Sec. 5.2.1) used for tuning PSO in parametric regression – where the true signal parameters are known – is not a practical one to use for tuning in non-parametric regression. This is because the latter is designed to capture a larger class of signals and it may not be possible to express all these signals in terms of a parametrized form with a fixed number of parameters. Moreover, one does not expect a non-parametric method to reproduce every feature of any given signal. Therefore, it is not possible to compare the fitness found by PSO with the fitness at the "true location" because the former will almost always be worse. On the other hand, a consistently better fitness from PSO may simply be an indication of overfitting and not necessarily a better fit to the true signal.

One straightforward option for tuning PSO for non-parametric regression is to increase N_{iter} and N_{runs} to the largest value one can afford based on available computational resources. This brute-force strategy works because we know that the performance of PSO becomes better as both these numbers go up. Although the relative influence of N_{runs} and N_{iter} is hard to pin down exactly and depends on the fitness function at hand, a smarter strategy that often works is to have N_{runs} be the same as the available number of parallel workers (usually 4 or 8 in high end desktops) and increase N_{iter} until a reasonable performance is achieved across a sufficiently broad range of signal shapes.

If one finds that increasing N_{iter} for a moderately large value of N_{runs} (≈ 12 or less) does not yield good performance, it is usually worth taking a second look at the variant of PSO being used or the formulation of the problem itself. If sources of degeneracies in the fitness function can be identified, it is best to reformulate the problem, using reparametrization for example, to remove them before pushing PSO to the extreme. Reparametrization in the context of the regression spline method is discussed next.

5.3.1 Reparametrization in regression spline

The parameters in the regression spline method that constitute the search space for PSO are the M breakpoints $\overline{b} = (b_0, b_1, \ldots, b_{M-1})$. The spline that is subtracted from the data when evaluating the least squares function is a linear combination of B-splines determined by $\overline{b}$ (see Eq. 1.11).

There is a clear source of parametric degeneracy if one were to use the components of $\overline{b}$ itself as the parameters for PSO to search over: Since particles can move in an unconstrained way in this space, it is possible for two particles to be at locations that simply correspond to a permutation of the elements of $\overline{b}$. (For example, consider the case where the two locations are

$(b_0 = 1, b_1 = 2, b_2 = 3, \ldots)$ and $(b_0 = 2, b_1 = 3, b_2 = 1, \ldots)$ with all b_i, $i \geq 3$ being equal.) However, any sequence of breakpoints must be sorted before it is used in the construction of B-splines, making the splines corresponding to the two locations above identical. Consequently, these two particles will have the same fitness values. Extending this simple example to all the particle locations in this space, it is easy to see that there is a huge scope for degeneracies in the fitness function. Such a fitness function would be extremely difficult for any stochastic optimization method to handle as it would be very rugged due to the numerous local minima with identical values.

To cure the source of degeneracy identified above, an obvious solution is to enforce monotonicity in the breakpoint sequence. That is, ensure that $b_{i+1} > b_i$. But this makes the shape of the search space non-hypercubical. As mentioned earlier in Sec. 4.5, PSO is known to work best when the shape of the search space is a hypercube. Hence, curing the degeneracy problem in this way will open a different, perhaps equally nasty, problem.

There is a different way to maintain monotonicity of the breakpoints while maintaining the hypercubical shape of the search space. The solution is to change from $\bar{b}$ to a new set of parameters, $\bar{\gamma}$, defined as follows.

$$\gamma_0 = \frac{b_0 - x_0}{T}, \tag{5.6}$$

$$\gamma_i = \frac{b_i - b_{i-1}}{x_{N-1} - b_{i-1}} \quad \text{for } i = 1, 2, \ldots, M-1, \tag{5.7}$$

where $T = x_{N-1} - x_0$. It follows that for $0 \leq \gamma_i \leq 1$, $b_{i+1} > b_i$ and $\bar{b}$ is monotonic. In addition, each γ_i can be varied independently of the others over the range $[0, 1]$, making the search space a hypercube. Hence, this reparametrization solves both the monotonicity and the search space shape requirements.

While it solves the two main problems identified above, it turns out that this particular reparametrization is not conducive to extracting a good performance from PSO. The reason is that one expects the uniformly spaced breakpoint sequence, where $b_{i+1} - b_i$ is a constant, to be of special importance in a spline-based smoothing method. In particular, if the true signal is not concentrated in a small section of the interval $[x_0, x_{N-1}]$, one would expect that the best fit spline would have nearly uniformly spaced points.

In the above reparametrization, a uniformly spaced breakpoint sequence corresponds to $\gamma_i \propto 1/(x_{N-1} - b_{i-1})$ and, hence, γ_i increases as i (and b_{i-1}) increase. This means that the uniformly spaced breakpoint sequence corresponds to a point in the new parameters that is displaced strongly from the center of the hypercubical search space. One would like to avoid this situation because it would result in a poor performance of PSO – particles near the boundary tend to exit the search space more often and convergence to a breakpoint sequence that is approximately uniform would be harder.

It turns out that the latter problem can also be solved by slightly altering the definition of $\overline{\gamma}$ as follows.

$$\gamma_0 = \frac{b_0 - x_0}{T} , \tag{5.8}$$

$$\gamma_i = \frac{b_i - b_{i-1}}{b_{i+1} - b_{i-1}} , \quad \text{for } 1 \leq i \leq M - 2 , \tag{5.9}$$

$$\gamma_{M-1} = \frac{b_{M-1} - b_0}{x_{N-1} - b_0} . \tag{5.10}$$

In the above reparametrization, setting $\gamma_i = 0.5$ for $1 \leq i \leq M - 2$ generates a breakpoint sequence where b_i, $1 \leq i \leq M - 2$, are uniformly spaced. Hence, the uniformly spaced sequence (at least up to the interior breakpoints) is now closer to the center of the search space (where $\gamma_i \in [0, 1]$, $\forall i$).

In order to evaluate the fitness function, $\overline{\gamma}$ must be inverted to give the actual breakpoint sequence $\overline{b}$. For Eqs. 5.6 and 5.7, this inversion can be performed iteratively starting with the inversion of b_0 from γ_0. For the parameters defined in Eqs. 5.8 to 5.10, the system of equations relating $\overline{\gamma}$ and $\overline{b}$ is of the form[2] $\mathbf{A}(\overline{\gamma}) \overline{b}^T = \overline{c}^T(\overline{\gamma})$, allowing $\overline{b}$ to be solved by inverting $\mathbf{A}(\overline{\gamma})$.

---

[2]The reader can easily work out the matrices $\mathbf{A}(\overline{\gamma})$ and $\overline{c}(\overline{\gamma})$ from the defining equations for $\overline{\gamma}$.

5.3.2   Results: Fixed number of breakpoints

Having reparametrized the breakpoints as defined by Eqs. 5.8 to 5.10, we can proceed to characterize the performance of the regression spline method. First, we will do this for a fixed, and the smallest possible, number of breakpoints for a cubic B-spline: $\overline{\gamma} = (\gamma_0, \gamma_1, \ldots, \gamma_{M-1})$ with $M = 5$. This matches the case of the true signal described in Eq. 1.12, which is simply a single B-spline.

In order to see if the optimization over breakpoints is essential, let us also consider the case where they are not optimized but spaced uniformly. B-splines constructed out of uniformly spaced breakpoints are called *cardinal* B-splines and we call a linear combination of such B-splines a *cardinal spline*. The fitness function for a cardinal spline remains the same as the one used in regression spline except that there is no minimization over breakpoints after that over the spline coefficients $\overline{\alpha}$.

Fig. 5.5 shows the estimated signals, for one data realization, obtained from the regression spline method and for a cardinal spline. The same number, $M = 5$, of breakpoints is used in both cases. It is clear from just a visual inspection that the solution obtained by optimizing the breakpoints is far better, in terms of its agreement with the true signal, than the one obtained using fixed breakpoints. Here, we have used $N_{\text{runs}} = 4$ and $N_{\text{iter}} = 200$ for PSO with all other settings kept as shown in Table 4.2 for lbest PSO.

A numerical issue arises in evaluating the fitness function when two or more breakpoints come closer than the spacing between the values of the independent variable. This can lead to the analytical minimization over the B-spline coefficients, $\overline{\alpha}$, becoming ill-posed. A simple way to mitigate this problem is to trap such anomalous breakpoint sequences before they are passed on to the fitness function and move apart the breakpoints that are too close to a preset minimum distance. If the minimum distance is set to be the minimum separation between the independent variable values, the ill-posedness is

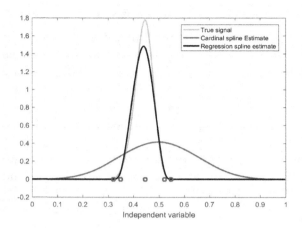

Figure 5.5 The true signal (gray) and the ones estimated by regression spline (black) and cardinal spline (blue) methods. The square markers show the breakpoints optimized by PSO.

resolved. This extra operation alters the fitness value slightly in those parts of the search space that contain the pathological breakpoint sequences, but it does not affect the overall performance of the method in the bulk of the search space.

5.3.3 Results: Variable number of breakpoints

In order to be of any use in the general situation where the true signal is not as simple as the one used in this book, the regression spline method must use a variable number of breakpoints. One way to incorporate a variable number of breakpoints is to perform model selection for which we use the *Akaike Information Criterion* (AIC) defined below.

Let $\widehat{L}_P(\mathbb{T})$ be the maximum value of the log-likelihood function, for given data $\mathbb{T}$, over the parameters $\overline{\theta} = (\theta_0, \theta_1, \ldots, \theta_{P-1})$ of a regression model with P parameters. Then

$$\text{AIC} = 2P - 2\widehat{L}_P(\mathbb{T}) . \tag{5.11}$$

With an increase in P, a model will be better able to fit the data and $\widehat{L}_P(\mathbb{T})$ will increase and AIC will decrease. However, at some point the improvement in the fit will slacken off, allowing the first term to take over and AIC will start to increase. The model selected is the value of P at which AIC attains a minimum.

In the context of regression spline, P is the number of breakpoints used. Thus, model selection in this case means applying regression spline to the *same* data realization with different numbers of breakpoints, $M \in \{M_1, M_2, \ldots, M_K\}$, and picking the optimal value of M using AIC. The set of K values of the number of breakpoints to use has to be specified by the analyst and, hence, is somewhat subjective. Nonetheless, it is much easier to specify a reasonable list of values for M rather than guess one particular optimal value. Moreover, it allows the regression spline method to adapt better to the true signal present in the data.

However, the direct use of AIC results in a new problem. As the number of breakpoints increases in the regression spline method, so does its tendency to cluster them together in order to fit outliers in the data rather than fit a smooth signal. Since the presence of noise can result in many such outliers, model selection alone tends to favor models with higher breakpoint numbers so that such outliers can be removed. In order to counter this and force the method to seek smooth solutions, which is the behavior we expect from the signals of interest, we need to regularize the method. This can be done by adding a term to the fitness function that penalizes undesirable solutions.

Since the clustering of breakpoints conspires with high values for the B-spline coefficients $\overline{\alpha}$ in order to fit outliers, one possibility is to impose a penalty,

$$R(\overline{\alpha}) \;=\; \overline{\alpha}\overline{\alpha}^T = \sum_{i=0}^{M-1} \alpha_i^2 \;. \qquad (5.12)$$

on the B-spline coefficients themselves. The fitness function to be minimized becomes

$$L_S^R = L_S + \lambda R(\overline{\alpha}) , \qquad (5.13)$$

where L_S is the original fitness function of the regression spline method. Here, $\lambda \geq 0$ is a constant called the *regulator gain* which governs the effect of the penalty term on the solution. For $\lambda = 0$, the original fitness function is recovered. As $\lambda \to \infty$, the solution tries to minimize only $R(\overline{\alpha})$, in which case $\alpha_i \to 0$, $\forall i$, leading to an estimated signal that is zero everywhere. (The derivation of l_S provided in Sec. C.2 can easily be extended to include the penalty term and is left as an exercise.)

Fig. 5.6 shows a comparison of the regularized regression spline method with the cardinal spline fit. It is clear that the optimization of breakpoints leads to better results again. The settings used for PSO here remain the same as in Sec. 5.3.2: $N_{\mathrm{runs}} = 4$ and $N_{\mathrm{iter}} = 200$ with all other settings kept as in Table 4.2 for lbest PSO.

The results in Fig. 5.6 for the cardinal spline fit method have been derived without regularization ($\lambda = 0$). One observes that the estimated signals have a systematic shift towards a lower peak amplitude. Using regularization will only make matters worse for this method as the estimated signals will be pushed down further in amplitude.

While we have chosen the regulator gain for regression spline somewhat arbitrarily here, there exist methods, such as *cross-validation*, that may be used to find an optimum value for it. We do not extend the discussion further on this topic as it will take us well outside the scope of this book.

Finally, Fig. 5.7 shows the distributions of the GLRT under H_0 and H_1. We see that the two distributions are well separated at the SNR used for the true signal.

5.4   NOTES AND SUMMARY

The terminology used in Sec. 5.1.2 for describing the simulated data is derived from the field of time-domain signal

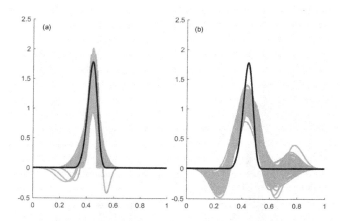

Figure 5.6 The estimated signals from (a) the regularized regression spline method, and (b) cardinal spline fit for 100 data realizations. The true signal in all the realizations is the one shown in black. It follows Eq. 1.12 and has SNR = 10. All the estimated signals are plotted together in gray. Model selection was used in both cases with the number of breakpoints $K \in \{5, 6, 7, 8, 9\}$. The regulator gain in the case of the regression spline method was set at $\lambda = 5.0$. For the cardinal spline fit, $\lambda = 0$.

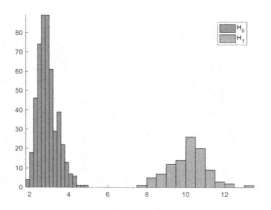

Figure 5.7 Histograms of the square root of the GLRT (L_G) for the regularized regression spline method with model selection. The number of breakpoints used were $M \in \{5, 6, 7, 8, 9\}$ and the regulator gain was set at $\lambda = 5.0$. The H_0 histogram was obtained with 500 and the H_1 histogram with 100 data realizations.

processing [21]. The minimal performance condition was proposed as a figure of merit for quantifying the performance of PSO in [48]. Its use in the metric for tuning defined in Eq. 5.5 is formalized in [35]. The reparametrization of breakpoints in Eq. 5.8 to 5.10 was introduced in [32]. A reference text for model selection is [6]. The penalty term in Eq. 5.12 is the basis of the penalized spline approach discussed in [42]. (This text also contains references to the literature on cross-validation.) An example of a regression method where both AIC-based model selection and cross-validation are used can be found in [33].

5.4.1 Summary

In this concluding chapter we have shown how a SI method such as PSO can help open up new possibilities in statistical analysis. The computational cost of minimizing the fitness function can become infeasible even for a small search space dimensionality when a grid-based strategy is used. Traditionally, the only option for statistical analysts in such situations has been to use methods under the random sampling metaheuristic, which are not only computationally expensive and wasteful in terms of fitness evaluations, but also fairly cumbersome to tune. This has prevented many a novice user of statistical methods from venturing too far from linear models in their analysis since these lead to convex optimization problems that are easily solved. However, non-linear models are becoming increasingly prevalent and unavoidable across many application areas as the problems being tackled become progressively more complex.

Using a three dimensional non-linear parametric regression example, we showed that PSO could easily solve for the best fit model with very small computational cost (in terms of the number of fitness evaluations) while admitting a fairly robust and straightforward tuning process. The example we chose was deliberately kept simple in many ways in order to let the reader reproduce it easily. However, PSO has already been applied to many real world non-linear parametric regression problems with higher dimensionalities and found to work quite well. Some of these applications do require special tweaking of PSO, such as changing the PSO variant that is used or imposing special boundary conditions, but the effort needed to get it to work well is generally far less than that for more traditional methods.

The problem of non-linear optimization has been a major roadblock in non-parametric regression. Here, more than in the case of parametric regression, the large number of parameters has generally forced the use of linear models. At some level, of course, the number of parameters can become so large that it is best to stick to linear models. However, as shown in this book, it is possible to extract much better performance in some cases by shifting to a moderate dimensional non-linear approach than a large dimensional linear one. PSO played a critical role in allowing this shift for the regression spline method considered in this book. In this method, the search space dimensionality (number of breakpoints) went up to a large value of 9 but PSO could still handle the optimization well.

We hope that the fairly simple illustrative problems considered in this book will not only allow readers to get familiar with PSO but also act as gateways to experimenting with and learning about other SI methods. We have not had the luxury of space and time to delve into a more detailed exposition of some important issues. One of them, for example, is that of non-hypercubical search spaces: a problem known in the optimization literature as constrained searches. However, with the working vocabulary in optimization theory and SI terminology acquired from this book, the reader is well placed to start exploring the technical literature on SI and EC methods and discovering new ways of solving statistical analysis problems.

Primer on Probability Theory

CONTENTS

This is a quick – and not a mathematically rigorous – tour of a limited set of concepts from probability theory that appear in this book. The reader interested in acquiring a deep foundation in probability theory is urged to use the many excellent texts that are available (e.g., [37] and [40]).

A.1   RANDOM VARIABLE

The first concept in probability theory is that of a *random variable* X. It is a quantity whose value cannot be predicted but all of whose possible values are known. A *trial* is the process of obtaining a value of X. Thus, the *outcome* of a trial is one of the possible values of X but we cannot predict in advance of the trial what this value is going to be. The value of a random variable obtained in a trial is commonly denoted by the corresponding lower case letter (e.g., x is a trial outcome of X).

A simple example is the number that shows up in throwing a six-sided die: we cannot predict the particular value that will appear in advance of a throw but we do know that it will be an element of the set $\{1, 2, 3, 4, 5, 6\}$. Here, the action of throwing the dice is a trial and the number that shows up is an outcome. (Deliberately placing the dice to show a particular number is not a trial in the probabilistic sense!) Another example is the temperature tomorrow at noon: We cannot predict it for sure but we believe that it will be a real number between the lowest and highest sea level temperatures ever measured on Earth. The temperature measurement process is a trial and the value we obtain is an outcome of the trial.

The set $\mathbb{S}$ of possible values that X can take is called its *sample space*. A subset $\mathbb{A}$ of the sample space is called an *event*. In the case of the dice, $\mathbb{S} = \{1, 2, 3, 4, 5, 6\}$ and $\mathbb{A} = \{2, 4, 6\}$ is the event that X takes an even value. In the case of temperature, the sample space is some interval on the real line. The former type of random variable, where the sample space is finite or may be infinite but countable, is called *discrete*. The latter type is an example of a *continuous* random variable.

A.2 PROBABILITY MEASURE

Probability theory starts with assigning a number $P(\mathbb{A})$, called a *probability measure*, to every event[1] $\mathbb{A} \in \mathbb{S}$. It helps to interpret $P(\mathbb{A})$ as the "probability of any element of event $\mathbb{A}$ occurring in a trial", or "probability that X takes a value in $\mathbb{A}$". Therefore, an equivalent notation would be $P(x \in \mathbb{A})$.

The assignment of the probability measure must follow the rules below.

- $P(\Phi) = 0$, where Φ is the empty set.

- $P(\mathbb{S}) = 1$.

---

[1] The mathematically rigorous approach requires that, before a probability measure is assigned to subsets, the subsets be chosen to form a *Borel Algebra* in order to guarantee that a countable union or intersection of events is an event and the empty set is also an event.

- $P(\mathbb{A} \cup \mathbb{B}) = P(\mathbb{A}) + P(\mathbb{B}) - P(\mathbb{A} \cap \mathbb{B})$.

The LHS of the last equation can be interpreted as "the probability of any element from $\mathbb{A}$ *or* $\mathbb{B}$ occurring in a trial is ...", while the term $P(\mathbb{A} \cap \mathbb{B})$ is subtracted in order to not double-count elements that belong to both $\mathbb{A}$ and $\mathbb{B}$.

In the *Frequentist* approach to probability theory, $P(\mathbb{A})$ is supposed to be objectively assigned by performing (or imagining) N_{trials} trials and counting the number $N(\mathbb{A})$ of those trials in which the event $\mathbb{A}$ occurred. Then $P(\mathbb{A})$ is given by the frequency of occurence of $\mathbb{A}$,

$$P(\mathbb{A}) \;=\; \lim_{N_{\text{trials}} \to \infty} \frac{N(\mathbb{A})}{N_{\text{trials}}} \,. \tag{A.1}$$

In the *Bayesian* approach to probability theory, $P(\mathbb{A})$ is assigned subjectively according to one's *degree of belief* in $\mathbb{A}$.

Consider the sample space of a single dice as an example. The so-called *fair dice* probability measure on the sample space is given by: $P(x) = 1/6$ for $x \in \{1, 2, \ldots, 6\}$. Once the probability measure is defined on these elementary subsets, the probability for any other event then simply follows from the rules stated above. For instance, for $\mathbb{A} = \{2, 4, 6\}$,

$$
\begin{aligned}
P(\mathbb{A}) \;&=\; P(\{2\} \cup \{4\} \cup \{6\})\,, \\
&=\; P(\{2\}) + P(\{4\}) + P(\{6\})\,, \\
&=\; \frac{1}{6} + \frac{1}{6} + \frac{1}{6} = \frac{1}{3}\,.
\end{aligned}
\tag{A.2}
$$

In the Frequentist interpretation of probability, the fair-dice measure implies that each number out of $\{1, 2, \ldots, 6\}$ appears the same number of times if the dice were, hypothetically, thrown an infinite number of times. For the same dice, everyone is supposed to agree with this measure since it is, at least in principle, verifiable.

In the Bayesian interpretation, the fair-dice measure represents our belief, independently of performing trials, that the die is constructed to be free of structural defects that favor one outcome more than the others. However, this degree of belief may not be shared by all people looking at the same dice. Some may assign an unfair dice measure for the same dice.

Once the probability measure is defined, whether in the Frequentist or Bayesian sense, the mathematical rule for obtaining the probability of any event, as illustrated in Eq. A.2, is the same in both the Frequentist and Bayesian interpretation.

A.3 JOINT PROBABILITY

The next elaboration of probability theory is the introduction of additional random variables. Let X_1 and X_2 be two random variables. If the sample space of X_i, $i = 1, 2$, is denoted by $\mathbb{S}_i$, then the sample space of X_1 and X_2 is the *Cartesian product* $\mathbb{S}_{12} = \mathbb{S}_1 \times \mathbb{S}_2 = \{(a_1, a_2); a_i \in \mathbb{S}_i, i = 1, 2\}$. Take the throw of two dice as an example. The outcome of the trial is two numbers a_1 and a_2 that belong to the respective sample spaces (in this case each sample space is $\{1, 2, \ldots, 6\}$). Thus, there are two random variables here and the outcome of each trial is taken from the set $\mathbb{S}_{12} = \{(a_1, a_2); a_i \in \{1, 2, \ldots, 6\}, i = 1, 2\}$.

As in the case of a single random variable, one can also impose a probability measure on subsets of $\mathbb{S}_{12}$. This probability measure is called the *joint probability* of X_1 and X_2. To distinguish between this measure and those for X_1 and X_2 taken separately, let us use P_{12}, P_1 and P_2 for the probability measure on the subsets of $\mathbb{S}_{12}$, $\mathbb{S}_1$, and $\mathbb{S}_2$ respectively. Since every event $\mathbb{A} \subseteq \mathbb{S}_{12}$ can be expressed as $\mathbb{A} = (\mathbb{A}_1 \subseteq \mathbb{S}_1) \times (\mathbb{A}_2 \subseteq \mathbb{S}_2)$, for some $\mathbb{A}_1$ and $\mathbb{A}_2$, $P_{12}(\mathbb{A}) = P_{12}(\mathbb{A}_1 \times \mathbb{A}_2)$ is the same as saying "probability that X_1 takes a value in $\mathbb{A}_1$ *and* X_2 takes a value in $\mathbb{A}_2$ **in the same trial**".

From the above statement, it follows that the probability $P_1(\mathbb{A}_1 \subseteq \mathbb{S}_1)$ is the "probability that X_1 takes a value in $\mathbb{A}_1$ and X_2 takes **any** value in $\mathbb{S}_2$ in the same trial" (i.e., we do not care what value X_2 takes in a trial). This means that $P_1(\mathbb{A}_1) = P_{12}(\mathbb{A}_1 \times \mathbb{S}_2)$. Similarly, one can derive $P_2(\mathbb{A}_2) = P_{12}(\mathbb{S}_1 \times \mathbb{A}_2)$. Obtaining the individual probability measure P_i of X_i from

P_{12} is called *marginalization* of the joint probability over the other random variable $X_{j \neq i}$.

Let us consider N_{trials} trials in the limit $N_{\text{trials}} \to \infty$. Then, the number of trials in which $X_1 \in \mathbb{A}$ is $N_{\text{trials}} P_1(\mathbb{A})$ and the number of trials in which $X_2 \in \mathbb{B}$ and $X_1 \in \mathbb{A}$ is $N_{\text{trials}} P_{12}(\mathbb{A} \times \mathbb{B})$. If we now confine ourselves to the subset of trials where $X_1 \in \mathbb{A}$, the measured probability of $X_2 \in \mathbb{B}$ in those trials is called the *conditional probability* $P_{2|1}(\mathbb{B}|\mathbb{A})$ of $X_2 \in \mathbb{B}$ given $X_1 \in \mathbb{A}$. Taking the ratio of the number of trials, we get

$$P_{2|1}(\mathbb{B}|\mathbb{A}) = P_{12}(\mathbb{A} \times \mathbb{B})/P_1(\mathbb{A}) . \tag{A.3}$$

Similarly, one obtains the conditional probability,

$$P_{1|2}(\mathbb{A}|\mathbb{B}) = P_{12}(\mathbb{A} \times \mathbb{B})/P_2(\mathbb{B}) , \tag{A.4}$$

of $X_1 \in \mathbb{A}$ given $X_2 \in \mathbb{B}$.

From Eq. A.3 and Eq. A.4, we get

$$
\begin{aligned}
P_{1|2}(\mathbb{A}|\mathbb{B})P_2(\mathbb{B}) &= P_{2|1}(\mathbb{B}|\mathbb{A})P_1(\mathbb{A}) , \\
\Rightarrow P_{1|2}(\mathbb{A}|\mathbb{B}) &= \frac{P_{2|1}(\mathbb{B}|\mathbb{A})P_1(\mathbb{A})}{P_2(\mathbb{B})} .
\end{aligned} \tag{A.5}
$$

Eq. A.5 is the famous *Bayes law* in statistical inference. Loosely speaking, when X_2 and X_1 represent "data" and the "quantity to be estimated from the data" respectively, Bayes law allows us to update our *prior degree of belief*, P_1, about X_1 to a *posterior degree of belief*, $P_{1|2}$, based on some model, $P_{2|1}$, called the *likelihood*, of how the probability of X_2 is affected by different values of X_1.

It is important to understand that the conditional probability of a random variable is **not** the same as its marginal probability in general: $P_{1|2}(\mathbb{A}|\mathbb{B}) = P_1(\mathbb{A})$ only when $\mathbb{B} = \mathbb{S}_2$. However, when two random variables are *statistically independent*, $P_{1|2}(\mathbb{A}|\mathbb{B}) = P_1(\mathbb{A})$ for any $\mathbb{B}$. (Similarly, $P_{2|1}(\mathbb{B}|\mathbb{A}) = P_2(\mathbb{B})$ for any $\mathbb{A}$.) This is the mathematically proper way of stating that the probability of $X_1 \in \mathbb{A}$ is not affected by the value that X_2 takes. From the definition of conditional probability in Eq. A.3 or Eq. A.4, it follows that when X1 and X2 are statistically independent,

$$P_{12}(\mathbb{A} \times \mathbb{B}) = P_1(\mathbb{A})P_2(\mathbb{B}) , \qquad (A.6)$$

in order for their respective conditional and marginal probabilities to be equal.

An example of statistical independence is given by two people in two separate rooms throwing a die each. Knowing the outcome of the throw in one room gives us no clue about the probability of getting a particular outcome in the other room. On the other hand, if one takes the temperature in a room and some measure of fatigue, such as efficiency in performing a given task, of a worker in that room, one would expect that the probability of obtaining a particular level of efficiency depends on the temperature in the room. This is an example where the value of one random variable (temperature) affects the probability of another (efficiency). Note that measuring a certain level of efficiency also tells us how probable it is for the room to be at a certain temperature. Of course, one does not expect measuring efficiency to be a good way of predicting temperature. On the other hand, measuring temperature may be a good way of predicting efficiency. This asymmetry is not an inconsistency because the two conditional probabilities are measures of different events.

A.4 CONTINUOUS RANDOM VARIABLES

From now on, we will focus on only continuous random variables. It will turn out that much of the mathematics of probability theory for discrete random variables is a special case of that for continuous ones.

When X is a continuous random variable that takes values in $\mathbb{R}$, it only makes sense to talk about the probability of an interval[2] in $\mathbb{R}$, no matter how small, rather than a single point in

---

[2]Formally, an interval in $\mathbb{R}$ is a set with the property that any number between two unequal numbers in the set is also in the set.

$\mathbb{R}$. Heuristically, this is easy to understand from the Frequentist perspective: one would have to wait an infinite number of trials before the exact real number x obtained in a given trial is obtained again. (Think of $x = \pi$ for example. One would have to do an infinite number of trials before one obtains π exact to each of its non-repeating decimal places.) Hence, in the notation of Eq. A.1, $P(\{x\}) = \lim_{N_{\text{trials}} \to \infty} N(\{x\})/N_{\text{trials}} = 0$.

Just as in the case of a die, where assigning the probability measure to elementary events $\{1\}$, $\{2\}$, etc., was sufficient to obtain the probability of any event (e.g., the subset of even numbers), it is sufficient to assign probability measures to infinitesimal intervals $[x, x + dx) \subset \mathbb{R}$ for a continuous random variable[3]. This is represented as,

$$P([x, x + dx)) \quad = \quad p_X(x)dx \ . \tag{A.7}$$

The function $p_X(x)$ is called the *probability density function* (pdf) of X.

Using the rules of probability theory, $P([a, b])$ for a finite interval $[a, b]$ can be obtained by breaking it up into contiguous disjoint discrete intervals and summing up the probabilities for each. This naturally leads to an integral,

$$P([a, b]) \quad = \quad \int_a^b p_X(x)dx \ . \tag{A.8}$$

Note that while $P([a, b]) \leq 1$, $p_X(x)$ can be greater than 1. Since the sample space of X is a subset of $\mathbb{R}$, $P(\mathbb{R}) = 1$ and

$$\int_{-\infty}^{\infty} p_X(x)dx = 1 \ . \tag{A.9}$$

Finally, since $P([a, b]) \geq 0$ for any interval, $p_X(x) \geq 0$ everywhere on $\mathbb{R}$.

From the properties of $p_X(x)$, it follows that if $P([a, b]) = 0$ then $p_X(x) = 0$ for $x \in [a, b]$. This allows us to treat the sample space of any continuous random variable to be the whole of $\mathbb{R}$

---

[3] $[x, x + dx)$ denotes an interval that includes x but not $x + dx$.

even if the actual sample space is a proper subset $\mathbb{S} \subset \mathbb{R}$. We simply set $p_X(x) = 0$ for $x \notin \mathbb{S}$, ensuring that X never gets a value outside $\mathbb{S}$ in any trial.

By introducing the *Dirac delta function*, $\delta(x)$, defined by

$$\int_a^b dx \delta(x - c) f(x) = \left\{ \begin{array}{ll} f(c) , & a < c < b , \\ 0 , & \text{otherwise} \end{array} \right. \quad \text{(A.10)}$$

one can subsume the case of a discrete random variable into the probability theory of a continuous one. If the probability measure for each element x_i, $i = 1, 2, \ldots$, of the sample space of a discrete random variable is $P(x_i)$, then $p_X(x) = \sum_{i=1}^N P(x_i) \delta(x - x_i)$ as can be verified directly by integrating $p_X(x)$ in any finite interval.

The next step is the introduction of the *joint probability density function* (joint pdf) $p_{X_1 X_2}(x, y)$ of two continuous random variables X and Y. The sample space of X and Y is $\mathbb{R}^2 = \mathbb{R} \times \mathbb{R}$, namely the 2-dimensional plane. If one takes an area $\mathbb{A} \subset \mathbb{R}^2$, then

$$P_{12}(\mathbb{A}) = \int_{\mathbb{A}} dx dy p_{XY}(x, y) . \quad \text{(A.11)}$$

From the joint pdf, one can derive the pdf of any one of the variables by marginalization.

$$\begin{aligned} P_1([a, b]) &= \int_a^b p_X(x) dx = P_{12}([a, b] \times \mathbb{R}) , \\ &= \int_a^b dx \int_{-\infty}^\infty dy p_{XY}(x, y) , \\ \Rightarrow p_X(x) &= \int_{-\infty}^\infty dy p_{XY}(x, y) . \end{aligned} \quad \text{(A.12)}$$

The conditional probability of $x \in \mathbb{A}$ given $y \in \mathbb{B}$ is given by

$$P_{1|2}(\mathbb{A}|\mathbb{B}) = \frac{\int_{\mathbb{A} \times \mathbb{B}} p_{XY}(x, y) dx dy}{\int_{\mathbb{B}} p_Y(y) dy} . \quad \text{(A.13)}$$

If $\mathbb{B} = [y, y + \epsilon)$, where ϵ is infinitesimally small, then the denominator is $p_Y(y)\epsilon$ and the numerator is $\epsilon \int_A p_{XY}(x, y)dx$, which gives

$$P_{1|2}(A|[y, y + \epsilon)) = \int_A dx \frac{p_{XY}(x, y)}{p_Y(y)} \qquad (A.14)$$

This motivates the definition of the conditional pdf,

$$p_{X|Y}(x|y) = \frac{p_{XY}(x, y)}{p_Y(y)}, \qquad (A.15)$$

giving $P_{1|2}(A|[y, y + \epsilon)) = \int_A dx p_{X|Y}(x|y)$.

All the above definitions for two continuous random variables extend easily to N random variables $\overline{X} = (X_0, X_1, \ldots, X_{N-1})$.

Joint pdf: $p_{\overline{X}}(\overline{x})$ defined by

$$P_{12\ldots N}(A) = \int_A d^N x p_{\overline{X}}(\overline{x}) . \qquad (A.16)$$

Conditional pdf: $p_{\overline{X}|\overline{Y}}(\overline{x}|\overline{y})$ where $\overline{x} \in \mathbb{R}^N$ and $\overline{y} \in \mathbb{R}^M$ and $N \neq M$ in general.

$$p_{\overline{X}|\overline{Y}}(\overline{x}|\overline{y}) = \frac{p_{\overline{X}\,\overline{Y}}(\overline{x}, \overline{y})}{p_{\overline{Y}}(\overline{y})} . \qquad (A.17)$$

In the special case where X_i is statistically independent of X_j, for $i \neq j$, the joint pdf becomes,

$$\begin{aligned} p_{\overline{X}}(\overline{x}) &= p_{X_0}(x_0)p_{X_1}(x_1)\ldots p_{X_{N-1}}(x_{N-1}) , \\ &= \Pi_{i=0}^{N-1} p_{X_i}(x_i) . \end{aligned} \qquad (A.18)$$

A.5 EXPECTATION

With the probabilistic description of a continuous random variable in hand, we can introduce some additional useful concepts.

Expectation: $E[f(X)]$, where $f(x)$ is a function, defined by

$$E[f(X)] = \int_{-\infty}^{\infty} dx f(x) p_X(x) . \qquad (A.19)$$

Some expectations have a special significance in probability theory. Among these is the nth *central moment* defined by $E[(X - E[X])^n]$ for $n > 1$. $E[X]$ itself is called the *mean* of the pdf, while the 2nd central moment is called the *variance*. The square root of the variance is called the *standard deviation* of the pdf.

The definition of expectation can be generalized to more than one random variables,

$$E[f(X, Y)] = \int_{-\infty}^{\infty} \int_{-\infty}^{\infty} dx\, dy\, f(x, y) p_{XY}(x, y) \,. \text{(A.20)}$$

$E[(X - E[X])(Y - E[Y])]$ for two random variables X and Y is called their *covariance*. Note that the covariance of a random variable with itself is simply its variance. It is easily shown that if X and Y are statistically independent, their covariance is zero. However, the opposite is not true: a covariance of zero does not necessarily mean statistical independence.

Another type of expectation that can be defined in the case of two random variables is the **conditional expectation**: $E_{X|Y}[f(X)|y]$ is defined by

$$E_{X|Y}[f(X)|y] = \int_{-\infty}^{\infty} dx\, f(x) p_{X|Y}(x|y) \,. \quad \text{(A.21)}$$

The different types of expectations discussed above simply differ in the pdf – marginal, joint, or conditional – inside the integrand.

A.6 COMMON PROBABILITY DENSITY FUNCTIONS

In this book, we encounter (i) the *normal* (also called *Gaussian*) pdf,

$$p_X(x; \mu, \sigma) = \frac{1}{\sqrt{2\pi}\sigma} \exp\left(\frac{1}{2\sigma^2}(x - \mu)^2\right) \,, \quad \text{(A.22)}$$

where the two parameters μ and σ correspond to the mean and standard deviation of X respectively, and (ii) the *uniform* pdf,

$$p_X(x; a, b) = \begin{cases} \frac{1}{b-a}, & a \le x \le b \\ 0, & \text{otherwise} \end{cases} . \quad (A.23)$$

The normal pdf is common enough that a special symbol, $N(x; \mu, \sigma)$, is often used to denote it. The symbol used for the uniform pdf is $U(x; a, b)$. (A random variable with a normal or uniform pdf is said to be *normally* or *uniformly* distributed.)

Two random variables X_0 and X_1 are said to have a *bivariate normal* pdf if,

$$p_{XY}(x, y) = \frac{1}{2\pi |\mathbf{C}|^{1/2}} \exp\left(-\frac{1}{2}\|\bar{x} - \bar{\mu}\|^2\right), \quad (A.24)$$

where for any vector $\bar{z} \in \mathbb{R}^2$,

$$\|\bar{z}\|^2 = \begin{pmatrix} z_0 & z_1 \end{pmatrix} \mathbf{C}^{-1} \begin{pmatrix} z_0 \\ z_1 \end{pmatrix}, \quad (A.25)$$

$\bar{\mu} = (\mu_0, \mu_1)$ contains the mean values of X_0 and X_1 respectively, and C_{ij}, for $i, j \in \{0, 1\}$, is the covariance of X_i and X_j. The symbol $|\mathbf{C}|$ denotes the determinant of $\mathbf{C}$.

The generalization of the bivariate normal pdf to more than two random variables is called the *multivariate normal* pdf. It is given by the same expression as Eq. A.24 but with $\bar{x} = (x_0, x_1, \ldots, x_{N-1})$, $\bar{\mu}_i = E[X_i]$, and $\mathbf{C}$ being an $N \times N$ matrix of covariances between each pair of random variables. In particular, for $\mathbf{C} = \sigma^2 \mathbf{I}$, where $\mathbf{I}$ is the identity matrix,

$$\begin{aligned} p_{\bar{X}}(\bar{x}) &= \frac{1}{(\sqrt{2\pi})^N \sigma^N} \exp\left(-\frac{1}{2} \sum_{i=0}^{N-1} \frac{(x_i - \mu)^2}{\sigma^2}\right), \\ &= \Pi_{i=0}^{N-1} N(x_i; \mu, \sigma), \quad (A.26) \end{aligned}$$

is the joint pdf of identically and independently distributed (*iid*) normal random variables.

A thorough description of useful and common univariate and multivariate pdfs can be found in [23] and [30] respectively. The properties of the multivariate normal pdf are discussed in the context of statistical analysis in [1].

Splines

CONTENTS

Splines form an important class of functions in non-parametric regression problems where the regression model needs to be flexible while satisfying a smoothness condition. Spline based models also allow some useful constraints to be imposed in addition to smoothness. This appendix summarizes some basic concepts related to splines that are sufficient to understand the material in the book. The reader may consult sources such as [10] and [42, 47] for an exhaustive review of splines and spline-based regression respectively.

B.1 DEFINITION

Consider data that is in the form $\{(b_i, y_i)\}$, $i = 0, 1, \ldots, M-1$, where $b_{i+1} > b_i$ are called *breakpoints*. A spline, $f(x)$, is a piece-wise polynomial function defined over the interval $x \in [b_0, b_{M-1}]$ such that for $j = 1, \ldots, M - 1$,

$$f(x) \;=\; g^{(j)}(x) \,, x \in [b_{j-1}, b_j] \,, \tag{B.1}$$

$$g^{(j)}(x) \;=\; \sum_{r=1}^{k} a_r^{(j)} (x - b_{j-1})^{r-1} \,, \tag{B.2}$$

$$g^{(j)}(b_j) \;=\; y_j \,, \tag{B.3}$$

$$g^{(j)}(b_{j-1}) \;=\; y_{j-1} \,, \tag{B.4}$$

where k is the *order* of the polynomial pieces, and for $m = 1, \ldots, M - 2$, and $n = 1, \ldots, k - 2$,

$$\frac{d^n g^{(m)}}{dx^n}\bigg|_{x=b_m} = \frac{d^n g^{(m+1)}}{dx^n}\bigg|_{x=b_m} . \tag{B.5}$$

Splines are more commonly referred to in terms of the degree of the polynomial pieces rather than their order. Thus, for $k = 4$ the polynomial pieces are cubic and the corresponding spline $f(x)$ is called a *cubic spline*. Similarly, for $k = 2$, one has the *linear spline*.

In words, Eq. B.1 states that $f(x)$ is composed of polynomials defined over the disjoint intervals $[b_{j-1}, b_j]$, $j = 1, 2, \ldots, M - 1$. Eq. B.2 defines a polynomial piece where we see the as yet undetermined coefficients $a_r^{(j)}$ defining the polynomial. Eqs. B.3 to B.5 are the conditions, all defined at the breakpoints, that determine the polynomial coefficients. The conditions enforce that $f(x)$ matches the data at each breakpoint, $f(b_j) = y_j$, and that all its derivatives up to order $k - 2$ are continuous across each interior breakpoint b_j, $j = 1, 2, \ldots, M - 2$. (The derivatives of $f(x)$ are automatically continuous at any point between breakpoints since $g^{(j)}(x)$ are polynomials.) Thus, $f(x)$ interpolates the supplied data points with some prescribed degree of smoothness that is determined by the order of the polynomial pieces.

The basic concept of a spline can be generalized to data where there are multiple identical breakpoints. In this case, $b_{i+1} \geq b_i$, and the sequence $(b_0, b_1, \ldots, b_{M-1})$ is called the *knot sequence*. In this book, we do not consider knot sequences that are not breakpoint sequences. However, knots are very useful in general because they allow splines to approximate functions that are discontinuous in their values or in their derivatives.

Counting all the conditions from Eq. B.3 to Eq. B.5, we get $2(M-1) + (k-2)(M-2) = kM - 2(k-1)$ conditions for a total of $k(M-1)$ unknowns, namely, the set of coefficients $\{a_r^{(j)}\}$, $j = 1, 2, \ldots, M - 1$, $r = 1, 2, \ldots, k$. Thus, there is a deficit of $k - 2$ conditions in getting a unique solution for the coefficients.

The extra conditions used to fill the deficit lead to different types of splines for the same data. For the case of cubic splines, where there are two extra conditions, some of the popular choices are as follows[1].

Natural spline: The second derivatives of $f(x)$ are zero at the two end breakpoints, $f''(b_0) = 0 = f''(b_{M-1})$.

Clamped spline: The first derivatives of $f(x)$ are set to some user-specified values, $f'(b_0) = a$ and $f'(b_{M-1}) = b$.

Periodic spline: Applicable when $y_0 = y_{M-1}$ with the conditions being $f'(b_0) = f'(b_{M-1})$ and $f''(b_0) = f''(b_{M-1})$.

Not-a-knot spline: The conditions are based on the continuity of third derivatives at the first and last interior breakpoints, $g^{(1)'''}(b_1) = g^{(2)'''}(b_1)$ and $g^{(M-2)'''}(b_{M-2}) = g^{(M-1)'''}(b_{M-2})$.

B.2   B-SPLINE BASIS

For a fixed set of M breakpoints $\bar{b} = (b_0, b_1, \ldots, b_{M-1})$, the set of all splines of order k defined by $\bar{b}$ and all possible values of $\bar{y} = (y_0, y_1, \ldots, y_{M-1})$ is a linear vector space. This simply follows from the fact that a linear combination of two splines from this set is another piecewise polynomial of the same order corresponding to the same breakpoints, hence another spline from the same set.

If no conditions are imposed on the polynomial in each interval $[b_i, b_{i+1}]$, $i = 0, 1, \ldots, M - 2$, the dimensionality of the space of piecewise polynomical functions for a given $\bar{b}$ is $k(M - 1)$. This is because the polynomial in each of the $M - 1$ intervals is a linear combination of k independent functions. Imposing continuity and differentiability conditions at the $M - 2$ interior breakpoints – with $k - 1$ conditions for a polynomial of order k – reduces the number of free parameters

---

[1]Here, a superscript prime on $f(x)$ denotes differentiation with respect to x: $f'(x) = df/dx$, $f''(x) = d^2f/dx^2$ and so on.

to $k(M-1)-(M-2)(k-1) = M+k-2$. Since these parameters uniquely label each element of the spline vector space, this is also its dimensionality if no further conditions are imposed. As discussed in the Sec. B.1, one is left with $k-2$ conditions over and above those given in Eqs. B.1 to B.5. If these conditions are imposed, usually in the form of boundary conditions such as the ones listed in Sec. B.1, the dimensionality of the space of splines drops to M.

The set of *Basis spline* (B-spline) functions, denoted by $B_{i,k}(x; \bar{b})$, $i = 0, 1, \ldots, M+k-3$, constitutes a very useful basis for the vector space of splines (without boundary conditions) described above. As such, any member, $f(x)$, of this space can be expressed as a linear combination,

$$f(x) = \sum_{i=0}^{M+k-3} \alpha_i B_{i,k}(x; \bar{b}) . \tag{B.6}$$

The B-spline basis functions can be derived from the Cox-de Boor recursion relations [9] given below. The recursions start with B-splines of order 1, which are piecewise constant functions, and build up B-splines of a higher order using the ones from the previous order. While the recursion relations can utilize any knot sequence, we will present them here for the special case where a breakpoint sequence $\bar{b} = (b_0 < b_1 < \ldots < b_{M-1})$ is given.

First, a knot sequence $\bar{\tau}$ is constructed by prepending and appending $k-1$ copies of b_0 and b_{M-1} respectively to $\bar{b}$. The recursion is initialized for $k' = 1$ by

$$B_{i,k'}(x; \bar{b}) = \begin{cases} 1, & \tau_i <= x < \tau_{i+1} \\ 0 & \text{else} \end{cases} . \tag{B.7}$$

For $2 \leq k' \leq k$,

$$B_{i,k'}(x; \bar{b}) = \frac{x - \tau_i}{\tau_{i+k'-1} - \tau_i} B_{i,k'-1}(x; \bar{b}) + \frac{\tau_{i+k'} - t}{\tau_{i+k'} - \tau_{i+1}} B_{i+1,k'-1}(x; \bar{b}) . \tag{B.8}$$

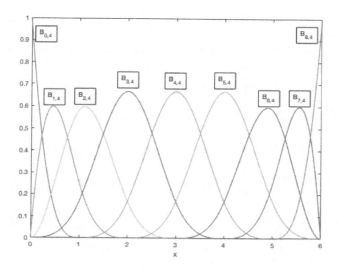

Figure B.1 The B-spline functions for an order 4 spline (cubic spline) with uniformly spaced breakpoints. The breakpoints are $\bar{b} = (b_0, b_1, \ldots, b_6)$ with $b_i = i$. If $B_{0,4}$ and $B_{8,4}$ are dropped from the basis set, the span of the remaining B-splines is a subspace of dimensionality 7 in the full 9-dimensional space of splines defined by $\bar{b}$.

For each k', $0 \le i \le 2(k-1) + M - (k' + 1)$. From Eq. B.7, $B_{i,1}(x; \bar{b}) = 0$ when $\tau_i = t = \tau_{i+1}$, which sets any term in Eq. B.8 that has a zero in the denominator (due to knot multiplicity) to zero. Fig. B.1 illustrates B-splines for the case $k = 4$ (cubic spline) and uniformly spaced breakpoints.

For $k = 4$ (cubic splines), we can reduce the dimensionality of the vector space of splines by $k - 2 = 2$. If this is done by dropping $B_{0,4}(x; \bar{b})$ and $B_{M+1,4}(x; \bar{b})$ from the basis set constructed as above, $f(x)$ will go continuously to zero at b_0 and b_{M-1} but will have discontinuous first and second derivatives at both the locations. The splines defined in Eq. 1.10 follow from this approach.

Analytical Minimization

CONTENTS

As discussed in Sec. 5.1.1, the minimization of the fitness function in a statistical regression problem should be nested in a way that, if applicable, the inner minimization is carried out analytically. Here, we derive the form of the fitness function in the outer minimization for the parametric and non-parametric regression examples in Sec. 1.3.1 and Sec. 1.3.2 respectively. In each case, we start with the least squares function, L_S, and minimize it analytically over the parameters that appear linearly. The resulting function, l_S, is the one that is then minimized numerically.

In the following, we will use the notation

$$\langle \overline{x}, \overline{y} \rangle = \sum_{i=0}^{N-1} x_i y_i = \overline{x}\,\overline{y}^T , \tag{C.1}$$

$$\|\overline{x}\|^2 = \langle \overline{x}, \overline{x} \rangle . \tag{C.2}$$

For $\overline{x} \in \mathbb{R}^N$ and $\overline{y} \in \mathbb{R}^N$.

C.1 QUADRATIC CHIRP

The parameters of the regression model are $\overline{\theta} = (A, a_1, a_2, a_3)$ and L_S is given by Eq. 1.8. Before proceeding further, it is convenient to replace A by a redefined parameter ρ as follows. Let us express the quadratic chirp as

$$q_c(X) = Ag(X), \tag{C.3}$$
$$g(X) = \sin(2\pi\phi(X)). \tag{C.4}$$

Using $\overline{q}_c$ and $\overline{g}$ to denote $(q_x(x_0), q_x(x_1), \ldots, q_c(x_{N-1}))$ and $(g(x_0), g(x_1), \ldots, g(x_{N-1}))$ respectively, we get

$$\overline{q}_c = A\|\overline{g}\|\frac{\overline{g}}{\|\overline{g}\|} = \rho\overline{u}, \tag{C.5}$$

where, by definition, $\|\overline{u}\| = 1$ and $\overline{u}$ only depends on $\overline{\theta}' = (a_1, a_2, a_3)$.

Expressing L_S in terms of the redefined quantities, we get

$$L_S = \frac{1}{2\sigma^2}\|\overline{y} - \rho\overline{u}\|^2,$$
$$= \frac{1}{2\sigma^2}\left[\|\overline{y}\|^2 - 2\rho\langle\overline{y}, \overline{u}\rangle + \rho^2\right]. \tag{C.6}$$

Let $\widehat{\rho}$ be the minimizer of L_S over ρ keeping the other parameters fixed. Using the standard condition for the extremum of a function, we get

$$\left.\frac{\partial L_S}{\partial\rho}\right|_{\widehat{\rho}} = 0 \Rightarrow \widehat{\rho} = \langle\overline{y}, \overline{u}\rangle. \tag{C.7}$$

Substituting $\widehat{\rho}$ into Eq. C.6 gives us the fitness function l_S,

$$l_S = \frac{1}{2\sigma^2}\left[\|\overline{y}\|^2 - \langle\overline{y}, \overline{u}\rangle^2\right] \tag{C.8}$$

that needs to be minimized over $\overline{\theta}'$.

C.2   SPLINE-BASED SMOOTHING

In this example, the regression model depends linearly on the parameters $\overline{\alpha} = (\alpha_0, \alpha_1, \ldots, \alpha_{M-1})$ defined in Eq. 1.10. Denoting $B_{j,4}(x_i; \overline{b})$ by $B_{j,4}[i]$ and $(B_{j,4}[0], B_{j,4}[1], \ldots, B_{j,4}[N-1])$ by $\overline{B}_{j,4}$, we can express L_S (c.f., Eq. 1.11) in the compact form

$$L_S \;=\; \frac{1}{2\sigma^2}\|\overline{y} - \overline{\alpha}\mathbf{B}\|^2 \;, \tag{C.9}$$

$$\;=\; \frac{1}{2\sigma^2}\left[\|\overline{y}\|^2 - 2\overline{y}\mathbf{B}^T\overline{\alpha}^T + \overline{\alpha}\mathbf{B}\mathbf{B}^T\overline{\alpha}^T\right] \;, \tag{C.10}$$

$$B_{ji} \;=\; B_{j,4}[i] \;. \tag{C.11}$$

Let $\widehat{\alpha}$ be the minimizer of L_S over $\overline{\alpha}$ keeping the other parameters fixed. Using the standard condition for the extremum,

$$\left.\frac{\partial L_s}{\partial \alpha_i}\right|_{\widehat{\alpha}} = 0 \;\Rightarrow\; \widehat{\alpha} = \overline{y}\mathbf{B}^T(\mathbf{B}\mathbf{B}^T)^{-1} \;. \tag{C.12}$$

Substituting $\widehat{\alpha}$ into Eq. C.10 gives the fitness function l_S,

$$l_S \;=\; \frac{1}{2\sigma^2}\left[\|\overline{y}\|^2 - \overline{y}\mathbf{B}^T(\mathbf{B}\mathbf{B}^T)^{-1}\mathbf{B}\overline{y}^T\right] \;. \tag{C.13}$$

that is to be minimized over the remaining parameters.

Bibliography

[1] T.W. Anderson. *An introduction to multivariate statistical analysis*. Wiley Series in Probability and Statistics. Wiley, 2003.

[2] Mauro Birattari and Marco Dorigo. How to assess and report the performance of a stochastic algorithm on a benchmark problem: mean or best result on a number of runs? *Optimization letters*, 1(3):309–311, 2007.

[3] Sergio Boixo, Troels F Rønnow, Sergei V Isakov, Zhihui Wang, David Wecker, Daniel A Lidar, John M Martinis, and Matthias Troyer. Evidence for quantum annealing with more than one hundred qubits. *Nature Physics*, 10(3):218, 2014.

[4] Daniel Bratton and James Kennedy. Defining a standard for particle swarm optimization. In *Swarm Intelligence Symposium, 2007. SIS 2007. IEEE*, pages 120–127. IEEE, 2007.

[5] Jie Chen, Bin Xin, Zhihong Peng, Lihua Dou, and Juan Zhang. Optimal contraction theorem for exploration-exploitation tradeoff in search and optimization. *IEEE Transactions on Systems, Man, and Cybernetics-Part A: Systems and Humans*, 39(3):680–691, 2009.

[6] Gerda Claeskens and Nils Lid Hjort. *Model selection and model averaging*, volume 330. Cambridge University Press, 2008.

[7] Maurice Clerc. *Particle swarm optimization*, volume 93. John Wiley & Sons, 2010.

[8] Maurice Clerc and James Kennedy. The particle swarm-explosion, stability, and convergence in a multidimensional complex space. *Evolutionary Computation, IEEE Transactions on*, 6(1):58–73, 2002.

[9] Carl de Boor. On calculating with b-splines. *Journal of Approximation Theory*, 6(1):50 – 62, 1972.

[10] Carl de Boor. *A practical guide to splines*, volume 27 of *Applied mathematical sciences*. Springer, 2001.

[11] R. C. Eberhart and Y. Shi. Comparing inertia weights and constriction factors in particle swarm optimization. In *Proceedings of the 2000 Congress on Evolutionary Computation*, volume 1, pages 84–88. IEEE, 2000.

[12] Russell C Eberhart, Yuhui Shi, and James Kennedy. *Swarm intelligence*. Elsevier, 2001.

[13] Bradley Efron. Bootstrap methods: another look at the jackknife. In *Breakthroughs in statistics*, pages 569–593. Springer, 1992.

[14] Andries P Engelbrecht. *Fundamentals of computational swarm intelligence*, volume 1. Wiley Chichester, 2005.

[15] Jerome Friedman, Trevor Hastie, and Robert Tibshirani. *The elements of statistical learning*, volume 1 of *Springer series in statistics*. Springer, 2001.

[16] Jerome H. Friedman. Multivariate adaptive regression splines. *The Annals of Statistics*, 19(1):1–67, 03 1991.

[17] Akemi Gálvez and Andrés Iglesias. Efficient particle swarm optimization approach for data fitting with free knot b-splines. *Computer-Aided Design*, 43(12):1683–1692, 2011.

[18] Andrew Gelman, John B Carlin, Hal S Stern, and Donald B Rubin. *Bayesian data analysis*. Chapman and Hall/CRC, 1995.

[19] Ian Goodfellow, Yoshua Bengio, and Aaron Courville. *Deep learning*. The MIT press, 2016.

[20] W. Hardle. *Applied Nonparametric Regression*, volume 19 of *Econometric Society Monographs*. Cambridge University Press, 1990.

[21] C. W. Helstrom. *Statistical Theory of Signal Detection*. Pergamon, 1968.

[22] Mark W Johnson, Mohammad HS Amin, Suzanne Gildert, Trevor Lanting, Firas Hamze, Neil Dickson, R Harris, Andrew J Berkley, Jan Johansson, Paul Bunyk, et al. Quantum annealing with manufactured spins. *Nature*, 473(7346):194, 2011.

[23] N.L. Johnson, S. Kotz, and N. Balakrishnan. *Continuous univariate distributions*, volume 2 of *Wiley series in probability and mathematical statistics: Applied probability and statistics*. Wiley & Sons, 1995.

[24] David LB Jupp. Approximation to data by splines with free knots. *SIAM Journal on Numerical Analysis*, 15(2):328–343, 1978.

[25] S. Kay. *Fundamentals of Statistical Signal Processing, Volume II: Detection Theory*. Prentice Hall, 1998.

[26] Steven Kay. *Fundamentals of Statistical Signal Processing, Volume I: Estimation Theory*. Prentice Hall, 1993.

[27] J. Kennedy and R. C. Eberhart. Particle swarm optimization. In *Proceedings of the IEEE International Conference on Neural Networks: Perth, WA, Australia*, volume 4, page 1942. IEEE, 1995.

[28] Scott Kirkpatrick, C Daniel Gelatt, and Mario P Vecchi. Optimization by simulated annealing. *Science*, 220(4598):671–680, 1983.

[29] Donald E Knuth. *The art of computer programming, 2: seminumerical algorithms.* Addison Wesley, 1998.

[30] S. Kotz, N. Balakrishnan, and N.L. Johnson. *Continuous multivariate distributions, volume 1: Models and applications.* Wiley series in probability and statistics. Wiley, 2004.

[31] Erich L Lehmann and Joseph P Romano. *Testing statistical hypotheses.* Springer Texts in Statistics. Springer, 2006.

[32] Calvin Leung. Estimation of unmodeled gravitational wave transients with spline regression and particle swarm optimization. *SIAM Undergraduate Research Online (SIURO)*, 8, 2015.

[33] Soumya D Mohanty. Spline based search method for unmodeled transient gravitational wave chirps. *Physical Review D*, 96(10):102008, 2017.

[34] Jorge Nocedal and Stephen J Wright. *Numerical optimization.* Springer Series in Operations Research and Financial Engineering. Springer, 2006.

[35] Marc E Normandin, Soumya D Mohanty, and Thilina S Weerathunga. Particle swarm optimization based search for gravitational waves from compact binary coalescences: performance improvements. *Physical Review D*, 98:044029, 2018.

[36] Yiannis G Petalas, Konstantinos E Parsopoulos, and Michael N Vrahatis. Memetic particle swarm optimization. *Annals of Operations Research*, 156(1):99–127, 2007.

[37] C.R. Rao. *Linear statistical inference and its applications.* Wiley Series in Probability and Statistics. Wiley, 2009.

[38] Christian Robert and George Casella. *Monte Carlo statistical methods.* Springer Texts in Statistics. Springer, 2005.

[39] Jacob Robinson and Yahya Rahmat-Samii. Particle swarm optimization in electromagnetics. *IEEE Transactions on Antennas and Propagation*, 52(2):397–407, 2004.

[40] Jeffrey S Rosenthal. *A first look at rigorous probability theory*. World Scientific Publishing Company, 2006.

[41] Sheldon M Ross. *Stochastic processes*. Wiley, New York, second edition, 1996.

[42] David Ruppert, Matthew P Wand, and Raymond J Carroll. *Semiparametric regression*, volume 12. Cambridge University Press, 2003.

[43] Yuhui Shi and Russell Eberhart. A modified particle swarm optimizer. In *The 1998 IEEE International Conference on Evolutionary Computation*, pages 69–73. IEEE, 1998.

[44] Francisco J Solis and Roger J-B Wets. Minimization by random search techniques. *Mathematics of Operations Research*, 6(1):19–30, 1981.

[45] Rainer Storn and Kenneth Price. Differential evolution– a simple and efficient heuristic for global optimization over continuous spaces. *Journal of Global Optimization*, 11(4):341–359, 1997.

[46] Rangarajan K Sundaram. *A first course in optimization theory*. Cambridge University Press, 1996.

[47] Grace Wahba. *Spline models for observational data*. SIAM, 1990.

[48] Y. Wang and S.D. Mohanty. Particle swarm optimization and gravitational wave data analysis: Performance on a binary inspiral testbed. *Physical Review D*, 81:063002, 2010.

[49] Wikipedia contributors. Errors-in-variables models — Wikipedia, the free encyclopedia, 2018. [Online; accessed 15-July-2018].

[50] Wikipedia contributors. Hertzsprung-Russell diagram — Wikipedia, the free encyclopedia, 2018. [Online; accessed 25-July-2018].

[51] Wikipedia contributors. Pearson correlation coefficient — Wikipedia, the free encyclopedia, 2018. [Online; accessed 10-August-2018].

[52] David H Wolpert and William G Macready. No free lunch theorems for optimization. *IEEE Transactions on Evolutionary Computation*, 1(1):67–82, 1997.

[53] Mauricio Zambrano-Bigiarini, Maurice Clerc, and Rodrigo Rojas. Standard particle swarm optimisation 2011 at CEC-2013: A baseline for future PSO improvements. In *2013 IEEE Congress on Evolutionary Computation (CEC)*, pages 2337–2344. IEEE, 2013.

Index

Akaike Information
    Criterion, 82
Algorithm condition, 27
Ant colony optimization, 42

Basis spline, 10, 104
Benchmark fitness function,
    30, 60
    Griewank, 21, 31, 63
    Rastrigin, 21
Benchmarking, 30, 32
Best-fit model, 1, 3
Best-of-M-runs, 34, 70
Big data, 2
Boundary condition, 56
    Let them fly, 56
Breakpoint, 78, 101

Cognitive
    Force, 51
    Term, 51
Conditional probability, 93,
    see also Statistical
    independence
Conditional probability
    density function, 97
Constraint space, 20
Convergence condition, 27,
    32
Convex function, 23
Convex set, 21

Data, 1, 3
    Realization, 3, 32, 67
    Simulation, 68
Density estimation, 3, 11
    Non-parametric, 4
    Parametric, 4
Dependent variable, 5, 14
Detection threshold, 13
Differential evolution, 41

EC, see Evolutionary
    computation
Evolutionary computation,
    37
    DE, see Differential
    evolution
    GA, see Genetic
    algorithm
Expectation, 97
Exploitation, 28, 54
Exploration, 28, 54, 63

False alarm, 13
False dismissal, 13
Feasible set, 20
Fitness function, 20, 46, 63,
    108
    Benchmark, see
    Benchmark fitness
    function
    Ruggedness, 28, 67, 79

Printed in the United States
by Baker & Taylor Publisher Services